Advances in Intelligent Systems and Computing

Volume 369

Series editor

Janusz Kacprzyk, Polish Academy of Sciences, Warsaw, Poland
e-mail: kacprzyk@ibspan.waw.pl

About this Series

The series "Advances in Intelligent Systems and Computing" contains publications on theory, applications, and design methods of Intelligent Systems and Intelligent Computing. Virtually all disciplines such as engineering, natural sciences, computer and information science, ICT, economics, business, e-commerce, environment, healthcare, life science are covered. The list of topics spans all the areas of modern intelligent systems and computing.

The publications within "Advances in Intelligent Systems and Computing" are primarily textbooks and proceedings of important conferences, symposia and congresses. They cover significant recent developments in the field, both of a foundational and applicable character. An important characteristic feature of the series is the short publication time and world-wide distribution. This permits a rapid and broad dissemination of research results.

Advisory Board

Chairman

Nikhil R. Pal, Indian Statistical Institute, Kolkata, India
e-mail: nikhil@isical.ac.in

Members

Rafael Bello, Universidad Central "Marta Abreu" de Las Villas, Santa Clara, Cuba
e-mail: rbellop@uclv.edu.cu

Emilio S. Corchado, University of Salamanca, Salamanca, Spain
e-mail: escorchado@usal.es

Hani Hagras, University of Essex, Colchester, UK
e-mail: hani@essex.ac.uk

László T. Kóczy, Széchenyi István University, Győr, Hungary
e-mail: koczy@sze.hu

Vladik Kreinovich, University of Texas at El Paso, El Paso, USA
e-mail: vladik@utep.edu

Chin-Teng Lin, National Chiao Tung University, Hsinchu, Taiwan
e-mail: ctlin@mail.nctu.edu.tw

Jie Lu, University of Technology, Sydney, Australia
e-mail: Jie.Lu@uts.edu.au

Patricia Melin, Tijuana Institute of Technology, Tijuana, Mexico
e-mail: epmelin@hafsamx.org

Nadia Nedjah, State University of Rio de Janeiro, Rio de Janeiro, Brazil
e-mail: nadia@eng.uerj.br

Ngoc Thanh Nguyen, Wroclaw University of Technology, Wroclaw, Poland
e-mail: Ngoc-Thanh.Nguyen@pwr.edu.pl

Jun Wang, The Chinese University of Hong Kong, Shatin, Hong Kong
e-mail: jwang@mae.cuhk.edu.hk

More information about this series at http://www.springer.com/series/11156

Álvaro Herrero · Bruno Baruque
Javier Sedano · Héctor Quintián
Emilio Corchado
Editors

International Joint Conference

CISIS'15 and ICEUTE'15

International Conference on

Computational Intelligence in
Security for Information Systems

International Conference on

EUROPEAN
Transnational Education

Springer

Editors

Álvaro Herrero
Department of Civil Engineering, Escuela
    Politécnica Superior
University of Burgos
Burgos
Spain

Bruno Baruque
Department of Civil Engineering, Escuela
    Politécnica Superior
University of Burgos
Burgos
Spain

Javier Sedano
Technological Institute of Castilla y León
Burgos
Spain

Héctor Quintián
University of Salamanca
Salamanca
Spain

Emilio Corchado
University of Salamanca
Salamanca
Spain

ISSN 2194-5357   ISSN 2194-5365   (electronic)
Advances in Intelligent Systems and Computing
ISBN 978-3-319-19712-8   ISBN 978-3-319-19713-5   (eBook)
DOI 10.1007/978-3-319-19713-5

Library of Congress Control Number: 2015940981

Springer Cham Heidelberg New York Dordrecht London
© Springer International Publishing Switzerland 2015

Printed on acid-free paper

Springer International Publishing AG Switzerland is part of Springer Science+Business Media
(www.springer.com)

Preface

This volume of *Advances in Intelligent Systems and Soft Computing* contains accepted papers presented at the *8th International Conference on Computational Intelligence in Security for Information Systems* (CISIS 2015) and the *6th International Conference on EUropean Transnational Education* (ICEUTE 2015). These conferences were held in the beautiful and historic city of Burgos (Spain), in June 2015.

The aim of the 8th CISIS conference is to offer a meeting opportunity for academic and industry-related researchers belonging to the various, vast communities of Computational Intelligence, Information Security, and Data Mining. The need for intelligent, flexible behavior by large, complex systems, especially in mission-critical domains, is intended to be the catalyst and the aggregation stimulus for the overall event.

After a through peer-review process, the CISIS 2015 International Program Committee selected 43 papers, written by authors from 16 different countries. These papers are published in present conference proceedings, achieving an acceptance rate of 39 %.

In the case of the 6th ICEUTE conference, the International Program Committee selected 12 papers (from seven countries), which are also published in these conference proceedings.

The selection of papers was extremely rigorous in order to maintain the high quality of the conference and we would like to thank the members of the International Program Committees for their hard work during the reviewing process. This is a crucial issue for creation of a high standard conference and the CISIS and ICEUTE conferences would not exist without their help.

CISIS'15 and ICEUTE'15 enjoyed outstanding keynote speeches by distinguished guest speakers: Prof. Senén Barro—University of Santiago de Compostela (Spain) and Prof. Hans J. Briegel—University of Innsbruck (Austria).

For this CISIS'15 edition, as a follow-up of the conference, we anticipate further publication of selected papers in a special issue of the prestigious Logic Journal of the IGPL Published by Oxford Journals.

Particular thanks go as well to the Conference main Sponsors, IEEE—Spain Section, IEEE Systems, Man and Cybernetics—Spanish Chapter, and The International Federation for Computational Logic, who jointly contributed in an active and constructive manner to the success of this initiative. We want also to extend our warm gratitude to all the Special Sessions chairs for their continuing support to the CISIS and ICEUTE Series of conferences.

We would like to thank all the Special Session organizers, contributing authors, as well as the members of the Program Committees and the Local Organizing Committee for their hard and highly valuable work. Their work has helped to contribute to the success of the CISIS 2015 and ICEUTE 2015 events.

June 2015

Álvaro Herrero
Bruno Baruque
Javier Sedano
Héctor Quintián
Emilio Corchado

CISIS 2015

Organization

General Chair

Emilio Corchado—University of Salamanca (Spain)

Honorary Chairs

Alfonso Murillo—Rector of the University of Burgos (Spain)
José Mª Vela—Director of the Technological Centre ITCL (Spain)

Local Chair

Álvaro Herrero—University of Burgos (Burgos)

International Advisory Committee

Ajith Abraham—Machine Intelligence Research Labs (USA)
Antonio Bahamonde—Universidad de Oviedo at Gijón (Spain)
Michael Gabbay—Kings College London (UK)

Program Committee

Emilio Corchado—University of Salamanca (Spain) (PC Co-Chair)
Álvaro Herrero—University of Burgos (Spain) (PC Co-Chair)
Bruno Baruque—University of Burgos (Spain) (PC Co-Chair)
Javier Sedano—Technological Institute of Castilla y León (Spain) (PC Co-Chair)
Alberto Peinado—Universidad de Malaga (Spain)
Amparo Fuster-Sabater—Institute of Applied Physics (CSIC) (Spain)

Jose L. Salmeron—University Pablo de Olavide (Spain)
Jose Luis Calvo-Rolle—University of A Coruña (Spain)
Jose Luis Imana—Complutense University of Madrid (Spain)
José Luis Casteleiro-Roca—University of Coruña (Spain)
Jose M. Molina—Universidad Carlos III de Madrid (Spain)
José M. Benítez—University of Granada (Spain)
José Manuel Molero Pérez—Austrian Academy of Sciences (Austria)
José Such—Lancaster University (UK)
Josep Ferrer—Universitat de les Illes Balears (Spain)
Juan Tena—University of Valladolid (Spain)
Juan Álvaro Muñoz Naranjo—University of Almería (Spain)
Juan Jesús Barbarán—University of Granada (Spain)
Juan Pedro Hecht—Universidad de Buenos Aires (Argentina)
Krzysztof Walkowiak—Wroclaw University of Technology (Poland)
Leocadio G. Casado—University of Almeria (Spain)
Luis Hernandez Encinas—Consejo Superior de Investigaciones Científicas (Spain)
Luis Alfonso Fernández Serantes—FH Joanneum—University of Applied Sciences (Austria)
Luis Enrique Sanchez Crespo—Universidad de Castilla-La Mancha (Spain)
Mª Belen Vaquerizo—University of Burgos (Spain)
Manuel Grana—University of Basque Country (Spain)
Michal Choras—ITTI Ltd. (Poland)
Nicolas Cesar Alfonso Antezana Abarca—San Pablo Catholic University (Perú)
Pino Caballero-Gil—University of La Laguna (Spain)
Rafael Alvarez—University of Alicante (Spain)
Rafael Corchuelo—University of Seville (Spain)
Rafael M. Gasca—University of Seville (Spain)
Raquel Redondo—University of Burgos (Spain)
Raúl Durán—Universidad de Alcalá (Spain)
Ricardo Contreras—Universidad de Concepción (Argentina)
Robert Burduk—Wroclaw University of Technology (Poland)
Rodolfo Zunino—University of Genoa (Italy)
Roman Senkerik—TBU in Zlin (Czech Republic)
Rosaura Palma-Orozco—CINVESTAV—IPN (Mexico)
Salvador Alcaraz—Miguel Hernandez University (Spain)
Simone Mutti—Università degli Studi di Bergamo (Italy)
Sorin Stratulat—Université de Lorraine (France)
Tomas Olovsson—Chalmers University of Technology (Sweden)
Tomasz Kajdanowicz—Wroclaw University of Technology (Poland)
Urko Zurutuza—Mondragon University (Spain)
Vincenzo Mendillo—Central University of Venezuela (Venezuela)
Wenjian Luo—University of Science and Technology of China (China)
Wojciech Kmiecik—Wroclaw University of Technology (Poland)
Zuzana Oplatkova—Tomas Bata University in Zlin (Czech Republic)

ICEUTE 2015

Organization

General Chair

Emilio Corchado—University of Salamanca (Spain)

Honorary Chairs

Alfonso Murillo—Rector of the University of Burgos (Spain)
José Mª Vela—Director of the Technological Centre ITCL (Spain)

Local Chair

Álvaro Herrero—University of Burgos (Burgos)

International Advisory Committee

Jean-Yves Antoine—Université François Rabelais (France)
Reinhard Baran—Hamburg University of Applied Sciences (Germany)
Fernanda Barbosa—Instituto Politécnico de Coimbra (Portugal)
Bruno Baruque—University of Burgos (Spain)
Emilio Corchado—University of Salamanca (Spain)
Wolfgang Gerken—Hamburg University of Applied Sciences (Germany)
Arnaud Giacometti—Université François Rabelais (France)
Helga Guincho—Instituto Politécnico de Coimbra (Portugal)
Álvaro Herrero—University of Burgos (Spain)
Patrick Marcel—Université François Rabelais (France)
Gabriel Michel—University Paul Verlaine—Metz (France)
Viorel Negru—West University of Timisoara (Romania)
Jose Luis Nunes—Instituto Politécnico de Coimbra (Portugal)

Mª Belen Vaquerizo—University of Burgos (Spain)
Wozniak, Michal—Wroclaw University of Technology (Poland)
Hujun Yin—University of Manchester (UK)
Daniela Zaharie—West University of Timisoara (Romania)

Organizing Committee

Álvaro Herrero—University of Burgos (Spain) (Chair)
Bruno Baruque—University of Burgos (Spain) (Co-Chair)
Javier Sedano—Technological Institute of Castilla y León (Spain) (Co-Chair)
Emilio Corchado—University of Salamanca (Spain)
Ángel Arroyo—University of Burgos (Spain)
Raquel Redondo—University of Burgos (Spain)
Leticia Curiel—University of Burgos (Spain)
Belkis Díaz—University of Burgos (Spain)
Belén Vaquerizo—University of Burgos (Spain)
Pedro Burgos—University of Burgos (Spain)
Juán Carlos Pérez—University of Burgos (Spain)
Héctor Quintian—University of Salamanca (Spain)
José Luis Calvo—University of La Coruña (Spain)
José Luis casteleiro—University of La Coruña (Spain)
Amelia García—Technological Institute of Castilla y León (Spain)
Mónica Cámara—Technological Institute of Castilla y León (Spain)
Silvia González—Technological Institute of Castilla y León (Spain)

Contents

CISIS 2015: Infrastructure and Network Security

CISIS 2015-SS02: User-Centric Security & Privacy

ICEUTE 2015: Domain Applications and Case Studies

ICEUTE 2015: Information Technologies for Transnational Learning

ICEUTE 2015: Teaching and Evaluation Methodologies

CISIS 2015: Applications of Intelligent Methods for Security

Performance Analysis of Vertically Partitioned Data in Clouds Through a Client-Based In-Memory Key-Value Store Cache

Jens Kohler and Thomas Specht

Abstract Data security and protection in Cloud Computing are still major challenges. Although Cloud Computing offers a promising technological foundation, data have to be stored externally in order to take the full advantages of public clouds. These challenges lead to our distribution approach that vertically distributes data among various cloud providers. As every provider only gets a small chunk of the data, the chunks are useless without the others. Unfortunately, the actual performance is disillusioning and the access times of the distributed data are indisputable. Thus, this lousy performance is now in the focus of this work. The basic idea is the introduction of a cache that stores the already joined tuples in its memory. Thus, not always the different cloud storages have to be queried or manipulated, but only the faster caches. Finally, we present the implementation and evaluation of a client-based In-Memory cache in this work.

Keywords Cloud computing security · Vertically partitioned data · Cloud abstraction · Database abstraction

1 Introduction

Data security and data protection issues in Cloud Computing environments are still major challenges especially for enterprises. Although Cloud Computing with its dynamic *pay as you go* billing models offers a promising technological foundation, critical enterprise data (e.g. customer data, accounting data, etc.) have to leave the secure enterprise network in order to take the full advantages of public clouds. Considering the four cloud deployment models (public, private, hybrid and community) [12], only the commonly shared usage of resources from public clouds avoids huge initial investments in physical hardware systems. However, these shared multi-tenant

J. Kohler (✉) · T. Specht
University of Applied Sciences Mannheim, Mannheim, Germany
e-mail: j.kohler@hs-mannheim.de

T. Specht
e-mail: t.specht@hs-mannheim.de

© Springer International Publishing Switzerland 2015   3
Á. Herrero et al. (eds.), *International Joint Conference*, Advances in Intelligent
Systems and Computing 369, DOI 10.1007/978-3-319-19713-5_1

systems and the storage of critical enterprise data at an external cloud provider location are the main reasons that prevent enterprises from using public cloud computing environments. Above that, legal regulations prohibit enterprises to store data (i.e. customer data) at an external provider location. All these challenges lead to the *SeDiCo* project (A Framework for a *Se*cure and *Di*stributed *C*loud Data Store) conducted at the Institute for Enterprise Computing at the University of Applied Sciences in Mannheim, Germany. The basic idea of *SeDiCo* is to create a data store that vertically distributes database data to various cloud providers. As every provider only gets a small chunk of the data that is logically separated from the other chunks, the respective chunks are useless without the others. The basic approach splits data into two vertical database partitions and stores these partitions at different clouds. Thus, our approach ensures both, data security and data protection. Considering the four cloud deployment models it does not matter if the partitions are stored on a public, private, hybrid or community cloud, as we encapsulated different cloud interfaces in the *SeDiCo* prototype such that all models are transparently usable. Moreover, the actual prototypical implementation supports Amazon EC2 (public cloud) and Eucalyptus (private cloud) programming interfaces. Figure 1 illustrates the approach in an exemplified use case.

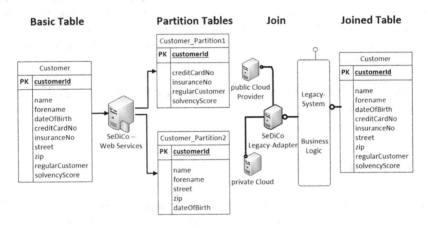

Fig. 1 *SeDiCo* Architecture

The remainder of this paper is structured as follows. Sect. 2 discusses the current state of the art and closely related other research works. We outline the performance problem of our approach in Sect. 3 in a concrete problem formulation. Then, Sect. 4 presents our caching approach in order to tackle the before-mentioned performance challenge. Afterwards, Sect. 5 points out the main topics of our cache implementation and this is followed by its evaluation in Sect. 6. After analyzing our evaluation results in Sect. 7 we give an outlook on our future work tasks in Sect. 8.

2 Related Work

The focus of this paper is the evaluation of a client-based proxy cache for the vertically distributed cloud databases that is implemented as an In-Memory key-value store. The presented performance evaluation results are based on a standard TPC-W [28] implementation. Another benchmark suite is SPECWeb [26], but they provide no specific database benchmarks.

Moreover, the analysis of cache loading strategies, such as the initial loading of all data into a centralized server-based In-Memory cache, such as DBProxy [1], RAM-Cloud [23] or even similar approaches as we presented in this paper with NoSQL caches (e.g. Redis, Voldemort, Hbase, etc.) [25] would go beyond the limits of this work.

Similar approaches that analyzed the performance of In-Memory databases and caches are stated in [5] and [27]. These works also prove the immense performance gain that comes with the usage of In-Memory technologies. Moreover, they give useful hints for our work, as we showed with this work that our vertical partitioning approach is also suitable for *Big Data* volumes when we use In-Memory caching architectures.

This work goes beyond the state of the art In-Memory caching architectures, as it introduces a cache that is based on vertically distributed database tables. Database partitioning is a common approach especially with the afore-mentioned NoSQL architectures [18, 24]. These architectures use horizontal partitioning to increase the overall database performance. They all have in common that the mixture of different database vendors is not possible. With respect to this, our *SeDiCo* framework supports different database vendors and they can even be used simultaneously. Hence, it is possible to partition a table into two partitions and store them in different database systems. Moreover, SeDiCo not only supports different database vendors but also different cloud providers. We achieved this with the encapsulation of different cloud application programming interfaces (APIs) into a generic interface and this encapsulation is based on the jclouds [3] framework.

Concerning the cloud abstraction frameworks besides jclouds, there are also worth mentioning libcloud [4] and deltacloud [2]. The ongoing research work in various cloud abstraction projects (e.g. CloudStack, OpenStack, FIWare, SLA@SOI, Eucalyptus, etc.) [21] aim to implement the OCCI standard [22]. This holds especially for the IaaS layer with the CIMI specification [7], as the offered services are relatively homogeneous compared to the upper cloud layers, such as PaaS and SaaS. This encapsulation of different provider application programming interfaces (APIs) is also necessary in SeDiCo, but as SeDiCo only deals with IaaS providers, its concrete implementation is less complex than the one proposed in the aforementioned standards and specifications.

All in all, these partitioning approaches are still a choice between performance (horizontal partitioning) and security (vertical partitioning). However, with the introduction of a cache as presented in this work, the gap between performance and security is getting smaller and smaller.

3 Problem Formulation

Currently, we implemented the above-mentioned *SeDiCo* approach as an Open Source prototype [9] as part of the project. Furthermore, we measured and evaluated its performance in [10, 11]. Unfortunately, the actual performance is disillusioning and the query and access times of the vertically distributed data is indisputable, as the query performance, for instance, decreased by 60 % in a MySQL database and by 90 % in an Oracle database [10]. This is the main reason why we have not integrated any de-/encryption of the data yet. However, the lousy performance is now in the focus of our ongoing research work and we present one of our auspicious approaches to increase the performance again in this work.

4 Approach

The basic idea of the presented approach is the introduction of a cache that stores the already joined tuples in its memory. Thus, not always the different cloud storages have to be queried or manipulated, but only the faster cache. Therefore, we present the implementation of a *client-based In-Memory cache* that acts as a proxy between the clients and the cloud databases. We further evaluate its performance for querying and manipulating data. The focus of this work is the evaluation of the performance of the In-Memory cache. Besides this performance measurement, we consider the key-value store cache architecture that stores the relational data schema that is partitioned across two clouds. Thus, this work describes the usage of a key-value store as a client-based cache for vertically distributed data in the cloud and its performance evaluation.

Cache Synchronization. As soon as a tuple is manipulated (i.e. updated, inserted, deleted) the *SeDiCo* client application ensures, that this tuple is manipulated immediately in the cache and in the cloud databases if it is already cached (i.e. cache hit). If a tuple is not in the cache, the application logic queries it (manipulates it if necessary) and adds it to the cache. Thus, the second access of the tuple is much faster. As soon as tuples are accessed, the cache is loaded with these tuples. This also holds for newly inserted tuples. Firstly, a new tuple is added to the cache and then immediately to the two partitions in the clouds. Finally, we discuss the last case where a tuple is deleted. If the tuple that should be deleted is in the cache, it is removed from the cache and from the clouds. If it is not in the cache, it is just removed from the clouds and not added to the client cache.

Especially the presented update, insert, delete operations raise the question of how to keep the data consistent throughout all client caches. This question is beyond the limits of this work, but here also a read-only approach like presented in [24] that just invalidates tuples but does not actually delete them is thinkable.

Cache Loading Strategies. Despite the fact that this approach promises access times to the data, that are comparable to general In-Memory architectures [24] with access ranges within milliseconds, another challenge is to initially load required tuples into the cache before they are accessed by the respective clients. Correctly identifying such tuples, that the user will use in the near future is not possible, although there exists various approaches such as *most frequently used* or *least frequently used.*

Finally, the above-mentioned challenges concerning *Cache Loading, Synchronization* and *Invalidation* are closely related to advanced *Computational Intelligence* approaches such as *Genetic Algorithms, Neural Networks*, etc. These methods and concepts are insofar important for our work here, as it is impossible with larger data volumes (*Big Data*) to hold entire data volumes in one or several caches. Thus, the next steps after improving the data access performance are *Intelligent Cache Loading* strategies that load possibly required data in the cache before they are actually used.

5 Implementation

In order to achieve comparable evaluation results, we use the same physical hardware infrastructure as presented in our previous works [10, 11]. Thus, the *SeDiCo* client (the so-called *Legacy Adapter*, Fig. 1) acts as the proposed client-based In-Memory cache. Basically it implements logic that acts as a proxy that coordinates the communication with the cache (in case of a cache hit) or with the cloud partitions (in case of a cache miss).

With this approach, we are able to use another feature of *SeDiCo* as it supports different databases systems. With the current implementation is it possible to use MySQL and Oracle Express. Moreover, it is possible to use them simultaneously, where one partition is in a MySQL database and the other one in an Oracle database. We chose this Open Source databases as they were generally available and because of their broad dissemination in both industrial and private environments [6]. Above that, we plan to integrate more relational databases such as PostgreSQL or MariaDB but also NoSQL databases such as HBase, Cassandra, MongoDB, etc. Here, we first have to determine a concept on how it is generally possible to integrate the respective NoSQL architectures [20] such as column, key-value, document stores or even graph databases).

We implemented our client-base cache with a cache size of 1 GB. Thus, we are able to ensure, that the entire data volume (288K tuples) fit entirely into the cache and that the operating systems, the SeDiCo client and the other applications have enough RAM left. However, if the cache is full we rely on the memcached standard invalidation method *least recently used* (LRU), that removes the oldest tuple that has not been used recently from the cache in order to get new cache storage space.

Based on that, the introduction of a key-value based In-Memory cache is a first step towards the integration of NoSQL into our framework and this is exactly where memcached [16] comes into play.

Memcached. Memcached is a database cache based on an In-Memory key-value storage architecture. It has the following advantages in the context of *SeDiCo.*

Firstly, we are able to easily adopt the NoSQL key-value concept to our relational architecture. This is because of the required primary key column that is replicated to the two partitions, in order to join them. Therefore, we are able to use this primary key as the *key* for the key-value cache. The rest of the distributed table is then stored as the *value*-part. Moreover, we are able to use it for arbitrary data structures, as all tuples are stored as serialized objects.

Secondly, memcached provides a tested and validated database caching technology, as recent works [5, 8, 19] and implementations (e.g. Craigslist, Wikipedia or Twitter show). Moreover, it is included in the software repositories of various operating systems like Debian, Ubuntu, RedHat and CentOS [16]. Thus, it is simple to install, to integrate and to maintain it in various heterogeneous software infrastructures. Moreover, it offers advanced redundant installations for load balancing or high availability recommendations.

Thirdly, it is possible to use memcached in various scenarios such as centralized/decentralized client and server-based caching [15].

Lastly, memcached is lightweight software component and therefore well suited for a client-based cache with a small amount of available memory.

Finally, all these advantages lead to the usage of memcached, although there are some restrictions that we now point out. It is worth mentioning, that memcached cannot be used with partitioned tables as presented in our *SeDiCo* approach [17]. Nevertheless, the *SeDiCo* client ensures, that all data loaded into the cache are already joined together. This is also the case for data manipulations where the application logic of the *SeDiCo* client guarantees that the manipulations are send to the cache as well as to the corresponding cloud partitions. Moreover, it is not possible with memcached to list all entries that are stored in the cache due to locking and performance issues [14]. This is similar to a "*SELECT * FROM...*" query, which is essential for a relational database schema. Therefore we implemented this as a feature in the *SeDiCo* client application logic and therefore a new performance evaluation is necessary.

6 Evaluation

According to our previous works [10, 11] we conducted our performance evaluation with all components (i.e. two cloud databases and the *SeDiCo* client) installed on one physical hardware machine with 2×2.9 GHz processors (with hyper threading), 8 GB RAM and 500 GB hard disk storage.

Above that, we use the same sample data volume as presented in [10]. Thus, the evaluation results range from 0 to 288K tuples. We used the database schema from the TPC-W benchmark [28] and distributed the *CUSTOMER* table vertically, as it mostly corresponds to the initial use case for the *SeDiCo* approach (Fig. 1). We performed every operation (i.e. query, update, delete) three times against the cache and present the average of these measurements. We further performed a *SELECT * FROM CUSTOMER* query in order to retrieve the respective number of tuples. The

Table 1 Experiment results - In-Memory cache performance in ms

# Tuples	Query time	Update time	Delete time
288,000	12,346	18,998	13,374
88,000	2,832	3,955	3,017
50,000	1,503	2,249	1,626
25,000	787	1,046	883
20,000	642	1,019	640
15,000	590	987	523
10,000	445	780	484
5,000	145	225	159
4,000	114	165	127
3,000	84	124	99
2,000	57	90	62
1,000	40	43	33
500	15	22	15
250	5	11	10
125	3	6	7
100	2	5	6
50	2	4	5
25	2	3	5
15	1	2	4
10	1	2	3
5	1	1	2
1	1	1	1

UPDATE operation contains updates for every value of the corresponding tuple. Lastly, the DELETE operation deletes the respective number of tuples from the cache. In order to extend our previous works [10, 11] and for a better comparability, we assume that the client has already accessed all existing tuples once. Therefore, we loaded all 228K tuples into the cache before the actual performance evaluation. With this approach, we are able to measure the pure cache performance and eliminate all side effects that could possibly emerge during the cache-loading phase (e.g. database I/O or network bottlenecks, etc.). These bottlenecks affect both, the cloud databases and the client caches. Eliminating these bottlenecks means, that we are able to completely focus on the In-Memory performance of memcached, to receive meaningful and reproducible evaluation results.

Table 2 now shows the pure cache query performance where all tuples are in the cache and thus, no cache misses occur, whereas Table 1 presents the pure database performance without any partitioning of the data. Finally, Fig. 2 visualizes all evaluation results.

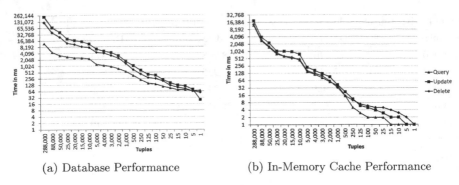

(a) Database Performance　　　　　　　(b) In-Memory Cache Performance

Fig. 2 Comparision - database and cache performance

Table 2 Experiment results - database performance in ms

# Tuples	Query time	Update time	Delete time
288,000	12,073	226,902	121,337
88,000	4,603	69,788	40,411
50,000	3,253	41,172	24,691
25,000	2,774	20,176	12,339
20,000	2,475	16,505	10,678
15,000	2,461	14,636	8,541
10,000	2,267	11,200	7,605
5,000	1,215	6,533	4,491
4,000	1,078	5,054	4,067
3,000	943	4,287	3,138
2,000	752	3,064	2,378
1,000	545	1,653	1,298
500	350	1,027	653
250	221	635	414
125	152	396	277
100	138	358	250
50	107	223	194
25	89	155	118
15	74	122	95
10	73	110	83
5	69	78	66
1	67	25	55

7 Conclusions

Setting the achieved values in relation to our previous works [10, 11], and Table 2, we recognize a remarkable performance gain for all evaluated operations. We did not measure the insert operation (i.e. cache loading times), as these values can also be derived from the query times presented in our previous works [10, 11]. For the sake of comparability, Table 1 now shows the performance evaluation of the database queries that we run against a locally installed MySQL database without vertically partitioned data. Our results show that the cache (Table 2) outperforms the database (Table 1).

We restricted the cache size during our evaluation to 1 GB in order to have enough memory available for the SeDiCo client, the operating system, the storage of the results, etc. Our analysis shows that the storage of all 288K tuples required 33 % of the 1 GB memory. Finally, these results show, that a client-based cache is a viable approach to tackle the lousy performance of our vertical data distribution approach. Besides the evaluation of other caching architectures and technologies, we are thus able to continue our research work and extend our approach towards an additional en-/decryption of the respective partitions or the vertical partitioning of *3* or even *n* partitions.

We also have to consider the fact that we used a rather small data volume that entirely fits into the cache memory. Thus, we are able to avoid side-effects as cache loading, synchronization and invalidation. These are important considerations in real world application scenarios. However, our goal with this work was to show the technical feasibility or our caching approach. This is insofar a new approach, as the cache is placed on top of two vertically distributed databases. Thus, we achieved promising results concerning the cache performance and this work is therefore an excellent foundation to consider the above-mentioned further challenges (i.e. larger data volumes, cache synchronization, invalidation, etc.) in our future work.

8 Outlook

With the evaluation in this paper we showed, that a client-based cache increases the data access performance in our vertically distributed scenario tremendously. Thus, we identified a strong demand for further research work that focuses on introducing other caching architectures such as centralized server-based caches or intercepting proxy caches, etc. Moreover, there are other technical cache implementations (DBProxy, RAMCloud, etc.) that need to be evaluated.

Closely related to the introduction of a cache are challenges that refer to cache synchronization and loading strategies. Here, further research work that evaluates approaches, like the CAS methods of memcached is necessary. This so-called *compare and set* feature is a mechanism that guarantees a consistent set of data throughout various client caches [13].

Finally, this work builds an excellent foundation to address the above-mentioned challenges. The presented evaluation results focus on the pure cache performance in order to achieve a performance metric that serves as an excellent foundation for our future work. Thus, we are able to rely on these results in our future work and to put our future evaluation results of the above-mentioned challenges in relation to these numbers.

References

1. Amiri, K., Park, S., Tewari, R., Padmanabhan, S.: DBProxy: a dynamic data cache for web applications. In: Proceedings of the 19th International Conference on Data Engineering, pp. 821–831. Bangalore, India (2003)
2. Apache Software Foundation: DeltaCloud Web Page. https://deltacloud.apache.org (2015)
3. Apache Software Foundation: jclouds Web Page. https://jclouds.apache.org (2015)
4. Apache Software Foundation: Libcloud Web Page. https://libcloud.apache.org (2015)
5. Carra, D., Michiardi, P.: Memory partitioning in memcached: an experimental performance analysis. In: ICC 2014, IEEE International Conference on Communications, pp. 1154–1159. Australia, Sydney (2014)
6. Db-engine: DB-Engines Ranking. http://db-engines.com/en/ranking (2014)
7. DMTF: Cloud infrastructure management interface—common information model (CIMI-CIM). In: Technical Report, Distributed Management Task Force (DMTF). http://dmtf.org/sites/default/files/standards/documents/DSP0264_1.0.0.pdf (2012)
8. Grolinger, K., Higashino, W.A., Tiwari, A., Capretz, M.A.: Data management in cloud environments: NoSQL and NewSQL data stores. J. Cloud Comput. Adv. Syst. Appl. 2(1), 22 (2013)
9. Kohler, J.: GitHub Page of SeDiCo Implementation. http://www.github.com/jenskohler
10. Kohler, J., Specht, T.: Vertical query-join benchmark in a cloud database environment. In: Proceedings of the 2nd World Conference on Complex Systems. IEEE Computer Society, Agadir, Marocco (2014)
11. Kohler, J., Specht, T.: Vertical update-join benchmark in a cloud database environment. In: WiWiTa 2014 Wismarer Wirtschaftsinformatiktage, pp. 159–175. Wismar, Germany (2014)
12. Mell, P., Grance, T.: The NIST definition of cloud computing. In: Technical Report 800–145, National Institute of Standards and Technology (NIST), Gaithersburg, MD (Sept 2011). http://csrc.nist.gov/publications/nistpubs/800-145/SP800-145.pdf
13. Memcached: Memcached CAS Operations. https://cloud.google.com/appengine/docs/python/memcache/#Python_Using_compare_and_set_in_Python (2014)
14. Memcached: Memcached FAQ. https://code.google.com/p/memcached/wiki/NewProgrammingFAQ (2014)
15. Memcached: Memcached Limits. https://cloud.google.com/appengine/docs/java/memcache/#Java_Limits (2014)
16. Memcached: Memcached Web Page. http://memcached.org (2014)
17. MySQL: MySQL Reference Guide. http://dev.mysql.com/doc/refman/5.7/en/innodb-memcached-porting-mysql.html (2014)
18. Neves, B.A., Correia, M.P., Bruno, Q., Fernando, A., Paulo, S.: DepSky: dependable and secure storage in a cloud-of-clouds. In: Kirsch, C.M., Heiser, G. (eds.) EuroSys '11 Proceedings of the Sixth Conference on Computer Systems, pp. 31–46. ACM, Salzburg, Austria (2011)
19. Nishtala, R., Fugal, H., Grimm, S.: Scaling memcache at facebook. In: NDSI'13 Proceedings of the 10th USENIX Conference on Networked Systems Design and Implementation, pp. 385–398 (2013)
20. NoSQL Archive: NoSQL Archive Web Page. http://nosql-databases.org/
21. OCCI: OCCI Implementing Projects. http://occi-wg.org/community/implementations

22. OCCI: OCCI (Open Cloud Computing Interface) Working Group Web Page. http://occi-wg. org/
23. Ousterhout, J., Parulkar, G., Rosenblum, M., Rumble, S.M., Stratmann, E., Stutsman, R., Agrawal, P., Erickson, D., Kozyrakis, C., Leverich, J., Mazières, D., Mitra, S., Narayanan, A., Ongaro, D.: The case for RAMCloud. Commun. ACM 54(7), 121–130 (2011)
24. Plattner, H.: A Course in In-Memory Data Management: The Inner Mechanics of In-Memory Databases. Springer, Berlin (2013)
25. Rabl, T., Gómez-Villamor, S.: Solving big data challenges for enterprise application performance management. Proc. VLDB Endow. 5(12), 1724–1735 (2012)
26. Standard Performance Evaluation Corporation: SPEC Web Page. https://www.spec.org
27. Tinnefeld, C., Kossmann, D., Boese, J.H., Plattner, H.: Parallel join executions in RAMCloud. In: Proceedings—International Conference on Data Engineering, pp. 182–190 (2014)
28. TPC: TPC Benchmark W (Web Commerce) Specification Version 2.0r. In: Technical Report, Transaction Processing Performance Council (TPC), San Francisco, USA (2003). http://www. tpc.org/tpcw/default.asp

Formalizing Agents' Beliefs for Cyber-Security Defense Strategy Planning

Karsten Martiny, Alexander Motzek and Ralf Möller

Abstract Critical information infrastructures have been exposed to an increasing number of cyber attacks in recent years. Current protection approaches consider the reaction to a threat from an operational perspective, but leave out human aspects of an attacker. The problem is, no matter how good a defense planning from an operational perspective is, it must be considered that any action taken might influence an attacker's belief in reaching a goal. For solving this problem this paper introduces a formal model of belief states for defender and intruder agents in a cyber-security setting. We do not only consider an attacker as a deterministic threat, but consider her as a human being and provide a formal method for reasoning about her beliefs given our reactions to her actions, providing more powerful means to assess the merits of countermeasures when planning cyber-security defense strategies.

Keywords Adaptive defense of network infrastructure · Semantic information representation · Situational awareness · Epistemic logic

1 Introduction

Critical information infrastructures have been exposed to an increasing number of cyber attacks in recent years. Cyber-physical systems in areas such as power plants or medical applications require special attention to defend them against any potential cyber attacks. It is important to note that, due to external constraints, established security measures—such as patching known vulnerabilities—are only applicable to a limited extent. Legal requirements might allow only the use of certified software

K. Martiny (✉)
Hamburg University of Technology, Hamburg, Germany
e-mail: karsten.martiny@tuhh.de

A. Motzek · R. Möller
University of Lübeck, Lübeck, Germany
e-mail: motzek@ifis.uni-luebeck.de

R. Möller
e-mail: moeller@ifis.uni-luebeck.de

© Springer International Publishing Switzerland 2015
Á. Herrero et al. (eds.), *International Joint Conference*, Advances in Intelligent Systems and Computing 369, DOI 10.1007/978-3-319-19713-5_2

versions, or new patches cannot be applied at a certain point in time because compatibility tests are still pending. A common effect is that external constraints leave critical information infrastructures exposed to known vulnerabilities, at least for some time.

An established approach to analyze threats from identified vulnerabilities is the use of attack graphs. Standard reactions to such an analysis include the proactive removal of identified vulnerabilities. However, for the intended application area of our work, these measures are usually not feasible without impairing the mission of the organization responsible for the critical infrastructure. Therefore, it is of utmost importance to carefully analyze potential consequences of applicable countermeasures or sequences of countermeasures (as part of defense strategies). Maintaining a model of an intruder's belief state provides the defender with improved means to assess the merits of potential defense strategies and novelly allows to analyze effects of taken actions on a human level.

To formalize the analysis of beliefs, we use *Probabilistic Doxastic Temporal (PDT) Logic* to represent the belief states of both the intruder and the defender. We assume that analysis starts at the last point in an attack graph where immediate consequences are pending, as described in [5, 10]. It is reasonable to react as late as possible in such scenarios, as intruders (or attackers) already spent significant efforts reaching into the network. As any action—including false alarms—might impact the network in the same way as a real attack, reactive measures have to be used very carefully. The approach presented in this paper aids the defender in selecting the best countermeasure by providing means to reason about the belief states of an attacker.

The remainder of this paper is structured as follows. An overview of related work is given in Sect. 2. After a summary of PDT Logic in Sect. 3, we show how the attacker's and defender's beliefs can be formalized in Sect. 4 and discuss a small example to show how beliefs evolve differently, depending on the respective observations. Finally, the paper concludes with Sect. 5.

2 Related Work

Automatic attack graph generation has been an active topic of research. Starting in about 1998 with [11], newer contributions such as MulVal [9] and, e.g., studies such as [4] pave the way for better cyber security. Formal methods of analyzing attack graphs are introduced in [5], which will serve as a base for our example.

Attack graphs provide an excellent base for creating corresponding defense strategies, which range from analysis of efficient placement of intrusion detection systems [8] over employing an integer optimization problem in selecting the current best countermeasure [13, 14] up to addressing situational awareness in quantitative scores such as [15] and validation of overall network defense as in [6]. However, indirect consequences—such as accidentally revealing information to an opponent—are not considered. We provide a formal method for analysis of indirect consequences and an automated consideration in defense planning.

Forms of intruder-defender-interactions have been studied on a personal level for instance in [12] and [16], but do only address the issue from a "psychological" point of view and do not provide means to formalize such behaviors. Studying intruders on a personal level comes along with developing attacker profiles and the distinction of different priorities and behaviors of attackers. Such profiles have notably been studied by Chiesa in [2, 3], but do not provide an assessment of consequences, or even formalisms of such profiles, on an (automated) process of defense planning. Ref. [1] proposes an automated approach considering behavioral models in cyber security, but focuses mostly on intruder detection, while a suitable countermeasure selection is not achieved.

While all those fundamental pieces for an assessment of agent beliefs in defense planning exist in the literature, their interaction is not covered. This paper proposes theoretic fundamentals for a formalized assessment of agent beliefs in cyber-security defense planning.

3 PDT Logic

In order to formally represent the belief states of both the intruder and defender, we use PDT Logic [7], a formalism to represent and reason about probabilistic beliefs and their temporal evolution in multi-agent systems. This section provides a summary of the key concepts of PDT Logic that are used utilized in this work.

Syntax We assume the existence of a first order logic language with finite sets of constant symbols $\mathcal{L}_{cons}$ and predicate symbols $\mathcal{L}_{pred}$, and an infinite set of variable symbols $\mathcal{L}_{var}$. Every predicate symbol $p \in \mathcal{L}_{pred}$ has an *arity*. Any member of the set $\mathcal{L}_{cons} \cup \mathcal{L}_{pred}$ is called a *term*. A term is called a *ground term* if it is a member of $\mathcal{L}_{cons}$. If $t_1, \cdots, t_k$ are (ground) terms, and p is a predicate symbol in $\mathcal{L}_{pred}$ with arity n, then $p(t_1, \cdots, t_k)$ with $k \in \{0, \cdots, n\}$ is a (ground) atom. If a is a (ground) atom, then a and $\neg a$ are (ground) *literals*. The former is called a *positive literal*, the latter is called a *negative literal*. The set of all ground literals is denoted by $\mathcal{L}_{lit}$. *Formulas* are built using $\wedge, \vee, \neg$ as usual. B denotes the Herbrand Base of $\mathcal{L}$, i.e., the set of all ground atoms that can be formed through from $\mathcal{L}_{pred}$ and $\mathcal{L}_{cons}$. Time is modeled in discrete steps. Generally, the set of agents $\mathcal{A}$ may be arbitrarily large, but for this work, we assume that the set of agents consists of a intruder I and a defender D.

Observation atoms To express that some group of agents $\mathcal{G} \subseteq \mathcal{A}$ observes some fact $F \in \mathcal{L}_{lit}$, we use the notion $Obs_{\mathcal{G}}(F)$. Note that F may be a negative literal and therefore we can explicitly specify observations of certain facts being false (such as "it is not raining"). We assume that the agents in $\mathcal{G}$ not only observe that l holds, but that each agent in $\mathcal{G}$ is also aware that all other agents in $\mathcal{G}$ make the same observation. The set of all observation atoms is denoted by $\mathcal{L}_{obs}$.

Possible Worlds The concept of possible worlds describes what combinations of events can actually occur in the modeled scenario. I.e., a world consists of a set of ground atoms and a set of observation atoms, describing what events actually hold and what is observed in this world, respectively. The set of all possible worlds is denoted by $W \subset 2^B \times 2^{\mathcal{L}_{obs}}$. If an agent is not able to differentiate between different possible worlds, we say that these worlds are *indistinguishable* to this agent. Namely, an agent i cannot distinguish two possible worlds w_1 and w_2, if both worlds contain exactly the same set of observations for agent i. We use $\mathcal{K}_i(w)$ to denote the set of worlds that agent i cannot distinguish from world w. Naturally, if i considers w as actually being possible (because it complies with all of i's observations), it also considers all worlds $\mathcal{K}_i(w)$ possible.

Threads To describe the temporal evolution of the modeled scenario, we use the concept of threads: A thread is a mapping $Th : \tau \to W$. Thus, a thread is a sequence of worlds and $Th(t)$ identifies the actual world at time t according to thread Th. The set of all possible threads is denoted by $\mathcal{T}$.

Subjective posterior probabilistic temporal interpretations Every possible thread in the modeled scenario can be associated with a probability value that describes how likely it is that the model evolves exactly according to the respective thread. Such a probability distribution across all possible threads is called a *probabilistic interpretation*. Initially, a probability distribution over the set of threads is given by the *prior probability assessment* $\mathcal{I}$, which is the same for all agents. With the occurrence of certain observations, agents will update their respective probability assessments over the set of threads. For instance, if some agent observes a specific fact, it will only consider threads possible, which actually contain this observation, i.e., the agent's probability assessments for all other threads will be updated to 0, while another agent, who did not make this observation, might still consider these threads possible. Thus, with the evolution of time, every agent maintains *subjective* interpretations. Since these interpretations depend on the occurrence of events in a specific thread, at a single time point different interpretations could be possible, depending on the actual thread. The subjective posterior probabilistic interpretation that an agent i associates to a thread Th at time t, given that the point-of-view thread is Th' is denoted by $\mathcal{I}_{it}^{Th'}(Th)$.

Subjective posterior probabilistic temporal interpretation In the beginning the probability distribution for the threads is given by the *prior probability assessment* $\mathcal{I}$. It is the same for all agents. With the observation of an event by one or a group of agents, the interpretation for every agent needs to be updated. With agent i, time point t, and *point of view thread Th* the update rule is

$$\mathcal{I}_{it}^{Th'}(Th) = \begin{cases} \dfrac{1}{\alpha_{it}^{Th'}} \cdot \mathcal{I}_{it-1}^{Th'}(Th) & \text{if } Th(t) \in \mathcal{K}_i(Th'(t)) \\ 0 & \text{if } Th(t) \notin \mathcal{K}_i(Th'(t)) \end{cases} \tag{1}$$

with

$$\alpha_{it}^{Th'} = \sum \mathcal{I}_{it-1}^{Th'}(Th) : Th(t) \in \mathcal{K}_i(Th'(t)).$$

The threads that were possible at time $t - 1$ are examined, if they are still possible at time t. $\alpha_{it}^{Th'}$ is the sum of the probabilities at time $t - 1$ of all possible threads at time t. These probabilities are divided by $\alpha_{it}^{Th'}$ and this leads to the new probability distribution, the *subjective posterior probabilistic temporal interpretation* $\mathcal{I}_{it}^{Th'}(Th)$ at time t of agent i. We assume a synchronous system, so the agents can distinguish between the worlds $Th(t)$ and $Th(t-1)$ even if they made no observation.

Belief in ground formulae $B_{it'}^{lu}(F_t)$ is a *belief formula* indicating that an agent i believes with a probability in a range of $[l, u]$ that a formula F, which was satisfiable at time t, still holds at time t'

$$\mathcal{I}_{it'}^{Th'} \vDash B_{it'}^{lu}(F_t) \quad \text{iff} \quad l \le \sum_{Th \in \mathcal{T}, Th(t) \vDash F} \mathcal{I}_{it'}^{Th'}(Th) \le u. \tag{2}$$

Nested beliefs A nested belief is the belief of an agent in another agent's belief. Agent i believes at time t' with a probability in the range $[l, u]$ that agent j believes at time t in a belief formula B with a probability in the range of $[l_j, u_j]$

$$\mathcal{I}_{it'}^{Th'} \vDash B_{it'}^{lu}(B_{jt}^{l_j u_j}(F)) \quad \text{iff} \quad l \le \sum_{\substack{Th \in \mathcal{T} \\ \mathcal{I}_{jt}^{Th} \vDash B_{jt}^{l_j u_j}(F)}} \mathcal{I}_{it'}^{Th'}(Th) \le u. \tag{3}$$

4 Formalizing Agents' Beliefs

4.1 Considerations on the Target Domain

As discussed in Sect. 1, we are concerned with situations where preventive security measures are not always an option. Thus, the network might be exposed to known vulnerabilities and the defender is left with choosing the best reactive countermeasure in case of an attack. Since we start our analysis at the last point in an attack graph, we have to assume that any attacker breaching this point is highly skilled (e.g., as described in [2]) and has already obtained extensive information about our network.

Any attack to the network consists of (at least) two actions: First, the attacker has to gain access to a target system with appropriate privileges. Then, custom code can be executed on this system to reach the attacker's actual goal. We do model the details of these steps, but abstractly represent the first step as an *attack* on a system resulting in a gained *shell* (e.g., through exploitation of known vulnerabilities), and the second step as some *code execution* on the target system. After having successfully obtained a shell on the target system, the attacker basically has two options: either she can proceed with the second stage of her attack (i.e., code execution) or she can try to gain access to further systems. Both options come with advantages and drawbacks for the intruder: continuing to attack further systems might result in additional compromised systems, but at the same time decreases the chance of performing an attack undetected. The choice of action depends on the attacker's actual goal; she might even attack another system without actually executing code there, but only to create distractions from her actual goal.

A network based intrusion detection system (IDS) can be used to detect attack actions on specific systems. However, in practice no IDS is perfect, i.e., both false alarms and missed attacks have to be considered when employing an IDS. This is an important point when planning defense strategies: if every detected attack is countered with a corresponding defense action, the lack of such a defense lets the attacker *know* that her attack went undetected and she might proceed with executing malicious code without having to fear any actions from the defender. Furthermore, deliberately letting the attacker execute code on a non-critical target host can provide valuable insights: an analysis of the executed code will reveal the actual goal of the attack and might further reveal the identity of the attacker. Another reason for refraining from a defense operation is that this action (e.g., unplugging a control server) might impact the mission success just as much as an attack. Consequently, deliberately letting an observed attack pass undefended might provide higher expected utility for the defender. By analyzing the potential evolution of the attacker's belief states, the choice of not defending can even be used to drive the attacker to false conclusions regarding her success.

Continuing these considerations, it might prove useful for the defender to maintain some kind of "honeypot" within the network. In its classical form, a honeypot is a system that has no productive meaning but is used instead to attract attackers and thereby provides means to analyze their goals and identities. However, since in our scenario we are dealing with highly skilled attackers, we have to assume that they would be able to identify such a honeypot immediately. Still, we can adapt the concept of honeypots to our model by maintaining backup devices of critical systems. These backup devices are disconnected from the physical world but otherwise indistinguishable from the actual productive system. This way, an intruder does not know which one the critical system is, but if she executes malicious code on the honeypot, she is not able to impact the mission success, but instead unknowingly provides the defender with the possibility to analyze the code and identify the attacker.

4.2 An Exemplary Domain Model

In the following, we introduce a small example to show how we can formally model potential attacks in a computer network and apply PDT Logic to analyze the evolution of the agents' belief states.

As explained in Sect. 3, our scenario contains two agents, the defender D and the intruder I. Following the considerations from the previous section, we assume that two systems are present in this network: some honeypot A and the corresponding critical system B. Possible actions on a system X are denoted by *attack(X)* and *defend(X)* with the obvious meanings. Furthermore, execution of malicious code on a system X is denoted by *exec(X)*. Finally, we have observation atoms such as $Obs_D(attack(A))$, indicating that the defender observed an attack on system A.

Building on these events, we can construct a set of threads representing all possible event sequences in this example. The resulting set is depicted in Fig. 1.

This model represents our considerations from the previous section: analysis starts at some time when no attack has occurred yet ($t = 0$). Possible subsequent events are then attacks on system A or system B) (represented through nodes 53 and 78 in the graph) or no attacks (node 1). If an attack has occurred on, say, system A, the IDS can detect this attack (i.e., an $Obs_D(attack(A))$ occurs, represented through the solid outgoing edges from node 53), or the attack is not detected (represented through the dashed line). For an undetected attack, the defender obviously has no options to defend against this. For an observed attack, the defender can choose between defending against this attack (node 55) or deliberately refrain from a defense (node 56). A defense forces the intruder to abort his attack (with potential downsides to the defender's mission). Lack of a defense action gives the intruder two options again: she can execute her malicious code on the attacked system (nodes 66 and 77), or she can proceed to attack the other system (nodes 57 and 68). After a second attack, possible subsequent events match the ones discussed for the first attack. Finally, if the first attack has not been defended, there are various options for the intruder to execute malicious code: if the second attack is defended, the attacker can execute the code only on the previously attacked system (nodes 63 and and 74), otherwise she can choose between executing code only on the previously attack system (e.g., node 61) or on both systems (e.g., node 62). If an attack has been detected by the IDS, the defender is able to observe these code executions (denoted in black),

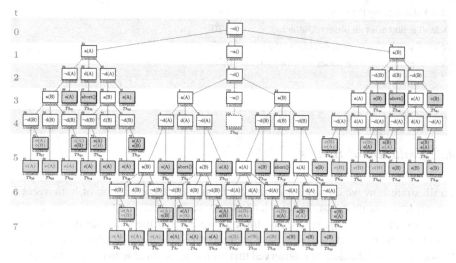

Fig. 1 Possible threads for the example domain. a(X), d(X), and e(X) denote attack, defend and code execution actions on node X, respectively. A defender's observation of an attack a(X) is marked through a solid edge (blue), a lack of an observation with a dashed edge. Unobserved code executions are marked in bold-red **e(X)**, otherwise they are marked roman-black e(X). Terminal nodes of threads are marked in gray

for undetected attacks, subsequent code executions will remain undetected as well (marked in bold red in the graph). If no attacks occur at all, the system continues in its normal state as indicated through node 104—allowing for equivalent branchings of the graph if attacks occur at later points in time.

Note that in most states of this model, none of the agents has complete knowledge of the world. For instance, if the defender does not observe an attack at time $t = 2$, he cannot distinguish between undetected attacks on either system or the actual absence of an attack (i.e., he considers nodes 54, 2, and 79 as actually being possible). If the intruder actually attacked a system and does not observe a defense action, she does not know whether her attack was actually undetected or whether it was observed and intentionally not defend (i.e., she is unable to distinguish between nodes 54 and 56, or 79 and 81, respectively) (Table 1).

To assign probabilities to every thread in this scenario, we start with assigning probabilities to single events. Then, we can determine the probability of a thread as a joint probabilities of the respective events contained in this thread. Reasonable values of single events corresponding to the considerations in Sect. 4.1 are given in the following table. The resulting probabilities of individual nodes are depicted in Fig. 1.

Table 1 Single event probabilities for the example

Event	Probability
Attack on a specific system at a specific time point	0.05
Attack detected by the IDS	0.80
Defend action after an observed attack	0.70
Second attack after a successful first attack	0.40
Code execution on both nodes after two attacks	0.65

4.3 Evolution of Beliefs

To illustrate how we can use this example to analyze the evolution of both agents' beliefs, let us assume that the intruder has attacked system A at time $t = 3$ (i.e., the actual world at $t = 3$ is represented through node 3). If the defender has observed this attack, he has to decide whether to defend the system against this attack or not. However, even if the defender observed this attack, he is not able to identify the actual node in the attack graph, because he is unable to distinguish between the situation where $attack(A)$ is the first action (node 3) and the situation where this was preceded by an undetected attack on B (node 82). To analyze the defender's expectations for the worst case (an unobserved code execution on the productive system B, i.e., $exec(B) \wedge \neg Obs_D(exec(B))$), one can analyze the threads containing nodes 5 or 84, and 6 or

85 in Fig. 1, respectively. By summing over the normalized probabilities of these respective threads, where the terminal node contains $exec(B) \wedge \neg Obs_D(exec(B))$, one can verify that the following holds:

$$\neg defend(A) \text{ at } t = 4 \quad \vDash \quad B_{D,4}^{0,\,0.05}(exec(B) \wedge \neg Obs_D(exec(B))), \text{ and}$$

$$defend(A) \text{ at } t = 4 \quad \nvDash \quad B_{D,4}^{0,\,0.05}(exec(B) \wedge \neg Obs_D(exec(B))).$$

I.e., the defender has lower expectations in the worst case actually happening if he choses to defend this attack. Consequently, we assume that the defender decides against a defend action, and we consider node 6 as the actual node for the following discussions. Next to protecting the system against the worst case, the defender is also interested in getting opportunities to analyze the intruder's malicious code. We can express his beliefs in observing a code execution (i.e., $Obs_D(exec(A)) \vee Obs_D(exec(B))$) as

$$\neg defend(A) \text{ at } t = 4 \quad \vDash \quad B_{D,4}^{0.9,\,1}(Obs_D(exec(A)) \vee Obs_D(exec(B))).$$

If the defender chooses not to take any defensive action, the intruder in turn is unable to distinguish between the situations where the defender deliberately took no action and where the defender simply missed the attack, i.e., the defender considers all threads possible that contain node 6 or node 4. Thus, the intruder has the following belief in actually being able to execute malicious code on a target system undetectedly:

$$\neg Obs_I(defend(A)) \text{ at } t = 4 \quad \vDash \quad B_{I,4}^{0,\,0.2}(\phi),$$

with $\phi = \neg Obs_D(exec(A)) \wedge \neg Obs_D(exec(B)) \wedge (exec(A) \vee exec(B))$

Since both the intruder and defender know the possible attack sequences, it is also possible to analyze belief states of the respective opponent: In the considered situation, the defender knows that the attacker has not observed any defense action and can therefore not distinguish between nodes 4 and 6. However, since the defender was not able to rule out a previous attack on B, from his point of view the intruder could still also consider nodes as 83, 85, 94, and 96 as possible (these are the nodes where B was attacked before and no defend action was taken). Still, the defender has a rather high belief in the intruder's actual belief state, as expressed in the following nested belief:

$$B_{D,4}^{0.8,\,1}(B_{I,4}^{0.0,\,0.2}(\phi))$$

5 Conclusion

In this paper we proposed a well-defined theory to formalize multi-agent beliefs in a security context. This formal representation enables the analysis of the adversary's belief evolution depending on specific actions. At a minimalistic example we demonstrated how a formal belief state analysis can be carried out. Next to formal representations of both the intruder's and defender's beliefs, this is especially useful to gain an opportunity to reason about nested beliefs. This provides novel means of assessing the expected utility of any action when planning a defense strategy: Along with analyzing the direct effect of any action on the network, we can also analyze how any action will influence the belief state of the opponent. With the use of more sophisticated attack models, this enables the defender to drive the intruder into desired safe states, where the intruder expects to achieve her goal, but is actually unable to cause real harm.

We only chose a minimalistic example because a manual analysis in complex attack graphs becomes infeasible by hand due to an excess amount of possible worlds. Still, PDT logic performs well in those and does not limit our approach. Future work includes a description of an—currently in implementation—autonomous system for analysis of complex attack graphs and an experimental evaluation of the demonstrated profound theoretic approach in complex settings.

References

1. Brdiczka, O., Liu, J., Price, B., Shen, J., Patil, A., Chow, R., Bart, E., Ducheneaut, N.: Proactive insider threat detection through graph learning and psychological context. In: Security and Privacy Workshops (SPW), pp. 142–149. IEEE (2012)
2. Chiesa, R.: Peering in the soul of hackers: HPP (the hacker's profiling project) v2.0 reloaded. In: 8.8 Security Conference, Santiago, Chile. 8dot8 (2012)
3. Chiesa, R., Ducci, S., Ciappi, S.: Profiling Hackers: the science of criminal profiling as applied to the world of hacking. CRC Press (2008)
4. Ingols, K., Lippmann, R., Piwowarski, K.: Practical attack graph generation for network defense. In: Computer Security Applications Conference, pp. 121–130. IEEE (2006)
5. Jha, S., Sheyner, O., Wing, J.: Two formal analyses of attack graphs. In: Computer Security Foundations Workshop, pp. 49–63. IEEE (2002)
6. Lippmann, R., Ingols, K., Scott, C., Piwowarski, K., Kratkiewicz, K., Artz, M., Cunningham, R.: Validating and restoring defense in depth using attack graphs. In: Military Communications Conference (MILCOM), pp. 1–10. IEEE (2006)
7. Martiny, K., Möller, R.: A probabilistic doxastic temporal logic for reasoning about beliefs in multi-agent systems. In: 7th International Conference on Agents and Artificial Intelligence (ICAART) (2015)
8. Noel, S., Jajodia, S.: Optimal IDS sensor placement and alert prioritization using attack graphs. J. Netw. Syst. Manag. 16(3), 259–275 (2008)
9. Ou, X., Govindavajhala, S., Appel, A.W.: Mulval: A logic-based network security analyzer. In: USENIX Security (2005)
10. Ou, X., Singhal, A.: Attack graph techniques. In: Quantitative Security Risk Assessment of Enterprise Networks, pp. 5–8. Springer (2011)

11. Phillips, C., Swiler, L.: A graph-based system for network-vulnerability analysis. In: Workshop on New Security Paradigms, pp. 71–79. ACM (1998)
12. Rogers, M.K.: A social learning theory and moral disengagement analysis of criminal computer behavior: An exploratory study. Ph.D. thesis, University of Manitoba (2001)
13. Roy, A., Kim, D.S., Trivedi, K.: Cyber security analysis using attack countermeasure trees. In: 6th Annual Workshop on Cyber Security and Information Intelligence Research, p. 28. ACM (2010)
14. Roy, A., Kim, D.S., Trivedi, K.: Scalable optimal countermeasure selection using implicit enumeration on attack countermeasure trees. In: Dependable Systems and Networks, pp. 1–12. IEEE (2012)
15. Sommestad, T., Ekstedt, M., Johnson, P.: Cyber security risks assessment with bayesian defense graphs and architectural models. In: 42nd Hawaii International Conference on System Sciences, pp. 1–10. IEEE (2009)
16. Theoharidou, M., Kokolakis, S., Karyda, M., Kiountouzis, E.: The insider threat to information systems and the effectiveness of ISO17799. Comput. Secur. **24**(6), 472–484 (2005)

Automatic Classification and Detection of Snort Configuration Anomalies - a Formal Approach

Amina Saâdaoui, Hajar Benmoussa, Adel Bouhoula
and Anas Abou EL Kalam

Abstract IDSs are core elements in network security. The effectiveness of security protection provided by an IDS mainly depends on the quality of its configuration. Unfortunately, configuring an IDS is work-intensive and error prone if performed manually. As a result, there is a high demand for analyzing and discovering automatically anomalies that can arise between rules. In this paper, we present (1) a new classification of anomalies between IDS rules, (2) three inference systems allowing automatic anomaly detection for discovering rule conflicts or redundancies and potential problems in IDS configuration, (3) optimization of IDS rules by removing automatically redundant rules and (4) formal specification and validation of these techniques and demonstration of the advantages of proposed approach on the sets of rules provided by open source Snort IDS. These techniques have been implemented and we proved the correctness of our method and demonstrated its applicability and scalability. The first results we obtained are very promising.

Keywords Anomalies · IDS · Snort · Snort configuration · Redundancies · Conflicts

1 Introduction

With the rapid increase of sophisticated attacks on computer systems, Intrusion detection systems (IDSs) remain at the center of intense research. IDSs allow detecting attack packets targeting network, hosts, services and applications. Basically,

A. Saâdaoui (✉) · A. Bouhoula
Higher School of Communication of Tunis, University of Carthage, Carthage, Tunisia
e-mail: amina.saadaoui@supcom.tn

A. Bouhoula
e-mail: adel.bouhoula@supcom.tn

H. Benmoussa · A.A.E. Kalam
ENSA Marrakesh, Cadi Ayyad University, Marrakesh, Morocco
e-mail: haj.benmoussa@gmail.com

A.A.E. Kalam
e-mail: a.abouelkalam@uca.ma

© Springer International Publishing Switzerland 2015
Á. Herrero et al. (eds.), *International Joint Conference*, Advances in Intelligent
Systems and Computing 369, DOI 10.1007/978-3-319-19713-5_3

Intrusion detection systems can use two main categories of detection methods: anomaly (or behavioral) detection and misuse (or signature) detection. The first one works by identifying a normal profile/behavior of a system or network and then a substantial deviation from this normal behavior is considered as an intrusion. The second one relies on constructing databases for known signatures of attacks. The signature detection is considered as the most commonly implemented techniques in modern intrusion detection systems (IDS) [12] to detect suspicious traffic and Snort [2] is one of the well-known, free and open source IDS that uses a set of signatures (rules). Nevertheless, these rules are generally manually written. Furthermore, it changes frequently due to network expansions and organizational changes. Deploying and managing IDS configuration in large networks with hundreds of devices makes it a time consuming, extensive process that is very error-prone. Gartner [7] has estimated that 65 % of cyber-attacks exploit systems with vulnerabilities introduced by configuration errors. Also, according to ISBS [1] 36 % of the worst security breaches in 2013 were caused by inadvertent human error. This is one of the major reasons why configuration analysis and anomalies management have been gaining significance in the security aspect of the network. Therefore, network security can be significantly improved if configuration errors can be automatically detected.

Given that Snort as the most of NIDSs uses a set of signatures (rules) to detect attacks and malicious packets, the accuracy of its detection depends on the quality of the used rule set and hence anomalies between them affect the performance of detection and create ambiguity in classification of network traffic.

In this context, we are interested in discovering Snort rules anomalies. In fact, the most existing works focuses on analyzing and managing firewall configurations [4, 6, 8, 10, 11]. These works proposed different methods to classify, detect and even correct firewall anomalies. However, a less attention was paid to study conflicts of the network IDS (NIDS) rules.

In [9], authors consider a general setting of network security, they proposed algorithms that allow analyzing relations between different component in Network (IDS, Firewall,...). But, this approach considers the same formalism for firewall and IDS without considering the specificity of each component. However, there is a difference between them especially in the manner of applying rules. The work in [13], presents a new approach for modeling configurations of different components in network without considering conflicts between IDS rules. Authors only optimize the rule-set by reducing the false alarms. In [14] authors introduce a method to analyze inconsistencies that could arise in configurations of different components in network. However, they talk about inconsistencies (conflicts) without proposing any method to classify and detect them. A validation and correctness of configurations approach is proposed in [5]. In fact, authors propose a new architecture that allows verifying the correctness of the intrusion detection policy with the global security policy, using an original meta-policy approach. But the problem is that they did not consider and analyze inconsistencies between IDS rules.

Another work has been done on IDS rule analysis [12], this work focused on classifying and discovering different types of anomalies that can arise between IDS rules. Nevertheless, each IDS has a specific default order of actions applied on rules and also a specific syntax to specify these rules. So, considering the same classification for all IDSs cannot always work efficiently.

Our approach is fundamentally different from what is already proposed. On one hand, we discover and distinguish different types of conflicts by considering combinations of rules regarding the order used to apply them. Also, our approach allows to automatically optimize the Snort configuration by eliminating all redundant rules. On the other hand, we formally prove the correctness and completeness of our approach.

In this paper, we propose our approach as inference systems allowing the automatic detection of IDS anomalies. We present also a novel anomaly management tool for IDSs configurations to facilitate more accurate anomaly detection.

The remainder of this paper is organized as follows. In Sect. 2, we provide information about Snort rules and we formally define some key notions. Then, in Sect. 3 we present our proposed classification of Snort rule anomalies and Sect. 4 presents inference systems which allow to discover existing anomalies. In Sect. 5, we address the implementation and evaluations of our tool. Finally, we present our conclusions and discuss our plans for future work.

2 Snort Rule

Our main goal is to classify and detect IDSs rules anomalies. In what follows, we define, formally, some key notions.

We consider a finite domain P containing all the headers of packets possibly incoming to or outgoing from a network. Generally, IDS configuration is a finite sequence of rules of the form $SC = \{r_j : \langle action_j, c_j \rangle\}_{1 < j < N+1}$. Each r_j has two parameters, the first one is the action $Action_j$ which defines the behavior of the IDS on filtered packets: according [2], there are six available default actions in Snort: [*Alert* rules generate an alert, and then log the packet - *Log* rules log the packet - *Pass* rules ignore the packet - *Activate* rules alert and then turn on another dynamic rule - *Dynamic* rules remain idle until activated by an activate rule - *Drop* rules block and log the packet]. The second parameter is the precondition c_j which is a region of the packets space. It consists usually of source address, destination address, protocol, source port, destination port and a set of options. This formal representation of rules ($\{r_j : \langle action_j, c_j \rangle\}_{1 < j < N+1}$) allows to define all IDSs rules even if the syntax of them varies across different intrusion detection devices. Figure 1 shows the syntax of some Snort rules.

Snort applies its rules to packets by the following default order: the pass rules are applied first, then the drop rules, then the alert rules, next activate rules, after dynamic rules and finally the log rules are applied. In our research work, we consider the default rule order cited above. Moreover, into each set of rules, packet inspection is performed in a sequential order starting from the first signature rule until a matching rule is found. So we consider rules with the same action as a set and we divide

```
r1: pass TCP 10.0.0.1 80 -> 172.16.0.0 * (content : "j0001 86 a5j " ; msg : "external mountd access";)

r2:alert TCP 192.168.0.0/23 80 -> 172.16.0.0 * (content : "j0001 86 a5j " ; msg : "external mountd access";)

r3:alert TCP 10.0.0.0/31 80 -> 172.16.0.0 * (content : "j0001 86 a5j " ; msg : "external mountd access";)

r4:alert TCP 192.168.0.0/24 80 -> 172.16.0.0 * (content : "j0001 86 a5j " ; msg : "external mountd access";)

r5:alert TCP 192.168.1.0/24 80 -> 172.16.0.0 * (content : "j0001 86 a5j " ; msg : "external mountd access";)

r6:drop TCP 10.0.0.1 80 -> 172.16.0.0/24 * (content : "j0001 86 a5j " ; msg : "external mountd access";)

r7: drop TCP 192.168.0.0/24 80 -> 172.16.0.0/24 * (content : "j0001 86 a5j " ; msg : "external mountd access";)
```

Fig. 1   Snort rule structure

rules in 6 sets A_i where the index i represents the priority order of each set of rules, this priority order is defined by respecting default order of actions. We also define the set A as follows:

$A = \{A_i\}_{\{i=1\to6\}} = \{A_1, A_2, A_3, A_4, A_5, A_6\} = \{rules(action=pass),\ rules (action =drop),\ rules(action=alert),\ rules(action=activate),\ rules(action=dynamic),\ ru-les(action=log)\}$.

Based on the priority order of Snort rules, we define a new set $C = \{ <pass, Alert> <pass, drop> <pass, activate>, <pass, log>, <drop, alert>, <drop, activate> <drop, dyanmic> <drop, log>, <alert, log>, <alert, log> and <activate, log> \}$ which represents the couple of sets $< A_i, A_j >$. Rules from these two sets may present different conflicts.

We consider the following functions: $dom(rule)$ which maps each rule into the subset of packets handled by this rule and $priority(r)$ which gives priority of the rule r with respect to other rules belonging to the same A_i.

Based on these rules, the efficiency of IDS detection heavily depends on the quality of the employed rule set. In this context, any conflicts or redundancies that arise between rules create ambiguity and possible misconfuration. In the followed section we classify and define, formally, possible anomalies.

3 Anomalies Classification

In this research work, we consider relations between rules in every set of rules working with the same action and also relations between rules between different sets. So we divide anomalies in two main types: Intra-set-redundancies and Inter-sets-conflicts. We define in the first one redundancy and in the second one three conflicts: correlation, generalization and shadowing.

3.1 Intra-Set-Redundancies

3.1.1 Redundancy

This type of anomaly is detected between one rule and set of rules having the same action when the domain of the first one is totally included in the domain of the other rules, so removing this rule will not affect the semantic of IDS configuration. This is the case for rule r_2 which is redundant to the union of rules r_4 and r_5 in the Snort configuration of Fig. 1.

Definition 1 *A Rule* $r_j \in A_i$ *is redundant to other rules iff* $\forall p \in P$, *if* $p \in r_j$ *then* $\exists r_k \in A_i$ *where* $p \in rk$.

3.2 Inter-Sets-Conflicts

Shadowing A rule r_n in A_i is shadowed if the previous rules to this rule (previous rules are defining as rules belonging to the same set A_i or other sets A_j which have a higher priority than A_i) match all the packets that match this rule and r_n which is always applied after these rules does not get the chance to match any packet. For instance, rule r_4 is simply shadowed by rule r_7 in the Snort configuration of Fig. 1 (*Drop* has a higher priority than *Alert*).

Definition 2 $\forall r_n \in A_i, r_n$ *is shadowed iff* $\forall p \in r_n \exists r_m \in A_j$ *such that* $p \in r_m$ *where* $(j < i$ *or* $(j == i$ *and priority* $(r_m) >$ *priority* $(r_n))$.

Correlation For each couple $<A_i, A_j>$ in C, A rule r_n from A_i is correlated with r_m from A_j if both of them match some common packets i.e. the rule r_n matches some packets, which are also matched by the rule r_m. For instance, Correlation happens for rules r_3 and r_6 in the configuration of Fig. 1.

Definition 3 *For* $<A_i, A_j> \in C, \forall r_n \in A_i$, r_n *is correlated with* r_m *where* $r_m \in A_j$ *iff* $r_n \cap r_m \neq \emptyset$ *and* $r_n \not\subseteq r_m$ *and* $r_m \not\subseteq r_n$.

Generalization For each couple $<A_i, A_j>$ in C, a rule q_j from A_j is said to be in generalization of $r_i \in A_i$ if the first rule matches all the packets which can be also matched by the second rule. For instance, rule r_6 is a generalization for rule r_1 in the configuration of Fig. 1.

Definition 4 *For* $<A_i, A_j> \in C, \forall r_n \in A_i$, r_m *is a generalization for* r_n *where* $r_m \in A_j$ *iff* $r_n \subseteq r_m$.

In the next section, we will present our automatic approach to identify all anomalies that could exist in an IDS configuration.

4 Inference Systems

In this section, we propose our approach as inference systems for examining the two classes of anomalies previously defined. These inference systems allow to analyze and manage all IDSs rules by using as input the formal representation of rules defined in Sect. 2.

4.1 Discovering and Removing Redundancies

In Fig. 2 we propose an inference system that presents necessary and sufficient steps to discover and then remove, automatically, Intra-set-redundant rules.

The rules of this inference system apply to quadruple (A, R_i, R_i^f, A^F). The first component A represents the set of A_i defined in Sect. 2, the second component R_i is a temporary variable used to parse all the rules of a given set A_i, the third component R_i^f is an updated version of A_i by removing all the redundant rules and the final component A^F is the final set of updated R_i^f.

The second inference rule **Parse** takes, at each iteration, a set A_i from A, then by using the other inference rules we will identify and remove, precisely, the set of redundant rules in this set.

$$
\begin{array}{ll}
\textit{Init} & \dfrac{}{\{A_1^{init}, .., A_6^{init}\}, \varnothing, \varnothing, \varnothing} \\[2ex]
\textit{Parse} & \dfrac{(A_i^{init} \cup A), \varnothing, \varnothing, A^F}{A, A_i^{init}, A_i^{init}, A^F} \\[2ex]
\textit{Redundancy} & \dfrac{A, (\{r_j\} \cup R_i), R_i^f, A^F}{A, R_i, R_i^f \setminus r_j, A^F} \quad \textit{if } dom(r_j) \subseteq \bigcup_{(k, k \neq j)} dom(r_k) \\[1ex]
& \qquad\qquad\qquad\qquad\qquad \textit{where } \begin{cases} r_k \in R_i^f \text{ and} \\ r_j \cap r_k \neq \varnothing \end{cases} \\[2ex]
\textit{Pass} & \dfrac{A, (\{r_j\} \cup R_i), R_i^f, A^F}{A, R_i, R_i^f, A^F} \quad \textit{if no other rule applies} \\[2ex]
\textit{Define} & \dfrac{A, \varnothing, R_i^f, A^F}{A, \varnothing, \varnothing, \{A^F, R_i^f\}} \\[2ex]
\textit{Success} & \dfrac{\varnothing, \varnothing, \varnothing, A^F}{A^F}
\end{array}
$$

Fig. 2 Inference system for discovering and removing Intra-set-redundancies

The main inference rule of this system is **Redundancy**, it allows discovering the redundant rules as defined in the previous section. In fact, it deals with each rule $r_j \in R_i$ and removes it if the domain of this rule is totally included by other rules in this set. In fact, each $dom(r_j)$ is compared with the union of $dom(r_k)(\forall k, k \neq j)$, if these two domains are equal or the first is included in the second domain then, the rule r_j is considered as redundant and will be removed. The fourth inference rule **Define** allows to define the new set A^F. The rule **Success** is applied when we parse all the A_i from the set A.

We write $C \vdash_A C'$: C' is obtained from C by application of one of the inference rules of Fig. 2 and we denote by $\vdash_A^*$ the reflexive and transitive closure of $\vdash_A$.

In order to prove the correctness of our approach, we start by the following definition:

Definition 5 *Let us consider* $A^1 = \{A_i^1\}_{\{i=1\to6\}}$ *and* $A^2\{A_i^2\}_{\{i=1\to6\}}$ *two distinct Snort configurations. We say that* A^1 *and* A^2 *are* **semantically equivalent** *if and only if* $\forall p \in P$, *if* $p \in A_i^1$ *then* $p \in A_i^2$.

Theorem 1 *If* $(A, \varnothing, \varnothing, \varnothing) \vdash^* A^F$ *(where* $A = \{A_1^{init}..A_6^{init}\}$ *and* $A^F = \{R_1^f..R_6^f\}$*) then A and* A^F *are semantically equivalent.*

Proof If $(A, \varnothing, \varnothing, \varnothing) \vdash^* A^F$ then either all steps and not last one are follow. In such case, $A_i^{init} = R_i^f \cup \bigcup_j dom(r_j)$ where $r_j \in R_i^f$. Suppose that $\exists p \in P, p \in A_i^{init}$ and $p \notin R_i^f$. $p \in A_i^{init}$ then $p \in \bigcup_j dom(r_j)$ where $r_j \in R_i^f$ then $p \in R_i^f$ which is a contradiction. Then, $p \in R_i^f$. Therefore, A and A^F are semantically equivalent.

Theorem 2 (termination) *The inference system shown in Fig. 2 is terminating.*

Proof For our inference system we have as input $A = \{A_1^{init}..A_6^{init}\}$ and as output $A^F = \{R_1^f..R_6^f\}$. Now let's define the ranking of A, $rank(A) = \sum_{i=1}^6 \left|A_i^{init}\right|$ where $|X|$ is the cardinality of the set X. Each $A_i^{init} = R_i^f \cup \bigcup_j dom(r_j)$, then $\left|A_i^{init}\right| > \left|R_i^f\right|$, then $\sum_{i=1}^6 \left|A_i^{init}\right| > \sum_{i=1}^6 \left|R_i^f\right|$. So $rank(A) > rank(A^F)$. It follows that our system is terminating.

4.2 Discovering Conflicts

Once A has been updated by removing all the redundant rules from all the sets $A_i \in A$, we can start the process of discovering Inter-sets-conflicts.

Discovering Shadowed Rules The rules of the system shown in Fig. 3 apply to triple (A, R_i, SH) whose first component A is a sequence of sets A_i, whose second component represents set of rules R_i and whose third component SH is the list of shadowed rules discovered. SH is initialized to an empty set.

Shadowing is the main inference rule for the inference system. It deals with each rule from the set A_i and verifies if the domain of this rule is totally included by the domain of other rules having a higher priority (A higher priority can be verified by one of these conditions. The first is when the compared rules are belonging to sets A_i

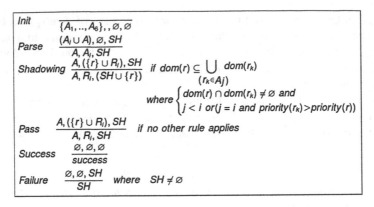

Fig. 3 Inference system for discovering shadowed rules

and A_j for example and A_j have a higher priority than A_i; the second condition when the two rules are belonging to the same set A_i and the priority of the second rule is higher than the rule under consideration). So, if this condition is verified, this rule will be added to the set SH. And we will indicate to the network administrator that this rule is totally masked and will never be applied.

The **Success** rule is applied when we parse all the rules and all the sets A_i of the set A without identifying shadowed rules, so in this case each rule in A is not totally masked and will be applied. And **Failure** is applied when at least one rule is identified as shadowed rule.

Theorem 3 *(Correctness-Success) If $(A, \varnothing, \varnothing) \vdash^* success$ then each rule in A is not totally masked and will be applied.*

Proof If $(A, \varnothing, \varnothing) \vdash^* success$ then we have $(A, \varnothing, \varnothing) \vdash (A^1, R_i^1, SH^1) \vdash \ldots \vdash (A^n, R_j^n, SH^n) \vdash \ldots \vdash success$ where $SH = \varnothing$. Suppose that $\exists r \in R_i$ where r is totally masked and will never be applied, then, $(dom(r) \subseteq \bigcup_{(r_k \in A_j)} dom(r_k))$ where $(j < i$ or $(j = i$ and $priority(r_k) > priority(r))$ and $dom(r) \cap dom(r_k) \neq \varnothing)$. So at this step the precondition of the inference rule *Shadowing* is verified, then $SH = SH \cup r$ then $SH \neq \varnothing$. Contradiction. It follows that each rule in A is not totally masked and will be applied.

Theorem 4 *(Correctness-Failure) If $(A, \varnothing, \varnothing) \vdash^* SH$ then all shadowed rules are identified.*

Proof If $(A, \varnothing, \varnothing) \vdash^* SH$ then we have $(A, \varnothing, \varnothing) \vdash (A^1, R_i^1, SH^1) \vdash \ldots \vdash (A^n, R_j^n, SH^n) \vdash \ldots \vdash SH$. Suppose that $\exists r \in R_i$ where r is totally masked and will never be applied and $r \notin SH$, then, according to the definition of shadowed rules presented in Sect. 3 $(dom(r) \subseteq \bigcup_{(r_k \in A_j)} dom(r_k))$ where $(j < i$ or $(j = i$ and $priority(r_k) > priority(r))$ and $dom(r) \cap dom(r_k) \neq \varnothing)$. So at this step the precondition of the inference rule *Shadowing* is verified, then $SH = SH \cup r$ so $r \in SH$. Which is a contradiction. It follows that all shadowed rules are identified.

Discovering Correlated and Generalized Rules The rules of the system shown in Fig. 4 apply to six components: $(C, <R_i, R_j>, R_j, r', CR, GN)$. The first component C represents the set of different combinations of the couple of sets $<A_i, A_j>$ as defined in Sect. 2. Rules from these two sets may present different anomalies (Correlation or generalization). Inference rules $Parse_C$ and $Parse_{A_i}$ are used to define sets used to verify intersection between different rules. In fact, at each iteration, the fourth element **r'** will contain a rule from the set R_i. This r' will be compared with other rules from the set R_j to check if it is correlated or generalized. CR and GN are the sets of correlated and generalized rules respectively.

Correlation and **Generalization** are the main inference rules for the inference system. The first one detects correlated rules. It deals with each rule from set R_j and verify if this rule is correlated with another rule from the set R_i, here the couple $<R_i, R_j>$ is belonging to the set of combination C. So, a rule is considered to be correlated with another one if and only if, they have different actions and their intersection is not empty. If it is the case, this couple of rules will be added to the set CR which contains the set of correlated couple of rules. The same for the second, **Generalization**, but here we will detect exceptions. I.e., rules that have different actions and the domain of rule belonging to the set R_i is totally included in the domain of the rule belonging to the set R_j.

The **Success** rule is applied when we parse all the rules and all the couples $< A_i, A_j >$ of the set C without identifying correlated or generalized rules, so in this case our configuration is considered to be conflict free. And **Failure** is applied when at least one of the sets CR and GN is not empty.

$$Init \quad \frac{}{\overline{C, \varnothing, \varnothing, \varnothing, \varnothing, \varnothing}}$$

$$Parse_C \quad \frac{(< A_i, A_j > \cup C), < \varnothing, A_i >, \varnothing, r', CR, GN}{C, < A_i, A_i >, \varnothing, r', CR, GN}$$

$$Parse_{Ai} \quad \frac{C, < r \cup R_i, R_j >, \varnothing, r', CR, GN}{C, < R_i, R_j >, R_j, r, CR, GN}$$

$$Correlation \quad \frac{C, < R_i, R_j >, (q \cup R), r', CR, GN}{C, < R_i, R_j >, R, r', (CR \cup < r', q >), GN} \quad if \begin{cases} r' \cap q \neq \varnothing \text{ and} \\ r' \not\subseteq q \text{ and } r' \not\supseteq q \end{cases}$$

$$Generalization \quad \frac{C, < R_i, R_j >, (q \cup R), r', CR, GN}{C, < R_i, R_j >, R, r', CR, (GN \cup < r', q >)} \quad if \ r' \subseteq q$$

$$Pass \quad \frac{C, < R_i, R_j >, (q \cup R), r', CR, GN}{C, < R_i, R_j >, R, r', CR, GN} \quad if \ no \ other \ rule \ applies$$

$$Success \quad \frac{\varnothing, < \varnothing, Rj >, \varnothing, r', \varnothing, \varnothing}{success}$$

$$Failure \quad \frac{\varnothing, < \varnothing, Rj >, \varnothing, r', CR, GN}{(CR, GN)} \quad if \ (CR \neq \varnothing \ or \ GN \neq \varnothing)$$

Fig. 4 Inference system for discovering conflicts

Definition 6 *A conflict is defined as a couple $<r, q> \in CR$ or GN where $<r, q> \in < A_i, A_j>$ and $<A_i, A_j> \in C$.*

Theorem 5 *(Correctness-Success) If $(C, \emptyset, \emptyset, \emptyset, \emptyset, \emptyset) \vdash^* success$ then we have a conflict free configuration.*

Proof If $(C, \emptyset, \emptyset, \emptyset, \emptyset, \emptyset) \vdash^* success$ then $CR = \emptyset$ and $GN = \emptyset$. Then $\forall < r, q> \in <A_i, A_j>$ where $<A_i, A_j> \in C$, $<r, q> \notin CR$ and $<r, q> \notin GN$. It implies that $dom(r) \cap dom(q) = \emptyset$, Then there is no conflict between r and q. Or $dom(q) \subseteq dom(r)$ then q is totally masked by r because $q \in A_j$ and A_i has a higher priority than A_j, so this anomaly is not a correlation or a generalization. Thus, there is no conflict between r and q. It follows that we have a conflict free configuration.

5 Implementation and Computer Experiments

To evaluate a practical value of our approach, we have implemented inference systems described in Sect. 4 using the C++ developing language to benefit from the execution speed provided by this language. We analyzed the effectiveness of our approach using the rule collections of Snort IDS and the ETOpen (Emerging Threats) IDS ruleset [3], where Emerging Threats is an open source community for collecting Suricata and Snort rules, firewall rules, and other IDS rulesets. The Table 1 describes the details of the discovered anomalies. But by default, all generated and distributed rules in [3] and [2] are **ALERT** rules because it is considered to be the safest way for distribution and it is up to each user to adjust them as needed for their network's needs. So anomalies discovered in table below are only redundancies.

In the following we describe the characteristics of Snort configuration used in our case study.

Table 1 Anomalies discovered

Rule set	Number of rules	Number of anomalies
emerging-pop3	9	8
emerging-dos	34	18
emerging-rbn-malvertisers	68	0
emerging-scada	14	9
protocol-voip	22	17

5.1 Step 1: Discovering and Removing Redundancies

Once we specify the Snort configuration, we proceed to the detection and elimination of redundant rules process, these rules are identified in each class of rules as explained in Sect. 4. In fact, based on the configuration shown in Fig. 5 and using the inference system for discovering and removing redundancies sets A_i will be initialized as follow:

$A = \{A_1^{init}..A_6^{init} =\}\{$ *rules(action=pass), rules(action=drop), rules(action= alert), rules(action=activate), rules (action=dynamic), rules(action=log)}* $= \{(R_1, R_2), (R_7, R_9), (R_3, R_4, R_5, R_6), \emptyset, \emptyset, (R_8)\}$.

Note that at each step only one inference rule could be applied.

Then, the inference rule **Parse** starting from set A_1 allows to recursively visit sets A_i. So first, we will consider the set A_1 (action = pass), then we verify whether this set contains redundant rules by using the third inference rule **Redundancy**. For this set rules R_1 and R_2 are not redundant, hence the precondition of the inference rule **Redundancy** is not verified and cannot be applied. Therefore, for these two rules **Pass** rule will be applied.

Then, we reapply **Parse** rule, and we parse the next set of the set A_i. Now, for example for the set A_3 (action = alert) we verify if the first rule of this set is redundant to other rules (it's the case of rule R_3 which is redundant to the union of rules R_5 and R_6) so the precondition of **Redundancy** is verified and we can apply it, therefore this rule (R_3) will be removed. Hence repeated application of our inference rules ensures that the new A^F (The updated version of A) obtained is redundancies-free.

5.2 Step 2: Extracting Shadowed Rules

Once ensured that all redundant rules are identified, we proceed to the shadowed rules discovering mechanism using the inference system described in Sect. 4. As explained before, our configuration is listed as follow:

$A = \{(R_1, R_2), (R_7, R_9), (R_4, R_5, R_6), \emptyset, \emptyset, (R_8)\}$ (R_3 has been removed).

So, to apply the inference system shown in Fig. 3 on A, we will first initialize our variables. Then we apply the **Parse** rule to extract the first set A_1, then by using the third inference rule we verify whether this set contains masked rules. So for example for the set A_3 (action = alert) the previous rules of the rule R_5 is the rule R_9, this rule is considered as previous rule because *Drop* is applied before *Alert* as explained in Sect. 2. Therefore, the precondition of the inference rule **Shadowing** is verified. So our inference system will notify the administrator that R_5 is shadowed by R_9. Hence repeated application of our inference rules ensures that all shadowed rules are extracted.

5.3 Step3: Discovering Correlated and Generalized Rules

After that shadowed rules discovering process have been established, we proceed to the detection of conflicting rules. We obtained the result displayed in Fig. 5. According to this outcome, our tool identifies 7 conflicts (indexes of rules are identified according to the initial configuration i.e., before removing redundant rules):

```
##################          ##################
IDS rules :                 Shadowed rules :
##################          ##################
R1 : pass TCP 10.0.0.0 80 -> 172.16.0.0 * <conten   rule number  5 is masqued by its previous rules
R2 : pass TCP 10.0.0.1 80 -> 172.16.0.0 * <conten
R3 : alert TCP 192.168.0.0/31 80 -> 172.16.0.0 *   ##################
R4 : alert TCP 10.0.0.0/31 80 -> 172.16.0.0 * <co   Conflicts :
R5 : alert TCP 192.168.0.0 80 -> 172.16.0.0 * <co   ##################
R6 : alert TCP 192.168.0.1 80 -> 172.16.0.0 * <co
R7 : drop TCP 10.0.0.1 80 -> 172.16.0.0/31 * <con   Generalization between R 4 and R 1
R8 : log TCP 10.0.0.1 * -> 172.16.0.0 * <content    Generalization between R 4 and R 2
R9 : drop TCP 192.168.0.0 80 -> 172.16.0.0/31 * <   Generalization between R 7 and R 2
                                                    Generalization between R 8 and R 2
##################
redundancies :                                      Correlation betweeen R 4 and R 7
##################                                  Correlation betweeen R 8 and R 7
      rule number  3 is redundant and will be removed   Correlation betweeen R 8 and R 4
```

Fig. 5   Case study

- Generalization: R_4 generalizes R_1, R_4 generalizes R_2, R_7 generalizes R_2, R_8 generalizes R_2.
- Correlation: R_4 is correlated with R_7, R_8 is correlated with R_7, R_8 is correlated with R_4.

In fact, based on this configuration we define the combinations C of conflicted couples $<A_i, A_j>$ as defined in Sect. 2. So, if we apply the inference system shown in Fig. 4 on this combination we first initialize our variables. Then we apply the $Parse_C$ rule to extract the first couple $<pass, alert>$. For each couple we use $Parse_{A_i}$ to parse rules of A_i, we compare each rule to others from set A_j if precondition is verified then we detect conflict (correlation or generalization). Otherwise we apply **Pass** rule. For example, for rules R_1 and R_4 we apply the inference rule **Generalization** and we add this conflict to the list of generalized rules.

6 Conclusion

The accuracy and effective of the detection in signature based IDS depends mainly on the quality of the rule configuration. Thus, the need of discovering anomalies between rules in an automatic manner. The work presented in this paper provides essentially two mechanisms. First, we classify anomalies that can arise between IDS rules, and second, we extract these anomalies by using inference systems. Our detection approach is totally automatic and allows an optimal optimization of IDS rules by removing automatically redundant rules. We proved also the correctness and completeness of our approach. Finally our method has been implemented using C++ language. The experimental results obtained are very promising.

As further work, we will apply our strategy to discover anomalies between multiple IDSs or with heterogeneous security equipment (Intrusion Detection systems /Firewall). Our second objective is to exploit the requirement of the global security policy to assist the network administrator to correct these discovered anomalies.

References

1. Information security breaches survey. Available from http://www.pwc.co.uk/assets/pdf/cyber-security-2013-technical-report.pdf (2013)
2. Snort Users Manual 2.9.3. Available from https://www.snort.org/documents (2014)
3. Emerging Threats.net Open rulesets. Available from http://rules.emergingthreats.net (2015)
4. Al-Shaer, E.S., Hamed, H.H.: Modeling and management of firewall policies. IEEE Trans. Netw. Serv. Manage. 1(1), 2–10 (2004)
5. Blanc, M., Briffaut, J., Clemente, P., El Rab, M.G., Toinard, C.: A collaborative approach for access control, intrusion detection and security testing, pp. 270–277 (2006)
6. Chomsiri, T., Pornavalai, C.: Firewall rules analysis, pp. 213–219 (2006)
7. Colville, R.J., Spafford, G.: Gartner ras core resarch note g00208328 (2010)
8. Cuppens, F., Cuppens-Boulahia, N., Garcia-Alfaro, J.: Detection and removal of firewall misconfiguration (2005)
9. Garcia-Alfaro, J., Cuppens, F., Cuppens-Boulahia, N.: Analysis of policy anomalies on distributed network security setups, pp. 496–511 (2006)
10. Hu, H., Ahn, G.-J., Ketan, K.: Detecting and resolving firewall policy anomalies. IEEE Trans. Dependable Secure Comput. 9(3), 318–331 (2012)
11. Mukkapati, N., Bhargavi, Ch.V.: Detecting policy anomalies in firewalls by relational algebra and raining 2d-box model. IJCSNS Int. J. Comput. Sci. Network Secur. 13(5), 94–99 (2013)
12. Stakhanova, N., Li, Y., Ghorbani, A.A.: Classification and discovery of rule misconfigurations in intrusion detection and response devices, pp. 29–37 (2009)
13. Uribe, T.E., Cheung, S.: Automatic analysis of firewall and network intrusion detection system configurations. In: Technical report, SRI international 9 (2004)
14. Zhang, D.: Inconsistencies in information security and digital forensics, pp. 141–146 (2010)

An Improved Bat Algorithm Driven
by Support Vector Machines
for Intrusion Detection

Adriana-Cristina Enache and Valentin Sgârciu

Abstract Today, the never-ending stream of security threats requires new security solutions capable to deal with large data volumes and high speed network connections in real-time. Intrusion Detection Systems are an omnipresent component of most security systems and may offer a viable answer. In this paper we propose a network anomaly IDS which merges the Support Vector Machines classifier with an improved version of the Bat Algorithm (BA). We use the Binary version of the Swarm Intelligence algorithm to construct a wrapper feature selection method and the standard version to elect the input parameters for SVM. Tests with the NSL-KDD dataset empirically prove our proposed model outperforms simple SVM or similar approaches based on PSO and BA, in terms of attack detection rate and false alarm rate generated after fewer number of iterations.

Keywords Feature selection · Bat algorithm · SVM and IDS

1 Introduction

Information security has become a major concern among most enterprises as their business depends on critical data stored on information systems. In spite of the recent technology advances and current concerns for assuring security, cyber threats are still proliferating and becoming more complex and dynamic. Intrusions are happening on a daily basis and adopting new types of hardware or software technologies creates a heterogeneous system and possibly opens new "doors" for attackers.

A.-C. Enache (✉) · V. Sgârciu
Faculty of Automatic Control and Computer Science, University Politehnica
of Bucharest, Romania, Bucharest, Europe
e-mail: adryanaenache@gmail.com

V. Sgârciu
e-mail: vsgarciu@aii.pub.ro

© Springer International Publishing Switzerland 2015
Á. Herrero et al. (eds.), *International Joint Conference*, Advances in Intelligent
Systems and Computing 369, DOI 10.1007/978-3-319-19713-5_4

To mitigate threats a multi-layered security system can be implemented. Intrusion Detection Systems have become an indispensable component of this multilayered approach as they offer several advantages such as: real-time detection, logging system events for further analysis and provide an additional line of defense. An **Intrusion Detection System** (IDS) monitors the activities in the system and determines if they indicate a possible attack or represent legitimate usage [1]. IDS can be classified based on the source of the analysed information (network and host), response to an intrusion (passive and active) or, most often used, their data analysis approach (*misuse* and *anomaly detection*). *Misuse* detection is simple and effective but, it can only uncover known intrusions. *Anomaly* detection constructs a normal profile of the system and identifies intrusions as deviations from it. Therefore, this second method can identify new type of attacks but, it can also generate a lot of false alarms [6].

Since 1980, when the IDS concept was introduced by James Anderson, researchers have proposed different approaches. Many have focused on anomaly detection and constructed two component models: preprocessing stage for feature selection and classification step using machine learning algorithms. The preprocessing step can impact intrusion detection because redundant or irrelevant features may hinder the classifier's accuracy or response time. Current IDS models implement *filter based* feature selection methods such as Correlation Feature Selection [11] or Information Gain [3]. These approaches are simple and independent of the classifier but, do not always improve the detection stage in terms of correctly generated alarms. Therefore, some researchers choose predictive models to determine the subset of features.

During the last decade, **Swarm Intelligence** (SI) algorithms have been adopted in many applications, intrusion detection being no exception. Their popularity is given by their capability to solve complex problems with simple and less intelligent agents which are auto-organized and can adapt to changing conditions. For example, Ma et al. [8] applied Binary Particle Swarm Optimization (BPSO) to create a hybrid intrusion detection model. For this, they joined BPSO with Support Vector Machines to select the improved feature subset and the input parameters for the classifier. Tests on the KDD-Cup99 dataset proved their method is accurate. Wang et al. [14] proposed a similar approach but, they used BPSO to determine the best subset of features and SPSO(Standard PSO) to search for improved SVM parameters. The fitness function for PSO is given by the accuracy of the classifier. Authors only report an improved detection (99.8438 %). Moreover, other SI algorithms have been combined with machine learning classifiers for feature selection such as: Ant colony [5], Artificial Bee Colny (ABC) [15], Bat Algorithm [4] or the Hybrid Bat Algorithm [7].

In this paper we propose an **anomaly based IDS** model combining a personal improved version of the **Bat Algorithm**(BA), we call BA(E), and Support Vector Machines (SVM) to create two components: a pre-processing phase implementing a wrapper feature selection method and a detection stage. For the feature selection we exploit the binary version of BA(E) (BBA(E)), while for detection we enhance the SVM classifier with input parameter selection based on BA(E). **SVM** has good generalization and learning abilities in noisy high dimensional data sets which make it a pertinent candidate for intrusion classification. While, **BA** is a promising novel SI algorithm for solving optimization problems and has outperformed PSO and GA [18] or the binary versions of PSO, Firefly Algorithm (FFA) or Gravitational Search Algorithm (GSA) [10]. The rest of the paper is organized as follows: first sections introduce the main algorithms used to construct our IDS model. Next, in Sect. 4 we show the proposed modified version of BA. Our approach for feature selection is described in Sect. 5. The model setup and test results are given in Sect. 6. Finally, the conclusions and future work.

2 Support Vector Machines

Support Vector Machines (SVM) is a binary classifier that conducts *structural risk analysis* of statistical learning theory to search for an optimum hyperplane to separate the two classes. In order to define this hyperplane, the algorithm computes some *support vectors* such that it obtains the maximum *margin* [2].

Let $X = [x_1, x_2, ..., x_N]$ be an input training datatset and $Y = [y_1, y_2, ..., y_N]$ the class label where $y_i \in \{-1, 1\}$. A new object x can be classified as $sign(f(x)) = y = sign(w \cdot x + b)$, where $f(x)$ is the separating hyperplane, w is the weight vector and b is the bias.

There are cases when the dataset is not *perfectly linearly separable*. To address this issue, we can introduce a *soft margin*, meaning the classifier will allow mislabeled data points, or/and we can use *kernel functions* to transform the nonlinear SVM into a linear problem by mapping the dataset into a higher-dimensional feature space. For our proposed model we will apply the **radial basis function (RBF)**, defined below:

$$K(x_i, x) = exp(-\frac{1}{2\sigma^2} \|x_i - x\|^2) \tag{1}$$

This kernel function offers some advantages such as: fewer controllable parameters and good nonlinear forecasting abilities. Therefore, the SVM classifier has two input parameters: C and σ, which influence its performances as follows:

- **the regularization parameter (C)** - influences the "softness" of the margin. *Smaller C* permits softer-margins and greater errors. While, *larger C* create a more accurate model with harder margins. However, the generalization of the classifier can be tampered.
- **kernel parameter** (σ) - is the constant variable for the kernel function. This parameter shows the correlation among support vectors that define the hyperplane and its choosing may induce overfitting or underfitting.

3 Bat Algorithm

The **Bat Algorithm** (BA) is a swarm intelligence algorithm proposed by Yang in 2010 [17]. The author was inspired by the echolocation of microbats, which emit a loud sound pulse to identify an obstacle or a prey. All individuals in the group have the same typology and *fly randomly* searching for their target. Their trajectory is defined by two internal variables: *position in space* $x_i = (x_{i,1}, x_{i,2}, ..., x_{i,d})$ and *flying velocity* $v_i = (v_{i,1}, v_{i,2}, ..., v_{i,d})$, where d is the dimension of the problem to be solved. A pseudocode of BA is given in Algorithm 1. At each iteration bats will update their internal frequency ($freq_i$), velocity (v_i) and position (x_i), according to the following equations:

$$freq_i = freq_{min} + (freq_{max} - freq_{min}) \cdot \beta \qquad (2)$$

$$v_{i,j}{}^t = v_{i,j}{}^{t-1} + (x_{i,j}{}^{t-1} - x\_best_j) \cdot freq_i \qquad (3)$$

$$x_{i,j}{}^t = x_{i,j}{}^{t-1} + v_{i,j}{}^t \qquad (4)$$

where $\beta \in [0, 1]$ is a random vector drawn from a uniform distribution. The position of the bat (x_i) denotes *the solution of the problem* and the best solution of the group (x_{best}) is determined and directly communicated to all individuals, after each iteration. The quality of the solution is given by its *fitness function*.

To zoom in and *exploit* a promising searching area, the bat will decrease its *loudness* (A_i) and increase its *rate of the pulse emission* (r_i) as follows:

$$A_i{}^{t+1} = \alpha \cdot A_i{}^t \qquad (5)$$

$$r_i{}^{t+1} = r_i{}^0 \cdot [1 - e^{-\gamma \cdot t}] \qquad (6)$$

where α ($0 < \alpha < 1$) and γ ($\gamma > 0$) are constants. To add *exploration* in the r space, Yang introduces a local search implemented with *random walks*:

$$x_{new} = x_{old} + \delta \cdot A^*{}_t \qquad (7)$$

where $\delta \in [-1, 1]$ is a random number and $A^*{}_t$ is the average loudness of all bats at iteration t. However, this new solution is generated only if the pulse rate emission of the bat satisfies the condition in line 7 from Algorithm 1.

Algorithm 1 Bat Algorithm

1: $INPUT$: $SN(population\ size), MAX\_IT(maximum\ nb.\ of\ loops)$
2: $OUTPUT$: x_{best} and $f_{best} = best(f(x_i))$ where $i = \overline{1, NS}$
3: **Initialization** : Generate initial population $(freq_i, x_i, y_i)$, compute fitness functions $(f(x_i))$ and determine the best solution of the group (x_{best}). Set $t = 0$.
4: **while** $t < MAX\_IT$ **do**
5:    **for** $i \leftarrow 1\ SN$ **do**
6:       $x_{new} = Generate\_new\_solution\ (x_i^{t-1})\ cf.\ eq.(2)(3)(4)$
7:       **if** $rand(0, 1) > r_i$ **then**
8:          $x_{new} = Improve\_candidate\_sol\ cf.\ (7)$
9:       **end if**
10:      **if** $rand(0, 1) < A_i$ AND $f(x_i) < f(x_{new})$ **then**
11:         $x_i = x_{new}$ (accept new solution)
12:         $r_i, A_i \leftarrow update\ cf.\ (5)(6)$
13:      **end if**
14:    **end for**
15:   $f_{best} = best(f(x_i)), \ i = \overline{1, NS}$ (determine best fitness and update current x_{best})
16:   $t \leftarrow t + 1$
17: **end while**

---

4 Proposed Bat Algorithm Improvement

All SI algorithms are based on two main concepts: *diversification*(exploration) which ensures the algorithm will carry out a global and hopefully efficient search for the solution and *intensification*(exploitation) that will try to improve a candidate solution. The balance between these two process plays an important role, as it may lead to a premature convergence with the probability to get trapped into local minima (if exploitation is too intensive) or a late convergence with a slow execution time (if exploration is too high).

BA has a quick start but, as the number of iterations grows the algorithm quickly looses exploration because condition $rand(0, 1) > r_i$ is hard to satisfy as r is increased exponentially. Furthermore, if the condition is true the new solution is generated near the best solution and with a small variation given by a decreasing loudness, meaning the algorithm can get trapped into local minima. To address these issues we improve the *exploration* component of BA by including an additional term in the new solution, denoted by the Euclidean distance between the current candidate solution and a solution with a better fitness value found by a neighbour (another individual in the group). This means equation (7) becomes:

$$x_{new} = x_{old} + u \cdot \sqrt{\sum_{i=1}^{d}(x_{old} - x_j)^2} + \delta \cdot A^*_{\ t} \qquad (8)$$

where $u \in [0, 1]$ is a random number and x_j is the position in space of bat j with a better fitness value. If we cannot find a neighbour with a superior fitness quality in a predefined number of trials then, we keep the original exploration of BA. This mean lines 7–9 from Algorithm 1 become:

Algorithm 2 Improved Bat Algorithm

1: **if** $rand(0, 1) > r_i$ **then**
2:    trial = SN*2 ; j = rand(1, SN)
3:    **while** $trial <> 0$ $AND f(x_j) <= f(x_i)$ **do**
4:       $j = rand(1, SN)$; trial = trial - 1;
5:    **end while**
6:    **if** $f(x_j) > f(x_i)$ **then**
7:       $x_{new} = Improve\_candidate\_sol(x_{old}, x_j)$ $cf.$ (8)
8:    **else**
9:       $x_{new} = Improve\_candidate\_sol\ (x_{old}) cf.$ (7)
10:   **end if**
11:   $x_i \leftarrow x_{new}$ (accept solution if $f(x_{new}) > f(x_i)$)
12: **end if**

This modification will ensure the algorithm converges to a searching area where good solutions are denser, before the pulse emission will inhibit further exploration. Moreover, if this new solution is better than the old one the bat will accept it. In terms of complexity level, we conserve the number of fitness computations and slightly increase the execution time in order to randomly choose an individual from the group with a higher fitness value. Our improvement might be seen as the attraction implemented in the Firefly Algorithm [16], where a higher fitness will attract other individuals. However, we preserve the original BA and only direct the individuals towards a position in the vicinity of a possibly finer location. We call this modified version **BA(E)**.

5 Feature Selection

Feature selection methods *are algorithms implemented in order to upgrade machine learning performances by removing irrelevant and/or redundant features and maintaining only the relevant attributes* [12]. In general, these methods are grouped based on how they work with the classifier as: *filter-based* (ranks features by correlating them with a class of features and its corresponding subset of features), *wrapper* (uses a predictive model to evaluate the subset and compute its importance) and *hybrid* (combines the previous two) [2]. The filter approach is more simple and does not depend on the classifier. On the other hand, the wrapper method is more complex as it requires training and testing the subset of features with a classifier. Usually, wrapper approaches are considered more reliable because they describe the problem for a specific classifier applied for detection hence, ensuring a good accuracy level.

5.1 Feature Selection Approach

The proposed feature selection method combines the modified Binary Bat Algorithm (BBA(E)) with SVM. To transform BA into the binary version we apply the sigmoid function:

$$S(v_{i,j}) = \frac{1}{1 + e^{-v_{i,j}}} \qquad (9)$$

as proposed in 2012 [10] to compute the new coordinates of an individual:

$$x_{i,j} = \begin{cases} -1 & if \ S(v_{i,j}) > \delta \\ 0 & otherwise \end{cases} \qquad (10)$$

where $\delta \in [0, 1]$ is a random number.

To simplify, all bats will fly inside a binary multi-dimensional grid searching for a position to attain a so-called optimal fitness value. In our feature selection problem, the dimension is given by the number of features and the best position found by the swarm will define the subset of features. Therefore, the solution can be viewed as an array of ones and zeroes that will render into the presence (if one) or the absence (if zero) of a feature from the subset.

6 Model Setup and Experiment Results

The proposed model is a network anomaly based IDS with two main components: *pre-processing* (wrapper feature selection that combines BBA(E) with SVM) and *detection* (SVM enhanced with parameter selection implemented by BA(E)). In order to rank our proposed BA(E), we compare it with the original BA and the well-known PSO by testing them for the same optimization problems. We implement in Java the standard and binary versions of the three SI algorithms and we combine them with the SVM classifier from weka version 3.6.10 [9].

Tests were conducted on a personal computer with 1.80 GHz Core (TM) 2 CPU and 2 GB of memory under Ubuntu 10.04.4. For the evaluations we used the NSL-KDD data set [13], an improved version of KDD-CUP. This dataset does not contain redundant records and has a lower complexity level of data. Each record is labeled as normal or attack and includes 41 features. The attacks from the dataset may belong to one of the following categories: Denial of service (DoS), Remote-to-Local(R2L), User-to-Root(U2R) or Probing. Also, the attributes can be classified into three groups: *connection based* (9 features), *content based* (13 features) and *time based* (19 features).

Our proposed model implies evaluations implemented with 10 or 5 folds cross validation, which might be time consuming for our test platform. Therefore, to simplify the process, we randomly select 9,566 records from the training dataset and

4,500 records from the test dataset. It is important to note that attacks in the test file are not incorporated in the training file thus, it will allow us to evaluate the proposed IDS model for new types of intrusions. Moreover, to better classification, we map the symbolic valued attributes (protocol_type, service, flag, class) to numerical values.

6.1 Model Setup

Our model includes two main components: *feature selection* and *detection*. For the first stage we use the training file and perform a 10 folds cross validation in order to avoid overfitting. In the case of detection we perform two evaluations. First we apply the training file and the test file with the selected attributes in order to evaluate the performance of our model for unknown attacks. In the second case, we use the test file to perform a 5 folds cross validation and apply SI for SVM parameter selection. To setup the IDS models we repeated the tests 50 times in order to obtain better results. For *BA, BA(E), BBA* and *BBA(E)* we ranged the *frequency* between 0.8 and 1.0, while the *maximum loudness* (A_0) is 0.5, *minimum pulse rate* (r_0) is 0.5, constants γ and α are set to 0.1 and respectively 0.9. In the case of *BPSO* and *PSO* the *inertia weight* is reduced from 0.9 to 0.5, c_1 and c_2 are equal to 2.3 and 1.8. All three binary SI algorithms have the problem dimension equal to the number of features in the original dataset $(d = 41)$, while standard versions have only two parameters to enhance $(d = 2)$ ranged as follows: $C \in [0, 2500]$ and $\sigma \in [0.0001, 25]$. The maximum number of iterations is 200.

Fitness Function. When evaluating IDS it is important to note the *Attack Detection Rate* and the *False Alarm Rate* as these two performance measures will indicate the number of correctly or incorrectly raised alarms:

- **attack detection rate** (ADR) - indicates if our IDS can detect intrusions by generating alarms. This performance measure will reveal if the classifier is capable to detect attacks, taking into account the selected subset of features.
- **false alarm rate** (FAR) - is the number of normal records that have generated false alarm. If too many improper alarms are raised, the model may become unreliable.

To establish the **fitness function** for the SI algorithms, we give each of the two previous performance measures a fraction and we add the number of features for the feature selection method and the execution time for the SVM parameter selection:

$$fitness_{FeatSelect} = 60\,\% \cdot ADR + 30\,\% \cdot \frac{1}{FAR} + 10\,\% \cdot \frac{1}{NbFeat} \tag{11}$$

$$fitness_{ParamSVM} = 70\,\% \cdot ADR + 10\,\% \cdot \frac{1}{FAR} + 20\,\% \cdot \frac{1}{Time_{execute}} \tag{12}$$

The aim of SI is to maximize the value of the fitness function.

6.2 Test Results and Analysis

Results from Table 1 show binary SI algorithms combined with SVM classifier can remove unimportant features while improving detection. Our proposed BBA(E) reduces the number of features with almost 60 % (for 5 individuals) or 53 % (for 2 individuals) and at the same time enhances the ADR (9.6 % for known intrusions and 7.9 % for unknown attacks) and the FAR (6.8 % for known intrusions and 5.6 % for unknown intrusions) of the simple SVM. When comparing the three feature selection approaches, BBA(E) outperforms them because it obtains a slightly higher fitness values after fewer number of iterations. This difference is visible for two individuals in the group (BBA(E) almost halves the required iterations when compared with the other two) but, it drastically reduces in the case of five individuals. The most expensive operation, in terms of execution time, performed by all feature models is fitness evaluation as it implies 10 folds cross validation. BBA(E) has the same number of fitness computations as BBA and might have a small delay due to the random search for a better neighbour. While, BPSO is actually BBA with A = 0 and r = 1. Therefore, a larger number of iterations results in a slower execution time for the feature selection process. Also, a notable remark is the relevance of selected attributes is confirmed by improved results even for unknown intrusions (test dataset in Table 1).

Having selected the subset of feature, we further improve our detection model by applying BA(E) to elect the two input parameters for SVM. Because the subsets from Table 1 obtain similar performances, we choose the one selected by BBA(E) with 5 individuals to reduce the test file and perform 5 folds cross validation to determine the fitness function. Similarly, we compare BA(E) with BA and PSO for the same optimization problem. Results from Table 2 show all SI algorithms can better the SVM classifier for detection but not significantly due to the good results obtained by the simple classifier. As in the feature selection case, for SVM parameter selection BA(E) obtains a fitness value close to BA and PSO, but after fewer iterations. Furthermore, the parameters selected by BA(E) give the highest ADR, lowest FAR and the fastest execution time, when performing 5 folds cross validation for SVM.

Table 1 Test results for feature selection methods

	Training dataset					Test dataset	
SI Alg.	Individ.	ADR	FAR	Nb. feat.	Iter.	ADR	FAR
BBA(E)	2	99.48	0.46	19	50	96.88	1.8
BBA	2	99.38	0.52	19	80	90.25	5.3
BPSO	2	99.27	0.61	21	100	97.17	1.6
BBA(E)	5	99.49	0.42	16	13	97.57	1.3
BBA	5	99.48	0.44	16	18	91.23	5.5
BPSO	5	99.54	0.40	17	25	97.11	1.5
Simple SVM		89.81	7.28	41		89.64	6.88

Table 2 Test results for SVM parameter selection

SI Alg.	C	σ	ADR	FAR	Time (s)	Iter.	Individ.
BA(E)	1	0.001	97.44	1.51	63.70	6	5
BA	1.26	0.001	96.75	2.0	70.39	11	5
PSO	1479.82	0.001	96.61	2.57	72.89	13	5
Simple SVM	1	0.5	93.8	4.97	77.45		

Although, the performance difference between the SVM parameters selected by the three SI algorithms is not high, the execution time required for this process, deducted from the number of iterations, is almost double in the case of BA and PSO. Hence, we can conclude BA(E) outperforms BA and PSO but, we claim our statement only for our selected dataset.

7 Conclusions and Future Work

The main contribution of this work is the personal improved Bat Algorithm. We apply this algorithm to a practical problem of intrusion detection by combining the binary version (BBA(E)) with SVM into a wrapper feature selection and the standard version (BA(E)) to select the input parameter for SVM. The proposed anomaly based IDS model is evaluated with the NSL-KDD dataset. Test results show BBA(E)-SVM reduces the dataset by 60 % obtaining higher ADR and lower FAR, while BA(E)-SVM for the selected subset only slightly improves detection. To rate our proposed BA(E), we compare it with the original BA and the popular PSO for the same optimization problems. We empirically prove BA(E) and BBA(E) outperforms them as it obtains almost the same fitness values, but after fewer iterations. In the future, further tests will consider combining BA(E) with other established classifiers. Also, we intend to compare our proposed algorithm with other standard optimization algorithms such as NSGA-II or SPEA2.

Acknowledgments The work has been funded by the Sectoral Operational Programme Human Resources Development 2007-2013 of the Ministry of European Funds through the Financial Agreement POSDRU/159/1.5/S/132395.

References

1. Debar, H., Dacier, M., Wespi, A.: Towards a taxonomy of intrusion-detection systems. Comput. Netw. **31**(9), 805–822 (1999)
2. Dua, S., Du, X.: Classical machine-learning paradigmsfor data mining. In: Data Mining and Machine Learning in Cybersecurity, pp. 23–56. Auerbach Publications Taylor and Francis Group (2011)

3. Enache, A.-C., Patriciu, V.V.: Intrusions detection based on support vector machine optimized with swarm intelligence. In: 9th IEEE International Symposium on Applied Computational Intelligence and Informatics, pp. 153–158 (2014)
4. Enache, A.-C., Sgarciu, V.: Enhanced intrusion detection system based on bat algorithm-support vector machine. In: 11th International Conference on Security and Cryptography, pp. 184–189. Vienna, Austria (2014)
5. Gao, H.-H., Yang, H.-H, Wang, X.-Y.: Ant colony optimization based network intrusion feature selection and detection. In: Proceedings of 2005 International Conference on Machine Learning and Cybernetics, pp. 3871–3875 (2005)
6. Kukielka, P., Kotulski, Z.: New unknown attack detection with the neural network-based ids. In: The State of the Art in Intrusion Prevention and Detection, pp. 259–284. Auerbach Publications (2014)
7. Laamari, M.A., Kamel, N.: A hybrid bat based feature selection approach for intrusion detection. In: Pan, L., Păun, G., Pérez-Jiménez, M.J., Song, T. (eds.) BIC-TA 2014. CCIS, vol. 472, pp. 230–238. Springer, Heidelberg (2014)
8. Ma, J., Liu, X., Liu, S.: A new intrusion detection method based on bpso-svm. Int. Symp. Comput. Intell. Des. 1, 473–477 (2008)
9. Mark, H., Eibe, F., Geoffrey, H., Bernhard, P., Peter, R., Ian, W.: The weka data mining software: an update. SIGKDD Explor. Newsl. 11, 10–18 (2009)
10. Nakamura, R., Pereira, L., Costa, K., Rodrigues, D., Papa, J., Yang, X.S.: Bba: a binary bat algorithm for feature selection. In: Proceedings of the 25th Conference on Graphics, Patterns and Images, pp. 291–297 (2012)
11. Nguyen, H., Franke, K., Petrovic, S.: Improving effectiveness of intrusion detection by correlation feature selection. In: ARES '10 International Conference on Availability, Reliability, and Security, 2010, pp. 17–24 (2010)
12. Sammut, C., Webb, G. I.: Feature selection. In: Encyclopedia of Machine Learning, pp. 429–433. Springer, New York (2010)
13. Tavallaee, M., Bagheri, E., Lu, W., Ghorbani, A.A.: A detailed analysis of the KDD CUP 99 data set. In: Proceedings of the IEEE Symposium on Computational Intelligence in Security and Defense Applications, pp. 1–6 (2009)
14. Wang, J., Hong, X., Ren, R., Li, T.: A real-time intrusion detection system based on pso-svm. In: Proceedings of the International Workshop on Information Security and Application, pp. 319–321. ACADEMY PUBLISHER (2009)
15. Wang, J., Li, T., Ren, R.: A real time IDSs based on artificial bee colony-support vector machine algorithm. In: Proceedings in the International Workshop on Advanced Computational Intelligence, pp. 91–96. IEEE (2010)
16. Yang, X.-S.: Firefly algorithms for multimodal optimization. In: Watanabe, O., Zeugmann, T. (eds.) SAGA 2009. LNCS, vol. 5792, pp. 169–178. Springer, Heidelberg (2009)
17. Yang, X.-S.: A new metaheuristic bat-inspired algorithm. In: González, J.R., Pelta, D.A., Cruz, C., Terrazas, G., Krasnogor, N. (eds.) NICSO 2010. SCI, vol. 284, pp. 65–74. Springer, Heidelberg (2010)
18. Yang, X.-S., He, X.: Bat algorithm: literature review and applications. Int. J. Bio-Inspired Comput. 5, 141–149 (2013)

A Formal Approach Based on Verification and Validation Techniques for Enhancing the Integrity of Concrete Role Based Access Control Policies

Faouzi Jaidi and Faten Labbene Ayachi

Abstract Our research works are in the context of verifying the integrity of access control policies in relational database management systems. This paper addresses the following question: In terms of security and particularly access control, *does an information system actually do what was planned for it to do*? Thus, an important aspect is to help *security architects* verifying the correspondence between the security planning and its real implementation. We present a synthesis of the problem of access control policy integrity within relational databases. Then, we introduce our proposal for addressing this issue. We especially focus on the verification and validation of the conformity of concrete role based access control policies in a formal environment. Finally, we illustrate the relevance of our contribution through a case of study.

Keywords Role based access control · Database security · Formal validation · Access control policy integrity · Conformity verification

1 Introduction, Problem Statement and Contribution

Generally, database systems are exposed to several security risks such as inner threats, unauthorized activities and accesses, logical and physical damages, performance constraints, design flaws and programming bugs, etc. Inner threats are security holes that can be exploited by legal users [3]. More, DataBase Management

F. Jaidi (✉) · F.L. Ayachi
Digital Security Research Unit (DSRU),
Higher School of Communication of Tunis (Sup'Com), Tunis, Tunisia
e-mail: faouzi.jaidi@gmail.com

F.L. Ayachi
e-mail: faten.labbene@supcom.rnu.tn

© Springer International Publishing Switzerland 2015
Á. Herrero et al. (eds.), *International Joint Conference*, Advances in Intelligent
Systems and Computing 369, DOI 10.1007/978-3-319-19713-5_5

Systems (DBMSs) function as network firewalls to control access to data, but unlike firewalls the access control policy is managed in the same place and way as the data it protects and consequently it is highly exposed to corruption attempts. In this context, we identify the following problems:

- A particular crucial problem is related to malicious use of administrative roles. If they are not used wisely, a malicious administrator can corrupt the policy and create other security breaches such as the following scenarios [9]:

  (a) Hidden users and hidden roles created and granted access rights by an administrator abusing his power.
  (b) Hidden access flow: users granted "*create any role*" privilege or granted roles "*with admin option*" privilege may delegate those roles to other users and therefore generate a new potential access flow invisible from outside the database.
  (c) Missed users, missed roles and missed access flow due to unintentional/ intentional incorrect use of already granted permissions via role misuse/abuse or partial implementation of the policy.

- The absence of restrictions that control the empowerment of an application role to reduce it to the minimum necessary.
- The coexistence in the DBMS of access control mechanisms based on different models may cause conflicts or redundancy in the expression of the policy.
- The violation of implicit negative authorizations since it is difficult to verify if a specified negative authorization is still enforced in the DBMS.

The described problems and scenarios cause non-compliance between the concrete policy and its specification. We derive four types of non-compliance anomalies [8]: inconsistency, contradiction, redundancy and partial implementation anomalies.

Protecting the database from insider threats requires sophisticated techniques such as anomaly detection tools able to build profiles of normal access and detect anomalous access with respect to those profiles [3]. More, specifying an access control policy, implementing it and monitoring its progress during its life cycle has emerged as a quite complex and confusing task. In this way, we defined a system that offers a global vision of the process of developing trusted policies [7, 8, 10]. It provides a complete solution for monitoring the compliance of a concrete policy and defines mechanisms for detecting possible attacks that can corrupt the policy. This paper focuses on the technical definition of the formal Validation and Verification (V&V) process that represents the core of our proposal. The paper contributions are as follows:

1. We define a methodology and process for identifying non-compliance anomalies in the concrete instances of access control policies.
2. We illustrate the basic characteristics of the verification and validation processes.
3. We discuss the security properties to be validated called validation properties.
4. We propose a formal representation of the defined validation process.

The remainder of the paper is structured as follows. Section 2, discusses related works. Section 3 details the proposed approach. Section 4 focuses on the V&V process. Section 5 illustrates the relevance of the proposal through a case of study. Finally, Sect. 6 concludes the paper.

2 Related Works

Several works treated the topic of verifying access control policies during the specification phase. Authors in [2] used to verify SecureUML models using the SecureMOVA tool which provides an evaluation of the security model through Object Constraint Language (OCL) requests. In [6], authors proposed to transform the specification realized with SecureUML to the Z language and to analyze the policy with the Jaza tool that allows animating the specification. Authors in [11] chose to transform the specification to the B notation using the B4Msecure tool and to analyze it with the ProB tool. Authors in [14] defined a logical framework to enforce the integrity of access control policies in relational databases. This framework focuses primarily on how to enforce and check constraints. The main goal of those works is to check the exactitude of the specified policy before proceeding to its implementation.

As for the validation of concrete access control policies, several researches propose representing roles in formalisms allowing the analysis, validation or optimization of the policy. Contributions deal with the following themes. (1) Validation of the implemented policy regarding the security constraints defined around that policy [4] using a finite model checking. (2) Detection of redundancy and inconsistency anomalies in the expression of a security policy [5] by adopting the formalism of graph of roles. The main objective of those works is to verify the correctness of the implemented policy regarding the defined constraints.

The aspect of checking the correspondence between the policy planning and its implementation according to our knowledge is not treated enough and needs more attention. To cover this gap, the main future of our proposal is to provide a global system allowing the verification, the optimization and helping the security architects validating the conformity of the implementation policy regarding its specification.

3 Our Approach for Validating and Verifying RBAC-Policies

Our approach [7, 8, 10] consists to validate and verify that a concrete role based access control (RBAC) [13] policy instantiates well a valid specification model. It helps detecting and correcting the discussed non-compliance anomalies. The V&V approach, defined in Fig. 1, consists of six basic steps that make up the body of the approach. *Phase 1* consists of specifying the policy during the specification of the database. It is based on SecureUML [12] as a modeling language. *Phase 2* concerns the encoding of the obtained functional and security models in a logic-like notation (we use in this paper the B notation [1]). It is performed via adopting the B4Msecure tool to our context. *Phase 3* defines reverse engineering techniques to extract the implemented policy from the Oracle DBMS. *Phase 4* defines the corresponding rules for transforming the obtained Data Definition Language (DDL) scripts relative to the extracted policy to the B notation. *Phase 5* consists to formally verify and validate the conformity of the concrete policy regarding its specification. Finally, *Phase 6* assures the adjustment and the optimization of the valid policy (Fig. 1).

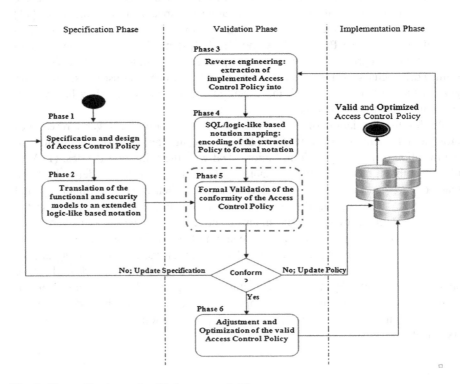

Fig. 1 The verification and validation approach [7]

4 The Verification and Validation Phase

4.1 Logic Formalization of RBAC Policies

The concepts and rules of a RBAC-based policy can be formalized as follows:

- **USERS**, represents the set of users authorized to access to the database schema.
- **ROLES**, is the set of roles defined in the database schema.
- **OBJECTS**, belongs to the set of resources defined in the database schema.
- **ACTIONS**, corresponds to the set of access modes.
- **PERMISSIONS**, is the set of permissions defined as possible actions on objects. It is noted: **PERMISSIONS $\subseteq$ ACTIONS $\times$ OBJECTS**.

AUR: USERS $\rightarrow$ ROLES, noted:

$$\text{AUR} \subseteq \text{USERS} \times \text{ROLES} \tag{1}$$

ARR: ROLES $\rightarrow$ ROLES, noted:

$$\text{ARR} \subseteq \text{ROLES} \times \text{ROLES} \tag{2}$$

APR: PERMISSIONS $\rightarrow$ ROLES, noted:

$$\text{APR} \subseteq \text{PERMISSIONS} \times \text{ROLES} \tag{3}$$

(1) **AUR** characterizes the users-roles assignments. (2) **ARR** describes the roles-roles assignments. (3) **APR** illustrates the assignment of permissions to roles.

Hence, we note **ACP = (USERS, ROLES, PERMISSIONS, AUR, ARR, APR)** the defined policy. The V&V process requires putting in duality two different notations. To do so, we consider that **ACP** denotes the formal specification of the policy and we note **ACP_IMP = (USERS_IMP, ROLES_IMP, PERMISSION_IMP, AUR_IMP, ARR_IMP, APR_IMP)** the formal representation of the concrete instance of that policy.

4.2 Verification of the Implemented Policy

The formal verification of a concrete policy concerns the check of the real implementation of the policy in order to control its exactitude. To do so, we proceed by verifying that the B machines obtained during the fourth phase are coherent, syntactically and semantically correct, well structured and the invariants are established on initialization and preserved during operations calls. The main proof activity comprises performing a number of demonstrations in order to prove the establishment and conservation of the invariants. For thus; by using the AtelierB tool; we verify the establishment of the invariants on initialization and during operations calls. This tool allows also type checking, generation and demonstration of proof obligations, etc.

4.3 Validation of the Concrete Policy

The validation of a concrete policy checks its conformity by comparing it to its specification. For thus, we define the validation properties and process as follows.

Validation Properties

We identify non-compliance anomalies based on predefined properties, called validation properties. They belong to predefined formulas used to detect inconsistency, contradiction, redundancy and partial implementation anomalies [8].

Anomalies of inconsistency belong to access control rules that are syntactically conform to the RBAC model, but not initially foreseen during the specification of the policy. It includes hidden users, hidden roles and hidden access flow.

Hidden users are visible when new users, not initially defined, are injected in the concrete instance. Logically, this is visible when $USERS_{IMP} - USERS \neq \emptyset$. Hence, we define the set of hidden users as described in (4) by the difference between the sets of implemented and specified users:

$$HiddenUsers = USERS_{IMP} - USERS \tag{4}$$

Hidden roles are observable when new roles, not initially planned, are introduced in the concrete policy. Logically, this is identifiable when $ROLES_{IMP} - ROLES \neq \emptyset$. Thus, we define the set of hidden roles as illustrated in (5) by the difference between the sets of implemented roles and the set of specified roles:

$$HiddenRoles = ROLES_{IMP} - ROLES \tag{5}$$

Hidden access flow is perceptible in the case of illegal assignments of: roles to roles, roles to users or permissions to roles. Logically, this is detectable when $ARR_{IMP} - ARR \neq \emptyset$ or $AUR_{IMP} - AUR \neq \emptyset$ or $APR_{IMP} - APR \neq \emptyset$
Therefore, we define the set of hidden access flow like presented in (6) by the union of possible hidden assignments of roles to roles, roles to users, and permissions to roles:

$$HiddenACFlow = (ARR_{IMP} - ARR) \cup (AUR_{IMP} - AUR) \cup (APR_{IMP} - APR) \tag{6}$$

Redundancy anomalies belong to access rules that express the same semantics and coexist in the target system. In (7) we express redundant access rules caused by transitivity. If $(U_k, R_i) \in \mathbf{AUR\_IMP}$ and $(R_j, R_i) \in \mathbf{APR\_IMP}$ then (U_k, R_j) must not be defined in $\mathbf{AUR\_IMP}$ to avoid redundancy. Otherwise (U_k, R_i) and (U_k, R_j) are redundant rules.

$$\mathbf{Redundancy} = \{(U_k, R_i) * (U_k, R_j) \mid (U_k, R_i): \mathbf{AUR\_IMP} \,\&\, (U_k, R_j): \mathbf{AUR\_IMP} \,\&\, (R_j, R_i): \mathbf{ARR\_IMP}\}$$

$$(7)$$

Another case of redundancy is caused by the coexistence of access control mechanisms based on different models. Thus, using RBAC, we assign permissions to users via roles, while using Dictionary Access Control (DAC) we directly assign the same permissions to the same users. To highlight this situation, we define in (8) a new function $\mathbf{APU\_IMP}$ that illustrates the direct assignment of permissions to users.

$$\mathbf{APU\_IMP}: \mathbf{PERMISSIONS\_IMP} \rightarrow \mathbf{USERS\_IMP}, \text{ noted also:}$$

$$\mathbf{APU\_IMP} \subseteq \mathbf{PERMISSIONS\_IMP} \times \mathbf{USERS\_IMP} \qquad (8)$$

Thus, we characterize in (9) redundant access rules. It relates the same permissions assigned via the RBAC model $((\mathbf{U, R}) * (\mathbf{P, R}))$ and through the DAC model $(\mathbf{U, P})$.

$$\mathbf{DacRedundancy} = \Big\{ (\mathbf{U, R})^* (\mathbf{P, R})^* (\mathbf{U, P}) \mid (\mathbf{U, R}): \mathbf{AUR\_IMP} \,\&\, (\mathbf{P, R}): \mathbf{APR\_IMP} \,\&\, (\mathbf{U, P}): \mathbf{APU\_IMP} \Big\}$$

$$(9)$$

Partially implemented policy anomaly is characterized by the absence of initially specified elements. It introduces non coherence and falsifies the behavior of the access control process. It includes missed users, missed roles and missed access flow.

Missed users are users initially specified but absent in the concrete policy. Logically, this is detectable when $\mathbf{USERS} - \mathbf{USERS\_IMP} \neq \varnothing$. Hence, we define missed users in (10) as the difference between the two sets of specified and implemented users:

$$\mathbf{MissedUsers} = \mathbf{USERS} - \mathbf{USERS\_IMP} \qquad (10)$$

Missed roles are roles initially identified but not implemented or removed. Logically, we recognize this when $\mathbf{ROLES} - \mathbf{ROLES\_IMP} \neq \varnothing$. Thus, we define missed roles in (11) by the difference between the two sets of specified and implemented roles:

$$\textbf{MissedRoles} = \textbf{ROLES} - \textbf{ROLES}_{\text{IMP}} \qquad (11)$$

Missed access flow is perceptible if planned roles-roles, roles-users or permissions-roles assignments are not implemented or removed. Logically, this is detectable when $\textbf{ARR} - \textbf{ARR}_{\text{IMP}} \neq \varnothing, \textbf{AUR} - \textbf{AUR}_{\text{IMP}} \neq \varnothing$ or $\textbf{APR} - \textbf{APR}_{\text{IMP}} \neq \varnothing$. Therefore, we define missed access flow in (12) by the union of missed assignments relations:

$$\textbf{MissedACFlow} = (\textbf{ARR} - \textbf{ARR}_{\text{IMP}}) \, \text{U} \, (\textbf{AUR} - \textbf{AUR}_{\text{IMP}}) \, \text{U} \, (\textbf{APR} - \textbf{APR}_{\text{IMP}})$$
$$(12)$$

Renamed users (resp. renamed roles) belong to a subset of users (resp. roles) that have their names changed but still have the same set of specified privileges. Logically, a renamed user (resp. a renamed role) is identified when a user (resp. a role) of the hidden users (resp. hidden roles) class share the same permissions with a user (resp. a role) of the missed users (resp. missed roles) class. For that, we define the functions:

- *PermissionsOfRole(R: Roles)* that returns the set of permissions of the role R.
- *RolesOfUser(U:Users)* that returns the set of roles assigned to the user U.
- *PermissionsOfUser(U: Users)* that returns the set of permissions associated to the user U: $\textbf{PermissionOfUser}(\textbf{U}) = \bigcup_{\text{Ri} \in \textbf{RolesOfUser}(\text{U})} \textbf{PermissionsOfRole}(\textbf{Ri})$

Hence, we define renamed roles in (13) as the set of couple of roles (R_i, R_j); where R_i is a missed role, R_j is a hidden role and both roles share the same permissions:

$$\textbf{RenamedRoles} = \{(\textbf{R}_i, \textbf{R}_j) \mid \textbf{R}_i : \textbf{MissedRoles} \, \& \, \textbf{R}_j : \textbf{HiddenRoles} \, \&$$
$$PermissionsOfRole(R_i) = PermissionsOfRole(R_j)\} \qquad (13)$$

We define renamed users in (14) as the set of couple of users (U_i, U_j); where U_i is a missed user, U_j is a hidden user and both users share the same permissions and roles:

$$\textbf{RenamedUsers} = \{(\textbf{U}_i, \textbf{U}_j) \mid \textbf{U}_i : \textbf{MissedUsers} \, \& \, \textbf{U}_j : \textbf{HiddenUsers} \, \&$$
$$PermissionsOfUser(U_i) = PermissionsOfUser(U_j) \, \& \, RolesOfUser(U_i) = \qquad (14)$$
$$RolesOfUser(U_j)\}$$

The Validation Process
The formal validation process, defined in Fig. 2, checks the equivalence between the two formal representations of the concrete policy and its specification. This comparison aims for detecting all kinds of non-compliance anomalies based on the predefined validation properties. It starts by checking the equivalence between: the sets of users, then the sets of roles and finally the sets of assignment relations. It stores the detected anomalies in a repository for a real time analysis and for further

usage. Depending on the obtained results, this process may propose, in an interactive way, some possible alternatives that help the *security architect* fixing the issues. This process is iterative in the sense that the changes introduced by the *security architect* must be checked until no anomaly is detected.

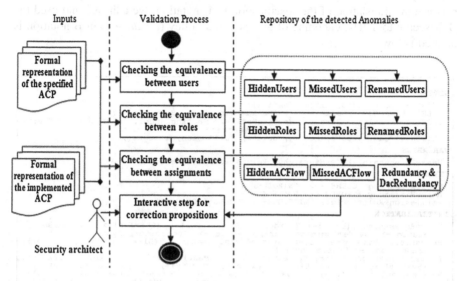

Fig. 2 The validation process

We note that intuitively, a policy based on $RBAC_1$ (*hierarchical RBAC*) model is a collection of finite sets and relations among them. The difference between initial and final states of that policy is evaluated as the difference between the initial and final sets/relations. Since the difference between sets is not commutative, the set of defined operators in this contribution is considered complete with respect to $RBAC_1$ model.

5   Case of Study

To illustrate the relevance of our proposal, we consider the "meeting scheduler" sample used in [6]. It defines four principal actors. A *system user* is able to create/modify/cancel meetings, add participants to a meeting and notify the participants about the meeting. The *system administrator* is responsible of managing persons. The *supervisor* is a special user who has the privilege to modify or cancel meetings he doesn't own. The *director* is both a user and an administrator.

5.1 Undergoing Phases 1 and 2

The specification of this system describes the assignments of users (Alice, Bob, Charles and David) to roles (system user, system administrator, supervisor and director), the hierarchy between roles and the permissions-roles assignments. To perform the translation of the specification to B notation, we adjusted and used the B4Msecure tool. The encoding of the specified policy (*instance 1*) in B notation is defined below:

```
...
SETS
    USERS = {Bob, David, Alice, Charles};
    ROLES={Director, Supervisor, SystemAdministrator, SystemUser};
    OBJECTS={Meeting, Person, MeetingNotify, MeetingCancel, MeetingModifyStart,
        MeetingModifyDuration, PersonModifyName};
    ACTIONS={read, create, modify, delete, fullAccess, execute}; ...
VARIABLES
    UsersRolesAssig, RolesHierarchy, PermissionsRolesAssig, ...
INVARIANT
    UsersRolesAssg: USERS --> POW(ROLES) &
    RolesHierarchy : ROLES <-> ROLES &
    PermissionsRolesAssig:PERMISSIONS--> (ROLES * OBJECTS)& ...
INITIALISATION
    UsersRolesAssg:={(Bob  |-> {Director, SystemUser}), (Charles|-> {SystemUser}),
        (David|-> {SystemAdministrator}), (Alice|-> {Supervisor, SystmUser})} ||
    RolesHierarchy:={Director|-> SystemAdministrator), (Director|-> SystemUser),
        (Supervisor|-> SystemUser)} ||
    PermissionsRolesAssig:={(SystemUser|->      (Meeting|->      {create,      read})),
        (SystemUser  |->  (Meeting|->  {delete,  modify})),  (SystemAdministrator|->
        (Meeting|->  {read})),    SystemAdministrator  |->  (Person|->  {fullAccess})),
        Supervisor  |->  (Meeting|->  {create,  read})),  (Supervisor|->  (Meeting|->
        {delete,    modify})),    (Supervisor|->    (MeetingCancel|->    {execute}),
        (Supervisor|->  (MeetingNotify|->  {execute}),  (Director|->  (Meeting  |->
        {create, read})), (Director|-> (Meeting|-> {delete, modify})), (Director|->
        (Meeting |-> {read})), (Director|-> (Person|-> {fullAccess})) } ||
...
```

5.2 Undergoing Phases 3 and 4

The extraction of the concrete policy is based on reverse engineering techniques that generates DDL statements describing the implemented policy. The encoding of the extracted policy (*instance 2*) to the target B notation is defined based on an appropriate SQL-B mapping as follows:

```
...
SETS
  USERS_IMP = {Alice,Bob, Charles, Marie, Paul};
  ROLES_IMP={Director,Supervisor, SystemAdministrator, SystemUser, Cosupervisor};
  OBJECTS_IMP={Meeting, Person, MeetingNotify, MeetingCancel, MeetingModifyStart,
    MeetingModifyDuration, PersonModifyName};
  ACTIONS_IMP={read, create, modify, delete, fullAccess, execute }; ...
VARIABLES
  UsersRolesAssig_IMP, RolesHierarchy_IMP, PermissionsRolesAssig_IMP, ...
INVARIANT
  UsersRolesAssg_IMP: USERS_IMP --> POW(ROLES_IMP) &
  RolesHierarchy_IMP : ROLES_IMP <-> ROLES_IMP &
  PermissionsRolesAssig_IMP: PERMISSIONS_IMP-->(ROLES_IMP* OBJECTS_IMP)& ...
INITIALISATION
  UsersRolesAssg_IMP:={ (Bob|->{Director,           SystemUser}),          Marie|->
    {SystemAdministrator}),   (Alice|->  {Supervisor,  SystemUser}),  (Charles|->
    {SystemUser}), (Paul|-> {Cosupervisor})} ||
  RolesHierarchy_IMP:={Director|->            SystemAdministrator),       (Director|->
    SystemUser), (Supervisor|-> SystemUser), (Cosupervisor|-> Supervisor)} ||
  PermissionsRolesAssig_IMP:={(SystemUser|->  (Meeting|->  {create,  read,  delete,
    modify})),(SystemAdministrator|->               (Meeting|->              {read})),
    (SystemAdministrator|-> (Person|-> {fullAccess})),(Supervisor|-> (Meeting|->
    {create,read,delete,modify})),(Supervisor|-> (MeetingCancel|-> {executeOp}),
    (Supervisor|-> (MeetingNotify|-> {executeOp}), (Cosupervisor|-> (Meeting|->
    {create,   read,   delete,   modify})),   (cosupervisor|->   (MeetingCancel|->
    {execute})),   (cosupervisor|->   (MeetingNotify|->   {execute})),   (Director|->
    (Meeting|->   {create,   read,   delete,   modify})),   (Director|->   (Meeting|->
    {read})),  (Director|-> (Person|-> {fullAccess})) } ||
  PermissionsUsersAssig_IMP:= { (Bob|->(Person|->{read}))}
...
```

5.3 Undergoing Phase 5

Applying the fifth phase of our approach, we detect the following abnormalities when checking the conformity between instances 1 and 2:

- **HiddenUsers** = {Marie, Paul}.
- **MissedUsers** = {David}.
- **HiddenRoles** = {Cosupervisor}.
- **MissedRoles** = ∅.
- **HiddenACFlow** = {Marie |- > {SystemAdministrator}), (Paul|- > {Cosupervisor}), (Cosupervisor|- > Supervisor), (Cosupervisor|- > (Meeting|- > {create, read, delete, modify})), (Cosupervisor|- > (MeetingCancel|- > {execute})), (Cosupervisor |- > (MeetingNotify|- > {execute}))}.
- **MissedACFlow** = {(David|- > {SystemAdministrator})}.
- **Redundancy** = {(Bob|- > Director) * (Bob|- > SystemUser), (Alice |- > Supervisor) * (Alice |- > SystemUser)}.
- **DacRedundancy** = {(Bob |- > Director) * ((Person |- > {fullaccess}) |- > Director) * (Bob |- > (Person|- > {read}))}.

Based on possible interpretations that emphasize for legal changes, the architect updates the specification and/or the implementation to reach the equivalence.

6 Conclusion and Future Work

This paper explores the notion of enhancing the integrity of access control policies in relational databases context. It probes the concept of formally validating and verifying the conformity of concrete RBAC-based policies. The main scope of the paper is basically to focus on the technical definition of the formal validation and verification process. This process aims for identifying all cases of non-compliance in concrete policies. Ongoing works address mainly the definition and formalization of an approach for assessing the risk introduced by the identified non-compliance anomalies.

References

1. Abrial J.-R.: The B-Book: Assigning Programs to Meanings. Press Syndicate of the University of Cambridge, Cambridge (1996)
2. Basin, D.A., Clavel, M., Doser, J., Egea, M.: Automated analysis of security-design models. Inf. Softw. Technol. **51**(5), 815–831 (2009)
3. Bertino, E., Ghinita, G., Kamra, A.: Access Control for Databases: Concepts and Systems. Found. Trends Databases **3**, 1–2, 1–148 (2010)
4. Hansen, F., Oleshchuk, V.: Conformance checking of RBAC policy and its implementation. In: 1st Information Security Practice and Experience Conference, pp. 144–155 (2005)
5. Huang, C., Sun, J., Wang, X., Si, Y.: Security policy management for systems employing role based access control model. Inf. Technol. J. **8**, 726–734 (2009)
6. Idani, A., Ledru, Y., Richier, J., Labiadh, M.A., Qamar, N., Gervais, F., Laleau, R., Milhau, J., Frappier, M.: Principles of the coupling between UML and formal notations. ANR-08-SEGI-018 (2011)
7. Jaidi, F., Labbene Ayachi, F.: An approach to formally validate and verify the compliance of low level access control policies. In: The 13th International Symposium on Pervasive Systems, Algorithms, and Networks, I-SPAN 2014 (2014)
8. Jaidi, F., Labbene Ayachi, F.: A formal system for detecting anomalies of non-conformity in concrete rbac-based policies. In: the International Conference on Computer Information Systems 2015 WCCAIS-2015- ICCIS (2015)
9. Jaidi, F., Labbene Ayachi, F.: The problem of integrity in rbac-based policies within relational databases: synthesis and problem study. In: ACM IMCOM 9th international conference on ubiquitous information management and communication: ICUIMC (2015)
10. Jaidi, F., Labbene Ayachi, F.: To summarize the problem of non-conformity in concrete rbac-based policies: synthesis, system proposal and future directives. In: NNGT Int. J. Inf. Secur. **2**, 1–12 (2015)
11. Ledru, Y., Idani, A., Milhau, J., Qamar, N., Laleau, R., Richier, J., Labiadh, M.A.: Taking into account functional models in the validation of is security policies. In: Advanced Information Systems Engineering (CAiSE) Workshops, vol. 83, pp. 592–606 (2011)
12. Lodderstedt, T., Basin, D., Doser, J.: SecureUML: a UML-based modeling language for model-driven security. In: 5th International Conference on the Unified Modeling Language, LNCS, vol. 2460, pp. 426–441. Springer, Heidelberg (2002)
13. Sandhu, R., Coynek, E.J., Feinsteink, H.L., Youmank, C.E.: Role-based access control models. IEEE. Computer **29**(2), 38–47 (1996)
14. Thion, R., Coulondre, S.: A relational database integrity framework for access control policies. J. Intell. Inf. Syst. **38**(1), 131–159 (2012)

Real-Time Hybrid Compression of Pattern Matching Automata for Heterogeneous Signature-Based Intrusion Detection

Ciprian Pungila and Viorel Negru

Abstract We are proposing a new hybrid approach to achieving real-time compression of pattern matching automata in signature-based intrusion detection systems, with particular emphasis on heterogeneous CPU/GPU architectures. We also provide details of the implementation and show how a hybrid approach can lead to improved compression ratios while performing real-time changes to the automata. By testing our methodology in a real-world scenario using sets taken from the ClamAV signature database the Snort rules database, we show that the approach we propose performs better than the current solutions, significantly reducing the storage required and paving the way for high-throughput CPU/GPU heterogeneous processing for such type of automata.

Keywords Hybrid compression · Heterogeneous architecture · Lempel-ziv-welch · Aho-corasick · GPU processing · Intrusion detection · Pattern matching · ClamAV · Snort

1 Introduction

Recent work in the field of GPGPU computing has shown that the future of intrusion detection systems is aimed to shift most computationally-intensive operations to heterogeneous architectures, where the CPU and GPU work together flawlessly towards achieving a common goal. The benefits of using a heterogeneous architecture are significant: the CPU is leveraged of tasks which are significantly CPU-bound and left to carry on more diverse tasks, while CPU-intensive operations will be parallelized (if at all possible of course and mostly in a SIMD manner) in high-throughput dedicated hardware, such as the GPU (or GPUs).

C. Pungila (✉) · V. Negru
West University of Timisoara, Blvd. V. Parvan 4, 300223 Timis, Romania
e-mail: cpungila@info.uvt.ro
url: http://info.uvt.ro

V. Negru
e-mail: vnegru@info.uvt.ro

© Springer International Publishing Switzerland 2015
Á. Herrero et al. (eds.), *International Joint Conference*, Advances in Intelligent
Systems and Computing 369, DOI 10.1007/978-3-319-19713-5_6

65

We are introducing an approach which combines a hybrid compression approach, inspired by the Lempel-Ziv-Welch algorithm [1, 2], with a highly efficient storage model (introduced in one of our earlier papers [3]) for GPGPU processing of pattern matching automata (specifically, for the Aho-Corasick algorithm [4]) and show how, through efficient data structures and efficient parallelization, we can achieve real-time compression of such automata and significantly reduce the storage space required for achieving signature-based intrusion detection. Related work in discussed in Sect. 2 of the paper, the approach we developed is presented in Sect. 3 and the experimental results obtained, along with the testing methodology are discussed in Sect. 4.

2 Related Work

Signature-based intrusion detection systems (IDS) work by keeping a database with patterns known to be a part of malicious activity (or software). Specifically, antivirus engines (such as the widely spread ClamAV antivirus [5]) work by keeping a database of virus signatures, which is constantly updated and allows detection of new threats. Other IDS, such as the popular Snort [6], work by sniffing packets and matching packet contents to a set of rules, which amongst other details, also contain one or more sets of strings which are known to belong to malicious threats. Both approaches share the common approach used in performing the pattern-matching process, through the implementation of the Aho-Corasick automaton. Other algorithms, such as Commentz-Walter [7] or Wu-Manber [8], have also been successfully used, but are not as popular as Aho-Corasick, for a few simple reasons: Aho-Corasick is easy to build and process (unlike Commentz-Walter), it uses a relatively modest storage space (unlike Wu-Manber) and has high-throughput when used in CPU/GPU architectures. Nevertheless, our approach is easily applicable to the Commentz-Walter automaton just as well - in fact, our previous work [9, 10] has shown how heterogeneous architectures can greatly improve the efficiency of building such type of automata.

Our paper's main contributions are: (a) combining a pruning approach of our memory-efficient model for the Aho-Corasick automaton with a Lempel-Ziv-Welch inspired approach to compressing the signature set and linking them to the automaton, together with an efficient storage format for the dictionary used by the compression algorithm; (b) showing how, after the initial stage, real-time compression of further patterns can be easily achieved in the GPU directly, with minimal impact on performance, providing an effective way of updating the entire dataset directly in RAM/V-RAM memory; (c) providing a real-world insight into how overall storage efficiency can be improved, by using datasets in two testing scenarios, one while using ClamAV virus signatures, and the other while using Snort rules.

2.1 The Aho-Corasick Automaton

The Aho-Corasick algorithm [4] is a common, fast approach to performing multiple pattern matching in intrusion detection systems. In fact, the two popular IDS systems, ClamAV [5] and Snort [6] are both using it to detect signatures in executable code (ClamAV) and network packet data (Snort). The algorithm works by constructing a trie tree out of the set of patterns it is supposed to locate, with each node storing a list of pointers to all children, for all accepted symbols, and a failure pointer, which gets called in case of a mismatch (an unaccepted symbol at current state). With the symbols being a part of the ASCII charset, the size of the alphabet used for the automaton implementation is 256 in most scenarios.

The failure pointer is computed as follows (see Figure 1 for a visual example):

- for the root and all children at the first level of the automaton, the failure always points back to the root.
- for all other states, the failure pointer is computed as follows:

  - we find the largest suffix of the word at the current level of the tree that is also matched by the automata, and we point back to it
  - if no such suffix of the word exists, the failure pointer links back to the root.

The automaton works by starting from the root and navigating through its nodes, until a leaf is reached, which means a match has been found.

Fig. 1 An Aho-Corasick automaton for the input set {abc, abd, ac, ad, bca, bce, cb, cd}. Dashes transitions are failure pointers [10]

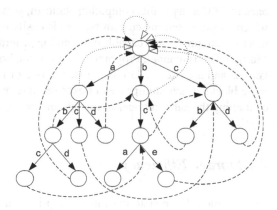

2.2 The Lempel-Ziv-Welch Compression Algorithm

The Lempel-Ziv-Welch algorithm [1, 2] is a lossless data compression approach using variable-length codes for achieving the compression. The algorithm's main idea lies in the fact that common symbols (or sequences of symbols) should be replaced with their corresponding codes, all of which have variable lengths, therefore reducing the storage space required. The algorithm builds a dictionary as the encoding process is on-going, assigning a unique code to each sequence that gets encoded. The process can be explained as follows:

1. initialize dictionary to contain all symbols in alphabet initially
2. find the longest word w in the dictionary that matches the current input
3. output the dictionary code for w and remove w from input
4. add word $w + c$ (the current symbol) to the dictionary
5. if input still exists, go to step 2.

The decompression routine works in a very similar manner: it reads a value from the encoded input and outputs the corresponding string from the dictionary. It then looks for the next value in the encoded input, adding to the dictionary a concatenation of the current word and the decoded value of the next input symbol. If next input symbol cannot be decoded, the current word with this symbol is added to the dictionary. In essence, the decoding process basically rebuilds the same dictionary as the encoding process, starting from the single symbols in the initial alphabet, which therefore does not require the distribution of the entire dictionary to the decoder.

2.3 The CUDA Architecture

The CUDA (Compute Unified Device Architecture) [11] was introduced by NVIDIA back in 2007 as a parallel computing platform, as well as a programming framework, which allows executing custom-built code easily on the GPU. Each GPU has one or more stream processors capable of executing programs in a SIMD (Single Instruction Multiple Data) manner. Programs sent from the host to the GPU (device) are called kernels and are executed on the device as one or more threads, organized in one or more blocks. Each multiprocessor executes one or more thread blocks, those active containing the same number of threads and being scheduled internally.

2.4 Storage Efficiency

As with most finite state machines comprised of a large number of nodes (in our experiments, for example, the ClamAV automaton had more than 2.9 million nodes), problems occur with efficient storage in memory. Additionally, since all programs running on the GPU must be first transferred from RAM to V-RAM, copying such a large automaton could take a significant amount of time (it took over 150 minutes with our automaton to copy all corresponding nodes from RAM to V-RAM). Additionally, allocating pointer nodes in RAM (or V-RAM) memory creates a cascade-effect of memory fragmentation (see Fig. 2), with huge gaps of unused space residing between two pointers allocated for consecutive nodes (leading to about double the amount of memory required to store the tree in our experiments).

Fig. 2 (a) Sparse, consecutive memory allocations and gaps that appear in memory fragmentation; (b) A single, sequentially allocated memory block without memory fragmentation

In recent work [3] we have presented an approach to efficiently storing the automaton, without compromising performance and allowing instant transfers between RAM and V-RAM through the PCI-Express architecture. The idea lies in the initial construction of the trie tree in RAM, after which a parsing of the tree is done (through pre-order traversal or any other type of traversal that produces a consecutive list of children for a node in the output), then building a stack of nodes, preserving the property of consecutive children in the stack. Each node will then just require a bitmap to indicate if a symbol is accepted or not, a failure offset indicating the node to jump to in case of a mismatch, and an offset to the first child in the stack, for situations when a symbol is accepted (the offset of the proper child to go to is computed by applying a population count in the bitmap). The approach has been successfully used in our earlier work [12] to significantly improve the construction stages of these automata, and still is at this time the most efficient storage format for heterogeneous automata processing.

3 Implementation

Virus signatures in the ClamAV database are stored as hexadecimally-encoded strings, e.g. *AA BB CC DD EE FF*. Signatures from the Snort database are stored in combined format, text and hexadecimally-encoded strings, e.g. *data|AA BB|some more data*, which we have transformed in the same format as used by ClamAV for inserting in the automaton (since they use the same alphabet, the entire ASCII charset). In order to reduce storage space in the GPU, we have implemented a pruned version of the Aho-Corasick automaton for both the ClamAV and Snort signatures used in the test. The idea is based on the observation that the entire automaton is not truly needed in memory, but a depth-limited version of it will suffice. The experiments we have carried on while analyzing a depth-limited implementation of the Aho-Corasick algorithm for virus signature scanning and network data sniffing respectively, over 100 MB of data analyzed, showed that a depth of 8 for the signatures is sufficient, with a precision of about 99.999 % (consistent to a similar test performed in [13], but only for a limited number of signatures and only for ClamAV). This shows that if the first 8 bytes of a signature are matched, there is a high probability that the entire signature will match also - but it also shows that we can easily reduce the computational time of the automaton and also the storage space by pruning.

3.1 Algorithm & Data Structures

The Aho-Corasick automaton we are using is built as a stack of nodes (see Fig. 3), as described in Sect. 2.4. Each node has a bitmap, occupying 256 bits, a failure offset and a child offset, each occupying 24 bits. In total, 38 bytes are used per node. By trimming the signatures and using the first 8 bytes to build the automaton, the rest of the signatures are compressed using the approach mentioned in Sect. 2.2 on the CPU. The dictionary built during the compression stage is stored as a trie, and the same serialization methodology as in Sect. 2.4 is applied: each node in the serialized trie uses a bitmap (of 32 bytes), an offset to first child of 3 bytes and an offset to the parent node, also 3 bytes in size. In total 38 bytes per node are used. The offset to the parent is used for the decompression routine, for reconstructing the unencoded string by going upwards in the dictionary when decompressing in case of a match. Additionally, a hash links all leaves in the dictionary to their proper codes.

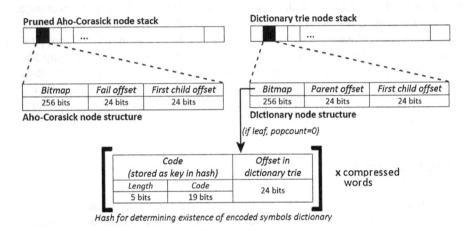

Fig. 3 Data structures used

All leaves in the pruned automaton are linked to all accepting patterns through a different two-level tree (see Fig. 4), since in case of a match the decompression routine will be enabled and the signature will be fully matched. The two-level tree is stored in the same serialized format described in Sect. 2.4. The full matching process works as follows:

1. search through pruned Aho-Corasick automaton
2. if match found on leaf, for each accepting pattern do

    (a) if current symbol is encoded and found in dictionary (hash in Fig. 3)
        i. locate Lempel-Ziv-Welch encoded pattern in dictionary trie
        ii. obtain offset of corresponding leaf in dictionary nodes stack
        iii. traverse trie upwards to determine unencoded symbol
    (b) otherwise match and if accepted, continue with next encoded symbol.

By applying all of the above, we ensure that transfers of the automaton, of the dictionary trie and of the hashes between RAM and V-RAM are instant, since all are stored as continuous, fragmentation-free memory blocks. The entire construction phase can be summarized as follows:

1. construct pruned Aho-Corasick trie on host (CPU)
2. compress remaining patterns on host and build dictionary trie (CPU)
3. move pruned automaton, dictionary and compressed patterns to V-RAM (instant transfers)
4. (optional) update in real-time all structures in case of updates/additions (CPU & GPU) [10].

Frequent scenarios in signature-based detection require updating of the signature dataset in real-time. With the approach described, the first 8 bytes of the signature are easily added to the Aho-Corasick automaton (performance has been thoroughly analyzed in [10]), while the remaining pattern will be easily compressed using the technique described in Sect. 2.2. In both scenarios, the additional simply requires the addition of a few nodes to the stacks of nodes for the automaton and dictionary trie.

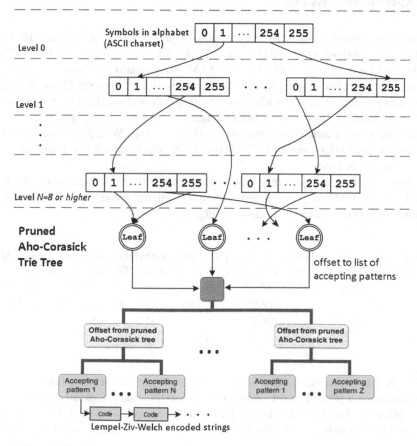

Fig. 4 Pruned Aho-Corasick tree and links to Lempel-Ziv-Welch encoded strings

3.2 Real-Time Processing

Adding patterns in real-time can be done both on the CPU or on the GPU and requires a few steps in order to update the structures used:

1. update pruned Aho-Corasick tree up to the depth level selected

   (a) adds additional nodes in the stack
   (b) update the two-level tree/hash used for linking to accepting patterns

2. update dictionary with remaining pattern

   (a) adds additional trie nodes in the stack
   (b) update the hash used for locating encoded symbols in dictionary

3. add compressed pattern to list.

4 Experimental Results

We have performed testing of the methodology proposed on an i7 Sandy-Bridge 2600K CPU, clocked at 4.3 GHz, backed up by a CUDA-enabled nVIDIA GTX 560Ti graphics card with 1 GB of DDR5 RAM. Additionally, we have tested using two datasets: one, using more than 38,400 virus signatures from the ClamAV database (with an average length of 82 bytes per signature), and another using more than 14,700 signatures from the Snort rules dataset (with an average length of 27 bytes per signature, and a minimum signature length of 6 bytes). We have tested multiple scenarios, to try to observe how LZW dictionary & Aho-Corasick automaton depths affect storage efficiency.

The storage used by the original automata, along with the storage obtained after applying the hybrid compression approach discussed, are presented in Table 1. The compression ratios obtained are shown in Fig. 5.

Table 1 Storage required (in MB) before and after applying the hybrid compression approach. **AC** denotes the Aho-Corasick pruning depth, while **LZW** denotes the LZW dictionary trie depth

	Uncompressed	AC=8, LZW=4	AC=8, LZW=6	AC=8, LZW=8	AC=8, LZW=16	AC=16, LZW=8
ClamAV	105.18	24.47	32.1	35.4	36.14	42.13
Snort	8.52	1.72	1.98	2.3	2.62	3.9

Best compression ratios were obtained for the Snort dataset, reaching 79.81 % for when using a pruned Aho-Corasick tree of depth 8, and a maximum dictionary trie depth of 4. The ClamAV dataset follows closely, with a ratio of 76.73 %. Just as expected, increasing the dictionary depth reduces the compression ratio, as more storage is required. In a similar manner, increasing the automaton depth also reduces the compression ratio significantly, since more nodes are stored in the serialized, uncompressed format.

The bigger the pruning depth of the automaton, the more accurate the matching process is. Nevertheless, results have shown that in practice, a pruning depth of 8 is sufficient (if a match is made of the first 8 bytes of such a signature, there is a very high probability that the entire signature will be matched completely). Anything lower would trigger too many matches, leading to additional computing cycles (and memory coalescing issues in SIMD hardware, e.g. GPUs) while anything higher does not significantly increase the accuracy of the matching. The LZW trie depth however affects compression differently: shorter patterns have a higher chance of repeating in the signature database once they are added to the dictionary (especially for viruses, where families share a very large portion of code), leading therefore to better compression ratios (since in memory at runtime, such repeating patterns would be replaced by their corresponding LZW codes).

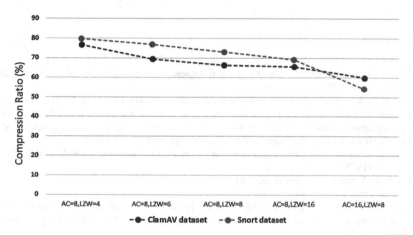

Fig. 5 The compression ratios obtained while applying our methodology to the two different datasets

The results prove that the storage has been reduced significantly in both scenarios, with the best performance obtained when the dictionary length is a maximum of 4. Adding new patterns is easily achievable by directly altering the stacks involved in the process, with minimal impact on performance (given the highly compact nature of storage used). Additionally, at the small expense of further processing in case of a full match (which seldom occurs in practice), the decompression works easily by parsing the dictionary trie upwards, allowing a multi-step matching process to take place (as symbols get decompressed, the matching occurs; therefore, if a mismatch occurs, no more symbols need to be decompressed and the process stops).

5 Conclusion

We have presented a new approach to achieving hybrid compression of commonly used pattern matching automata in signature-based intrusion detection systems, and tested our approach in a real-world scenario on two popular datasets, belonging to the ClamAV antivirus and the Snort firewall. Results have shown that the storage efficiency is greatly improved in both scenarios, while still preserving a high throughput while performing the matching process, in both cases. Additionally, with the high efficiency of the storage models chosen for the data structures used within, which resolve potential memory fragmentation issues, transfers between RAM and V-RAM are instantaneous, offering great potential of the approach in heterogeneous CPU/GPU hardware implementations.

Acknowledgments This work was partially supported by the Romanian national grant PN-II-ID-PCE-2011-3-0260 (AMICAS).

References

1. Ziv, J., Lempel, A.: Compression of individual sequences via variable-rate coding. IEEE Trans. Inf. Theor. **24** (1978)
2. Welch, T.: A technique for high-performance data compression. Computer **17**(6), 8–19 (1984)
3. Pungila, C., Negru, V.: A highly-efficient memory-compression approach for GPU-accelerated virus signature matching. In: Proceedings of the 15th Information Security Conference (ISC), Lecture Notes in Computer Science, pp. 354–369. Springer, Berlin (2012)
4. Aho, A., Corasick, M.: Efficient string matching: an aid to blbiographic search. CACM **18**(6), 333–340 (1975)
5. Clam AntiVirus. http://www.clamav.net
6. Snort. http://www.snort.org/
7. Commentz-Walter, B.: A string matching algorithm fast on the average. In: Maurer (ed.) Proceedings 6th International Coll. on Automata, Languages, and Programming, pp. 118–132. Springer (1979)
8. Wu, S., Manber, U.: A fast algorithm for multi-pattern searching. Technical Report TR, pp. 94–17. University of Arizona (1994)
9. Pungila, C., Negru, V.: Towards building efficient malware detection engines using hybrid CPU/GPU-accelerated approaches. Architectures and Protocols for Secure Information Technology Infrastructures. IGI Global, pp. 237–264. doi:10.4018/978-1-4666-4514-1.ch009 (2014)
10. Pungila, C., Negru, V.: Real-time polymorphic Aho-Corasick automata for heterogeneous malicious code detection. In: International Joint Conference SOCO'13-CISIS'13-ICEUTE'13. Advances in Intelligent Systems and Computing, vol. 239, pp. 439–448 (2014)
11. The CUDA Architecture. http://www.nvidia.com/object/cuda_home_new.html
12. Pungila, C., Reja, M.,Negru, V.: Efficient parallel automata construction for hybrid resource-impelled data-matching. Future Generation Computer Systems, vol. 36, pp. 31–41. Special section: intelligent big data processing (2014). doi:10.1016/j.future.2013.09.008
13. Vasiliadis, G., Ioannidis, S.: GrAVity: A massively parallel antivirus engine. Recent advances in intrusion detection. Lecture Notes in Computer Science, vol. 6307, pp. 79–96 (2010)

Adaptive Watermarking Algorithm of Color Images of Ancient Documents on YIQ-PCA Color Space

Mohamed Neji Maatouk and Najoua Essoukri Ben Amara

Abstract Ancient documents present a very precious cultural heritage, hence the need to preserve and protect it and fight against illegal copying and pirating in its numerical form. The watermarking presents a promising solution for the copyright protection of these documents. Considering the importance in the number of color images, we propose in this paper a watermarking approach of color images from ancient documents, which adapts to the characteristics and content of each image. It is also based on the principal component analysis, the wavelet packet decomposition, and the singular value decomposition. The recorded results have shown that our algorithm has a good invisibility and an excellent robustness against different attacks relative to ancient documents.

Keywords Watermarking · Ancient documents · Color image · YIQ · WPD · PCA · SVD

1 Introduction

Watermarking numerical documents refers to inserting a datum, called signature, mark or even watermark, in an imperceptibly digital content, which must be robust against the attacks that can incur the considered document. The insertion domain of the watermark can be spatial, frequency or multiresolution [1].

The main problem of a watermarking algorithm is to have a good compromise between invisibility and robustness. This implies a good choice of the insertion domain, the signature's carrier points and its coding – if it exists. For color-image watermarking, the invisibility problem is not easy to be solved essentially because

M.N. Maatouk (✉) · N.E.B. Amara
UR: SAGE, National Engineering School of Sousse, University of Sousse, Sousse, Tunisia
e-mail: maatoukneji@gmail.com

N.E.B. Amara
e-mail: najoua.benamara@eniso.rnu.tn

© Springer International Publishing Switzerland 2015
Á. Herrero et al. (eds.), *International Joint Conference*, Advances in Intelligent
Systems and Computing 369, DOI 10.1007/978-3-319-19713-5_7

of the appearance of false colors in the watermarked image (Fig. 1). Thus, it is necessary to choose a good color space (primary spaces (RGB, XYZ), achromatic spaces (YIQ, YUV, YCbCr) or perceptual (HSL, HSV) ones) as well as insertion components to ensure a good invisibility of the color image watermarking.

a. original image   b. watermarked image

Fig. 1 Illustration of false-color appearance on (b) watermarked image

As shown in Table 1, the choice of the color space depends on the application and the characteristics (contrast, texture and luminance) of the considered image.

Table 1 Selection of algorithms of watermarking color images

Ref.	Test images	Color space	Insertion component	Insertion domain/ Transformation
[1]	Standard images	RGB	Blue component (B)	Spatial/Amplitude modulation
[2]				Multi-resolution/DWT (Discrete Wavelet Transform)
[3]			Three components	Frequency/DCT (Discrete Cosine Transform)
[4]		YCbCr		
[5]		YIQ	Luminance component (Y)	Multi-resolution/IWT (Integer Wavelet Transform)/DWT/
[6]		YUV PCA		
[7]		YUV	Two chromatic components (U, V)	
[8]	Medical images	RGB-PCA	PC1 (first principal component)	DCT
[9]				DWT/WPD (Wavelet Packet Decomposition)
[14]	Ancient documents	YIQ-PCA		

The literature reveals that a good robustness of a color component is related to a low invisibility independently from the selected color space, and vice versa.

The primary and achromatic spaces are frequently exploited for color-image watermarking. For the primary space, the green-based components present an opportunity of robust signature insertion, especially against the JPEG compression, but they provoke a degradation of the watermarked image. However, the blue-based components are frequently used for color image watermarking since they respect the invisible criterion, despite their fragility against signal processing attacks.

Besides, the achromatic space is decomposed into one luminance component and two chromatic ones. The literature has shown that the luminance component is robust against a JPEG compression but fragile against a noise with a medium invisibility. It has also proved that the chromatic component provides a good invisibility with a better robustness against a low-pass filter and fragility against the JPEG compression in decreasing the quality factor. Indeed, the human eye is more sensitive to the luminance than to the chrominance of an image. Consequently, it is acceptable to degrade the chrominance of the image while keeping a good quality of a watermarked image.

In what follows, we present in Sect. 2 the problematic related to watermarking ancient documents. We describe in Sect. 3 the main steps of our approach and we present the used techniques. In Sect. 4, we expose the insertion and detection algorithms. In Sect. 5, we present the experimental results. A conclusion is given in Sect. 6.

2 Problematic Related to Watermarking Ancient Documents

The literature has shown that the majority of the watermarking algorithms have been applied on medical and standard images (classical algorithms). To the best of our knowledge, despite the major issue of preservation and copyright protection of ancient documents, no watermarking algorithm has been developed for images of ancient documents outside the work of our group [13, 14].

Watermarking the images of ancient documents is not trivial, especially because of the problems related to the particular structure of documents due to their different specificities and frequent mediocre quality (Fig. 2).

a.   b.   c.   d.   e.

Fig. 2 Examples of intrinsic problems of ancient documents: a. visible back from the front, b. ink fading, c. parasite points, d. splash due to dampness, e. weak contrast

In fact, a step of preprocessing is often performed on these documents to improve their quality, such as contrast enhancement and intensity distribution on the whole range of possible values. Also, taking into account their huge size of digital versions, a compression operation is often necessary. Furthermore, the ancient documents can be affected by noise during their transmission and exploitation. When watermarking the images of ancient documents, the latter mentioned operations are considered as attacks which could affect the good detection of the signature. In [13], we have presented our first contribution of watermarking gray-scale images in the multi-resolution domain, based on Wavelet Packet Decomposition (WPD). The Singular Value Decomposition (SVD) is applied on each sub-band of the best insertion wavelet packet base of the original image to retain its Singular Values (SV) as signature carrier points. The main results have shown the performance of this approach in terms of invisibility and robustness.

Watermarking color images requires considering other information mainly related to the color criterion. Accordingly, we propose in this paper an algorithm of watermarking color images of ancient documents. As a first step, some classical watermarking algorithms have been tested on some images of ancient documents [1, 2, 6]. According to the experimental results, we have noticed, as anticipated, that the classical algorithms are not adapted to this type of images. Actually, we have detected a weakness in both invisibility and robustness. This weakness is caused by different specificities and the important degradation of the ancient documents. Indeed, we are always noting a confusion problem of carrier points of the water-mark with the noise components from which the ancient documents often suffer, beside the problem of choosing the color space.

The principle of the algorithm suggested for watermarking color images of ancient documents consists in adding the SV of the signature to the stable SV of each sub-band of the best packet base of the first principal component (PC1) of the original image in the YIQ mode.

3   Proposed Approach

We propose a non-blind adaptive algorithm of watermarking color images of ancient documents. This approach is an adaptation of the one suggested in [13], using the marginal strategy. Next, we detail the main steps of the insertion and detection phases. Figure 3 shows the block diagram (insertion phase) of the suggested approach.

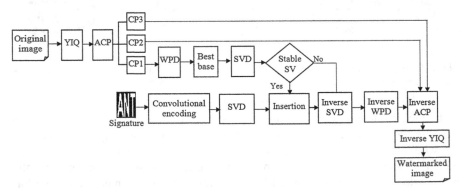

Fig. 3 Schema of the proposed watermarking approach: Insertion phase

3.1 Selection of Color Space and Insertion Component

The first step consists in choosing the color space and selecting the insertion component to get the performance of our approach. We suggest exploiting a decorrelated component as an insertion component obtained by merging three components of the used color space of the image. The objective of this measure is to benefit of each color component in terms of robustness and keep the visual properties of the signature marking. Consequently, we opt for the Principal Components Analysis (PCA) and exploit its PC1 as an insertion component [10]. Actually, this component is closely related to the statistical properties of the original image and has the advantage of a maximum-energy image concentration. A comparative study allows concluding that the YIQ space is the best insertion one, which shows a more pertinent contribution than others in the context of color-image watermarking of ancient documents [14].

3.2 Selection of the Best Wavelet Packet Insertion Base

The choice of the insertion domain and base is the second step. In this work, we retain the multi-resolution domain as an insertion domain that is more robust to signal processing attacks related to ancient documents (compression, filter, histogram equalization...). The WPD is selected after having studied the DWT, the dual-tree complex wavelet, the contourlet and the curvelet transform [12]. This decomposition guarantees an information-rich insertion base in the high-frequency sub-bands, i.e. a good localization of information compared to noises that dominate the images of ancient documents. Actually, this decomposition results in a redundant base, where for any "L" decomposition level, each sub-band (approximations and details) is itself decomposed into four sub-bands at the level "L + 1".

As an insertion base, we have selected from the redundant packet base the best sub-bands that are stable and robust against attacks and that ensure a good invisibility. All of these sub-bands construct the best insertion base.

Since we process images that are generally noisy, we have chosen to insert the signature into the less noisy sub-bands (weak entropic value). The provision of the criterion based on the minimization of the entropy is twofold. On the one hand, it reduces the confusion between the signature carrier points and the noise components. On the other hand, it allows taking advantage of a masking effect by watermarking the sub-bands that have little information.

The strategy for selecting this base is to run through the entropic tree of the last decomposition at an "L-1" level towards a "0" level. Then, we compare the entropy value of a father node with those of its four sons. The algorithm of the proposed strategy has been presented in [13]. The major advantage of this base is its adaptation to the contents and characteristics of ancient documents; i.e., it changes from one image to another. Figure 4 shows an example of the best image insertion base.

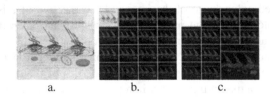

a.   b.   c.

Fig. 4 (a) Original image (b) Two levels of decomposition in wavelet packets (c) Selection of the best base

3.3 Selection of Signature Carrier Points

In our approach, we suggest applying the SVD to each insertion sub-band to exploit its SV as carrier points [15]. The number of exploited SV varies from one image to another. Indeed, it is equal to the minimum number of stable SV for all insertion sub-bands of each image.

In the insertion phase, after having coded the signature by a convolutif Error Correcting Code (ECC), we propose adding its SV to stable SV of each insertion sub-band. Finally, the inverse SVD is performed for reconstructing each watermarked sub-band and the inverse WPD is performed for reconstructing the watermarked PC1. The inverse ACP and inverse YIQ are performed consecutively for reconstructing the watermarked image.

In the detection phase, only the stable sub-bands of the best base of the original and watermarked images are used to extract the SV of each detected signature. The stability of the SV as carrier points and the stability of the insertion base are defined in [13].

4 Algorithm

The main steps of insertion and detection processes of our approach are detailed respectively by Algorithm1 and Algorithm2.

Algorithm1 insertion phase

Input: I : original image, W : signature
Output: I': watermarked image
<u>begin</u>
1. Transform the I to YIQ space
2. Apply ACP to image in YIQ space
3. Apply 3 level of WPD to PC1
4. Select the best base of wavelet packet B={SB$_k$}, k∈[1..N]
5. Perform SVD on each sub-band of the obtained best base:

$$SB=U_B*S_B*V_B^T$$

6. Code W by the convolutional ECC
7. Perform SVD on the coded signature: $W=U_w*S_w*V_w^T$
8. Modify the stable SV of each sub-band of the best basis with the SV of S_w:

for each j from 1 to R step r

for each i from 1 to r

$$S_w'(j)=S_B(j)+\propto S_w(i)$$

where R is the number of selected SV of S_B, r is the number of selected SV of signature, and α is a factor of a good compromise invisibility/ robustness.
9. Reconstruct each watermarking sub-band by performing the inverse SVD
10. Apply the inverse WPD to reconstruct the watermarked PC1
11. Apply the inverse ACP by using PC2 and PC3 to obtain watermarked image in YIQ space.
12. Perform YIQ inverse to reconstruct I'
<u>End</u>

Algorithm2: detection phase

Input : I: original image, I$^"$: attacked watermarked image
Output : W': detected signature
<u>Begin</u>
1. Transform the I and I$^"$ from RGB space to YIQ space
2. Apply PCA to two images in YIQ space
3. Apply 3 level of WPD on PC1 and PC1$^"$
4. Select the two best wavelet packet bases B and BT respectively for CP1 and CP1$^"$
5. Select the stable sub-band between B and BT
6. Apply the SVD to each stable sub-band of B and BT:

$$SB=U_B*S_B*V_B^T, SB^"=U_A*S_A*V_A^T$$

7. Detect the SV of watermark for each sub-band:

for each j from 1 to R step r

for each i from 1 to r

$$S_w'(i)=(S_A(j)-S_B(j))/\alpha$$

8. Reconstruction of each W' : $W'=U_w*S_w'*V_w^T$
9. Normalization: it consists in binarizing the values of the W'
10. Perform the Viterbi decoder on W'
<u>End</u>

5 Experimentations and Results

In this section, we describe the main experiments and the recorded results. In this work, we test the robustness of our algorithm against attacks related to the ancient documents, such as the JPEG compression with a 30 % quality factor, the 5×5 and 7×7 median filter, the histogram equalization, the 10 % Gaussian noise, and the salt and pepper noise with a density of 0.01. We have used the Peak Signal to Noise Ratio (PSNR) to determine the variation between the watermarked image and the original one [11]. We have also exploited the correlation to calculate the similarity between the original watermark and the extracted one [16].

5.1 Test Database

The different tests have been conducted upon color images of 512×512 size from the ancient documents of the National Archives of Tunisia. Our test database is composed of 100 color images in the RGB mode. In this paper, we illustrate the different results on four images (Fig. 5). We have kept, as a watermark, a binary image of 32×16 size, which contains the first letters of the "Archives Nationales de Tunisie".

Img1   Img2   Img3   Img4   Watermark

Fig. 5 Examples of images extracted from our test database, watermark

5.2 Results

After an experimental study on the images of our database, a PSNR value superior or equal to 39 dB allows having a good invisibility. In Fig. 6, we present the PSNR value evaluation depending on the insertion factor α, which ensures a compromise between invisibility and robustness for the four images in Fig. 5. According to the four curves in Fig. 6, the standard deviation of the PSNR values is very high between images. For example, the difference margin of the PSNR value of Img1 is in the order of 10 db compared to Img4, and 5 db compared to Img3. Therefore, this margin makes the problem of selecting a single value of α for all images, unlike the watermarking gray-scale images of ancient documents [13].

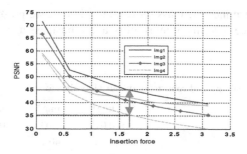

Fig. 6 PSNR values for four images

To resolve this problem, an adaptive α is used in the algorithm to obtain the best compromise. Here, α takes into account various degradations of the color images of ancient documents. The value of α is determined based on the entropy criterion that corresponds to the least PSNR value (39 db). In fact, we calculate the entropy value of PC1 of the original image into the YIQ mode and we select the highest value of α that offers the best invisibility. Selecting the highest α value ensures the robustness against recurrent attacks. The results are illustrated in Fig. 7 on 50 images of various specifics (structure, content, quality...).

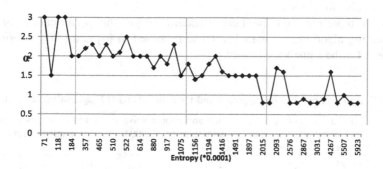

Fig. 7 α values according the entropy variations

According to Fig. 7, we can classify the entropy values into four classes depending on each image's α values. Table 2 shows the registered α value for each class.

Table 2 α value for each interval

Entropy value (e)	e ≤ 0.017	0.017 < e ≤ 0.1	0.1 < e ≤ 0.2	e > 0.2
Classes	Class 1	Class 2	Class 3	Class 4
α	3	2	1.5	0.8

After having determined the class for each test image, we present in Table 3 the approach performance in terms of invisibility and robustness against attacks of four different classes.

Table 3 Results of invisibility and robustness of our algorithm for four test images

	Img1	Img2	Img3	Img4
PC1 entropy	0.012	0.034	0.18	0.97
PSNR	39.6	40.11	41.56	39.19
Histogram equalization	1	1	1	0.78
JPEG compression with 30 % quality factor	0.96	1	0.74	0.85
5 × 5 median filter	0.99	1	0.94	0.99
7 × 7 median filter	0.76	0.99	0.81	0.87
Salt and pepper noise with a density of 0.01	0.96	0.88	0.96	0.91
10 % Gaussian noise	1	0.84	0.56	0.94

As shown in Table 3, the PSNR value varies from 39 to 41 db, which shows the invisibility of our approach. We note also a good robustness against different attacks for classes1, 2 and 4, but a fragility against the JPEG compression and the Gaussian noise for class 3.

Table 4 represents, on the Lena standard image, the performances of our watermarking algorithm compared to those proposed in [6] and [17], in terms of robustness for the same tested attacks.

Table 4 Comparison between our approach and those of [6] and [17] against the same attacks

	3 × 3 median filter	6 % Gaussian noise	Salt and pepper noise with density of 0.003	JPEG compression with 50 % quality factor
Our approach	0.99	1	1	0.99
[6]	0.57	0.68	0.65	0.85
[17]	0.93	0.96	0.95	0.94

6 Conclusion

In this paper, a watermarking schema of color images of ancient documents is put forward. We have used the PCA to retain PC1 as an insertion component of the original image in the YIQ color space and the WPD-SVD to select the signature carrier points. The signature is coded by a convolutif encoder. The difficulty of having a compromise between invisibility and robustness leads to categorize the test-image base into four classes according to the entropy variations. For each class, the recorded results show the best compromise between invisibility and robustness against most attacks, frequently undergone by ancient documents. As a future work, we will refine the number of classes of color images of ancient documents, using a more significant image base. In addition, we propose to exploit the image self-embedding techniques to improve the invisibility level for all the classes [3].

References

1. Kutter, M., Jordan, F., Bossen, F.: Digital signature of color images using amplitude modulation. In: Proceeding Storage and Retrieval for Image and Video Databases, SPIE, pp. 518–526 (1997)
2. Dharwadkar, N., Amberker, B., Gorai, A.: Non-blind Watermarking scheme for color images in RGB space using DWT-SVD. In: IEEE International Conference on Communications and signal processing, pp. 489–493 (2011)
3. Song, Q., Zhang, H.: Color image self-embedding and watermarking based on DWT. In: IEEE International Conference on Measuring Technology and Mechatronics Automation, vol. 1, pp. 796–799 (2010)
4. Lo-varco, G., Puech, W., Dumas, M.: Content based watermarking for securing color images. J. Imaging Sci. Technol. 49(6), 464–473 (2005)
5. Su, Q., Liu, X., Yang, W.: A Watermarking algorithm for color, image based on YIQ color space and integer wavelet transform. In: IEEE International Conference on Image and Signal Processing, pp. 70–73 (2009)
6. Sinha, S., Bardhan, P., Pramanick, S., Jagatramka, A., Kole, D., Chakraborty, A.: Digital video watermarking using discrete wavelet transform and principal component analysis. Int. J. Wisdom Based Comput. 1(2), 7–12 (2011)
7. Munawer, H., Al Sowayan, S.: Color image watermarking based on self-embedded color permissibility with preserved high image quality and enhanced robustness. Int. J. Electron. Commun. 65(7), 619–629 (2011)
8. Xinde, S.: A novel digital watermarking algorithm for medical color image. In: Proceedings of the International Conference on Computer and Information Application, pp. 1385–1388. Atlantis Press (2012)
9. Xinde, S., Shukui, B.: A blind digital watermarking for color medical images based on PCA. In: IEEE International Conference on Wireless Communications, Networking and Information Security, pp. 421–427 (2010)
10. Hien, T.D., Chen, Y.W., Nakao, Z.: Robust digital watermarking technique based on principal component analysis. Int. J. Comput. Intell. Appl. 4(2), 138–192 (2004)
11. Jin, C., Pan, L., Ting S.: A blind watermarking scheme based on visual model for copyright security. In: Proceeding Multimedia Content Analysis and Mining, pp. 454–463 (2007)
12. Maatouk, M., Jedidi, O., Essoukri Ben Amara, N.: Watermarking ancient documents based on wavelet packets. In: Document Recognition and Retrieval XVI Conference, vol. 7247, pp. 7238–7247, SPIE (2009)
13. Maatouk, M., Essoukri Ben Amara, N.: Intelligent hybrid watermarking ancient-document wavelet packet decomposition-singular value decomposition-based schema. J. IET Image Process. 8(12), 708–717 (2014)

14. Maatouk, M., Noubigh, Z., Essoukri Ben Amara, N.: Contribution of watermarking color images of ancient documents. In: IEEE International Conference Reasoning and Optimization in Information Systems, pp. 155–162 (2012)
15. Liu, R., Tan, T.: An SVD-based watermarking scheme for protecting rightful ownership. IEEE Trans. Multimedia 4(1), 121–128 (2002)
16. Douak, F., Benzid, R., Benoudjit, N.: Color image compression algorithm based on the DCT transform combined to an adaptive block scanning. Int. J. Electron. Commun. 65(1), 16–26 (2011)
17. Niu, P., Wang, X., Yang, Y., Lu, M.: A novel color image watermarking scheme in non sampled contourlet domain. Expert Syst Appl 38(3), 2081–2098 (2011)

Planar Multi-classifier Modelling-NN/SVM: Application to Off-line Handwritten Signature Verification

Imen Abroug Ben Abdelghani and Najoua Essoukri Ben Amara

Abstract We present in this paper a new approach to modeling an image in a pattern recognition context. It relates to a generalization of the planar Markov modeling, which we shall call Planar Multi-Classifier Modeling (PMCM). These models allow exploring various classifier types, among others, Markov ones, to model delimited bands of an image associated with a given pattern. Determining the different parameters of the proposed planar model is made explicit. Various PMCM architectures are proposed; in this paper we present the PMCM-NN/SVM architecture. The validation has been performed in a security context based on off-line handwritten signature verification. The different experiments, carried out on two public databases: the SID-Signature database and the GPDS-160 Signature database, have led to promising results.

Keywords Planar multi-classifier models · NN/SVM · Parallel classifier combination · Off-line handwritten signature verification

1 Introduction

The growing evolution of communications and the increase in identity fraud attempts has generated a highly important need for the identification of individuals. Biometric security technologies are among the most widely used methods to secure persons and information. Our research works focus on the identity verification by biometric features. We are particularly interested in modeling images forms in a biometric context such us face recognition [1] fingerprint and palm print

I.A.B. Abdelghani (✉)
Higher Institute of Applied Sciences and Technology of Kairouan,
UR: SAGE-ENISo, University of Sousse, Sousse, Tunisia
e-mail: abrougimen@yahooo.fr

N.E.B. Amara
National Engineering School of Sousse, UR: SAGE-ENISo,
University of Sousse, Sousse, Tunisia
e-mail: najoua.benamara@eniso.rnu.tn

© Springer International Publishing Switzerland 2015
Á. Herrero et al. (eds.), *International Joint Conference*, Advances in Intelligent Systems and Computing 369, DOI 10.1007/978-3-319-19713-5_8

recognition [2]. In this paper we are interested in handwritten signature verification; we present a new approach to modeling an image in the context of pattern recognition, handwritten signature in this case.

The literature shows different types of approaches to pattern recognition by images processing based on 1D and 2D models. The 1D models consider the image as a one-dimension signal, thus overshadowing its bi-dimensional properties. The 2D models process the image in its two dimensions, which leads to a complexity of exponential calculations [3]. The pseudo 2D models, still known as planar, based on Hidden Markov Models (HMM), provide an intermediate solution between the 1D and 2D models. They bring the calculations to a complexity lower than those of the 2D, considering the image in its two dimensions without applying real 2D models. These models have been proven efficacious in several application areas.

The basic idea of Planar Hidden Markov Models (PHMMs) is to divide the image into homogeneous zones (horizontal or vertical, depending on the chosen orientation) and is to model each one by a 1D HMM, which is properly called a secondary model. A model, known as principal, is defined in the orthogonal direction; the corresponding states are called super-states. As applied to a given shape, the main model of a PHMM performs a one-direction analysis, and its secondary models do that according to an orthogonal direction.

The definition of the secondary models implies firstly choosing their direction, segmenting the pattern image into homogeneous zones, and also choosing for each zone its descriptors and the type of the classifier that will be associated with. The main model, generally 1D Markov Bakis type, is fed by the outputs of the secondary models. Through specific choices of the used planar definition, the main model takes into account the correlation naturally existing between the different delimited zones (duration model, additional observations, ...). We invite the reader to take into consideration the basic reference of Levin et al. [4] for the detailed theoretical definitions of the PHMMs.

Following the study that we have conducted, we found that the variety of the different planar architectures proposed in the literature are based on the hidden Markov models hence their PHMM designation [5–8]. Our various works based on the PHMMs in the handwritten and printed Arabic script [3], have shown encouraging results and let us be convinced that defining the planar Markov models can be generalized to other classification techniques. The basic idea of our purpose is to extend the planar modeling by proposing other architectures based on the principle of multiple classifications. It is on the one hand to explore other classification techniques, apart from the HMMs, and on the other hand to bring the choice of observations, at the level of secondary models, to classically extracting a set of features from a local, global or hybrid characterization approach. Likewise, the zone segmentation is considered differently, which is adapted to the morphology of the shape to model.

In the next section we propose an extension of the planar Markov modeling to a planar multi-classifier modeling, while specifying the choice of various parameters associated with it. In Sect. 3, we validate the modeling proposed in the case of off-line handwritten signature verification. The experimental results are presented in Sect. 4. The conclusion is addressed in Sect. 5.

2 Planar Multi-classifier Modeling

We propose in this section a generalization of the PHMMs, which we shall call Planar Multi-Classifier Modeling (PMCM) whose operating principle is reminiscent of the parallel combination of classifiers. The proposed modeling is justified by the importance of taking into account the real dimensionality of a given shape— bi-dimensional in our case—in a simple and practical way without being restricted to Markov models. Having set the main direction, we associate with previously delimited bands classifiers that can be of different types—Markov, inter alia. The type of the main model is chosen in complementarily with the retained secondary models, more exactly in adequacy with the type of their decision (binary, probability, distance ...). We also put forward a step of conventional characterization of a pattern recognition system, which will be applied across each of the bands.

2.1 Segmentation into Zones and Characterization

The idea of zone segmentation remains the same as that of the planar Markov modeling. In the case where the shape morphology facilitates defining a segmentation criterion in accordance with a predetermined direction, the image is segmented into a set of zones whose number and size will be defined in keeping with the morphological content of the considered shape. In the event of a random variation in morphology, we suggest delimiting the image in a fixed number of zones having similar or dissimilar dimensions.

The different bands being delimited, their characterization is performed independently of each other. This phase follows the conventional characterization steps in a pattern recognition system. Therefore, each band will be characterized by a specific set of features which make up the observations of the secondary models.

2.2 Choice of Main and Secondary Models

The number of secondary models corresponds to the number of bands delimited beforehand. The secondary model is a classifier that can be of different types (NN, HMM, SVM ...). As for the main model, it is fed by the outputs of the secondary models. Choosing the classifier of the main model depends mainly on the output type of the secondary models. According to the decision as to highlight the correlation between bands, some additional information could feed the main model. Depending on the type and scale of the values of the outputs of secondary models, a normalization step is necessary to homogenize the data.

2.3 Learning Procedure

Learning the PMCM requires a procedure adapted to its architecture (Fig. 1). We propose stacked generalization learning inspired from classifier-combination learning [9]. The learning data set is divided into two subsets D_1 and D_2. Learning the secondary models is carried out in parallel with D_1. Once the secondary models are created, learning the main model is activated using the results of the secondary models, operating in recognition mode of the subset D_2.

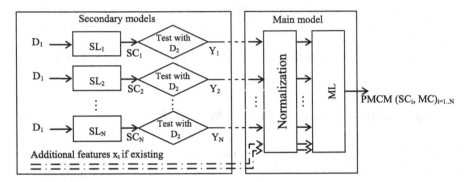

Fig. 1 Stacked generalization learning of planar multi-classifier model

The principle is given by the following algorithm.

```
Inputs:
N: Number of secondary classifiers
D₁={V₁,ᵢ|i=1..N}:N characteristic vectors of Learning set D₁
D₂={V₂,ᵢ|i=1..N}:N characteristic vectors of Learning set D₂
X={xₜ|t=1..T}:T eventual additional features
SLᵢ,i=1..N:N secondary classifiers learning algorithms
ML: Main classifier learning algorithm
Output:
SCᵢ,i=1..N: N Secondary classifiers
MC: Main classifier
PMCM(SCᵢ,MC),i=1..N: Planar multi-classifier model
Begin
   For i: 1...N do
     Construct secondary models: SCᵢ=SLᵢ(V₁,ᵢ)
       Evaluate secondary models: Yᵢ=SCᵢ(V₂,ᵢ)
   End
Construct main model: MC=ML(Yᵢ,X)
Return PMCM(SCᵢ,MC),i=1..N
End
```

2.4 Planar Multi-classifier Models Versus Parallel Combination of Classifiers

As defined, the main model of a PMCM makes the role of a fusion module (also called aggregation module) which allows merging the outputs of the various secondary models with the additional data, if they exist. Thus, a planar multi- classifier modeling could be compared to architecture of a parallel combination of classifiers. Indeed, in the case of a parallel combination of classifiers, according to the chosen application, firstly several sources are processed—each separately by a classifier—and secondly the responses for each classifier (rank, degree ...) are merged. A normalization step is often needed prior to the merger. The approach of parallel combination requires the activation of all classifiers of the system that are to be part of it concurrently and independently. The final decision is taken with the utmost knowledge made available by each classifier. According to the taxonomy of the methods of parallel combination of classifiers [10], particularly the method of merging combination (parametric, non-parametric), we note a great similarity with the PMCM concepts we have put forward. In fact, a PMCM may be represented by a set of different types of classifiers fed by various typical vectors. The corresponding outputs are then merged with an aggregation operator: the main model. The secondary models are represented each by the 1D classifier, and the main model corresponds to a 1D meta-classifier or to non-parametric techniques (vote, fixed rules, ...) which receives as an input the outputs of the secondary models and additional features to provide the final decision.

3 Handwritten Signature Modeling

The PMCM, we have put forward, can be applied for modeling, a priori, any 2D-shaped type such as writing, face, signature, fingerprint, ...In [1] we have validated a PMCM-NN/NN defined by $(NN_i, NN)_{i=1...3}$ in the case of face recognition. In this paper we validate our concepts in the case of off-line handwritten signature. We present the PMCM-NN/SVM architecture defined by PMCM $(NN_i, SVM)_{i=1...3}$ (Fig. 2). We also suggest a PMCM-HMM/HMM architecture defined by PMCM $(HMM_i, HMM)_{i=1...3}$ in order to compare the proposed architectures with the basic planar Markov modeling. The proposed architecture follows the design steps detailed in Sect. 2.

3.1 Definition of the Main and Secondary Models

The secondary models are defined according to the horizontal direction. Indeed, the random morphological nature of the signature has led us to determine the number

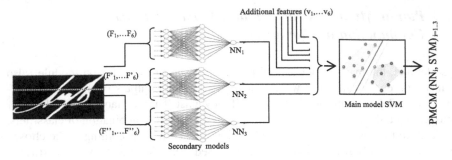

Fig. 2 Implemented planar multi-classifier PMCM-NN/SVM architecture

and size of the information bands for all classes of signatures. Accordingly, the previously-preprocessed signature image (binarization, filtering using median filter, and size standardization) is segmented into three horizontal bands having the same height. Each band is modeled by a horizontal secondary model. The vertically-oriented main model enables modelling the image according to the vertical direction while correlating the three delimited bands. The choice of classifiers used at the level of the main and secondary models is guided by the type of outputs of the secondary models feeding the inputs of the main model. In the first architecture MPMC-NN/SVM, we have considered the multilayer NN (MLP) in secondary models and SVM with an RBF kernel in main model. In the second architecture PMCM-HMM/HMM, secondary models are of ergodic topology HMM and main model is an HMM with left-right topology.

3.2 Definition of Observations

We have retained two types of global features: geometric and texture characteristics from the application of the wavelet transformation. Each secondary model is fed by a secondary input vector, which contains six features:

- The number of black pixels in the considered band (F_1),
- The mean and standard deviation of the coefficients of the approximation image resulting from applying the wavelet transformation (F_2, F_3),
- The standard deviations of the coefficients of the horizontal, vertical and diagonal detail images (F_4, F_5, F_6).

The main model is fed by the outputs of the three secondary models, to which we have added six other geometric characteristics to reinforce the description in the vertical direction; these characteristics have been shown to be effective in describing the interclass variations:

- The maximum of the vertical histograms of the three bands (V_1, V_2, V_3);
- The orientation of the signature image compared to the horizontal position (V_4).

- The number of intersections between the signature image lines (V_5).
- The number of closed areas of the signature image (V_6).

3.3 Learning and Recognition Procedure

According to the considered PMCM architecture, learning and evaluating main and secondary models has been carried out through the specific techniques of the classifiers taken into account. In the case of NNs, we have implemented the algorithm of the error retro-propagation for training. As to the HMMs, learning has been performed by means of the Baum-Welch algorithm and recognition has been realized by the Viterbi algorithm. Learning the three secondary models is made in parallel. Once the secondary models are created, learning the main model is activated.

4 Experiments and Results

We consider in this paper the results achieved in the case of the offline handwritten signature verification. The different experiments have been carried out on two public handwritten-signature databases: the SID-Signature database [11] and the GPDS-160 Signature database [12]. The different developed architectures have been evaluated by estimating the False Reject Rate (FRR) the False Acceptance Rate (FAR) and the Equal Error Rate (ERR).

4.1 Validation with the SID-Signature Database

The SID-Signature is a public database that has been developed for the study of Tunisian origin signatures and the impact of the cultural aspect on verification performance. The different experiments performed to validate the proposed architectures were carried out on a training set containing only genuine and random forgeries samples (Table 1).

Several experiments were carried out in order to fix the structure of the PMCM classifiers for each class of signatures. The comparison of performances of explored architectures shows that the PMCM-NN/SVM is more efficient than the basic

Table 1 Learning and test subsets of the SID-signature database		Genuine	Random	Simple	Skilled
	Learning	20	20	–	–
	Test	20	20	20	20

Table 2 Performances of PMCM architectures validated on the SID-signature database

	EER (%)		
Forgeries types	Random	Simple	Skilled
PMCM-NN/SVM	8.85	8.35	9.10
PMCM-HMM/HMM	28.45	29.5	37.05

PMCM-HMM/HMM and confirms that the classifier combination improve considerably the performance of planar models (Table 2).

4.2 Validation with the GPDS-160 Signature Database

We retain the PMCM-NN/SVM architecture for the validation on the GPDS database. The state of the art presents different works based on this database. To compare the PMCM architecture with existing systems, we consider only systems using the same experimental protocol: learning and test subsets (Table 3).

The results of previous works are presented with the AER which is the average of FRR and FAR. Table 4 give results and details corresponding to the considered works.

Table 3 Learning and test subsets of the GPDS-160 signature database

	Genuine	Random	Skilled forgeries
Learning	12	12	–
Test	12	30	30

Table 4 Performances results of selected works validated on the GPDS-160 Signature database

	Method		FRR (%)	FAR (%)		AER
	Characteristics	Classifiers		Random	Skilled	(%)
[13]	Local features	Combined global and modular SVM	15.41	–	15.41	15.35
[14]		SVM	22.08	–	18.07	20.07
		RBF	13.19	–	31.94	22.57
		MLP	27.29	–	25.66	26.48
[15]	Local/ global features	SVM	17.25	0.08	17.25	17.25
[16]	Local features	Combined classifiers	19.19	9.81	47.25	25.42
[17]			27.25	0.0031	18.17	15.24
Proposed approach	Global features	PMCM-NN/SVM	21.72	4.72	31.89	19.45

Table 5 Execution time of selected works validated on the GPDS-160 signature database

		Execution time (s)					
		Characterization	Classification				
[13]	Hog-polar	0.6	Modular	10^{-2}	Global	0.6
	Hog-grid	0.04		10^{-4}		0.25
	Lbp-grid	0.1		6×10^{-3}		0.75
[17]		0.27	10^{-5}			
Proposed approach		0.26	3.5×10^{-4}			

Besides, we present in Table 5 the execution times per signature. Among the work presented in Table 4, only [13] and [17] gives the execution time details.

The performance analysis shows that the proposed architecture gives encouraging results. The achieved AER is better than those performed in [14] and [16] which are based on local features. The execution time of the proposed approach is comparable or better than considered execution times. Compared with [16], which is based on combined classifiers and a complex dynamic selection classifiers; our model showed better performances.

Compared with [15] and [17] which conduct a better verification results, the time of characterization of [17] is comparable to the proposed model. The fusion of global and local features used in [15] may explain the little exceeding performance of this work.

The classifier presented in [13] performed better than the proposed model however the PMCM-NN/SVM is much better in both characterization and test execution times. These results confirm the contribution of the proposed modeling in handwritten signature verification as in face recognition [1].

5 Conclusion

In this paper we have presented an extension of the planar Markov modelling by proposing a planar multi-classifier modeling in the context of pattern recognition. First, we have presented a review of the planar models recalling the main steps of their design. The literature analysis shows that the various proposed planar modelings are based on the HMMs, which differ from each other depending on the morphological complexity of the considered shapes and on the choices made at the different design stages. We have proposed a new vision of the conception of the planar modeling inspired by the principle of the parallel combination of classifiers. Our idea is based on exploring several types of classifiers at the level of the main and secondary models, without being limited to the HMMs. The analogy of the planar architecture, suggested by the parallel combination of classifiers, allows us to take advantage of the theoretical and experimental results—greatly developed—of the combination of classifiers. The proposed planar multi-classifier modeling can be used, a priori, for modeling any shape type. We have validated that with the offline handwritten signatures verification. Our exploratory work has been validated in the case of a PMCM architecture based on the combination of NN/SVM and compared with a basic PMCM-HMM/HMM. The experiments have been carried out on the SID-Signature and GPDS-160 Signature databases. The recorded results confirm the different proposed concepts. Other work is being in progress to validate our concepts on other biometric modalities using other classifiers and characteristics.

References

1. Abroug Ben Abdelghani, I., Essoukri Ben Amara, N.: Planar multi-classifier modelling-NN/NN for face recognition. In: IEEE International Multi-Conference on Systems, Signals and Devices, Tunisia (2015)
2. Ben Khalifa, A., Essoukri BenAmara, N.: Adaptive score normalization: a novel approach for multimodal biometric systems. Int. J. Comput. Inf. Sci. Eng. **7**(3), 18–26 (2013)
3. Essoukri Ben Amara, N.: Utilisation des modèles de Markov cachés planaires en reconnaissance de l'écriture arabe imprimée. Ph.D. Thesis, Department of Electrical Engineering, University Tunis II, Tunisia, February (1999)
4. Levin, E., Pieraccini, R.: Dnamic planar warping for optical character recognition. In: International Conference on Acoustic, Speech and Signal processing, pp. III-149-III-152 (1992)
5. Kuo, S., Agazzi, O.E.: Keyword spotting in poorly printed documents using pseudo 2D hidden Markov models. IEEE Trans. Pattern Anal. Mach. Intell. **16**(8), 842–848 (1994)
6. Wang, J., Xu, C., Chng, E.: Automatic sports video genre classification using pseudo-2D-HMM. In: International Conference on Pattern Recognition, vol. 4, pp 778–781 (2006)
7. Hui, L., Jean-Claude, T., Legand, B., Chunmei, L.: A rapid 3D protein structural classification using pseudo 2D HMMs. In: IEEE International Conference on Granular Computing, pp. 742–745 (2012)
8. Tashk, A., Helfroush, M.S., Kazemi, K.: Automatic fingerprint matching based on an innovative ergodic embedded hidden markov model (E2HMM) approach. In: Iranian Conference on Electrical Engineering, pp. 265–269 (2010)
9. Wolpert, D.H.: Stacked generalization. Neural Networks **5**, 241–259 (1992)
10. Duda, R.O., Hart, P.E., Stork, D.G.: *Pattern Classification*, 2nd edn. (Wiley, New York, 2012)
11. Abroug Ben Abdelghani, I., Essoukri Ben Amara, N.: SID-signature database: a tunisian handwritten signature database. In: Petrosino A., Maddalena L., Pala P. (eds.) New Trends in Image Analysis and Processing, Lecture Note on Computer Science, vol. 8158, pp. 131–139 Springer-Verlag, Heidelberg (2013)
12. Vargas, F., Ferrer, M.A., Travieso, C.M., Alonso, J.B.: Off-line handwritten signature GPDS-960 corpus. In: International Conference on Document Analysis and Recognition, pp. 764–768 (2007)
13. Yilmaz, M.B., Yanikoglu, B., Tirkaz, C., Kholmatov, A.: Offline signature verification using classifier combination of hog and lbp features. In: International Conference on Biometrics, IEEE Computer Society, pp. 1–7 (2011)
14. Nguyen, V., Blumenstein, M., Muthukkumarasamy, V.M., Leedham, G.: Off-line signature verification using enhanced modified direction features in conjunction with neural classifiers and support vector machines. In: International Conference on Document Analysis and Recognition, pp. 734–738 (2007)

15. Nguyen, V., Blumenstein, M., Leedham, G.: Global features for off-line signature verification problem. In: International Conference on Document Analysis and Recognition, pp 1300–1304 (2009)
16. Batista, L., Granger, E., Sabourin, R.: Dynamic selection of generative-discriminative ensembles for off-line signature verification. Pattern Recogn. **45**(4), 1326–1340 (2012)
17. Eskander, G.S., Sabourin, R., Granger, E.: Hybrid writer-independent–writer-dependent offline signature verification system. In: IET Biometrics, pp. 1–13 (2013)

Email Spam Filtering Using the Combination of Two Improved Versions of Support Vector Domain Description

Mohamed EL Boujnouni, Mohamed Jedra and Noureddine Zahid

Abstract Email is fast and cheap message transfer way, it has become widely used form of communication, due to the popularization of Internet and the increasing use of smart devices. However, an unsolicited kind of email known as spam has been appeared and caused major problems of the today's Internet, by bringing financial damage to companies and annoying individual users. In this paper we present a new method to classify automatically legitimate email from spam, based on the combination of two improved versions of Support Vector Domain Description (SVDD), the first one aims to adjust the volume of the minimal spheres, while the second replaces the standard decision function of SVDD by a new improved one. An experimental evaluation of the proposed method is carried out on a benchmark spam email dataset. The experimental results demonstrate that the proposed method achieves high recognition rate with good generalization ability.

Keywords Spam filtering · Support vector domain description · Parametric volume · Decision function

1 Introduction

The internet occupies an integral part of everyday life, and electronic mail has become a popular and powerful tool intended for idea and information exchange. It attracts lots of users since its low cost send, simple delivery mode and convenient

M.E. Boujnouni (✉) · M. Jedra · N. Zahid
Mohammed V—Agdal University, Faculty of Sciences, Laboratory of Conception and Systems (Microelectronic and Informatics), Avenue Ibn Battouta B.P 1014, Rabat, Morocco
e-mail: med_elbouj@yahoo.fr

M. Jedra
e-mail: jedra@fsr.ac.ma

N. Zahid
e-mail: zahid@fsr.ac.ma

© Springer International Publishing Switzerland 2015
Á. Herrero et al. (eds.), *International Joint Conference*, Advances in Intelligent Systems and Computing 369, DOI 10.1007/978-3-319-19713-5_9

usage. It is estimated that 247 billion email messages were sent per day in 2009 [15]. Its applications range vary from basic informal communication to an indispensable business platform, however there are few issues that spoil the efficient usage of emails, like Spam.

E-mail spam also called, unsolicited bulk Email (UBE), junk mail, or unsolicited commercial email (UCE) is the abuse of electronic messaging systems to send unwanted email messages, frequently with commercial content, in large quantities to an indiscriminate set of recipients. It can originate from any location across the globe where internet access is available (Fig. 1).

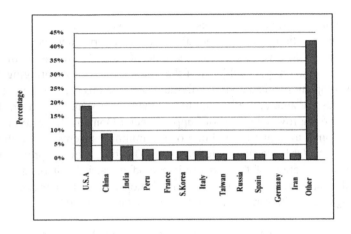

Fig. 1 Spam rate (December 2012 to Feb 2013)

Spam causes huge and various damage like:

- Wasting the network bandwidth by receiving massive quantity of spam;
- Consuming user's time and efforts by distinguishing between spam and legitimate messages;
- Spending storage space in the email servers;
- Crashing of mail-servers;
- Filling user's mailboxes;
- Delivering malicious software like virus, or immoral contain;
- Financial loss and violation of laws by broadcasting prohibited materials, the overall cost of spam in 2009 was estimated at 130 billion U.S. dollars [8];
- Hunting for sensitive information (passwords, credit card numbers, etc.) by imitating official requests from a trusted authorities, such as banks, server's administration;
- Delivering a whole range of privacy and security threats. Phishing scams, spyware, adware, Trojans and viruses could all be unwittingly unleashed by just clicking on a link or attachment contained within a spam message.

Due to the damage cited above spam filter must be associated to the network or computer; it will separate spam and legitimate email before entering it into the mail box of any E-mail user. To do that many solutions for fighting spam have been proposed. Below some of them:

- Adopting legislation against spam, although it is limited by the fact that many such messages are sent from various countries, beside it is difficult to track the actual senders of spam;
- Creating Whitelists (The opposite of blacklists), it contains lists of emails senders who are trusted to send ham and not spam. Emails from someone listed on a whitelist will normally not be marked as spam, no matter what the content of their email;
- Using of spam filters, although spammers began to use several tricky ways to overcome the filtering methods like: using random sender addresses, appending random characters to the beginning or the end of the message subject line, obfuscating text in the spam, inserting random space or word in the header;
- Classifying emails by adopting machine learning techniques such as Support Vector Machine, neural networks, naïve Bayesian classifier, K-Nearest Neighbor algorithm... etc.

In this paper we proposed a new approach to filter spam, based on the forth solution cited above which uses machine learning techniques. Our approach is to filter spam using the combination of two improved versions of the conventional Support Vector Domain Description (SVDD), the first one aims to adjust the volume of the minimal hyperspheres, it is called a small sphere and parametric volume for SVDD (SSPV-SVDD) [11], while the second replaces the standard decision function of SVDD by a new improved one [12].

The rest of this paper is organized as follows. Section 2 discusses several approaches to classify emails. Section 3 provides an overview of the proposed approach. Experimental results are provided in Sect. 4 and we conclude the paper in the last section.

2 Related Work

In literature, many researchers have discussed spam filtering, and proposed several approaches to classify emails. H. Drucker et al. [9] used Support Vector Machines (SVM) algorithm for classifying the emails as spam or legitimate and compared the performance of SVM to three others classification algorithms (Ripper, Rocchio and Boosting decision trees). B. Cui et al. [4] proposed a model based on the neural network preprocessed by the principal component analysis (PCA) to classify personal emails. The goal behind using PCA is to reduce the data in terms of both dimensionality and size. Naïve Bayes network algorithms [13, 14] were used and achieved a good performance in filtering spam e-mails. In [20] X. Carreras et al.

proposed boosting trees algorithm for anti-spam email filtering, A.G. López-Herrera et al. [1] developed a multi-objective evolutionary algorithm for filtering spam. A. Hotho et al. [2] and T. Fawcett [16] involved the deployment of data mining techniques to filtering emails, J. Chen et al. [10] and T. Oda et al. [17] proposed an artificial immune system for spam filtering.

3   The Two Improved Versions of Support Vector Domain Description

3.1   A Small Sphere and Parametric Volume for Support Vector Domain Description

Support Vector Domain Description has been developed by Tax et al. [5, 7], to solve the one-class classification problem based on Vapnik's Support Vector Machine learning theory [19]. The goal of SVDD is to describe the group of data by searching a spherically shaped boundary around the target dataset. To avoid accepting outliers the volume of the sphere is minimized. Mainly it is used to carry out single-class classification and remove the noises or the outliers. To utilize the information of the negative examples, additionally, Tax et al. [6] proposed SVDD with negative examples (NSVDD) method. Positive examples were enclosed in the minimum sphere and negative examples were rejected out of it (Fig. 2). SVDD is very flexible due to the incorporation of kernel functions. The latter allows to model nonlinearity, by mapping data into a higher dimensional feature space where a hyperspherical description can be found.

Fig. 2 A conventional SVDD classifier applied to an artificial dataset

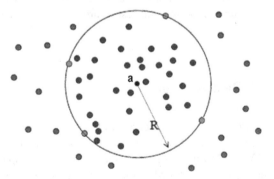

The algorithm called A Small Sphere and Parametric Volume for Support Vector Domain Description (SSPV-SVDD) [11]. is an improvement of the conventional SVDD [5–7], this approach aims to minimize the volume of the sphere describing the dataset, but following the value of a parameter called p, which controls the volume of the sphere and plays a compromise between the outlier's acceptance and the target's rejection. In what follows a summarized description of this algorithm is presented.

3.1.1 Mathematical Formulation

The goal of SSPV-SVDD is to integrate a new parameter p in the conventional SVDD, which controls the volume of the sphere describing data. To do that the SVDD problem was reformulated as follows:

Suppose we are given a training set of instance-label pairs $(x_i, y_i), i = 1, \ldots, N$ with $x_i \in R^d$ and d is the dimensionality of the vector. The value y_i is $+1$ for the target (positive) samples, and -1 for the negative ones. We aim to find the smallest sphere of radius R that encloses only the positive samples, which is described by the following constraints:

$$\|x_j - a^2\| \leq R^2 - p.y_j + \epsilon_j \forall j \text{ with } y_j = +1 \tag{1}$$

$$\|x_j - a^2\| \geq R^2 - p.y_j - \epsilon_j \forall j \text{ with } y_j = -1 \tag{2}$$

The ϵ_j are slack variables that measure the amount of violation of the constraints, when p is equal to 0 the results are the same as the conventional SVDD, to solve this problem the Lagrangian was introduced:

$$L(R, \epsilon, a) = R^2 + C \sum_{i=1}^{N} \epsilon_i - \sum_{i=1}^{N} \alpha_i y_i \left(R^2 - \|x_i - a\|^2 - p.y_i \right) - \sum_{i=1}^{N} \epsilon_i \mu_i \tag{3}$$

where α_i and μ_i are Lagrange multipliers, Setting the partial derivatives of L with respect to R, a, ϵ_i to zero gives the following constraints:

$$\frac{\partial L}{\partial R} = 0 \Rightarrow \sum_{i=1}^{N} \alpha_i y_i = 1 \tag{4}$$

$$\frac{\partial L}{\partial a} = 0 \Rightarrow a = \sum_{i=1}^{N} \alpha_i x_i y_i \tag{5}$$

$$\frac{\partial L}{\partial \epsilon_i} = 0 \Rightarrow \alpha_i = C - \mu_i \tag{6}$$

The dual optimization problem becomes:
Max:

$$L(\alpha) = - \sum_{i=1}^{N} \sum_{j=1}^{N} \alpha_i \alpha_j y_i y_j x_i x_j + \sum_{i=1}^{N} \alpha_i (y_i x_i x_i + p) \tag{7}$$

Subject to

$$0 \le \alpha_i \le C \quad et \quad \sum_{i=1}^{N} \alpha_i y_i = 1 \tag{8}$$

3.1.2 Standard Decision Function

For multi-class classification problems, multiple optimized hyperspheres which described each class of dataset were constructed separately. To classify a new test point z, we just investigate whether it is inside the hypersphere (a_k, R_k) constructed during the training and associated to the class k [3, 5–7] . The decision function is calculated by the equation (Eq. 9), if its value is positive for the kth class and negative for the others, then we conclude that z belong to the class k.

$$f(z) = sgn\left(R_k^2 - \|z - a_k\|^2\right) \tag{9}$$

where R_k is the radius associated to the kth class, it can be written as follows:

$$R_k^2 = \|x_l - a_k\|^2 + y_l.p \tag{10}$$

$$R_k^2 = x_1.x_1 - 2 \sum_{j=1}^{N} x_j x_1 y_j \alpha_j + \sum_{i=1}^{N} \sum_{j=1}^{N} x_i x_j y_i y_j \alpha_i \alpha_j + y_l.p \tag{11}$$

$x_l \in SV$, the set of Support Vectors having $0 < \alpha_l < C$, R_k and a_k are respectively the radius and the center of the kth class.

3.2 SVDD with a New Decision Function

In multiclass problem, the conventional SVDD associate to each class a minimal enclosing sphere, by consequence a new sample z will belong to the kth class, if z is inside the kth corresponding sphere and outside all of the others, unfortunately the spheres can overlap and thus produce common region(s), in that region(s) a new sample can't be classified using the conventional decision function (Eq. 9), because it will belong simultaneously to many spheres. To deal with this problem a new decision function is presented in [12], which aims to evaluate the membership degree in the common region(s), this new function can be evaluated in the feature space and thus represented by kernels functions. It reduces significantly the effects of overlap and improves greatly the classification accuracy [12].

The new decision function proposed in [12], can be written as:

$$s_k = 1 - \sqrt{\frac{\|z - a_k\|^2}{R_k^2 + \delta}} \tag{12}$$

where z is a sample existing in the overlapped region(s) (i.e. inside more than one hypersphere), a_k and R_k represent the centre and the radius of the kth sphere defined respectively by equations (Eq 5) and (Eq 11), δ is a strictly positive constant to avoid the case $s_k = 0$, After calculating the values of these functions for all spheres candidates, in which z exist simultaneously, z is classified into the class which maximize s_k:

$$f^{new}(z) = argmax_k(s_k) \tag{13}$$

4 Email Spam Filtering Using SSPV-SVDD with the New Decision Function

4.1 Datasets and Experimental Setting

Spam e-mail dataset was used to demonstrate the performance of the proposed method, it was downloaded from UCI repository [18]. This dataset contains 4601 samples each one is represented by 58 attributes (the first 57 attributes are continuous values, and the last one is the class label), out of which 1813 samples are related to spam while the rest of the samples are related to non-spam (legitimate). In all experiments we used the Gaussian kernel with the spread parameter σ, since it is a popular and powerful kernel used frequently in pattern recognition.

Training a Support Vector Domain Description requires the solution of a quadratic programming problem (Eq. 7), which is complex and time consuming, specifically when dealing with large data sets. To do that more quickly, we propose using a new SVDD learning algorithm called Sequential Minimal Optimization (SMO) [21], SMO breaks this large QP problem into a series of smallest possible QP problems. These small QP problems are solved analytically, which avoids using a time-consuming numerical QP optimization. The amount of memory required for SMO is linear in the training set size, which allows SMO to handle very large training sets.

The proposed experiments are performed as follows.

Step 0: Choose the value of the parameter $p*$.
Step 1: Choose the value of the spread parameter $\sigma*$.
Step 2: As said before the benchmark dataset used in the experiment is divided into two parts spam and legitimate, in this step we aim to construct two subsets training

and testing, the first one will contain 80 % of legitimate email and 80 % of spam (randomly), the second one will contain the remaining samples from each part.

Step 3:
Run: The conventional SVDD using $\sigma = \sigma^*$, with the standard decision function.
Run: SSPV for SVDD using $p = p^*$ and $\sigma = \sigma^*$, with the standard decision function.
Run: SSPV for SVDD using $p = p^*$ and $\sigma = \sigma^*$, with the new decision function.

Step 4: Calculate and store the recognition rate for both training and testing datasets, and repeat the experiment from the second step 20 times.

Step 5: Calculate the average of recognition rate and the standard deviation for (σ^*, p^*).

Step 6: Change the values of σ by going to Step 1.

Step 7: Change the values of p by going to Step 0.

4.2 Numerical Results

In the experiment we fix the value of the parameter $C = 1$, the set of $\sigma = \{1, 2, 3, 4, 5\}$, and $p = 0.01$.

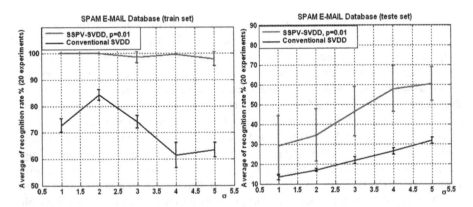

Fig. 3 Comparison between the conventional SVDD and SSPV-SVDD, using the standard decision function

In the first experiment (Fig. 3) we compare the average recognition rate (20 runs), for the two algorithms SSPV-SVDD and the conventional SVDD, using different values of σ. The figure is divided into two parts:

- The first one (left side) concerns the training dataset, it can be seen that, when we run the conventional SVDD, the average recognition rate varies between 62 % and 84 %, this results increases greatly when running SSPV-SVDD and achieves overall classification accuracy above 97 %. We conclude that

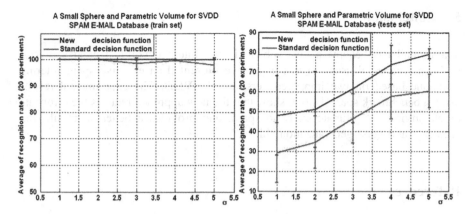

Fig. 4 Comparison between standard decision function, and the new decision function, using SSPV-SVDD with the same p = 0.01

SSPV-SVDD learns approximately all of the training dataset, and is more stable since it does not varies significantly with σ (3 % against 22 %).

- The second part of the figure (right side) corresponds to the testing dataset, in which we evaluate the generalization ability of both algorithms; we observe that their classification accuracy increases with the increase of σ, until a maximum values 60 % for SSPV-SVDD against 32 % for conventional SVDD. This result shows that SSPV-SVDD outperforms the conventional SVDD in detecting new spam emails.

In the second experiment (Fig. 4) we aim to improve the SSPV-SVDD by using the new decision function, to do that we kept the results of SSPV-SVDD found in the first experiment (red curve), and we run this algorithm (20 times) with different values of σ, using the proposed function. Also the figure is divided into two parts:

- The first one (left side) shows that the average recognition rate becomes much better when using the new decision function; it reaches 100 % whatever the value of σ.
- The second part of the figure (right side) shows that the average recognition rate becomes much better, it increases as the value of sigma is raised until a maximum value (80 %).

Both experiments demonstrate that the combination of the method SSPV-SVDD with the new decision function, achieves high recognition rate with good generalization ability.

5 Conclusion

Spam email has become a big problem in the Internet world, it creates several problems for email users. In this paper, we presented an effective spam filtering technique based on two improved versions of SVDD. The experimental results using a benchmark spam dataset, show that the new approach reduces the error rate of classifying a legitimate email to spam, and provides a better spam filtering accuracy, and good generalization ability.

References

1. López-Herrera, A.G., Herrera-Viedma, E., Herrera, F.: A multiobjective evolutionary algorithm for spam e-mail filtering. In: Proceedings of the 3rd International Conference on Intelligent System and Knowledge Engineering, vol. 1, pp. 366–371. Xiamen, China (2008)
2. Hotho, A., Staab, S., Stumme, G.: Ontologies improve text document clustering. In: Proceedings of 3rd IEEE International Conference on Data Mining, pp. 541–544. Melbourne, Florida, USA (2003)
3. Tavakkoli, A., Nicolescu, M., Bebis, G., Nicolescu, M.: A support vector data description approach for background modeling in videos with quasi-stationary backgrounds. Int. J. Artif. Intell. Tools 17(4), 635–658 (2008)
4. Cui, B., Mondal, A., Shen, J., Cong, G., Tan, K.: On effective email classification via neural networks. In: Proceedings of DEXA, pp. 85–94. Copenhagen, Denmark (2005)
5. Tax, D.M.J., Duin, R.P.W.: Data domain description using support vectors. In: Proceedings of European Symposium on Artificial Neural Networks, Bruges (Belgium), pp. 251–256 (1999)
6. Tax, D.M.J., Duin, R.P.W.: Support Vector Data Description. Mach. Learn. 54, 45–66 (2004)
7. Tax, D.M.J., Duin, R.P.W.: Support vector domain description. Pattern Recogn. Lett. 20 (11–13), 1191–1199 (1999)
8. Ferris Research: 2008. Industry Statistics-Ferris Research. http://www.ferris.com/research-library/industry-statistics/, Washington, DC, U.S.A, 2003
9. Drucker, H., Wu, S., Vapnik, V.N.: Support vector machines for spam categorization. IEEE Trans. Neural Netw. 10(5), 1048–1054 (1999)
10. Chen, J., Xiao, G., Gao, F., Zhang, Y.: Spam filtering method based on an artificial immune system. In: Proceedings of the IEEE International Conference on Multimedia and Information Technology (MMIT), pp. 169–171. Three Gorges, (2008)
11. El Boujnouni, M., Jedra, M., Zahid, N.: A small sphere and parametric volume for support vector domain description. J. Theor. Appl. Inf. Technol. 46(1), 471–478 (2012)
12. El Boujnouni, M., Jedra, M., Zahid, N.: New decision function for support vector data description. J. Inf. Syst. Manage. 2(3), 105–115 (2012)
13. Sahami, M., Dumais, S., Heckerman, D., Horvitz, E.: A Bayesian approach to filtering junk e-mail. In: Proceeding of the AAAI Workshop on Learning for Text Categorization, pp. 98–105. Madison, Wisconsin (1998)
14. Pantel, P., Lin, D.: Spamcop: a spam classification and organization program. In: Proceeding of the AAAI Workshop on Learning for Text Categorization, pp. 95–98. Madison, Wisconsin (1998)
15. Radicati, Email statistics report, 2009–2013
16. Fawcett, T., In vivo spam filtering: a challenge problem for data mining. In: Proceedings of the 9th ACM SIGKDD International Conference on Knowledge Discovery and Data Mining Explorations, vol. 5, No.°2. Washington, DC, USA (2003)

17. Oda, T., White, T., Developing an immunity to spam. In: Genetic and Evolutionary Computation Conference, pp. 231–242. Chicago, IL (2003)
18. UCI repository of machine learning databases. http://archive.ics.uci.edu/ml/
19. Vapnik, V.: Statistical Learning Theory. Wiley, New York (1998)
20. Carreras, X., Marquez, L., Salgado, J.G.: Boosting trees for anti-spam email filtering. In: Proceedings of the 4th International Conference on Recent Advances in Natural Language Processing, pp. 58–64. Tzigov Chark, BG (2001)
21. Platt, J.C.: Fast training of support vector machines using sequential minimal optimization. In: Schoelkopf B., Burges C., Smola A. (eds.) Advances in Kernels Methods: Support Vector Learning, MIT Press, Cambridge, Mass (1998)

Performance of the Cryptanalysis over the Shrinking Generator

Sara D. Cardell and Amparo Fúster-Sabater

Abstract The shrinking generator is a decimation-based nonlinear sequence generator with cryptographic application. Its output sequence can be modelled as one of the sequences generated by a linear cellular automata. Taking advantage of this linear structure, in this work a cryptanalysis of the shrinking generator has been introduced. The algorithm here developed recovers the secret key of the shrinking generator.

Keywords Shrinking generator · Cellular automata · Linearity · Security · Cryptanalysis

1 Introduction

Nowadays stream ciphers are the fastest encryption procedures to protect information [1]. Given a short key and a public algorithm (the sequence generator) stream ciphers create a long sequence of seemingly random bits, called keystream sequence. Bits are encrypted individually XOR-ing the message or plaintext and the keystream, giving rise to the ciphertext. Decryption is performed by executing the same XOR operation between the ciphertext and the keystream. The security of the transmitted information depends on the strength of the keystream generation algorithm.

Maximum-length Linear Feedback Shift Registers (LFSRs) [2] generate PN-sequences with good cryptographic properties, but their linearity make them vulnerable. This is the reason why these PN-sequences are not used as keystreams, but as basis for other more complex generators, for instance, decimation-based generators such as the shrinking and the self-shrinking generators [3, 4].

S.D. Cardell (✉)
Departamento de Estadística e Investigación Operativa,
Universidadde Alicante, E-03080, Alicante, Spain
e-mail: s.diaz@ua.es

A. Fúster-Sabater
Instituto de Tecnologías Físicas y de la Información (CSIC), Serrano 144,
20006 Madrid, Spain
e-mail: amparo@iec.csic.es

© Springer International Publishing Switzerland 2015
Á. Herrero et al. (eds.), *International Joint Conference*, Advances in Intelligent
Systems and Computing 369, DOI 10.1007/978-3-319-19713-5_10

On the other hand, it was found that some one-dimensional linear cellular automata (CA) generate exactly the same PN-sequences as those generated by maximum length LFSRs [5]. Thus, CA can be considered as alternative generators to LFSRs. Moreover, some keystream generators can be modelled in terms of linear CA [6–8]. In [6], authors showed that the output sequence of a cryptographic sequence generator, the shrinking generator, can be obtained from linear CA. In this work, we take advantage of such linear structures and their properties to design a cryptanalysis against the shrinking generator. We see that this generator can be easily broken, which attempts against the security of the information encoded using a structure based on it.

This paper is organised as follows. In Sect. 2, some basic concepts are introduced. In Sect. 3, special properties of the shrunken sequence are reminded. Besides, it is shown how the shrunken sequence is obtained as the output sequence of a linear periodic CA. In Sect. 4, the main idea of this work, that is, an algorithm to cryptanalyze the shrinking generator is proposed as well as numerical results are provided. Finally, conclusions in Sect. 5 end the paper.

2 Preliminaries

This **shrinking generator** was first introduced in [3]. It has good cryptographic properties and is easy to implement. It is composed of two maximum-length LFSRs R_1 and R_2, with lengths L_1, L_2 and periods $T_1 = 2^{L_1} - 1$, $T_2 = 2^{L_2} - 1$, respectively. The PN-sequence $\{u_i\}$ generated by the register R_1 decimates the PN-sequence $\{v_i\}$ produced by the register R_2. Given two bits u_i and v_i, $(i = 0, 1, 2, \ldots)$ from both PN-sequences, the output sequence $\{s_j\}$ of the generator is computed as follows:

$$\begin{cases} \text{If } u_i = 1 \text{ then } s_j = v_i. \\ \text{If } u_i = 0 \text{ then } v_i \text{ is discarded.} \end{cases}$$

The output sequence $\{s_j\}$ is called the **shrunken sequence**. This sequence has period $T = (2^{L_2} - 1)2^{L_1-1}$ and its number of 1s is $2^{L_1+L_2-2}$. Moreover, its characteristic polynomial has the form $p(x)^m$, with $2^{L_1-2} < m \leq 2^{L_1-1}$ and $p(x)$ being a primitive polynomial of degree L_2 [7]. The key of this generator is the initial state of each register.

Example 1 Consider the register R_1 with characteristic polynomial $p_1(x) = 1+x+x^2$ and initial state $\{1, 0\}$. Consider now R_2 with characteristic polynomial $p_2(x) = 1 + x + x^3$ and initial state $\{1, 0, 0\}$. The shrunken sequence can be computed in the following way:

$$R_1 : 1\ 0\ 1\ 1\ 0\ 1\ 1\ 0\ 1\ 1\ 0\ 1\ 1\ 0\ 1\ 1\ 0\ 1\ 1\ 0\ 1$$
$$R_2 : 1\ \cancel{0}\ 0\ 1\ \cancel{0}\ 1\ 1\ \cancel{1}\ 0\ 0\ \cancel{0}\ 0\ 1\ \cancel{1}\ 1\ 0\ \cancel{0}\ 1\ 0\ \cancel{1}\ 1$$
$$\mathbf{1\quad 0\ 1\quad 1\ 1\quad 0\ 0\quad 0\ 1\quad 1\ 0\quad 1\ 0\quad 1}$$

It has period 14 and it is easy to check that its characteristic polynomial is $p(x)^2 = (1 + x^2 + x^3)^2$, thus the linear complexity of this sequence equals 6. ∎

Cellular automata (CA) are devices composed by a finite number of cells whose content (binary in this work) is updated following a function of k variables [9]. Thus, the value of the cell in position i at time $t+1$, x_i^{t+1}, depends on the value of the k neighbour cells at time t. If these rules are composed exclusively by XOR operations, then the CA is said to be **linear**. In this work, the CA considered are **regular** (every cell follows the same rule), **periodic** (extreme cells are adjacent) and one-dimensional. For $k = 3$, the rule 102 is given by:

$$\textbf{Rule 102: } x_i^{t+1} = x_i^t + x_{i+1}^t$$

111	110	101	100	011	010	001	000
0	1	1	0	0	1	1	0

The number 01100110 is the binary representation of the number 102. In Table 1, we can find an example of a linear regular periodic CA with rule 102.

Due to their speed and the randomness of their sequences, CA are good building blocks for stream ciphers. In fact, their hardware implementation is simple and their regular structure makes possible an efficient software implementation. Many authors have proposed stream ciphers based on CA [10, 11].

Table 1 Linear regular periodic CA with rule 102

102	102	102
1	1	0
0	1	1
1	0	1
1	1	0
⋮	⋮	⋮

3 Previous Results

Due to the lack of space, we will show the properties of the shrunken sequence and its relation with the PN-sequences generated by R_1 and R_2 with an example. More details can be found in [6].

3.1 Properties of the Shrunken Sequence

Consider the characteristic polynomials $p_1(x) = 1 + x^2 + x^3$ and $p_2(x) = 1 + x^3 + x^4$ of R_1 and R_2, respectively. The corresponding shrunken sequence $\{s_j\}$ have period 60 and characteristic polynomial $p(x)^m = (1 + x + x^4)^4$. If we decimate such a sequence by $2^{L_1-1} = 4$, then we notice that the sequence is made out of 4 PN-sequences called **the interleaved PN-sequences** of the shrunken sequence. In our example, we obtain the following interleaved PN-sequences:

$$
\begin{array}{c}
\quad\; v_0\; v_7\; v_{14}\; v_6\; v_{13}\; v_5\; v_{12}\; v_4\; v_{11}\; v_3\; v_{10}\; v_2\; v_9\; v_1\; v_8 \\
\quad\; \uparrow\; \uparrow\; \uparrow\; \uparrow\; \uparrow\; \uparrow\; \uparrow\; \uparrow\; \uparrow\; \uparrow\; \uparrow\; \uparrow\; \uparrow\; \uparrow\; \uparrow \\
\{s_{4j}\} \;\rightarrow\; \mathbf{1}\; \mathbf{1}\; \mathbf{0}\; \mathbf{1}\; \mathbf{0}\; \mathbf{1}\; \mathbf{1}\; \mathbf{1}\; \mathbf{1}\; \mathbf{0}\; \mathbf{0}\; \mathbf{0}\; \mathbf{1}\; \mathbf{0}\; \mathbf{0} \\
\{s_{4j+1}\} \;\rightarrow\; 0\; 0\; 0\; 1\; 0\; 0\; 1\; 1\; 0\; 1\; 0\; 1\; 1\; 1\; 1 \\
\{s_{4j+2}\} \;\rightarrow\; 1\; 1\; 0\; 0\; 0\; 1\; 0\; 0\; 1\; 1\; 0\; 1\; 0\; 1\; 1 \\
\{s_{4j+3}\} \;\rightarrow\; 1\; 1\; 1\; 1\; 0\; 0\; 0\; 1\; 0\; 0\; 1\; 1\; 0\; 1\; 0
\end{array}
\tag{1}
$$

The characteristic polynomial of these PN-sequences is $p(x) = 1 + x + x^4$. Therefore, all of them are the same PN-sequence but shifted. If the first PN-sequence $\{w_i\}$ (the one in bold) is decimated by d, with $T_1 d = 1 \bmod T_2$, then the PN-sequence $\{v_i\}$ generated by the register R_2 is obtained. In this example, $d = 13$ and the PN-sequence $\{v_i\}$ is $\{1,0,0,0,1,1,1,1,0,1,0,1,1,0,0\}$.

The remaining interleaved PN-sequences of the shrunken sequence will have the form $\{w_{d_1+i}\}, \{w_{d_2+i}\}, \ldots, \{w_{d_{2^{L_1-1}-1}+i}\}$. Let $\{0, i_1, i_2, \ldots, i_{2^{L_1-1}-1}\}$ be the set of indices of the 1s in $\{u_i\}$. We know that $d_j = d \cdot i_j \bmod T_2$, for $j = 1, 2, \ldots, 2^{L_1-1} - 1$. According to the position of the last three PN-sequences in (1), we have that $d_1 = 9$, $d_2 = 7$ and $d_3 = 5$. Now, we can search for the indices of the 1s in $\{u_i\}$ solving the expressions $13 \cdot i_1 = 9 \bmod 15$, $13 \cdot i_2 = 7 \bmod 15$ and $13 \cdot i_3 = 5 \bmod 15$. It is easy to check that $i_1 = 3$, $i_2 = 4$ and $i_3 = 5$. Therefore, the set of indices is given by $\{0, 3, 4, 5\}$ and the PN-sequence $\{u_i\}$, generated by R_1, is $\{1,0,0,1,1,1,0\}$.

Recall that if $L_2 = L_1 + 1$, then the polynomial $p(x)$ is the reciprocal polynomial of $p_2(x)$ and $d = T_2 - 2$ (see [6]).

3.2 Modelling Sequences

We start this section with the concept of Zech logarithm. Let $\alpha \in \mathbb{F}_{2^L}$ be a primitive element. The **Zech logarithm** with basis α is the application $\mathcal{Z}_\alpha : \mathbb{Z}_{2^L-2} \rightarrow \mathbb{Z}^*_{2^L-2} \cup \{\infty\}$, such that each element $t \in \mathbb{Z}_{2^L-2}$ corresponds to $\mathcal{Z}_\alpha(t)$, attaining $1 + \alpha^t = \alpha^{\mathcal{Z}_\alpha(t)}$.

Modelling an LFSR: Consider a PN-sequence $\{a_i\}$ with characteristic polynomial $p(x)$ of degree L. The sequence $\{a_i + a_{i+1}\}$ is generated adding the PN-sequence $\{a_i\}$ with itself. As a consequence, this new sequence is the same PN-sequence but starting in some position D. According to the definition of Zech logarithm, for every PN-sequence $\{a_i\}$ generated by a primitive polynomial $p(x)$, there exists a unique number $D = \mathcal{Z}_\alpha(1)$, such that $a_i + a_{i+1} = a_D$.

Consider a linear periodic CA that uses rule 102. Assume the PN-sequence $\{a_i\}$ appears in the zero column. It is not difficult to check that every column in the CA is the same PN-sequence $\{a_i\}$ but shifted $D, 2D, 3D, \ldots$ (mod $2^L - 1$) positions, respectively. Besides, since the CA is periodic, the length of the CA must be $(2^L - 1)/\gcd(2^L - 1, D)$.

Example 2 Consider, for example, the polynomial $p(x) = 1 + x + x^2$. In this case $D = 2$, then the length of the CA will be 3. In Table 1, we can see a CA that generates the PN-sequence generated by $p(x)$ with initial state $\{1, 0\}$. The PN-sequence appears three times but shifted 2 and 4 positions, respectively.

Table 2 CA that generates the shrunken sequence in Example 3

102	102	102	102	102	102	102	102	102	102	102	102	102	102
1	1	0	1	0	0	1	0	0	1	1	0	1	1
0	1	1	1	0	1	1	0	1	0	1	1	0	0
1	0	0	1	1	0	1	1	1	1	0	1	0	0
1	0	1	0	1	1	0	0	0	1	1	1	0	1
1	1	1	1	0	1	0	0	1	0	0	1	1	0
0	0	0	1	1	1	0	1	1	0	1	0	1	1
0	0	1	0	0	1	1	0	1	1	1	1	0	1
0	1	1	0	1	0	1	1	0	0	0	1	1	1
1	0	1	1	1	1	0	1	0	0	1	0	0	1
1	1	0	0	0	1	1	1	0	1	1	0	1	0
0	1	0	0	1	0	0	1	1	0	1	1	1	1
1	1	0	1	1	0	1	0	1	1	0	0	0	1
0	1	1	0	1	1	1	1	0	1	0	0	1	0
1	0	1	1	0	0	0	1	1	1	0	1	1	0

Modelling the Shrunken Sequence: If we consider now the shrunken sequence $\{s_j\}$, whose characteristic polynomial is $p(x)^m$, $2^{L_1-2} < m \leq 2^{L_1-1}$. There exists a unique number $\widehat{D} = 2^{L_1-1}D$, such that $s_j + s_{j+2^{L_1-1}} = s_{j+\widehat{D}}$, where $D = \mathcal{Z}_\alpha(1)$, with $\alpha \in \mathbb{F}_{2^{L_2}}$ root of $p(x)$. The CA generates 2^{L_1-1} sequences, including the shrunken sequence, before generating the shrunken sequence again shifted $\widehat{D}$ positions. According to this, the CA must have length $L = T2^{L_1-1}/\gcd(\widehat{D}, T)$. More information about these results can be found in [6].

Example 3 The characteristic polynomial of the shrunken sequence in Example 1 is $p(x)^2 = (1 + x^2 + x^3)^2$. The number associated to $p(x)^2$ is $\widehat{D} = 2 \cdot D = 2 \cdot \mathcal{Z}_\alpha(1) = 2 \cdot 5 = 10$, with $\alpha \in \mathbb{F}_{2^3}$ root of $p(x)$. The shrunken sequence can be generated by a periodic CA of length 14 (see Table 2). This CA generates two different sequences, the shrunken sequence and another sequence with the same period 14. The shrunken sequence appears 7 times in columns 0, 2, 4, 6, 8, 10 and 12. The successive shrunken sequences start in positions 10, 20 mod 14, 30 mod 14 ... regarding the first one. The other sequence appears 7 times as well, in columns, 1, 3, 5, 7, 9, 11, 13 with the same shifts.

When the shrunken sequence appears in the zero column of the CA, 2^{L_1-1} different sequences are generated by the CA, including the shrunken sequence. All of them have the same characteristic polynomial $p(x)^m$ and are composed of 2^{L_1-1} interleaved PN-sequences with characteristic polynomial $p(x)$.

In Example 3, two sequences were generated by the CA, the shrunken sequence and another sequence with the same characteristic polynomial $p(x)^2 = (1 + x^2 + x^3)^2$ and the same period. If we decimate both sequences by 2, we can see that both sequences are composed of two interleaved PN-sequences whose characteristic polynomial is $p(x) = 1 + x^2 + x^3$. This means that all these PN-sequences are the same but shifted.

As a consequence of the previous statements, knowing some bits the CA sequences can help us to recover the shrunken sequence. In fact, we can recover the first interleaved PN-sequence of the shrunken sequence using the other interleaved PN-sequences of the same shrunken sequence and the interleaved PN-sequences of the other sequences generated by the CA. The knowledge of this first interleaved PN-sequence allows us to determine the PN-sequences $\{u_i\}$ and $\{v_i\}$ generated by the registers R_1 and R_2 and, consequently, their corresponding initial states.

Assume the interleaved PN-sequences of the shrunken sequence are denoted by $\{w_{d_0+i}\}, \{w_{d_1+i}\}, \{w_{d_2+i}\}, \ldots, \{w_{d_{2^{L_1-1}-1}+i}\}$, where $d_0 = 0$. Remember that the shifts d_j depend on the location of the 1 s in the PN-sequence $\{u_i\}$ generated by the first register R_1. If the interleaved PN-sequences of the next sequence in the CA are denoted by $\{w_{d_0^1+i}\}, \{w_{d_1^1+i}\}, \{w_{d_2^1+i}\}, \ldots, \{w_{d_{2^{L_1-1}-1}^1+i}\}$, these new positions can be computed as follows

$$d_j^1 = \mathcal{Z}_\alpha(d_j - d_{j+1}) + d_{j+1}, j = 0, 1, \ldots, 2^{L_1-1} - 2,$$

$$d_{2^{L_1-1}-1}^1 = \mathcal{Z}_\alpha(d_{2^{L_1-1}-1} - 1) + 1.$$

In a similar way, we can compute the shifts d_i^k, for $i = 0, 1, 2, \ldots, 2^{L_1-1} - 1$ for the next sequences in the CA using the same expressions and the previous shifts d_i^{k-1}, for $i = 0, 1, 2, \ldots, 2^{L_1-1} - 1$ and $k = 2, 3, \ldots$

4 Cryptanalysis

The main contribution of this work is to propose a cryptanalysis that takes advantage of the properties of the shrunken sequence observed in Sect. 3.1 and those of the other sequences obtained in the CA proposed in Sect. 3.2.

4.1 The Algorithm

If we intercept n bits of the shrunken sequence, $s = \{s_0, s_1, \ldots, s_{n-1}\}$, Algorithm 2 tests whether an initial state $u = \{u_0, u_1, \ldots, u_{L_1-1}\}$ for the register R_1 is the correct one or not. On the other hand, Algorithm 1 tests every possible initial state of length L_1 for R_1 applying Algorithm 2 to each one of them. In the end, a set of possible correct initial states for R_1 is provided by Algorithm 1.

Algorithm 1. Crypto: Search through the initial states of R_1

Input: $p_1(x), p(x)$ and s
function $S =$**Crypto**$(p_1(x), p(x), s)$

    Initialise S;
    for $k = 1$ to $2^{L_1} - 1$
        Compute k the binary representation of k;
        $[M, Stop]=$**SubCrypto**$(p_1(x), p(x), s, k)$;
        if Stop$=1$
            Store k in the set S;
        end if
    end for
end function
Output: S: Set of initial states of R_1 considered correct

Algorithm 2. SubCrypto: Test of each initial state for R_1

Input: $p_1(x)$, $p(x)$, s and u
function $[M, Stop]$ =**SubCrypto**$(p_1(x), p(x), s, u)$

01:    Compute $\{u_i\}$ using $p_1(x)$ and u until finding $length(s)$ ones;
02:    Store in P the positions of the 1s in the generated bits of $\{u_i\}$;
03:    Compute d, such that $d \cdot (2^{L_1} - 1) = 1 \bmod (2^{L_2} - 1)$;
04:    Store in P the new positions computed as $P_i \cdot d \bmod (2^{L_2} - 1)$;
05:    Store $[P_i, s_i]$ in a matrix M;
06:    $Stop = 1$;
07:    **while** $Stop = 1$ and $length(s) > 1$
08:       Update P with the new positions;
09:       Update s with $\{s_0 + s_1, s_1 + s_2, \ldots, s_{n-1} + s_n\}$;
10:       Store [m,n]=size(M);
11:       **for** $j = 0$ to $m - 1$
12:          **for** $k = 0$ to $length(P) - 1$
13:             **if** $M_{j1} = P_k$
14:                **if** $M_{j2} \neq s_k$
15:                   Initialise M;
16:                   $Stop = 0$;
17:                **end if**
18:             **end if**
19:          Store $[P_k, s_k]$ in M;
20:          **end for**
21:       **end for**
22:   **end while**
end function
Output:
M: Recovered bits and their position in the first interleaved PN-sequence.
$Stop$: 1 if the initial state is considered correct and 0 otherwise.

4.2 A Numerical Example

Consider two registers R_1 and R_2 with characteristic polynomials $p_1(x) = 1 + x + x^6$ and $p_2(x) = 1 + x^3 + x^7$, respectively.

Assume we intercept 6 bits of the shrunken sequence: $s = \{1, 0, 1, 0, 0, 0\}$.

Notice that, in this case, the period of the sequence is $2^6(2^7 - 1) = 4064$.

We apply Algorithm 2 in order to check if the initial state $u = \{1, 1, 1, 1, 0, 1\}$ is correct for R_1.

Since $L_2 = L_1 + 1$, we know that $p(x)$ is the reciprocal polynomial of $p_2(x)$, that is, $p(x) = 1 + x^4 + x^7$.

Input: $p_1(x) = 1 + x + x^6$, $p(x) = 1 + x^4 + x^7$, $u = \{1,1,1,1,0,1\}$ and $s = \{1,0,1,0,0,0\}$.

We compute the PN-sequence generated by R_1 using u until we find 6 ones: $\{1,1,1,1,0,1,0,0,0,1\}$.

The position of the 1 s is $pos = \{0,1,2,3,5,9\}$.

In this case, we have that the distance of decimation is $d = T_2 - 2 = 2^7 - 3 = 125$, since $L_2 = L_1 + 1$.

The positions of the intercepted 6 bits in the first interleaved PN-sequence are: $pos = \{0, 125, 123, 121, 117, 109\}$.

We store this information in the matrix:

$$M^T = \begin{bmatrix} 0 & 125 & 123 & 121 & 117 & 109 \\ 1 & 0 & 1 & 0 & 0 & 0 \end{bmatrix}$$

The new positions are computed as follows:

$$pos_0 = \mathcal{Z}_\alpha(0 - 125) + 125 = \mathcal{Z}_\alpha(2) + 125 = 65$$
$$pos_1 = \mathcal{Z}_\alpha(125 - 123) + 123 = \mathcal{Z}_\alpha(2) + 123 = 63$$
$$pos_2 = \mathcal{Z}_\alpha(123 - 121) + 121 = \mathcal{Z}_\alpha(2) + 121 = 61$$
$$pos_3 = \mathcal{Z}_\alpha(121 - 117) + 117 = \mathcal{Z}_\alpha(4) + 117 = 124$$
$$pos_4 = \mathcal{Z}_\alpha(117 - 109) + 109 = 1\mathcal{Z}_\alpha(8) + 109 = 123$$

The new bits to locate are: $\{1, 1, 1, 0, \mathbf{0}\}$.

The position 123 appeared already in the matrix M with 1 as associated bit. In this new round we are trying to store the bit 0 in position 123.

Output: $STOP = 0$ The initial state $u = \{1,1,1,1,0,1\}$ is not correct since we found a contradiction.

4.3 Discussion of the Algorithm

After analysing a considerable number of examples, we can conclude that the period of the shrunken-sequence grows much faster than the number of intercepted bits needed for its cryptanalysis. Below we introduce several examples to illustrate this conclusion.

In Fig. 1a, the performance of the algorithm for a shrinking generator with characteristic polynomials $p_1(x) = 1 + x + x^6$ and $p_2(x) = 1 + x^3 + x^7$ is depicted. It is easy to check that the more intercepted bits we have the smaller the set of possible correct initial states is. Intercepting 16 bits or more, this set is reduced to the correct one. In this case, the period of the shrunken sequence is 4064, so we need almost the 0.4 % of the bits of the sequence to recover the initial state of R_1.

In Fig. 1b, a case with higher degrees is considered. The characteristic polynomials are $p_1(x) = 1 + x^2 + x^3 + x^4 + x^8$ and $p_2(x) = 1 + x^4 + x^9$. Intercepting 17 bits or more, the algorithm just returns the correct initial state. The percentage in this example is

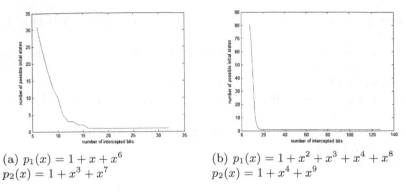

(a) $p_1(x) = 1 + x + x^6$
   $p_2(x) = 1 + x^3 + x^7$

(b) $p_1(x) = 1 + x^2 + x^3 + x^4 + x^8$
   $p_2(x) = 1 + x^4 + x^9$

Fig. 1 Examples of the performance of the algorithm

even better, since we only need 0.026 % of 65408 bits. According to Algorithm 2 and the size of the resultant matrix M, we recover 24 bits of the first interleaved PN-sequence of the shrunken sequence. This quantity is enough to recover the rest of the interleaved PN-sequence and then, recover the complete shrunken sequence.

In Table 3, we can see the minimum number of intercepted bits we need to recover the initial state, for several examples. The higher the degrees of the characteristic polynomial are, the more bits we need. However, the number of bits grows in a linear way and the period grows exponentially. This means that the number of necessary intercepted bits is very small compared with the period of the shrunken sequence.

Table 3 Minimum number of intercepted bits needed to recover the R_1 initial state

$p_1(x)$	$p_2(x)$	Intercepted bits	T
$1 + x^2 + x^5$	$1 + x + x^6$	11	1008
$1 + x^2 + x^5$	$1 + x^3 + x^7$	14	2032
$1 + x + x^6$	$1 + x^3 + x^7$	16	4064
$1 + x + x^7$	$1 + x^2 + x^3 + x^4 + x^8$	16	16320
$1 + x^2 + x^3 + x^4 + x^8$	$1 + x^4 + x^9$	17	65408
$1 + x + x^7$	$1 + x^4 + x^9$	16	32704
$1 + x^4 + x^9$	$1 + x^3 + x^{10}$	18	261888

The requirements of memory for Algorithm 2 are reasonably low. Assume we intercept n bits of the shrunken sequence; for the matrix M we store, in the worst case, $\sum_{i=1}^{n} 2i = n^2 + n$ integers. Besides, to store the positions d_i^k, we need n additional integers. In total, $n^2 + 3n$ integers to be stored.

5 Conclusions

In cryptography, decimation-based sequence generators, e.g. the shrinking generator, were designed in order to break the linearity of the PN-sequences generated by LFSRs. However, we have seen that the output sequence of this generator can be reduced to a succession of a unique PN-sequence and can be modelled as the output sequence of a linear CA. Due to the previous characteristics, this paper presents a cryptanalysis of the shrinking generator that exploits the inherent linearity of the shrunken sequence. Given a small number of intercepted bits of such a sequence, the proposed algorithm recovers the initial state of the first register R_1 and then from it, we can easily recover the initial state of the second register R_2. The number of intercepted bits is very small compared with the period of the shrunken sequence. Numerical results and a discussion on the performance of this algorithm complete the work.

Acknowledgments This work was supported by Generalitat Valenciana under grant with reference APOSTD/2013/081 and by Comunidad de Madrid under project S2013/ICE-3095-CIBERDINE-CM.

References

1. Paar, C., Pelzl, J.: Understanding Cryptography. Springer, Berlin (2010)
2. Golomb, S.W.: Shift Register-Sequences. Aegean Park Press, Laguna Hill (1982)
3. Coppersmith, D., Krawczyk, H., Mansour, Y.: The shrinking generator. In: Advances in Cryptology—CRYPTO '93, vol. 773, pp. 23–39. Springer (1993)
4. Meier, W., Staffelbach, O.: The self-shrinking generator. In Cachin, C., Camenisch, J. (eds.) Advances in Cryptology—EUROCRYPT 1994. vol. 950, pp. 205–214. Springer (1994)
5. Cattell, K., Muzio, J.C.: One-dimensional linear hybrid cellular automata. IEEE Trans. Comput. Aided Des. **15**(3), 325–335 (1996)
6. Cardell, S.D., Fúster-Sabater, A.: Modelling the shrinking generator in terms of linear CA. Submitted to Advances in Mathematics of Communications (2014)
7. Fúster-Sabater, A., Caballero-Gil, P.: Linear solutions for cryptographic nonlinear sequence generators. Phys. Lett. A **369**, 432–437 (2007)
8. Fúster-Sabater, A., Pazo-Robles, M.E., Caballero-Gil, P.: A simple linearization of the self-shrinking generator by means of cellular automata. Neural Netw. **23**(3), 461–464 (2010)
9. Wolfram, S.: Cellular automata as simple self-organizing system. Caltrech preprint CALT, pp. 68–938 (1982)
10. Das, S., RoyChowdhury, D.: Car30: a new scalable stream cipher with rule 30. Crypt. Commun. **5**(2), 137–162 (2013)
11. Jose, J., Das, S., RoyChowdhury, D.: Inapplicability of fault attacks against trivium on a cellular automata based stream cipher. In: 11th International Conference on Cellular Automata for Research and Industry, ACRI 2014. vol. 875, pp. 427–436. Springer (2014)

On the (Non)Improvement
of an Authenticated GKT Protocol

Ruxandra F. Olimid

Abstract Harn and Lin proposed in 2010 a secret sharing-based group key transfer protocol. One year later, Nam et al. showed their construction is vulnerable to a replay attack and proposed a way to fix it. Recently, Yuan et al. analyzed the same protocol, proved that it is also vulnerable to a man-in-the middle attack and considered a countermeasure. First, we slightly modify Yuan et al.'s attack to make it simpler to implement and harder to be detected. Second, we show that the improved version of the protocol remains susceptible to a man-in-the-middle attack.

Keywords Group key transfer · Secret sharing · Man-in-the-middle attack · Insider attack · Cryptanalysis

1 Introduction

A *group key transfer* (GKT) protocol permits multiple users to share a common secret key that is subsequently used for cryptographic purposes. A privileged party called *Key Generation Center* (KGC) selects a fresh random key and securely distributes it to the qualified users.

A particular and efficient way to build GKT protocols is to use *secret sharing schemes* [1, 8–10]. A secret sharing scheme splits a secret into multiple shares such that only *authorized* set of shares lead to the reconstruction of the secret. However, most of the existing protocols leak a proper security analysis and hence become susceptible to attacks [2–5, 9].

Harn and Lin proposed in 2010 a GKT protocol [1] based on Shamir's secret sharing scheme [7]. Soon after, Nam et al. proved their construction is susceptible to a replay attack and considered a countermeasure to stand against this vulnerability [3]. Recently, Yuan et al. showed that Harn and Lin's protocol is also vulnerable to a man-in-the-middle attack mounted from inside [9]. Although their attacks successfully fulfills its goal and breaks key confidentiality, it can easily be detected during the

R.F. Olimid (✉)
Department of Computer Science, University of Bucharest, Bucharest, Romania
e-mail: ruxandra.olimid@fmi.unibuc.ro

© Springer International Publishing Switzerland 2015
Á. Herrero et al. (eds.), *International Joint Conference*, Advances in Intelligent
Systems and Computing 369, DOI 10.1007/978-3-319-19713-5_11

execution of the protocol. We slightly modify their attack to prevent detection in the sense that no user can suspect a strange behavior. More, we show that to succeed the attack, the adversary actively intercepts and prevents from reaching their destination a lower number of messages than in the original attack. The improvement on Yuan et al.'s attack represents our first contribution.

In the same paper, Yuan et al. gave a new version of the protocol. We show that their proposal remains vulnerable to a man-in-the-middle attack mounted from inside. The attack allows the adversary to make the users of the session (except the user that the adversary impersonates) end up with a key he already knows and hence to ruin key secrecy. The vulnerability against the Yuan et al.'s proposal represents our second contribution.

The next section gives the preliminaries. Section 3 describes Harn and Lin's protocol. Section 4 presents Yuan et al.'s man-in-the-middle attack and specifies how to improve it. Section 5 explains the countermeasure Yuan et al. proposed for Harn and Lin's protocol to stand against the attack. Section 6 proves their construction remains vulnerable to a man-in-the-middle attack mounted from inside. Finally, we conclude in Sect. 7.

2 Preliminaries

Confidentiality assures that a session group key is available to authorized participants only and no other party can recover it, even if the protocol runs for several times, called *sessions*.

To satisfy confidentiality, a GKT protocol must stand against both *outsiders* and *insiders*. An outsider attack is mounted by an entity that it is not registered as a group member, does not share any secret information with the KGC and cannot initiate protocol sessions; the goal of an outsider is to reveal session keys. An insider attack is mounted by an entity that is registered as a group member, shares a long-term key with the KGC and may legitimate initiate protocol sessions by requesting to share a common key with a set of other group members; the goal of an insider is to determine keys of sessions he is unauthorized to know.

A secure GKT protocol must also prevent *active* attacks. An active adversary has full control over the communication channel: he can intercept, modify, inject or drop messages. Hence, he can mount a *replay attack*, meaning that he resends messages he had previously eavesdropped, usually with the intention to impersonate other parties. Another particular case of an active attack is the *man-in-the-middle attack*, where the attacker interposes on the communication link between the users and the KGC. This work focuses on man-in-the-middle attacks mounted by an insider.

3 Harn and Lin's Protocol

Harn and Lin introduced in 2010 a GKT protocol [1] based on Shamir's secret sharing scheme [7]. Figure 1 describes the protocol in detail.

We use the following notations: m the number of all possible users, $\{U_1, \ldots, U_m\}$ the set of all users, $\{U_1, \ldots, U_t\}$ the set of participants to a given session (after a possible reordering), h a collision-resistant hash function, $\leftarrow^R$ a random choice from a specified set of values. We denote by $A \rightarrow B : M$ a message M sent by an entity A to an entity B and by $A \rightarrow^* : M$ a broadcast message M originating from an entity A.

The original paper mentions two possible enhancements of the protocol to achieve *users authentication* and *key confirmation*. We ignore them here to stick to the same version Yuan et al. referred to in their paper [9].

4 Man-in-the-Middle Attack Against Harn and Lin's Protocol

Yuan et al. showed that Harn and Lin's protocol is susceptible to a man-in-the-middle attack [9]. Figure 2 explains the attack in detail. U_a denotes an active insider who can intercept, modify, drop or inject messages over the communication channel.

The attacker's goal is to impersonate a user U_i in a session U_a is unauthorized for and hence to share a common key with the rest of the participants to the session. We highlight that the users $\{U_1, \ldots, U_{i-1}, U_{i+1}, \ldots, U_t\}$ believe to share a key within the group $\{U_1, \ldots, U_t\}$, while the KGC thinks he generates a session key for $\{U_1, \ldots, U_{i-1}, U_a, U_{i+1}, \ldots, U_t\}$.

The attack becomes possible because the first two rounds of the protocol lack authentication; therefore, any insider or outsider may impersonate a valid user or the KGC and send a list of participants in their behalf. Furthermore, the nonce R_i is also unauthenticated; hence U_a can impersonate U_i in Round 3 also. Note that in the original protocol, Harn and Lin mention a way to authenticate the origin of R_i that would avoid the simple impersonation of U_i [1], but Yuan et al. skip it in their work [9]. Also, Yuan et al. do not consider the key confirmation phase, a general technique used to assure that all parties own the same key at the end of the protocol.

We remark an important drawback of the attack: U_i might suspect a malicious action, since he is invited to take part to a protocol session that never finalizes. To eliminate this weakness, we propose a slightly modified version of the attack that makes U_i unaware that the session is taking place.

To avoid repetition, we skip the complete description and only emphasize the differences from the original attack as follows:

Step 4. U_a broadcasts the original list $\{U_1, \ldots, U_t\}$ to all the group members except U_i (as originating from the *KGC*).

Step 5. It is no longer performed.

Initialization
  The KGC selects 2 large safe primes p and q (i.e. $p' = \frac{p-1}{2}$ and $q' = \frac{q-1}{2}$ are also primes) and computes $n = pq$;

Users Registration
  Each user $U_i, i = 1, \ldots, m$, shares a long-term secret $(x_i, y_i) \in \mathbb{Z}_n^* \times \mathbb{Z}_n^*$ and the value n with the KGC;

Round 1
  User U_1:
  1.1. sends a key generation request:
    $U_1 \rightarrow KGC : \{U_1, \ldots, U_t\}$

Round 2
  The KGC:
  2.1. broadcasts:
    $KGC \rightarrow^* : \{U_1, \ldots, U_t\}$

Round 3
  Each user $U_i, i = 1, \ldots, t$:
  3.1. chooses $R_i \leftarrow^R \mathbb{Z}_n^*$;
  3.2. sends:
    $U_i \rightarrow KGC : R_i$

Round 4
  The KGC:
  4.1. selects a group key $k \leftarrow^R \mathbb{Z}_n^*$;
  4.2. generates the polynomial $f(x)$ of degree t that passes through the $t + 1$ points $(0, k)$, $(x_1, y_1 \oplus R_1), \ldots, (x_t, y_t \oplus R_t)$;
  4.3. computes t additional points $P_1, \ldots, P_t$ on $f(x)$;
  4.4. computes the authentication message
    $Auth = h(k, U_1, \ldots, U_t, R_1, \ldots, R_t, P_1, \ldots, P_t)$;
  4.5. broadcasts:
    $KGC \rightarrow^* : (R_1, \ldots, R_t, P_1, \ldots, P_t, Auth)$

Key Computation
  Each user $U_i, i = 1, \ldots, t$:
  5.1. computes the group key $k = f(0)$ by interpolating the points $P_1, \ldots, P_t$ and $(x_i, y_i \oplus R_i)$;
  5.2. checks if $Auth = h(k, U_1, \ldots, U_t, R_1, \ldots, R_t, P_1, \ldots, P_t)$;
    If the equality does not hold, he quits.

Fig. 1 Harn and Lin's group key transfer protocol [1]

There is no point for U_a to forward the original list $\{U_1, \ldots, U_t\}$ to U_i in Step 4 because U_a does not use messages originating from U_i to succeed the attack. So, U_a no longer invites U_i to participate to the session. Since no subsequent message will ever reach U_i, this simple modification makes him unaware of the execution of the

Step 1. U_a intercepts the message $\{U_1, \ldots, U_t\}$ sent in Round 1 of the protocol and prevents it from reaching the KGC;

Step 2. U_a replaces U_i by U_a in the message from Step 1 and sends to the KGC $\{U_1, \ldots, U_{i-1}, U_a, U_{i+1}, \ldots, U_t\}$;

Step 3. U_a intercepts the broadcast message $\{U_1, \ldots, U_{i-1}, U_a, U_{i+1}, \ldots, U_t\}$ sent in Round 2 of the protocol and prevents it from reaching the group members;

Step 4. U_a broadcasts the original list $\{U_1, \ldots, U_t\}$ to all the group members (as originating from the KGC);

Step 5. U_a intercepts the message R_i sent in Round 3 of the protocol and prevents it from reaching the KGC;

Step 6. U_a chooses $R_a \xleftarrow{R} \mathbb{Z}_n^*$ and sends it to the KGC (instead of R_i), but allows the other messages R_j, $j = 1, \ldots, t$, $j \neq i$ to reach the KGC;

Step 7. U_a intercepts $\{R_1, \ldots, R_{i-1}, R_a, R_{i+1}, \ldots, R_t, P_1, \ldots, P_t, Auth\}$ sent in Round 4 of the protocol and prevents it from reaching the group members;

Step 8. U_a computes $k = f(0)$ by interpolating the public points $P_1, \ldots, P_t$ and $(x_a, y_a \oplus R_a)$, forges $Auth' = h(k, U_1, \ldots, U_t, R_1, \ldots, R_t, P_1, \ldots, P_t)$ and sends $(R_1, \ldots, R_t, P_1, \ldots, P_t, Auth')$ to $U_1, \ldots, U_{i-1}, U_{i+1}, \ldots, U_t$ as originating from the KGC;

Step 9. For all $j = 1, \ldots, t$, $j \neq i$, U_j, recovers the group key $k = f(0)$ by interpolating the public points $P_1, \ldots, P_t$ and $(x_j, y_j \oplus R_j)$, checks that $Auth' = h(k, U_1, \ldots, U_t, R_1, \ldots, R_t, P_1, \ldots, P_t)$ holds and therefore accepts k as the correct group key.

Fig. 2 Man-in-the-middle attack against Harn and Lin's protocol [9]

protocol. In consequence, U_i will not send a nonce R_i to the KGC and thus Step 5 does not make sense anymore. This reduces the active implication of the adversary during the attack, since U_a must intercept and prevent fewer messages from reaching their destination.

The modified attack maintains the original goal: it allows U_a to learn a common key k with the users $\{U_1, \ldots, U_{i-1}, U_i, \ldots U_t\}$, while they all believe to share it within the group $\{U_1, \ldots, U_t\}$. In addition, our modification introduces two advantages for the adversary: it prevents detection by U_i during the execution of the protocol and it is easier to mount.

Initialization
    The KGC selects 2 large safe primes p and q (i.e. $p' = \frac{p-1}{2}$ and $q' = \frac{q-1}{2}$ are also primes) and computes $n = pq$;
    Then, it selects $e \in \mathbb{Z}_n^*$ s.t. $(e, \Phi(n)) = 1$ and computes $d \in \mathbb{Z}_n^*$ s.t. $ed = 1 \pmod{\Phi(n)}$;

Users Registration
    Each user $U_i, i = 1, \ldots, m$, shares a long-term secret $(x_i, y_i) \in \mathbb{Z}_n^* \times \mathbb{Z}_n^*$ and the public key (e, n) with the KGC;

Round 1
    User U_1:
    1.1. sends a key generation request:
        $U_1 \rightarrow KGC : \{U_1, \ldots, U_t\}$

Round 2
    The KGC:
    2.1. computes $v = (h(U_1, \ldots, U_t))^d \pmod{n}$;
    2.2. broadcasts:
        $KGC \rightarrow^* : (\{U_1, \ldots, U_t\}, v)$

Round 3
    Each user $U_i, i = 1, \ldots, t$:
    3.1. checks if $h(U_1, \ldots, U_t) = v^e \pmod{n}$;
        If the equality does not hold, he quits.
    3.2. chooses $R_i \leftarrow^R \mathbb{Z}_n^*$;
    3.3. sends:
        $U_i \rightarrow KGC : R_i$

Round 4
    The KGC:
    4.1. selects a group key $k \leftarrow^R \mathbb{Z}_n^*$;
    4.2. generates the polynomial $f(x)$ of degree t that passes through the $t+1$ points $(0, k)$, $(x_1, y_1 \oplus R_1), \ldots, (x_t, y_t \oplus R_t)$;
    4.3. computes t additional points $P_1, \ldots, P_t$ on $f(x)$;
    4.4. computes the authentication message
        $Auth = h(k, U_1, \ldots, U_t, R_1, \ldots, R_t, P_1, \ldots, P_t)$;
    4.5. broadcasts:
        $KGC \rightarrow^* : (R_1, \ldots, R_t, P_1, \ldots, P_t, Auth)$

Key Computation
    Each user $U_i, i = 1, \ldots, t$:
    5.1. computes the group key $k = f(0)$ by interpolating the points $P_1, \ldots, P_t$ and $(x_i, y_i \oplus R_i)$;
    5.2. checks if $Auth = h(k, U_1, \ldots, U_t, R_1, \ldots, R_t, P_1, \ldots, P_t)$;
        If the equality does not hold, he quits.

Fig. 3 Yuan et al.'s improvement on Harn and Lin's group key transfer protocol [9]

5 Yuan et al.'s Improvement

Yuan et al. claim that they improved Harn and Lin's protocol to stand against the previous attack [9]. Figure 3 describes their proposal.

We highlight the modifications from the original construction. The KGC generates a RSA public-private key pair in the Initialization Phase. Let n be the RSA modulus, $\Phi(n) = (p-1)(q-1)$ the Euler's totient function, e the signature verification exponent and d the corresponding signature generation exponent. We skip more details of RSA here, but invite the reader to address the original paper [6]. The KGC makes his public RSA key (e, n) available to users during the Registration Phase, as they will later need it for signature verification purposes. In Round 2, the KGC signs $h(U_1, \ldots, U_t)$ and broadcast the signature v along with the list of participants. This aims to authenticate the origin of the message such that it prevents an attacker to send a different message on his behalf. In Round 3, each user that identifies himself in the list verifies the origin of the message by using the public key of the KGC.

6 Man-in-the-Middle Attack Against Yuan et al.'s Improvement

Yuan et al. claim that in their improved version that no intermediate entity can play the role of a participant in a session without being detected (due to the signature of the *KGC* in Round 2 and the verification performed by the users in Round 3 [9]). We prove next that they are wrong.

First, we remark that the proposed solution does not exclude a replay attack: the adversary U_a can eavesdrop on a message $(\{U_1, \ldots, U_t\}, v)$ in one session of the protocol and reuse it to impersonate the KGC in another session that intends to establish a common key for the same set of users. Even more, if U_a is an insider, he can legitimately initiate protocol sessions and therefore learn valid signatures for sets of players on his own choice. As the key generation request in Round 1 is unauthenticated, U_a can impersonate any user U_i to obtain the signature of the *KGC* on any group of players U_i belongs to. The replay attack is possible because the signature is deterministic on the list of participants to a given session and the signing exponent remains unchanged for a long period of time (it is generated during the Initialization Phase and maintained for multiple sessions). Once the adversary learns the signature v on the set of users, U_a can mount the attack presented in Sect. 4.

Second, we show that Yuan et al.'s proposal remains vulnerable to a man-in-the-middle attack even in the absence of message replay. This is desirable because replay attacks can be easy avoided by standard techniques like nonce usage. For example: in Round 1, the initiator sends as key generation request the list of players $\{U_1, \ldots, U_t\}$ together with a nonce N_i; in Round 2, the *KGC* verifies that N_i has not been used before and if this holds, he signs $\{U_1, \ldots, U_t\}$ as $v = (h(U_1, \ldots, U_t, N_i))^d \pmod{n}$.

Step 1. U_a does not interfere in Round 1 of a genuine session, but he intercepts the response $(\{U_1, \ldots, U_t\}, v)$ sent in Round 2 and prevents it from reaching the group members.

Step 2. U_a impersonates all participants to the session by sending R'_j on behalf of U_j, $j = 1, \ldots, t$;

Step 3. U_a intercepts the message $(R'_1, \ldots, R'_t, P'_1, \ldots, P'_t, Auth')$ sent in Round 4 and prevents it from reaching the group members.

Step 4. U_a initiates a new protocol session by sending to the KGC $\{U_1, \ldots, U_{i-1}, U_a, U_{i+1}, \ldots, U_t\}$;

Step 5. U_a intercepts the message $(\{U_1, \ldots, U_{i-1}, U_a, U_{i+1}, \ldots, U_t\}, v')$ sent in Round 2 of the protocol and prevents it from reaching the group members;

Step 6. U_a forwards the response in Step 1 to all group members except U_i; For all $j = 1, \ldots, t$, $j \neq i$, U_j checks that $h(U_1, \ldots, U_t) = v^e \pmod{n}$ holds and therefore cannot detect the attack;

Step 7. U_a chooses $R_a \leftarrow^R \mathbb{Z}_n^*$, sends it to the KGC and allows the other messages R_j, $j = 1, \ldots, t$, $j \neq i$ to reach the KGC;

Step 8. U_a intercepts $(R_1, \ldots, R_{i-1}, R_a, R_{i+1}, \ldots, R_t, P_1, \ldots, P_t, Auth)$ sent in Round 4 of the protocol and prevents it from reaching the group members;

Step 9. U_a computes the group key $k = f(0)$ by interpolating the public points $P_1, \ldots, P_t$ and $(x_a, y_a \oplus R_a)$, forges $Auth'' = h(k, U_1, \ldots, U_t, R_1, \ldots, R_{i-1}, R_a, R_{i+1}, \ldots, R_t, P_1, \ldots, P_t)$ and sends $(R_1, \ldots, R_{i-1}, R_a, R_{i+1}, \ldots, R_t, P_1, \ldots, P_t, Auth'')$ to $U_1, \ldots, U_{i-1}, U_{i+1}, \ldots, U_t$ as originating from the KGC.

Step 10. For all $j = 1, \ldots, t$, $j \neq i$, U_j, recovers the group key $k = f(0)$ by interpolating the public points $P_1, \ldots, P_t$ and $(x_j, y_j \oplus R_j)$, checks that $Auth'' = h(k, U_1, \ldots, U_t, R_1, \ldots R_{i-1}, R_a, R_{i+1}, \ldots, R_t, P_1, \ldots, P_t)$ holds and therefore accepts k as the correct group key.

Fig. 4 Man-in-the-middle attack against Yuan et al.'s improvement

Our attack is similar to the attack in Sect. 4 in the sense that U_a maintains the same abilities (he is an active insider with full control over the communication channel that shares a long-term key (x_a, y_a) with the KGC) and has the same goal (to impersonate a victim U_i and share a common key within a set of participants $\{U_1, \ldots, U_{i-1}, U_a, U_{i+1}, \ldots, U_t\}$, while the other users believe to share a common key within $\{U_1, \ldots, U_t\}$). Figure 4 describes the attack in detail.

To fulfill his goal, the attacker initiates a second session and requests a key for $\{U_1, \ldots, U_{i-1}, U_a, U_i, \ldots, U_t\}$. During the attack, U_a makes the *KGC* believe that the initial session has successfully finished by impersonating all users and sending the values R_j in their behalf; this is possible because the origin of the messages sent in Round 3 is unauthenticated. Then, U_a initiates a new protocol session and uses the message received from the *KGC* to make $\{U_1, \ldots, U_{i-1}, U_i, \ldots, U_t\}$ finalize.

For clearness, we specify the view of the players on the sessions they participate to:

- the *KGC*: believes to participate to a session for sharing a key between $\{U_1, \ldots, U_t\}$ in Step 1 to Step 3 and believes to participate to a second session for sharing a key between $\{U_1, \ldots, U_{i-1}, U_a, U_{i+1}, \ldots, U_t\}$ starting from Step 4;
- the victim U_i: believes that there is no active session;
- any user $U_j, j = 1, \ldots, t, j \neq i$: believes to participate to a single session for sharing a key between $\{U_1, \ldots, U_t\}$; all messages in Step 2 to Step 5 are hidden to U_j.

We remark that neither the victim U_i nor the KGC can suspect a malicious behavior during the execution of the protocol: U_i is not aware that a protocol session is taking place and the KGC considers that both sessions have successfully finished. The rest of the users $\{U_1, \ldots, U_{i-1}, U_{i+1}, \ldots, U_t\}$ are only aware of the execution of a single session, which properly terminates.

By comparison to the attack in Sect. 4, the adversary U_a must impersonate all users $U_1, \ldots, U_t$ instead of only the victim U_i and must intercept and prevent from reaching their destination 4 broadcast message instead of 2. Nevertheless, this is feasible.

Our attack shows that the usage of a signature in Round 2 to attest the origin of the list of players can be easily overcome by an attacker that runs one additional session (even in the absence of a replay attack).

7 Conclusions

Yuan et al. recently mounted a man-in-the-middle attack against Harn and Lin's GKT protocol [9]. However, the attack can be detectable by the user that the adversary impersonates during the attack. We slightly modified the attack to decrease the probability of detection by preventing the user from knowing that the protocol session is taking place. The modification brings an additional advantage: it requires less implication of the adversary in the protocol execution.

In the same paper, Yuan et al. also proposed a countermeasure that they claim to stand against the attack [9]. Our work proved that their proposal remains vulnerable to a man-in-the-middle attack that permits the adversary to inject a known session key to all players except the one the adversary impersonates during the attack.

Table 1 Attacks and countermeasures

Protocol	Attacks
Harn and Lin [1]	Replay Attack [3]
	Man-in-the-Middle Attack [9]
	Improved Man-in-the-Middle Attack [Sect. 4]
Yuan et al.'s Improvement [9]	Man-in-the-middle Attack [Sect. 6]

Table 1 synthesizes the vulnerabilities against the original protocol and its modification that the current paper focuses on. To conclude, we highlight that simple attacks arise natural against protocols that lack a security proof.

Acknowledgments This work was supported by the strategic grant POSDRU/159/1.5/S/137750, Project Doctoral and Postdoctoral programs support for increased competitiveness in Exact Sciences research cofinanced by the European Social Found within the Sectorial Operational Program Human Resources Development 2007–2013.

References

1. Harn, L., Lin, C.: Authenticated group key transfer protocol based on secret sharing. IEEE Trans. Comput. **59**(6), 842–846 (2010)
2. Kim, M., Park, N., Won, D.: Cryptanalysis of an authenticated group key transfer protocol based on secret sharing. In: Park, J.J(Jong Hyuk), Arabnia, H.R., Kim, C., Shi, W., Gil, J.-M. (eds.) GPC 2013. LNCS, vol. 7861, pp. 761–766. Springer, Heidelberg (2013)
3. Nam, J., Kim, M., Paik, J., Jeon, W., Lee, B., Won, D.: Cryptanalysis of a group key transfer protocol based on secret sharing. In: Kim, T., Adeli, H., Slezak, D., Sandnes, F.E., Song, X., Chung, K., Arnett, K.P. (eds.) FGIT 2011. LNCS, vol. 7105, pp. 309–315. Springer, Heidelberg (2011)
4. Olimid, R.F.: On the security of an authenticated group key transfer protocol based on secret sharing. In: Mustofa, K., Neuhold, E.J., Tjoa, A.M., Weippl, E., You, I. (eds.) ICT-EurAsia 2013. LNCS, vol. 7804, pp. 399–408. Springer, Heidelberg (2013)
5. Olimid, R.F.: A chain of attacks and countermeasures applied to a group key transfer protocol. In: Proceedings of International Joint Conference SOCO13 CISIS13 ICEUTE13, AISC 239, pp. 333–342, (2014)
6. Rivest, R., Shamir, A.: A method for obtaining digital signatures and public-key cryptosystems. Commun. ACM **21**(2), 120–126 (1978)
7. Shamir, A.: How to share a secret. Commun. ACM **22**(11), 612–613 (1979)
8. Sun, Y., Wen, Q., Sun, H., Li, W., Jin, Z., Zhang, H.: An authenticated group key transfer protocol based on secret sharing. Int. Workshop Inform. Electron. Eng. Procedia Eng. **29**, 403–408 (2012)
9. Yuan, W., Hu, L., Li, H., Chu, J.: Security and improvement of an authenticated group key transfer protocol based on secret sharing. Appl. Math. Inf. Sci. **7**(5), 1943–1949 (2013)
10. Yuan, W., Hu, L., Li, H., Chu, J.: An effcient password-based group key exchange protocol using secret sharing. Appl. Math. Inf. Sci **7**(1), 145–150 (2013)

A Heterogeneous Fault-Resilient Architecture for Mining Anomalous Activity Patterns in Smart Homes

Ciprian Pungila, Bogdan Manate and Viorel Negru

Abstract We are presenting a massively parallel heterogeneous cloud-based architecture oriented towards anomalous activity detection in smart homes. The architecture has very high resilience to both hardware and software faults, it is capable of collecting activity from various data sources and performing anomaly detection in real-time. We corroborate the approach with an efficient checkpointing mechanism for data processing which allows the implementation of hybrid (CPU/GPU) fault-resilience and anomaly detection through pattern mining techniques, at the same time offering high throughput.

Keywords Anomaly detection · Pattern mining · Smart home · Fault resiliency · Heterogeneous architecture · Graphics processing unit

1 Introduction

Anomaly detection has become quite important nowadays in various scenarios: helping determine abnormal situations in smart homes, which could potentially lead to disastrous scenarios (e.g. building catching fire, uncontrolled power consumption, burglary, etc.), helping patients in need of healthcare (by tracking location, determining abnormally long stays in one location, measuring the heart rate, etc.), etc. [1, 2].

Numerous architectures developed for smart homes are using interconnected sensors that transmit information wirelessly, from the sensor node to a receiver node and from there onwards to a data aggregator. Such information is collected by a central

C. Pungila (✉) · B. Manate · V. Negru
West University of Timisoara, Blvd. V. Parvan 4, 300223 Timisoara, Timis, Romania
e-mail: cpungila@info.uvt.ro
url: http://info.uvt.ro

B. Manate
e-mail: bogdan.manate@info.uvt.ro

V. Negru
e-mail: vnegru@info.uvt.ro

© Springer International Publishing Switzerland 2015
Á. Herrero et al. (eds.), *International Joint Conference*, Advances in Intelligent Systems and Computing 369, DOI 10.1007/978-3-319-19713-5_12

machine (usually stored in a cloud, where it can benefit from useful features such as failure protection or data preservation) and it is either stored for archival purposes, or analyzed for producing better recommendation systems for the future.

We are proposing an architecture for mining anomalous patterns of activity in smarthomes by analyzing and classifying activities based on sequences of relevant events, offering fault resilience, processing multiple subjects at once while being backed up by a custom checkpointing mechanism. We construct a methodology for classifying activities into different tasks, as well as provide the means to mine anomalous activity patterns by identifying sequences which do not fit a specific profile or which deviate from those already existing. We discuss related work in Sect. 2, our approach is presented in Sect. 3 and the experimental results obtained are discussed in Sect. 4.

2 Related Work

2.1 Vector Space Model

The vector space model (VSM) [3] has been used successfully in the past as a technique for information retrieval and text mining. In literature, it is also known as term vector model (TVM). It relies on a corpus comprised of a set of vectors as follows:

$$v_i = (w_{1,i}, w_{2,i}, \ldots, w_{n,i}) \tag{1}$$

Each element of the vector, $w_{j,i}$ represents a weight computed as a measure of relevance that the j-th term, t_j, has in the object v_i. In the original approach, the terms and the objects have been used as text words and text documents, respectively, with the weight $w_{j,i}$ measuring therefore the importance of the word t_i in the document v_i. As a method of clustering or classification, the approach is being commonly used with the cosine of the angle of two vectors.

As statistical indicator, VSM uses term frequency-inverse document frequency (also known as tf-idf), computed as:

$$tf\_idf(w_{j,i}) = tf(w_{j,i}) \cdot idf(t_j) \tag{2}$$

The term frequency, $tf(w_{j,i})$ measures the number of occurrences of t_j in v_i, and the inverse term frequency $idf(t_j)$ measures how rare (or common) t_j is across all documents in the corpus. A higher value of the $tf\_idf(w_{j,i})$ statistical indicator represents shows not just that the corresponding term appears many times in the document, but also that it is less frequent in other documents, making the indicator suitable for filtering out common terms in the documents of the corpus.

2.2 Cosine Similarity Measurement

The cosine similarity measurement allows tracking down similarities between one or more objects in the corpus. Assuming the objects are p_i and r_j, the cosine similarity can be computed as follows:

$$sim(p_i, r_j) = \frac{\sum_{k=1}^{n} p_{k,i} r_{k,j}}{\sqrt{\sum_{k=1}^{n} p_{k,i}^2 \cdot \sum_{k=1}^{n} r_{k,j}^2}} \tag{3}$$

Possible values for $sim(p_i, r_j)$ range from -1 to 1, with values of 1 representing identical documents, -1 representing complete dissimilarity and 0 a certain independence between the two.

3 Implementation

As a general layout, in a smart home, the design relies on a data aggreggator, which collects information from the various sensors mounted in the home and sends them to a central hub for further processing and recommendations, or processes them locally. We design our approach based on the general consensus that security should be at the heart of all data transmissions [4–6], therefore assuming all encryption of the data received and sent is accomplished through an encryption algorithm. We assume encryption happens both on a sensor-level, as well as between the data aggregator and the analyzer module recipient. Our architecture's layout is presented in Fig. 1.

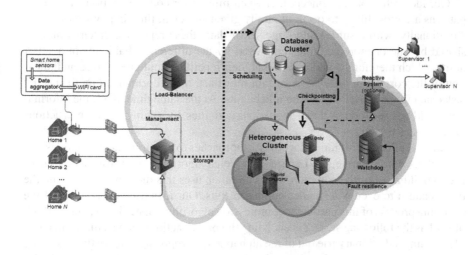

Fig. 1 The proposed architecture's overall layout

Our proposed cloud configuration makes use of two different types of watch-dogs: a localized (per instance) watchdog, whose purpose is to analyze CPU and GPU resources and decide the extent of computations it can perform when the need arises, as well as communicate with the global (per cloud) watchdog and inform him of its status.The GPU memory is organized as to hold several structures for different households. This ensures efficient resource usage. It is highly desirable that tree data structures use as little memory as possible, which is why we have opted for the bitmapped model we have presented in [7]. The overall architecture for the watchdog monitoring system is presented in Fig. 2.

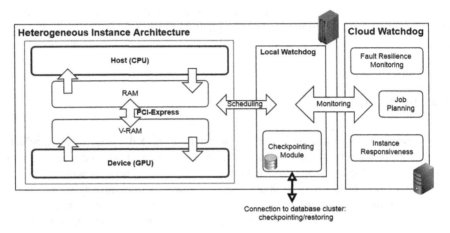

Fig. 2 The proposed architecture for a heterogeneous instance in the cloud

Our idea relies on the concept that when mining for common actions, repetitive patterns are most likely to occur naturally given sufficient time in any smart home. Additionally, with a sufficiently large set of data, these repetitive patterns may be shared by one or more activities employed by the different inhabitants in a smart home, which therefore would ensure that their repetition could be used as an indication of the correlation between a certain set of activities, corresponding to specific tasks, and those carried out by the inhabitants. At the same time, the same information could be used to track down anomalous activities, by mining uncommon actions or sequences of actions.

Other applications of more 'localized' approaches to data mining are compression algorithms. For example, the Lempel-Ziv-Welch algorithm [8, 9] uses a dictionary trie to build a list of the common words and then uses this dictionary to encode the input pattern to occupy a smaller number of bits. Our idea uses the concept of the encoding process of this particular compression algorithm. Specifically, our methodology has the following steps: (1) encoding the proper actions using a finite alphabet, (2) building a dictionary trie out of the alphabet, (3) proposing a heuristic for mining activities and classifying them accordingly, based on the keywords from the dictionary. The idea may be considered to some extent as being similar to that of using a variable n-gram approach, but differs through the encoding mechanism, the heuristic proposed and the massively data-parallel character of the architecture (making it perfectly suitable for heterogeneous CPU/GPU architectures).

3.1 Encoding

The encoding process looks at the different types of sensors and their list of possible states, and builds an alphabet capable of supporting each possible combination of that sensor, based on their possible states. For example, a motion sensor only has two possible states, on and off. Therefore, the partial alphabet for this particular type of sensor has a cardinality of 2. The final alphabet is comprised of the collection of all partial alphabets from all sensor types.

It is worth noticing that we do not consider specific sensor values in this scenario, e.g. if a running water sensor is measuring the amount of water consumed, this information is not taken into account in the encoding stage. It may however be used when applying heuristics over the dictionary obtained.

In the tests we have performed, using datasets from the CASAS project [11], we have used a set of 100 ADL activities, corresponding to 20 participants performing 5 different tasks. A total of 5,312 sensor events have been recorded in this dataset. The alphabet cardinality for this specific dataset turned out to be 53, with the different sensors used including: motion sensors, item sensors (medicine container, pot, phone book, burner, etc.) and miscellaneous activity sensors (e.g. running water, phone usage).

3.2 Dictionary of Variable-Length n-grams

Once the proper encoding has been constructed for each sensor type and their possible states, the next step was to build the adequate dictionary tree for the encoded actions. The encoding process was inspired by the compression algorithm proposed by Lempel-Ziv-Welch [8, 9]. The approach builds a dictionary trie by parsing the entire sequences of actions and storing sequences of actions on-the-go in the dictionary, as follows:

1. we initialize the dictionary to contain all encoded symbols in the alphabet
2. find the longest sequence of actions s in the dictionary that matches the current input
3. remove s from input
4. add sequence $s + c$ (the current symbol) to the dictionary
5. if not reached end of input yet, repeat by going to step 2.

In order to store the dictionary trie (and to that extent, any tree at all) efficiently in GPU memory, we have used an approach we have presented in our previous work

[7]. The primary problem of transferring trees node-by-node from RAM memory to V-RAM memory lies in the overhead involved by the actual API used for the transfers, as well as the huge amount of memory which is wasted as a result of memory fragmentation (see Fig. 3). Our methodology involves parsing the dictionary tree in pre-order (or any other parsing methodology which would offer a list of nodes, with all children of a specific node being listed sequentially), building a stack of nodes out of the dictionary, and transferring the entire tree in a single, instant burst through the PCI-Express architecture to V-RAM, avoiding any overheads and benefitting of maximum throughput.

Additionally, all changes to the tree may be executed in real-time in the GPU directly, as presented in our earlier work [12]. With this in mind, operating changes over an existing dictionary may happen directly in GPU memory, with minimal impact on performance.

3.3 Heuristics

Once the encoding process has happened, we propose a heuristic based on the Vector Space Model (VSM) [3] text mining model presented earlier in Sect. 2.1, and the cosine similarity measurement discussed in Sect. 2.2. The corpus of elements is represented by all sequences of activities which have been encoded in the dictionary trie, as variable n-grams. In our experiments, with different minimum sequence lengths, we have obtained a corpus of 268 to 1,667 elements (see Fig. 5a), which means that each specific task (out of the 5 tested) will have a vector of this size. However, these are mostly sparse vectors, therefore a small storage than normal is required.

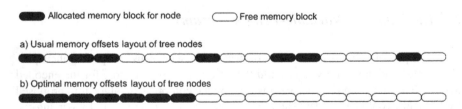

Fig. 3 **a** The usual, non-optimal memory offsets layout for pointers to nodes in a tree. **b** The optimal layout of memory offsets for pointers to nodes in the same tree

Just like presented, we use $tf(w_{j,i})$ as the term frequency, meaning the number of times the sequence of activities t_j is found in the task v_i. We then computed $idf(t_j)$ as follows:

$$idf(t_j) = ln(\frac{T}{df_j}) \tag{4}$$

Here, T is the total number of tasks and df_j is the total number of tasks which contain the sequence of activities t_j. To classify an unknown sequence of activities, we proceed as follows:

1. encode incoming data using the alphabet developed initially
2. find longest sequence of activities s which exists in dictionary trie
3. build a suffix tree of the sequences encoded (for storing precedence information)
4. for each task identified

    (a) for each sequence in the initial alphabet
        i. compute $tf_i df$ as presented in Sect. 2.1
    (b) compute similarity using the cosine measurement discussed in Sect. 2.2.

The suffix tree is useful for storing precedence information. For example, let's assume the sequence of activities found in the incoming data is decomposed, after parsing the dictionary trie as presented, in the following sequences from the internal alphabet: s_2, s_4, s_3, s_5. Each of these sequences would be stored in a suffix tree where s_2 would a parent for s_4, s_4 a parent for s_3, and so on. The storing methodology in V-RAM for this tree is identical to the dictionary storage approach presented in Sect. 3.2, so that transfers between RAM and V-RAM are instantaneous (Fig. 4).

We use the suffix tree mentioned for anomalous/abnormal activity detection. Specifically, whenever decomposing a new activity into its sequences for analysis through the VSM proposed, we track down consecutive groups of sequences with a length smaller than the minimal length used when building the dictionary trie. This allows us to track down groups of sequences which do not exist in the dictionary, and which are being sent to the reactive component of the entire architecture for decision-making or reporting. Additionally, future implementations planned are looking at using the precedence information in correlation with the VSM for classification.

As a measure of classification (or clustering), we use the $dist(p, r)$ attribute:

$$dist(p, r) = 1 - sim(p, r) \tag{5}$$

As a result, after the VSM is computed, we choose the one which is closer to 0 and consider that the sequence of activities analyzed belongs to that specific task.

3.4 Data Storage and Replication

The checkpointing mechanism is employed regularly every few cycles of the computation process presented in Sect. 3.3. Each GPU on each instance in the cloud uses a continuous block of memory for all data structures involved, as presented in Sect. 3.2, making instant transfers between RAM and V-RAM possible, therefore allowing the checkpointing to occur uninterrupted.

Fig. 4 The Riak cluster ring

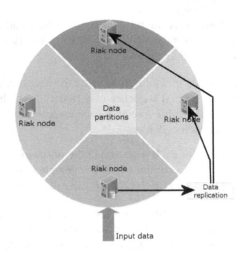

Our architecture uses a Riak cluster for real-world deployment [10] and NoSQL databases for stories time-series data. Riak offers high availability, operational simplicity, scalability and it is masterless, so there is no need to direct all the requests to a single server.

The Riak cluster is divided in nodes which usually represent a single server. Data is stored using a key/value scheme and multiple key/value pairs are stored in buckets. For data retrieval, Riak uses a hash generated from the bucket and the key where the data was stored. The resulted hash value maps into a 160-bit integer space. When new data is inserted on a Riak node it automatically replicates on neighboring nodes, which are designed to take over the first node's functionality in case of a failure, therefore maintaining data integrity and availability. In case of a recovery of the failed node, lost updates (if any) will be delivered by the neighboring nodes.

4 Experimental Results

We have used a real-world dataset from the CASAS project [11, 13], comprised of a total of 5,312 sensor events containing activities of daily living (ADL), from a total of 20 participants, each involved in performing 5 different tasks. The encoding process was discussed in Sect. 3.1 and the dictionary trie construction has been presented in Sect. 3.2. We then applied the heuristics described in Sect. 3.3 on the initial dataset, and verified it by re-employing the same analysis on the same set after the initial learning stage has taken place, just like the experiments performed in the original paper [11]. In the CASAS project, the authors use a Bayes algorithm for classification (which achieved 91 % accuracy) and a Markov model (98 % accuracy).

In our experiment, we have varied the minimum sequence length that was to be inserted in the dictionary, from 1 to 6, in order to see how this affects the error in clas-

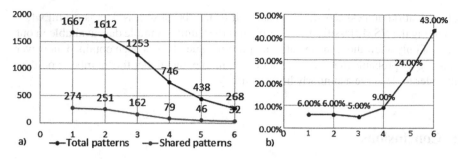

Fig. 5 **a** The experimental results on the CASAS dataset [13]. **b** The classification error obtained when applying different minimum sequence lengths to the dictionary trie construction, using the CASAS dataset [13]

sification. The best accuracy, of 95 %, was obtained for a minimum sequence length of 3 (Fig. 5b). The number of shared patterns (items in the list of sequences which exist in at least two daily activities) decreases with the minimum sequence length as well. Just as expected, the accuracy drops significantly when the minimum n-gram length increases, since there are fewer samples to use for an accurate classification. Nonetheless, the accuracy is slightly worse when values of 1 and 2 are used, most likely because of the high density of the smaller repeating patterns (e.g. a motion sensor turning on several times in different tasks and activities).

For testing the checkpointing approach, we configured 3 nodes in a Riak cluster on Amazon EC2, using 64-bit m1.large instances. Our simulation included 3 replication nodes and 100 concurrent connections, configured to execute 1 GET for every 100 UPDATE operations. Figure 6 shows that the entire data transfer reaches 1.5 GB/s. Every update operation had a fixed key value generator of 10 KB.

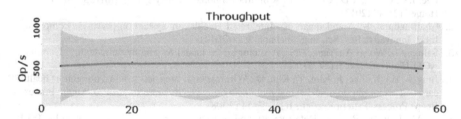

Fig. 6 Data replication benchmark during stress-testing through the checkpointing mechanism

Accuracy of the classification process (of daily activities into tasks) is directly affected by the minimum sequence length. This is however normal: shorter events comprising an activity are more likely to be found in multiple different tasks, causing the classification to lose the dependency on certain unique sequences, specific to them. The longer the events comprising an activity are, the more they are tied to that specific activity. Nevertheless, choosing sequences which are too long makes the classification process too vague, causing it to fail, because less data is useful when applying the VSM classification. With the rather limited sample data available in our test, of only a few thousand events, an optimal value of 3 has been obtained, however it is highly probable that the more data is available, the higher this number would be in order to preserve a similarly high accuracy.

5 Conclusions

We have proposed an efficient heterogeneous architecture for data mining using a massively parallel fault-resilient approach. Starting with a custom encoding process, followed by an idea inspired from the encoding process of an efficient compression algorithm and ending with our own variable n-gram heuristic implementation based on the vector space model (VSM), we have shown how the approach can be used to efficiently mine activities in a smart home in a cloud-based environment. Our approach delivers good performance when an adequate minimum sequence length is chosen, achieving an accuracy of 95 % in the tests performed with a real-world dataset obtained from the CASAS project [11] and providing high throughput in massively parallel implementation in CPU/GPU hybrid hardware.

Acknowledgments This work was partially supported by the Romanian national grant PN-II-ID-PCE-2011-3-0260 (AMICAS).

References

1. Lee, J.V., Chuah, Y.D., Chai, C.T.: A multilevel home security system (MHSS). Int. J. Smart Home **7**(2), 49 (2013)
2. Chairmadurai, K., Manikannan, K.: Integrated Health care system on pervasive computing. Int. J. Innovative Res. Sci. Eng. Technol. **3**(1) (2014)
3. Salton, G., Wong, A., Yang, C.S.: A vector space model for automatic indexing. Commun. ACM **18**(11), 613–620 (1975)
4. Jung, J., Ha, K., Lee, J., Kim, Y., Kim, D.: Wireless body area network in a ubiquitous health-care system for physiological signal monitoring and health consulting. Int. J. Signal Process. Image Process. Pattern Recogn. **1**(1), 47–54 (2008)
5. Li, M., Lou, W., Ren, K.: Data security and privacy in wireless body area networks. IEEE Wireless Commun. **17**(1), 51–58 (2010)
6. Lim, S., Oh, T.H., Choi, Y.B., Lakshman, T.: Security issues on wireless body area network for remote healthcare monitoring. In: Proceedings of the 2010 IEEE International Conference on Sensor Networks, Ubiquitous, and Trustworthy Computing, pp. 327–332. IEEE Computer Society (2010)
7. Pungila, C., Negru, V.: A highly-efficient memory-compression approach for GPU-accelerated virus signature matching. In: Gollmann, D., Freiling, F.C. (eds.) ISC 2012. LNCS, vol. 7483, pp. 354–369. Springer, Heidelberg (2012)
8. Ziv, J., Lempel, A.: Compression of individual sequences via variable-rate coding. IEEE Trans. Inform. Theor. **24** (1978)
9. Welch, T.: A technique for high-performance data compression. Computer **17**(6), 8–19 (1984)

10. Riak. http://basho.com/riak/
11. Cook, D.J., Schmitter-Edgecombe, M.: Assessing the quality of activities in a smart environment. Methods Inf. Med. **48**(5), 480–485 (2009). doi:10.3414/ME0592
12. Pungila, C., Reja, M., Negru, V.: Efficient parallel automata construction for hybrid resource-impelled data-matching. Future Gener. Comput. Syst. **36**, 31–41 (2014) Special Section: Intelligent Big Data Processing 2014. doi:10.1016/j.future.2013.09.008
13. The CASAS project. http://ailab.wsu.edu/casas/datasets/

Investigating the Impacts of Brain Conditions on EEG-Based Person Identification

Dinh Phung, Dat Tran, Wanli Ma and Tien Pham

Abstract Person identification using electroencephalogram (EEG) as biometric has been widely used since it is capable of achieving high identification rate. Brain conditions such as epilepsy and alcohol are some of problems that cause brain disorders in EEG signals, and hence they may have impacts on EEG-based person identification systems. However, this issue has not been investigated. In this paper, we perform person identification on two datasets, Australian and Alcoholism EEG, then compare the classification rates between epileptic and non-epileptic groups, and between alcoholic and non-alcoholic groups, to investigate the impacts of such brain conditions on the identification rates. Shannon (SEn), Spectral (SpEn), Approximate entropy (ApEn), Sample (SampEn) and Conditional (CEn) entropy are employed to extract features from these two datasets. Experimental results show that both epilepsy and alcohol actually have different impacts depending on feature extraction method used in the system.

1 Introduction

Person identification is the process of recognizing the identity of a given person out of a closed pool of N persons [1]. Applications are found in video surveillance (public places, restricted areas) and information retrieval (police databases). In general, people can be identified by their biometrics such as voice, face, iris, retina, and fingerprint. It has been shown that electroencephalogram (EEG) can also be used as biometric for person identification [1].

D. Phung (✉) · D. Tran · W. Ma · T. Pham
Faculty of Education, Science, Technology and Mathematics,
University of Canberra, ACT 2601, Canberra, Australia
e-mail: Dinh.Phung@canberra.edu.au

© Springer International Publishing Switzerland 2015
Á. Herrero et al. (eds.), *International Joint Conference*, Advances in Intelligent
Systems and Computing 369, DOI 10.1007/978-3-319-19713-5_13

1.1 Electroencephalogram

Electroencephalogram (EEG) is a measurement of the brain signals containing information generated by brain activities [2]. EEG signals are captured by using multiple electrodes either from inside the brain (invasive methods), over the cortex under the skull, or certain locations over the scalp (non-invasive methods) [2]. In fact, there is a connection between genetic information and EEG of an individual [3], and brain wave patterns are unique to individuals [1, 4]. Moreover, EEG features are universal as all living and functional persons have recordable EEG signals [5]. Therefore, EEG data can be suitably used for person identification [1, 3, 6, 7].

Accuracy is one of the crucial requirements of any person identification system, including EEG-based. Factors which may affect the accuracy of an EEG-based identification system can be signal noises, feature extraction methods, and/or classification algorithms. Brain conditions (such as epilepsy and alcohol) may also have some effects on the performance of person identification. As entropies have shown their ability in reflecting the changes in the chaotic level and degree of complexity in time series as well [2, 8, 9], we employ five different entropy methods including Shannon, Spectral, Approximate, Sample and Conditional, to investigate the impacts of epilepsy and alcohol on EEG-based person identification. These methods are used as feature extraction on two sets of EEG database, namely Australian EEG and Alcoholism, to obtain different features for EEG-based person identification systems.

1.2 Epilepsy

Epilepsy is chronic neurological disorder which is generally characterized the sudden and the recurrent seizures [8]. Several research results have shown that brain state changes from less to more ordered state, from more to less chaotic, or from more to less complexity during epileptic seizures. So far, there have been many studies on detecting epilepsy such as in [7–12], and most of the methods used are entropies. In particular, [8] showed that Sample, Approximate, and Spectral entropies for epileptic EEG signals were lower than those of normal EEG signals. [10] and [11] concluded that value of the ApEn drops sharply during an epileptic seizure, and [7] indicated that entropies of epileptic activity are less than those of non-epileptic activity. The same trend is observed in both spectral and embedding (using time series directly for estimations) entropies. It also reported that epileptic seizures are emergent states with reduced dimensionality compared to non-epileptic activity. This means that the number of independent variables required to describe the system during epileptic seizures is smaller than at other times [7].

1.3 Alcohol

Alcohol is one of the most common used substance in the world. Alcoholic beverages have been a part of social life for many years. So far several researches have shown that alcohol effects on neurophysiologic parameters which can be detected through EEG signals. For the resting EEG, [13] reports that the alcohol-dependent people have higher resting theta power at all scalp locations. This reflects a deficiency in the information processing capacity of the central nervous system. Unstable of poor alpha rhythm is found in alcoholics. An increase power in beta in frontal brain regions. While in the active brain conditions, [14] states that alcohol makes an increase in absolute power (amplitude) of delta and theta, and of alpha rhythm in both the frontal mid-line and parietal regions of the brain in both high and low load task conditions. Similarly, the pre-frontal region of brain decreases an absolute power with increase in the amount of alcohol intake [14]. In addition, [13] demonstrates that both evoked delta and theta waves are deficient in alcoholics in the cognitive process. Alcoholics are also depleted in the production of evoked gamma during the processing of target stimuli. [15] reports that moderate doses of alcohol have an increase in amplitude and a decrease in the dominant frequency of alpha. Lower frequencies increase with larger doses of alcohol. In addition, topographic maps of have been affected in alcoholics as [13] indicates that they have less distinct spatial-temporal patterns during various tasks.

In brief, we expect in this paper that if epilepsy and alcohol have impacts on EEG-based person identification systems, there should be a difference between the rates of epileptic and non-epileptic, and between alcohol and non-alcohol groups.

2 Entropy

Originally, entropy is a thermodynamic quantity measuring the amount of disorder in the system. From a perspective of information theory, entropy is described as the information amount stored in a more general probability distribution. Recently, a number of different entropy estimators have been applied to quantify the complexity of signals. [7] Entropy is a measure of uncertainty. In brain-computer interface systems, entropy can be used to measure the level of chaos of the system [2]. It is a non-linear measure quantifying the degree of complexity in a time series [8].

The advantage of using entropy methods for EEG feature extraction is that EEG signals are complex, non-linear, non-stationary, and random in nature [2, 3, 7–9, 16–18]. Therefore, approaches for non-linear analysis such as entropy would be appropriate for EEG signals [7]. Entropy is one of several approaches for non-linear analysis that has been proposed for EEG feature extraction as randomness of non-linear time series data is well embodied by calculating entropies of the time series data [19]. Entropy reflects how well one can predict the behavior of each respective part of the trajectory from the other. Basically, higher entropy indicates more complex or chaotic systems, thus, less predictability [7].

So far some entropy methods has been utilized in EEG feature extraction for EEG-based applications including epilepsy detection such as in [7, 8, 10, 11, 20]. Specifically, it has been shown that there are significant differences between epileptic and normal EEG, that is, entropy values of for epileptic are lower than those of non-epileptic EEG signals [8]. Epileptic EEGs are more regular and less complex than the normal, and entropies of epileptic are less as compared to that of non-epileptic activity [7]. Furthermore, entropy methods have recently been used for EEG-based person identification and shown promising results such as Shannon [21] and the effects of epilepsy on EEG-based person identification are also investigated in [22] using Approximate entropy.

3 Datasets

Our experiments are conducted on the Australian EEG (AEEG) and Alcoholism datatsets. The AEEG dataset was collected in the John Hunter Hospital, New South Wales, Australia, over a period of 11 years [23]. The recordings were made by using 23 electrodes (channels) placed on the scalp of a subject with the sampling rate of 167 Hz for about 20 min. The subset of the data used for our experiments consists of the EEG data of 80 subjects.

The Alcoholism dataset comes from a study to examine EEG correlates of genetic predisposition to alcoholism [24]. The dataset contains EEG recordings of control and alcoholic subjects. Each subject was exposed to either a single stimulus (S1) or to two stimuli (S1 and S2) which were pictures of objects chosen from the 1980 Snodgrass and Vanderwart picture set. When two stimuli were shown, they were presented in either a matched condition where S1 was identical to S2 or in a non-matched condition where S1 differed from S2. The 64 electrodes placed on the scalp sampled at 256 Hz for 1 s. The Alcoholism full dataset contains 120 trials for 122 subjects.

4 Feature Extraction and Classification

We extract EEG features by using Shannon Entropy (SEn), Spectral entropy (SpEn), Approximate entropy (ApEn), Sample entropy (SampEn) and Conditional entropy (CEn) methods. First, EEG signals are filtered into six wavebands, namely delta, theta, alpha, beta, gamma, and mix (includes five above bands). Features are then extracted and classified separately for each waveband. While each subject in the Alcoholism datatset has 120 separate 1 s trials, there is only one single trial of about 1200 s per person in the AEEG (the actual length is different for each individual); therefore, one further step is taken to cut all AEEG's trials into the same length of

900 s. Then each 900-s trial is divided into 60 15-s segments, so that each subject has 60 trials of 15 s. Next, entropy methods are used to extract features on each trial. Features from all EEG channels will be joined together to form a feature vector. Apart from CEn, of which each feature vector has 253 (for AEEG dataset) or 2016 features (for Alcoholism dataset), other entropy methods have 23 (for AEEG dataset) or 64 features (for Alcoholism dataset). The selected parameters for ApEn and SampEn are $m = 4$ and $r = 0.5 * std$.

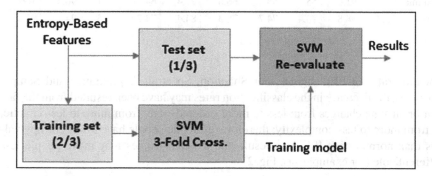

Fig. 1 Classification process

The extracted features are used to train Linear Support Vector Machine (SVM) classifiers for person identification as described in [25, 26]. SVM uses a regularization parameter C that enables accommodation to outliers and allows errors on the training set [25]. Originally, SVM was designed for binary classification; therefore, it cannot deal with multi-class classification directly [27]. However, binary SVM classifiers can be combined to form multi-class SVM to handle the multi-class case [26].

During the classification, two-third of the datasets are used building up training model, and one-third are used for testing. In the training phase, Linear SVM classifiers are trained in 3-fold cross-validation with parameter C ranging from 1 to 1000 in 5 steps [26]. The selection of parameters C is conducted by using a Weka's meta-classifier named CVParameterSelection. SVM's models are then built on the training data based on the best-found parameters of C. In the test phase, the supplied test data will be evaluated against the trained models for person identification (see Fig. 1).

5 Results

Our experimental results show that both epilepsy and alcohol have impacts on EEG-based person identification.

As shown in Table 1, the epileptic group has lower entropies in all bands. These results have supported the that of [22] that epilepsy decreases the person iden-

Table 1 Person identification rates (%) between Epileptic (Epi.) and Non-epileptic (Non.) groups

Band	Shannon		Spectral		Approximate		Sample		Conditional	
	Epi.	Non.	Epi.	Non.	Epi.	Non.	Epi.	Non.	Epi.	Non.
Delta	78.8	84.6	36.7	41.1	42.6	45.4	35.6	40.7	92.5	93.8
Theta	85	86.3	37.8	43.8	39	40.7	26.1	32	95.3	96.3
Alpha	76.9	89.7	43	44.2	52.9	58.4	41.3	52.1	96.2	100
Beta	90.8	91.8	48.2	56	67.8	74.1	66.9	79.2	95.8	97.5
Gamma	91.3	91.6	55	58	66.6	77.4	64	76.1	96.5	100
Mix	86.3	88.8	70.4	74.7	78.3	84.4	73.7	81.5	94.1	97.1

tification rates of EEG signals. For Shannon, Spectral, Approximate and Sample entropies, the decrease in the classification rates may have been resulted from the fact that brain state changes from less to more ordered state, from more to less chaotic, or from more to less complexity; therefore, epileptic people have lower entropy values than normal ones [8]. As a result, the lower entropies may make people less differentiable (for example, see Fig. 2).

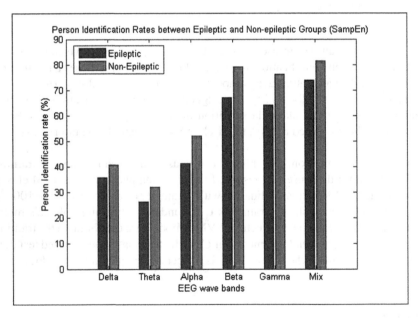

Fig. 2 Person Identification rates between Epileptic and Non-epileptic groups (Sample entropy)

The lower in Conditional entropy values of the epileptic group in comparison with the non-epileptic one also support that epileptic seizures have reduced dimensionality compared to non-epileptic activity. This means that the number of independent

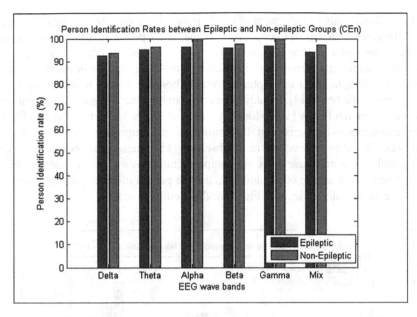

Fig. 3 Person Identification rates between Epileptic and Non-epileptic groups (Conditional entropy)

variables required to describe the system during epileptic seizures is smaller than at other times [7]. This might make the lower classification rates of epileptic group compared with the normal group, and CEn can reflect this difference respectively (see Fig. 3).

Table 2 Person identification rates (%) between Alcoholic (Alc.) and Non-alcoholic (Non.) groups

Band	Shannon		Spectral		Approximate		Sample		Conditional	
	Alc.	Non.	Alc.	Non.	Alc.	Non.	Alc.	Non.	Alc.	Non.
Delta	43.1	38.1	35	34.1	35.6	34.8	35.9	33.4	60.7	72.2
Theta	52.1	48.4	33.4	32.9	35.8	33	34.4	33.3	32.8	42.8
Alpha	60	49.9	33.6	32.6	36	34.4	33.6	33.4	50.3	52.3
Beta	75.5	69.8	36	33.8	59.9	58	59.9	58	68.9	79.1
Gamma	79	72.1	43	42.1	54.1	50.1	54.1	50.1	69.2	79.4
Mix	68.1	61.5	59.2	55	70.9	63.4	73	69.5	63.9	74.1

The person identification rates between Alcoholic and Non-alcoholic groups are presented in Table 2 which shows that there is a separation between the rates of the alcoholic and non-alcoholic people. That is, features extracted from Shannon, Spectral, Approximate and Sample entropy of the alcoholic group provide higher classification rates than those of the non-alcoholic one. The results support the findings that alcohol makes changes to all bands of EEG signals. For example, during the resting states, the alcohol-dependent people have higher theta power in all brain locations, unstably poor alpha rhythm and an increase power in beta in frontal brain regions [13]. While in the active brain conditions, alcohol make an increase in absolute power (amplitude) of delta, theta and alpha rhythms in both the frontal mid-line and parietal regions of the brain [14]. In addition, evoked delta theta and gamma oscillations are deficient in alcoholics [13]. Moderate doses of alcohol also have an amplitude increase of alpha and a decrease of the dominant alpha frequency. With larger doses, an increase in the lower frequencies is observed [15]. These changes tend to make EEG signals more nonlinear, thus more unpredictable; as a result, features of SEn, SpEn, ApEn and SampEn of alcoholics make the person identification rates higher than those of non-alcoholics (see Fig. 4 for this demonstration).

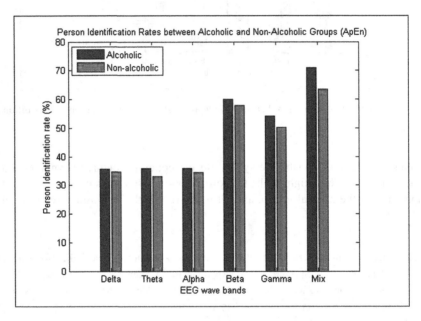

Fig. 4 Person Identification rates between Alcoholic and Non-alcoholic groups (Approximate entropy)

However, alcoholic group has lower performance than the non-alcoholic group for CEn. This can support the findings that alcoholics have less distinct spatial-temporal patterns during various tasks [13], thus this would affect on the poorer classification rates of CEn's features (see Fig. 5).

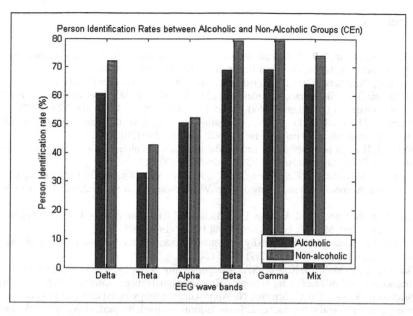

Fig. 5 Person Identification rates between Alcoholic and Non-alcoholic groups (Conditional entropy)

6 Conclusions

We have found that brain conditions such as epilepsy and alcohol have impacts on EEG-based person identification systems. Entropy methods show the separation in the rates of person identification between different brain conditions. The highest difference between Epileptic and Non-epileptic groups is 12.8 %, and between Alcoholic and Non-alcoholic groups is 14.7 %. These big gaps show that brain conditions such as epilepsy and alcohol can have significant influence on the performance of a EEG-based person identification system. In case of epilepsy, Conditional Entropy would be recommended to improve the EEG-based person identification rates since it has shown to be the best performance among the five investigated entropy methods, and it has the smallest gap in the rates between Epileptic and Non-epileptic groups. Shannon Entropy would be recommended in case of alcohol because, like Spectral, Approximate and Sample Entropy, it performs better on Alcoholic than Non-alcoholic groups, and it is the best method of all in terms of the identification rates for Alcoholic group. Our future works are to investigate a wider range of datasets and impacts of other brain conditions such as emotion on EEG-based person identification.

References

1. Marcel, S., Millán, Jl.R.: Person authentication using brainwaves (EEG) and maximum a posteriori model adaptation. Technical Report Idiap-RR-81-2005, IDIAP, 2005
2. Sanei, S., Chambers, J.A.: EEG Signal Processing. Wiley-Interscience (2007)
3. Mohammadi, G., Shoushtari, P., Ardekani, B.M., Shamsollahi, M.B.: Person identification by using ar model for EEG signals. World Acad. Sci. Eng. Technol. 1(11), 918–923 (2007)
4. Pham, T., Ma, W., Tran, D., Nguyen, P., Phung, D.: A study on the feasibility of using EEG signals for authentication purpose. In: ICONIP, pp. 562–569 (2013)
5. Shedeed, H.A.: A new method for person identification in a biometric security system based on brain EEG signal processing. WICT 1, 205–1210 (2011)
6. Abdullah, M.K., Subari, K.S., Loong, J.L.C., Ahmad, N.N.: Analysis of the EEG signal for a practical biometric system. Biomed. Eng. World Acad. Sci. Eng. Technol. 4(8), 944–949 (2010)
7. Kannathal, N., Choo, M.L., Acharya, U.R., Sadasivan, P.K.: Entropies for detection of epilepsy in EEG. Comput. Methods Prog. Biomed. 80(3), 187–194 (2005)
8. Kumar, Y., Dewal, M.L.: Complexity measures for normal and epileptic EEG signals using ApEn, SampEn and SEN. IJCCT 2(7), 6–12 (2011)
9. Acharya, U.R., Molinari, F., Sree, S.V., Chattopadhyay, S., Ng, K.-H., Suri, J.S.: Automated diagnosis of epileptic EEG using entropies. Biomed. Signal Process. Control 7, 401–408 (2012)
10. Srinivasan, V., Eswaran, C., Sriraam, N.: Approximate entropy-based epileptic EEG detection using artificial neural networks. IEEE Trans. Inform. Technol. Biomed. 11(3), 288–295 (2007)
11. Vukkadala, S., Vijayalakshmi, S., Vijayapriya, S.: Automated detection of epileptic EEG using approximate entropy in elman networks. Int. J. Recent Trends Eng. 1(1), 307–312 (2009)
12. Zandi, A.S., Javidan, M., Dumont, G.A., Tafreshi, R.: Automated real-time epileptic seizure detection in scalp EEG recordings using an algorithm based on wavelet packet transform. IEEE Trans. Biomed. Eng. 57(7), 1639–1651 (2010)
13. Porjesz, B., Begleiter, H.: Alcoholism and human electrophysiology. Alcohol Res. Health 27(2), 153–160 (2003)
14. Kendre, S., Janvale, G., Mehrotra, S.: Mental and behavioural disorders related to alcohol and their effects on EEG signals an overview. ProcediaSoc. Behav. Sci. 133(0), 116–121 (2014). International Conference on Trade, Markets and Sustainability (ICTMS-2013)
15. Kahkonen, S., Wilenius, J., Nikulin, V.V., Ollikainen, M., Ilmoniemi, R.J.: Alcohol reduces prefrontal cortical excitability in humans: a combined tms and eeg study. Neuropsychopharmacology 28(4), 747–754 (2003)
16. Chaovalitwongse, W., Prokopyev, O., Pardalos, P.: Electroencephalogram (EEG) time series classification: applications in epilepsy. Ann. OR 148(1), 227–250 (2006)
17. Hope, A., Rosipal, R.: Measuring depth of anesthesia using electroencephalogram entropy rates. Technical report, Department of Theoretical Methods, Slovak Academy of Sciences, Slovak Republic, 2001
18. Litscher, G.: Electroencephalogram-entropy and acupuncture. Anesth. Analg. 102(6), 1745–1751 (2006)
19. Song, Y., Liò, P.: A new approach for epileptic seizure detection: sample entropy based feature extraction and extreme learning machine. Biomed. Sci. Eng. 3, 556–567 (2010)
20. Vollala, S., Gulla, K.: Automatic detection of epilepsy EEG using neural networks. Int. J. Internet Comput. (IJIC) 1(3), 68–72 (2012)
21. Phung, D., Tran, D., Ma, W., Nguyen, P., Pham, T.: Using shannon entropy as EEG signal feature for fast person identification. In: Proceedings of 22nd European Symposim on Artificial Neural Networks, Computational Intelligence and Machine Learning (ESANN), vol. 4, pp. 413–418 (2014)
22. Phung, D., Tran, D., Ma, W., Nguyen, P., Pham, T.: Investigating the impacts of epilepsy on EEG-based person identification systems. In 2014 International Joint Conference on Neural Networks (IJCNN), pp. 3644–3648. IEEE (2014)

23. Hunter, M., Smith, R.L.L., Hyslop, W., Rosso, O.A., Gerlach, R., Rostas, J.A.P., Williams, D.B., Henskens, F.: The Australian EEG database. Clin. EEG Neurosci. **36**(2), 76–81 (2005)
24. Begleiter, H.: EEG database
25. Lotte, F., Congedo, M., Lécuyer, A., Lamarche, F., Arnaldi, B.: A review of classification algorithms for EEG-based BCIs. J. Neural Eng. **4**(2) (2007)
26. Nguyen, P., Tran, D., Huang, X., Sharma, D.: A proposed feature extraction method for EEG-based person identification. In: International Conference on Artificial Intelligence (2012)
27. He, Z.Y., Jin, L.W.: Activity recognition from acceleration data using AR model representation and SVM. In: Machine Learning and Cybernetics, January 2008

Conditional Entropy Approach to Multichannel EEG-Based Person Identification

Dinh Phung, Dat Tran, Wanli Ma and Tien Pham

Abstract Person identification using electroencephalogram (EEG) as biometric has been widely used since it is capable of achieving high identification rate. Since single-channel EEG signal does not provide sufficient information for person identification, multi-channel EEG signals are used to record brain activities distributed over the entire scalp. However extracting brain features from multi-channel EEG signals is still a challenge. In this paper, we propose to use Conditional Entropy (CEN) as a feature extraction method for multi-channel EEG-based person identification. The use of entropy-based method is based on the fact that EEG signal is complex, non-linear, and random in nature. CEN is capable of quantifying how much uncertainty an EEG channel has if the outcome of another EEG channel is known. The mechanism of CEN in correlating pairs of channels would be a solution for feature extraction from multi-channel EEG signals. Our experimental results on EEG signals from 80 persons have shown that CEN provides higher identification rate, yet less number of features than the baseline Autoregressive modelling method.

Keywords EEG · Conditional Entropy · Autoregressive model · Multi-channel EEG · Person identification

1 Introduction

The identification of a person interacting with computers plays an important role for automatic systems in the area of information retrieval, automatic banking, control of access to security areas, buildings, etc. [1]. Person identification is the process of recognising the identity of a given person out of a closed pool of N persons [2]. Biometric recognition, which includes physiological (fingerprint, iris pattern, facial feature, etc.) or behavioral characteristics (signature, speech pattern, etc.) [3], is now a common and reliable way to identify a person.

D. Phung (✉) · D. Tran · W. Ma · T. Pham
Faculty of Education, Science, Technology and Mathematics,
University of Canberra, ACT 2601, Canberra, Australia
e-mail: Dinh.Phung@canberra.edu.au

© Springer International Publishing Switzerland 2015
Á. Herrero et al. (eds.), *International Joint Conference*, Advances in Intelligent
Systems and Computing 369, DOI 10.1007/978-3-319-19713-5_14

Electroencephalogram (EEG) has been emerged as a new biometric for person identification. EEG is a measurement of the brain signals containing information generated by brain activities [4]. EEG signals are captured using multiple electrodes either from inside the brain or certain locations over the scalp [4]. EEG signals carry genetic information of an individual [5], and recent publications show that EEG patterns are probably unique for individuals [5]. Moreover, EEG features are universal as all living and functional persons have recordable EEG signals [6]. Therefore, EEG data can be suitably used for person identification [2, 5, 7, 8].

EEG-based person identification can be utilized as an additional reliable biometric-based person identification method when the physiology-based biometrics do not offer satisfactory recognition rates [3]. It can also be employed in health clinics to either automatically identify patients being brain-diagnosed by EEG systems or identify unknown EEG records.

Single-channel recordings are normally used for typical sleep research as noticeable changes in brainwave activity are detected during the different stages of sleep. However, single-channel EEG cannot provide sufficient information; therefore, multi-channel EEG is recorded and analysed in order to detect brain state changes for other problems since brain activities distributed over the entire scalp [9]. The common feature vectors are constructed from single channel features on their own or multi-channel features concatenated together [10].

In this paper, we propose a Condition Entropy (CEN)-based feature extraction method for multi-channel EEG signals and apply the proposed method to EEG-based person identification. The use of CEN is based on the fact that EEG signal is complex, non-linear, and random in nature [4, 5, 8, 11–15]. Therefore, entropy-based approach to non-linear analysis would be appropriate for EEG signals [8]. In addition, some entropy methods have been utilized in EEG feature extraction for EEG-based applications including epilepsy detection (Sample, Approximate, Spectral entropy [14]), and motor imagery (Approximate [16], Kolmogorov [17], and Spectral entropy [18]). Especially, entropy methods have recently been used for EEG-based person identification and shown promising identification results such as Shannon Entropy (94.9 % on 40 subjects) [19] and Approximate Entropy (70.8 % on 40 subjects) [20]. Moreover CEN computes entropy values based on the correlation between channels, and as a result, the proposed method would be a better solution than the traditional way of feature vector construction from multi-channel EEG signals as referred in [10].

We also use Autoregressvive model (AR) for feature extraction as a baseline method to evaluate the performance of CEN. The reason for choosing AR model as it has been widely used in various applications ranging from identification, prediction, and control of dynamical systems [21]. It has also been a popular feature extraction for EEG-based person identification as in [5, 22, 23].

2 Conditional Entropy

CEN measures the amount of information contributed by one random variable about a second random variable [24]. Consider two random variables $X, Y, (X, Y) \sim p(x, y)$ with $\sum_X = \{x_1, x_2, \ldots, x_m\}$ and $\sum_Y = \{y_1, y_2, \ldots, y_n\}$, which are not necessarily probabilistic independent. The conditional probability mass function, $p(y|x)$, is defined as $p(y|x) = Pr\{Y = y | X = x\}$, $x \in X$, $y \in Y$. In this case $p(x, y) = p(x)p(y|x)$, and $p(x)$, $x \in X$, is a marginal distribution, which is probability of x regardless the occurrence of y. It is given as:

$$p(x) = \sum_{y \in \sum_Y} p(x, y) \tag{1}$$

For the conditional probabilities at $p(x) > 0$ and $x \in X$, we have:

$$p(y|x) = \frac{p(x, y)}{p(x)} \tag{2}$$

Conditional entropy $H(Y|X)$, which is the entropy of Y calculated on the assumption that X has occurred, is defined as:

$$H(Y|X) = - \sum_{x \in \sum_X} \sum_{y \in \sum_Y} p(x, y) ln(p(y|x)) \tag{3}$$

$H(X|Y)$ can be defined in the similar way.

3 Autoregressive Model

The Autoregressive model (AR) predicts the current values of a time series from the previous values of the same series [10] to find a set of model parameters that best describe the signal generation systems [4]. The AR model, with the order p, is defined to be linearly related with respect to a number of its previous samples [4], i.e.

$$x(n) = - \sum_{k=1}^{p} a_k x(n - k) + y(n) \tag{4}$$

where $x(n)$ is the data of the signal at the sampled point n, $a_k, k = 1, 2, \ldots, p$ are the AR coefficients, and $y(n)$ is the noise input.

AR modeling is an alternate for EEG spectral estimation. However, the AR model is only applicable to stationary signals. In contrast, EEG signals are non-stationary

[4], and they are considered stationary only within short intervals [25, 26]. Therefore, EEG signals are normally segmented into short intervals prior to AR modeling process.

4 Datasets

Our experiments are conducted on the Australian EEG (AEEG) and Alcoholism datasets. The AEEG dataset was collected in the John Hunter Hospital, New South Wales, Australia, over a period of 11 years [27]. The recordings were made by using 23 electrodes (23 channels) placed on the scalp of a subject with the sampling rate of 167 Hz for about 20 min. The subset of the data used for our experiments consists of the EEG data of 80 subjects.

Table 1 Datasets descriptions

Dataset	No. of subjects	No. of channels	No. of trials	No. of sessions	Trial length (s)
AEEG	80	23	1	1	$\approx$1200
Alcoholism	122	64	120	1	1

The Alcoholism dataset comes from a study to examine EEG correlates of genetic predisposition to alcoholism [28]. The dataset contains EEG recordings of control and alcoholic subjects. Each subject was exposed to either a single stimulus (S1) or two stimuli (S1 and S2) which were pictures of objects chosen from the 1980 Snodgrass and Vanderwart picture set. When two stimuli were shown, they were presented in either a matched condition where S1 was identical to S2 or a non-matched condition where S1 differed from S2. The 64 electrodes placed on the scalp sampled at 256 Hz for 1 s. The Alcoholism full dataset contains 120 trials for 122 subjects. The summary of these datasets is described in Table 1.

Although the datasets were recorded for different purposes, i.e. for analysing epilepsy or alcohol, they were used in our experiments for this paper as the generic EEG records for person identification only. This means that the impacts of the brain conditions (epilepsy or alcohol) are not considered in this paper.

5 Feature Extraction

We extract EEG features based on CEN and AR methods. For the AEEG dataset, all 23 channels are used, while only 13 of 64 Alcoholism's channels are selected to avoid overfitting as there are only 120 trials for each Alcoholism subject. The channels,

Table 2 Feature vector descriptions

Dataset	No. of features per channel			Total no. of features for all channels			Number of feature vectors
	CEN	AR6	AR14	CEN$((n(n-1)/2))$	AR6	AR14	per subject
AEEG	1 each pair	n/a	14	253	n/a	322	1200
Alcoholism	1 each pair	6	n/a	78	78	n/a	120

FP1, FP2, F3, Fz, F4, C3, Cz, C4, P3, Pz, P4, O1, and O2, are selected based on their locations such as frontal, central, parietal and occipital lobes respectively (see Fig. 1a).

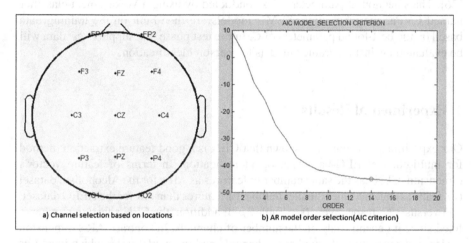

a) Channel selection based on locations   b) AR model order selection(AIC criterion)

Fig. 1 Channel and AR model order selection

EEG signals are then filtered to achieve those of 8–45 Hz using a bypass filter. This helps not only to achieve the signals that cover three EEG's main rhythms, such as alpha (8–13 Hz), beta (14–26 Hz), and gamma (30–45 Hz), but also to keep the baseline noise (50 Hz) off the range. Prior to feature extraction, the AEEG's 1200-s trials are truncated into 1200 1-s samples. After that, feature extractions are conducted using CEN or AR order 14 (for AEEG) or 6 (for Alcoholism). The model orders are automatically selected using Akaike's information criterion (AIC) [29]. As demonstrated in Fig. 1b, the order 14 is sufficient for the AEEG dataset since there is very little change in the AIC beyond that value. Finally, each feature vector is formed by joining CEN values or AR's coefficients from all computed channels together (see Table 2).

6 Classification

The extracted features are used to train Linear SVM classifiers for person identification as described in [25, 26]. An advantage of SVM is the hyperplane selection that maximizes the margins, which is known to increase the generalization capabilities in classification. SVM also uses a regularization parameter C that enables accommodation to outliers and allows errors on the training set [25]. Originally, SVM was designed for binary classification; therefore, it cannot deal with multi-class classification directly [30]. However, binary SVM classifiers can be combined to form multi-class SVM to handle the multi-class case [26].

During the classification, two-third of the datasets are used for training models, and one-third are used for testing. In the training phase, Linear SVM classifiers are trained in 3-fold cross-validation with parameter C ranging from 1 to 1000 in 5 steps [26]. The selection of parameters C is conducted by using a Weka's meta-classifier named CVParameterSelection. SVM's models are then built on the training data based on the best-found parameters of C. In the test phase, the supplied test data will be evaluated against the trained models for person identification.

7 Experimental Results

Our experimental results have shown that CEN is a good feature extraction method for multi-channel EEG-based person identification. In terms of feature vector's dimension, CEN has the same number of features as AR's for the Alcoholism dataset (78 features) but has much smaller number of features than AR for the AEEG dataset, 253 versus 322, respectively (see Table 2). In addition, the CEN's feature number is fixed since it depends only on the number of channels. In contrast, AR's dimension varies as it relies on both number of channels and the model order which increases as data expands.

Table 3 Experimental results

Dataset	CEN		AR	
	3 Fold X-validation (%)	Test (%)	3 Fold X-validation (%)	Test (%)
AEEG	99.0	95.4	97.8	83.9
Alcoholism	72.7	66.9	39.9	36.1

Moreover, CEN's identification rates are significantly better than AR's (see Table 3 and Fig. 2). That is, in the test phase, the CEN's accuracy is 12.1 % higher than AR's for the AEEG dataset (95.4 % versus 83.9 % respectively). Furthermore, CEN achieves an accuracy of 66.9 % for the Alcoholism dataset, while AR gains only 36.1 % which is almost half of CEN's rate. There may be some reasons why the identification rates for the Alcoholism dataset is much lower than those for the AEEG dataset. Firstly, the number of Alcoholism subjects, 122, is higher than that of AEEG, 80. Secondy, the number of trials of the Alcoholism (120 trials) is ten times less than that of the AEEG (about 1200 trials). This especially limits the order of the AR model to meet the requirement of reliability: $N * M \ll L$ [10], where N, M, and L are number of channels, model order, and number of data sample respectively. As a result, it leads to very poor results of AR for the Alcoholism dataset.

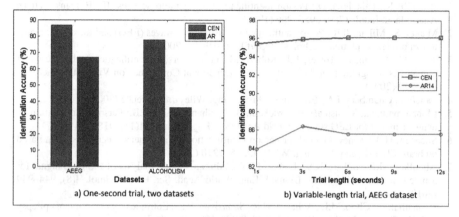

a) One-second trial, two datasets

b) Variable-length trial, AEEG dataset

Fig. 2 Identification accuracy comparson between CEN and AR

Finally, AEEG dataset's variable sample lengths, $N = 3, 6, 9$, and 12 s (overlapping $N - 1$ s), are also tested to find out if the length of trials affects the identification rates. As a result, CEN's identification rate increases as trials are lengthened from 1 to 6 s and remains stably high (at 96.1 %) beyond that length. This could suggest that window sizes bigger or equal to 6 s are likely long enough to maintain significant chaotic levels of the EEG signals for the effective of entropy [13]. On the other hand, AR14's accuracy reaches the peak of 86.4 % then decreases and remains stable at 85.6 % beyond 6 s (see Fig. 2b). This may be explained as AR modelling is a linear method, and EEG signals are considered stationary only within short intervals [4]; therefore, AR gains better results with shorter window sizes.

8 Conclusions

Our proposed method of Conditional Entopy (CEN) has proven to be a good feature extraction method for multi-channel EEG-based person identification. The results have shown that CEN gains better identification accuracy than Autoregressive (AR) model regarding to all datasets and sample sizes. Meanwhile, CEN's channel correlation helps to retain number of features smaller than or equal to those of AR. Our future work will extend to investigate effects of mental conditions, i.e., epilepsy and alcohol, on EEG-based person identification. We will also conduct experiments on other feature extraction methods and on larger datasets to evaluate the use of CEN for EEG-based feature extraction.

References

1. Brunelli, R., Falavigna, D.: Person identification using multiple cues. IEEE Trans. Pattern Anal. Mach. Intell. **17**(10), 955–966 (1995)
2. Marcel, S., Millán, Jl.R.: Person authentication using brainwaves (EEG) and maximum a posteriori model adaptation. Technical Report Idiap-RR-81, 2005
3. Tisse, C., Martin, L., Torres, L.l., Robert, M., et al.: Person identification technique using human iris recognition. In: Proceedings of International Conference on Vision Interface, pp. 294–299 (2002)
4. Sanei, S., Chambers, J.A.: EEG Signal Processing. Wiley-Interscience (2007)
5. Mohammadi, G., Shoushtari, P., Ardekani, B.M., Shamsollahi, M.B.: Person identification by using ar model for EEG signals. World Acad. Sci. Eng. Technol. **1**(11), 918–923 (2007)
6. Shedeed, H.A.: A new method for person identification in a biometric security system based on brain EEG signal processing. WICT **1**, 205–1210 (2011)
7. Abdullah, M.K., Subari, K.S., Loong, J.L.C., Ahmad, N.N.: Analysis of the EEG signal for a practical biometric system. Biomed. Eng. World Acad. Sci. Eng. Technol. **4**(8), 944–949 (2010)
8. Kannathal, N., Choo, M.L., Acharya, U.R., Sadasivan, P.K.: Entropies for detection of epilepsy in EEG. Comput. Methods Prog. Biomed. **80**(3), 187–194 (2005)
9. Anderson, C.W., Stolz, E.A., Shamsunder, S.: Multivariate autoregressive models for classification of spontaneous electroencephalographic signals during mental tasks. IEEE Trans. Biomed. Eng. **45**(3), 277–286 (1998)
10. Hytti, H., Takalo, R., Ihalainen, H.: Tutorial on multivariate autoregressive modelling. J. Clin. Monit. Comput. **20**(2), 101–108 (2006)
11. Chaovalitwongse, W., Prokopyev, O., Pardalos, P.: Electroencephalogram time series classification: applications in epilepsy. Ann. OR **148**(1), 227–250 (2006)
12. Acharya, U.R., Molinari, F., Sree, S.V., Chattopadhyay, S., Ng, K.-H., Suri, J.S.: Automated diagnosis of epileptic EEG using entropies. Biomed. Signal Process. Control **7**, 401–408 (2012)
13. Hope, A., Rosipal, R.: Measuring depth of anesthesia using electroencephalogram entropy rates. Technical report, Slovak Academy of Sciences, Slovak Republic, 2001
14. Kumar, Y., Dewal, M.L.: Complexity measures for normal and epileptic EEG signals using ApEn, SampEn and SEN. IJCCT **2**(7), 6–12 (2011)
15. Litscher, G.: Electroencephalogram-entropy and acupuncture. Anesth. Analg. **102**(6), 1745–1751 (2006)
16. Fang, Y., Chen, M., Zheng, X., Harrison, R.F.: Feature extraction of motor imagery in BCI with approximate entropy. JCIS **8**(6), 2485–2491 (2012)
17. Gao, L., Wang, J., Zhang, H., Xu, J., Zheng, Y.: Feature extraction and classification of event-related EEG based on kolmogorov entropy. In: 4th International Congress on Image and Signal Processing, vol. 5, pp. 2650–2653 (2011)
18. Zhang, A., Yang, B., Huang, L.: Feature extraction of EEG signals using power spectral entropy. In: BMEI 2, pp. 435–439. IEEE Computer Society (2008)
19. Phung, D., Tran, D., Ma, W., Nguyen, P., Pham, T.: Using shannon entropy as EEG signal feature for fast person identification. In: Proceedings of 22nd European Symposim on Artificial Neural Networks, Computational Intelligence and Machine Learning (ESANN), vol. 4, pp. 413–418 (2014)

20. Phung, D., Tran, D., Ma, W., Nguyen, P., Pham, T.: Investigating the impacts of epilepsy on eeg-based person identification systems. In: Proceedings of The annual International Joint Conference on Neural Networks (IJCNN), vol. 6 (2014)
21. Palaniappan, R., Raveendran, P., Nishida, S., Saiwaki, N.: Autoregressive spectral analysis and model order selection criteria for EEG signals. In: Proceedings of TENCON 2000, vol. 2, pp. 126–129 (2000)
22. Paranjape, R.B., Mahovsky, J., Benedicenti, L., Koles', Z.: The electroencephalogram as a biometric. In: 2001 Canadian Conference on Electrical and Computer Engineering, vol. 2, pp. 1363–1366 (2001)
23. Poulos, M., Rangoussi, M., Alex, N., Evangelou, A.: Person identification from the eeg using nonlinear signal classification. In: Methods of Information in Medicine, pp.41–64 (2001)
24. Klan, P.: Entropy handbook - definitions, theorems, m-files
25. Lotte, F., Congedo, M., Lécuyer, A., Lamarche, F., Arnaldi, B.: A review of classification algorithms for EEG-based BCIs. J. Neural Eng. 4(2) (2007)
26. Nguyen, P., Tran, D., Huang, X., Sharma, D.: A proposed feature extraction method for EEG-based person identification. In: International Conference on Artificial Intelligence (2012)
27. Hunter, M., Smith, R.L.L., Hyslop, W., Rosso, O.A., Gerlach, R., Rostas, J.A.P., Williams, D.B., Henskens, F.: The Australian EEG database. Clin EEG Neurosci. 36(2), 76–81 (2005)
28. Begleiter, H.: EEG database
29. Bozdogan, H.: Akaike's information criterion and recent developments in information complexity. J. Math. Psychol. 44(1), 62–91 (2000)
30. He, Z.Y., Jin, L.W.: Activity recognition from acceleration data using AR model representation and SVM. In: Machine Learning and Cybernetics, Jan 2008

On the Selection of Key Features
for Android Malware Characterization

Javier Sedano, Camelia Chira, Silvia González, Álvaro Herrero,
Emilio Corchado and José Ramón Villar

Abstract Undoubtedly, mobile devices (mainly smartphones and tablets up to now)
have become the new paradigm of user-computer interaction. The use of such
gadgets is increasing to unexpected figures and, at the same time, the number of
potential security risks. This paper focuses on the bad-intentioned Android apps, as it
is still the most widely used operating systems for such devices. Accurate detection
of this malware remains an open challenge, mainly due to the ever-changing nature
of malware and the "open" distribution channel of Android apps through Google
Play. Present work uses feature selection for the identification of those features that
may help in characterizing mobile Android-based malware. Maximum Relevance
Minimum Redundancy and genetic algorithms guided by information correlation

J. Sedano (✉) · S. González
Instituto Tecnológico de Castilla Y León, C/López Bravo 70, Pol. Ind. Villalonquejar,
09001 Burgos, Spain
e-mail: javier.sedano@itcl.es

S. González
e-mail: silvia.gonzalez@itcl.es

C. Chira
Department of Computer Science, University of Cluj-Napoca, Baritiu
26-28, Cluj-Napoca 400027, Romania
e-mail: camelia.chira@cs.utcluj.ro

Á. Herrero
Department of Civil Engineering, University of Burgos, Avenida de Cantabria S/N,
09006 Burgos, Spain
e-mail: ahcosio@ubu.es

E. Corchado
Department of Computer Science and Automation, University of Salamanca,
Plaza de La Merced, S/N, 37008 Salamanca, Spain
e-mail: escorchado@usal.es

J.R. Villar
Computer Science Department, ETSIMO, University of Oviedo, 33005 Oviedo, Spain
e-mail: villarjose@uniovi.es

© Springer International Publishing Switzerland 2015   167
Á. Herrero et al. (eds.), *International Joint Conference*, Advances in Intelligent
Systems and Computing 369, DOI 10.1007/978-3-319-19713-5_15

measures have been applied to the Android Malware Genome (Malgenome) dataset, attaining interesting results on the most informative features for the characterization of representative families of existing Android malware.

Keywords Feature selection · Max-Relevance Min-Redundancy criteria · Information correlation coefficient · Android · Malware

1   Introduction

Since the first smartphones came onto the market (late 90s), sales on that sector have increased constantly until present days. Among all the available operating systems, Google's Android actually is the most popular mobile platform, according to [1]. 250.06 million of Android-run units were sold in Q3 2014 worldwide, out of 301.01. Similarly, the number of apps available at Android's official store has increased constantly from the very beginning, up to around 1,541,500 apps [2] available nowadays. Moreover, Android became the top mobile malware platform as well. Recent news [3] confirm this trend as 99 % of the new threats that emerged in Q1 2014 were run on Android. This operating system is an appealing target for bad-intentioned people, reaching unexpected heights, as there are cases where PC malware is now being transfigured as Android malware [3].

To fight against such a problem, it is required to understand the malware and its nature. Otherwise, it will not be possible to practically develop an effective solution [4]. Thus, present study is focused on the characterization of Android malware families, trying to reduce the amount of app features needed to distinguish among all of them. To do so, a real-life publicly-available dataset [5] has been analyzed by means of several feature selection strategies. From the samples contained in such dataset, several alarming statistics were found [4], that motivate further research on Android malware:

- Around one third (36.7 %) of the collected samples leverage root-level exploits to fully compromise the Android security.
- More than 90 % turn the compromised phones into a botnet controlled through network or short messages.
- 45.3 % of the samples have the built-in support of sending out background short messages (to premium-rate numbers) or making phone calls without user awareness.
- 51.1 % of the samples harvested user's information, including user accounts and short messages stored on the phones.

To improve the characterization of the addressed malware, this paper proposes the use of feature selection. To more easily identify the malware family an app belongs to, authors address this feature selection problem using a genetic algorithm guided by information theory measures. Each individual encodes the subset of selected features using the binary representation. The evolutionary search process is guided by crossover and mutation operators specific to the binary encoding and a fitness function that evaluates the quality of the encoded feature subset. In the current study, this fitness function can be the mutual information or the information correlation coefficient.

Feature selection methods are normally used to reduce the number of features considered in a classification task by removing irrelevant or noisy features [6, 7]. Filter methods perform feature selection independently from the learning algorithm while wrapper models embed classifiers in the search model [8, 9]. Filter methods select features based on some measures that determine their relevance to the target class without any correlation to a learning method. The Minimum-Redundancy Maximum-Relevance (MRMR) feature selection framework [8] is a well-known filter method. Besides the maximal relevance criteria, MRMR requires selected features to further be maximally dissimilar to each other (the minimum redundancy criteria). On the other hand, wrapper models integrate learning algorithms in the selection process and determine the relevance of a feature based on the learning accuracy [10]. Population-based randomized heuristics are normally used to guide the search towards the optimal feature subset. Wrapper methods require a high computational time and present a high risk of overfitting [10] but they are able to model feature dependencies and the interaction of the search model with the classifier [11]. Although MRMR was previously applied to the detection of malware [12] and machine learning has also been applied to the detection of android Malware [13, 14], present study differentiates from previous work as feature selection is now applied from a new perspective, trying to ease the characterization of different Android malware families.

The MRMR method is further used in this study to compare or confirm the subsets of selected features related to Android malware. The results obtained for the considered problem are extensively analysed, describing their relevance that probes the positive aspects of gaining deep knowledge of malware nature.

The structure of the paper is as follows: the MRMR method and the proposed GA-based feature selection algorithm are described in Sect. 2, the experiments for the Android Malware Genome dataset are presented in Sect. 3, the results obtained are discussed in Sect. 3.1 and the conclusions of the study are drawn in Sect. 4.

2 Feature Selection Methods

Since the number of features to be analysed in present study is small (see Sect. 3), the various feature subsets can be extensively evaluated using different methods. The result of these methods can then be aggregated in a ranking scheme. It is proposed to determine an ordered list of selected features using (i) Minimum-Redundancy Maximum-Relevance criteria [8] and (ii) a genetic algorithm based on information theory measures as fitness function. The methods described in this section assume a matrix X of N feature values in M samples and an output value y for each sample.

2.1 Minimum-Redundancy Maximum-Relevance Criteria

The Minimum-Redundancy Maximum-Relevance (MRMR) [8] feature selection method aims to obtain maximum relevance to output and in the same time minimum redundancy between the selected features.

Defined by means of their probability distribution, the mutual information between two variables has a higher value for higher degrees of relevance between the two features. Let I(X, Y) be the mutual information between two features, given by:

$$I(X, Y) = \iint p(x, y) * \log\left(\frac{p(x, y)}{p(x) * p(y)}\right) dx dy \tag{1}$$

In the first step, the MRMR approach selects one feature out of the N input features in the set X which has the maximum value of $I(x, y)$. Let this feature be x_k. Next, one of the features in $X - x_k$ is chosen according to the MRMR criteria.

Let us suppose that we have $m - 1$ features selected already in the subset S_{m-1} and the task is to select the mth feature from $X - S_{m-1}$. This will be the feature that maximizes the following formula:

$$\max_{x_j} \in X - S_{m-1}\left[I(x_j, y) - \frac{1}{m-1}\sum_{x_i \in S_{m-1}} I(x_j, x_i)\right] \tag{2}$$

This MRMR scheme can be run for m = 1, 2, 3... resulting in different feature subsets.

2.2 A Genetic Algorithm Using Information Theory Measures for Feature Selection

The proposed Genetic Algorithm (GA) encodes in each individual the feature selection by using a binary representation of features. The size of each individual equals the number of features and the value of each position can be 0 or 1, where 1 means that the corresponding feature is selected (the number of features is N).

It is proposed to evaluate feature selection results using the following two measures from information theory [15] as fitness functions: mutual information(I) and information correlation coefficient (ICC).

Let $H(X, Y)$ denote the joint entropy of the two features, and by $I(X, Y)$ the mutual information between X and Y (see Eq. 1).The information correlation coefficient $ICC(X, Y)$ is calculated based on the Eq. 3 for all the features selected in individual X and the output Y. ICC measures how independent two features are from each other (the higher the ICC value the more relevant the relationship is).

$$ICC(X, Y) = \frac{I(X, Y)}{H(X, Y)} \tag{3}$$

If $ICC(X, Y) = 1$ then the two variables X and Y are strictly dependent whereas a value of 0 indicates that they are completely irrelevant to each other.

The resulting genetic algorithm (called GA-INFO, where INFO can be either I or ICC) is outlined below. The population size is denoted by N, the maximum number of generations is denoted by G and t represents the current generation.

Algorithm: GA-INFO Feature Selection

Require: X the input variables data set
Require: Y the output vector
P ← a vector of N Individual objects
t ← 0
Generate the initial population P(t): randomly initialize the value of each individual
while t <G do
   Evaluate each individual IND in P(t): calculate I(IND, Y) or ICC(IND, Y) value
   P(t +1) ← roulette wheel selection from P(t)
   for all individuals IND in P(t + 1) **do**
     Select mate J from P(t + 1)
     K ←two-point crossover (IND, J)
     if fitness(K) > fitness(IND) **then**
       IND ← K
    end if
     L ← mutation(IND)
     if fitness(L) > fitness(IND) **then**
       IND ← L
    end if
   end for
   t ← t+1
end while
Return Best Individual in P(t)

The GA follows a standard scheme in which roulette wheel selection, two-point crossover and swap mutation are used to guide the search. Each individual is evaluated based on the correlation between the current subset of selected features and the output. This correlation is given by either I or ICC used to evaluate the fitness. Therefore, depending on the fitness function used, two GA variants result:

GA-I is the GA using mutual information as fitness, while GA-ICC denotes the GA based on ICC fitness function.

3 Experiments

As previously mentioned, the Malgenome dataset [4], coming from the Android Malware Genome Project [5] has been analysed in preset study. It was the first large collection of Android malware (1,260 samples) that was split in 49 different malware families. It covered the majority of existing Android malware, collected from their debut in August 2010.

Data related to many different apps were accumulated over more than one year from a variety of Android Markets, and not only Google Play. Additionally, malware apps were thoroughly characterized based on their detailed behavior breakdown, including the installation, activation, and payloads.

Collected malware was split in 49 families, that were obtained by "carefully examining the related security announcements, threat reports, and blog contents from existing mobile antivirus companies and active researchers as exhaustively as possible and diligently requesting malware samples from them or actively crawling from existing official and alternative Android Markets" [4]. The defined families are: ADRD, AnserverBot, Asroot, BaseBridge, BeanBot, BgServ, CoinPirate, Crusewin, DogWars, DroidCoupon, DroidDeluxe, DroidDream, DroidDreamLight, DroidKungFu1, DroidKungFu2, DroidKungFu3, DroidKungFu4, DroidKungFu-Sapp, DoidKungFuUpdate, Endofday, FakeNetflix, FakePlayer, GamblerSMS, Geinimi, GGTracker, GingerMaster, GoldDream, Gone60, GPSSMSSpy, Hippo-SMS, Jifake, jSMSHider, Kmin, Lovetrap, NickyBot, Nickyspy, Pjapps, Plankton, RogueLemon, RogueSPPush, SMSReplicator, SndApps, Spitmo, TapSnake, Walkinwat, YZHC, zHash, Zitmo, and Zsone. Samples of 14 of the malware families were obtained from the official Android market, while samples of 44 of the families came from unofficial markets.

Bad-intentioned apps were then aggregated into 49 malware families [4], and information on those families is considered in present study. Thus, the analysed dataset consists of 49 samples (one for each family) and each sample has 26 different features. The features are divided into six categories; installation (repackaging, update, drive-by download, standalone), activation (BOOT, SMS, NET, CALL, USB, PKG, BATT, SYS, MAIN), privilege escalation (exploit, RATC/zimperlich, ginger break, asroot, encrypted), remote control (NET, SMS), financial charges (phone call, SMS, block SMS), and personal information stealing (SMS, phone number, user account). The values of those features are 0 (that feature is not present in that family) and 1 (the feature is present).

3.1 Results

MRMR, GA-ICC and GA-I algorithms were used for the selection of the best four features to characterize the above described Android malware families. The GA parameter setting used is the following:

- population size: 100.
- number of generations: 100
- number of runs for the algorithm in both cases (GA-ICC and GA-I): 50.

Firstly, the four first features have been selected by MRMR method, sorted by relevance and minimal redundancy. The selected features after the experiments are: Installation - Repackaging, Activation - SMS, Activation - BOOT, and Remote Control - NET.

Secondly, GA-INFO has been also applied to the same dataset for comparison purposes. Figure 1 displays the values obtained by ICC and mutual information (I) when running the GA for each one of the features. The Y axis shows the values of ICC and I respectively.

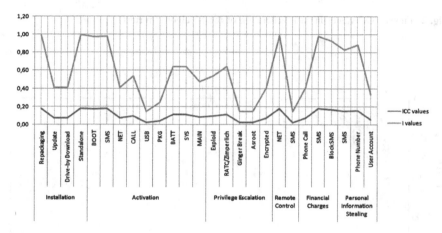

Fig. 1 ICC and I results for each feature

The GA-ICC and GA-I select two different individuals ([Installation - Repackaging, Installation - Standalone, Remote Control - NET, Activation - SMS] and [Installation - Repackaging, Installation –Standalone, Remote Control - NET, Financial Charges- SMS]) respectively, with the same fitness value for ICC and MI (0.18 and 0.99 respectively).

It should be noted that the GA methods were able to reach the optimum values in the population very early in the search process – around generation 11 (see Figs. 2 and 3). Each line represents a run of the algorithm, some lines overlap in some executions that were similar - and that is why 50 lines can not be identified. This is due to the relatively small number of features that had to be considered in the search, leading to an individual size easy to handle and quickly explore many feature subsets.

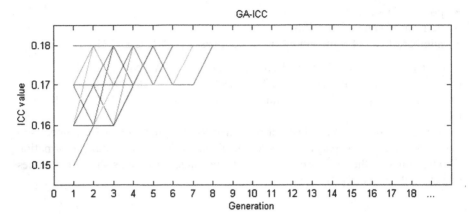

Fig. 2 Fitness ICC values in each generation of the 50 algorithm runs

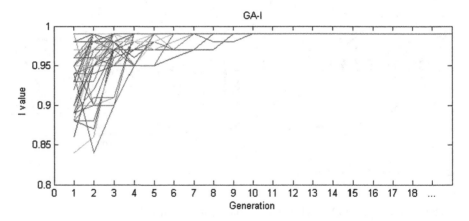

Fig. 3 Fitness I values in each generation of the 50 algorithm runs

The features selected by the three different algorithms are shown in Table 1 and, according to their relevance (being included in the selected subset), they are ordered in Table 2.

Table 1 Selected features by each one of the applied methods

Feature	MRMR	GA-I	GA-ICC
Installation - repackaging	√	√	√
Installation - standalone		√	√
Activation - SMS	√	√	
Activation - BOOT	√		
Remote control - NET	√	√	√
Financial charges - SMS			√

Table 2 Ordered list of selected features

Feature	Relevance
Installation - repackaging	100 %
Remote control - NET	100 %
Installation - standalone	66 %
Activation - SMS	66 %
Activation - BOOT	33 %
Financial charges - SMS	33 %

According to results shown in Tables 1 and 2, features 'Installation – Repackaging' and 'Remote Control – NET' have been selected by the three algorithms. Hence, it can be concluded that those features are the most relevant ones for the characterization of Android malware families.

The repackaging way of installation was defined by the authors of the dataset [4] as "one of the most common techniques malware authors use to piggyback malicious payloads into popular applications (or simply apps). In essence, malware authors may locate and download popular apps, disassemble them, enclose malicious payloads, and then re-assemble and submit the new apps to official and/or alternative Android Markets." Furthermore, from the collected samples, dataset authors found that 1,083 of them (or 86.0 %) were repackaged versions of legitimate applications with malicious payloads.

Regarding the remote control feature, dataset authors stated [4] that 93.0 % of the samples turn the infected phones into bots for remote control. Moreover, 1, 171 of the samples use the HTTP-based web traffic to receive bot commands.

In a second order of importance, 'Installation – Standalone' and 'Activation – SMS', together with 'Activation – BOOT' and 'Financial Charges – SMS' have been identified as key features for characterizing malware families.

4 Conclusions and Future Work

This paper has proposed several methods for selecting those features that best characterize malware families in Malgenome dataset. A genetic algorithm using a binary representation and a fitness function based on information theory measures (mutual information and information correlation coefficient) has been developed for the selection of the optimal subset of features in the considered problem. Experimental results show that the applied methods agree on the selection of 3 of the 4 major features.

Future work will extend these methods to consider other measures as fitness functions in evolutionary search or other population-based search heuristics. A hybridization of such methods and MRMR-based approaches will also be investigated. Additionally, the applicability of these methods to more datasets for the characterization of Android malware families will be further explored.

Acknowledgments This research has been partially supported through the project of the Spanish Ministry of Economy and Competitiveness RTC-2014-3059-4. The authors would also like to thank the BIO/BU09/14 and the Spanish Ministry of Science and Innovation PID 560300-2009-11.

References

1. Statista - The Statistics Portal, http://www.statista.com/statistics/266219/global-smartphone-sales-since-1st-quarter-2009-by-operating-system/
2. AppBrain Stats, http://www.appbrain.com/stats/stats-index
3. F-Secure: Q1 2014 Mobile Threat Report (2015)
4. Yajin, Z., Xuxian, J.: Dissecting android malware: characterization and evolution. In: 2012 IEEE Symposium on Security and Privacy **5**, 95–109 (2012)
5. Malgenome Project, http://www.malgenomeproject.org/
6. Guyon, I., Elisseeff, A.: An introduction to variable and feature selection. J. Mach. Learn. Res. **3**, 1157–1182 (2003)
7. Larrañaga, P., Calvo, B., Santana, R., Bielza, C., Galdiano, J., Inza, I., Lozano, J.A., Armañanzas, R., Santafé, G., Pérez, A.: Machine learning in bioinformatics. Brief. Bioinform **7**(1), 86–112 (2006)
8. Ding, C., Peng, H.: Minimum redundancy feature selection from microarray gene expression data. J. Bioinform. Comput. Biol. **3**(02), 185–205 (2005)
9. Liu, H., Liu, L., Zhang, H.: Ensemble gene selection by grouping for microarray data classification. J. Biomed. Inform. **43**(1), 81–87 (2010)
10. Saeys, Y., Inza, I., Larrañaga, P.: A review of feature selection techniques in bioinformatics. Bioinformatics **23**(19), 2507–2517 (2007)
11. Hatami, N., Chira, C.: Diverse accurate feature selection for microarray cancer diagnosis. Intell. Data Anal. **17**(4), 697–716 (2013)
12. Vinod, P., Laxmi, V., Gaur, M.S., Naval, S., Faruki, P.: MCF: MultiComponent Features for malware analysis. In: 27th International Conference on Advanced Information Networking and Applications Workshops (WAINA), 2013, pp. 1076–1081 (2013)
13. Sanz, B., Santos, I., Laorden, C., Ugarte-Pedrero, X., Bringas, P.G.: On the automatic categorisation of android applications. In: 2012 IEEE Consumer Communications and Networking Conference (CCNC), pp. 149–153 (2012)
14. Sanz, B., Santos, I., Laorden, C., Ugarte-Pedrero, X., Bringas, P., Álvarez, G.: PUMA: Permission Usage to Detect Malware in Android. In: Herrero Á., Snášel V., Abraham A., Zelinka I., Baruque B., Quintián H., Calvo J.L., Sedano J., Corchado E. (eds.) International Joint Conference CISIS'12-ICEUTE´12-SOCO´12 Special Sessions, vol. 189. Springer, Berlin, Heidelberg. pp. 289–298 (2013)
15. Cover, T.M., Thomas, J.A.: Elements of Information Theory. Wiley, New York (1991)

Performance Analysis of User to Root Attack Class Using Correlation Based Feature Selection Model

Shilpa Bahl and Sudhir Kumar Sharma

Abstract Intrusion detection system (IDS) research field has grown tremendously over the past decade. Most of the IDSs employ almost all data features for detection of intrusions. It has been observed that some of the features might not be relevant or did not improve the performance of the system significantly. Objective of this proposed work is to select a minimal subset of most relevant features for designing IDS. A minimal subset of features is chosen from the features commonly selected by correlation based feature selection with six search methods. Further, the performance comparison among seven selected subsets and complete set of features is analyzed. The simulation results show better performance using the proposed subset having only 12 features in comparison to others.

Keywords Correlation based feature selection · Feature selection · Intrusion detection system · Machine learning · User to root attack class

1 Introduction

Intrusion detection system (IDS) is a complement of traditional computer networks protection techniques namely user authentication, data encryption, and firewall. An ID has been recognized as an intense research area in the past decade owing to the rapid increase of sophisticated attacks on networks [1]. The objective of IDS is to detect any anomalous or unusual activity as an attempt of breaking the security policy of computer networks. The designing of IDS can be formulated into any one of the following three distinct approaches in the machine learning (ML) problem domain:

S. Bahl (✉)
KIIT, College of Engineering, Gurgaon, India
e-mail: gerashilpa@gmail.com

S.K. Sharma
School of Engineering and Technology, Ansal University, Gurgaon, India
e-mail: sudhir.sharma@aitgurgaon.org

© Springer International Publishing Switzerland 2015
Á. Herrero et al. (eds.), *International Joint Conference*, Advances in Intelligent Systems and Computing 369, DOI 10.1007/978-3-319-19713-5_16

(1) signature-based misuse detection system; (2) anomaly-based system and (3) clustering-based system. A signature-based system can be treated as a pattern classification problem. In the supervised classification approach, we construct the model by the training data for classifying network traffic data into different attack types during testing phase [2]. An anomaly-based system is a supervised/ semi-supervised approach where we build a model of normal usage of computer networks and flag exceptions to that model [3–5]. This paper addresses classification approach for building effective and efficient IDS.

The KDD Cup'99 dataset is publically available benchmark for evaluation of IDS methods [6]. This dataset has a large number of duplicate examples. In this paper, we used publically available NSL-KDD Cup 99 dataset [7]. This dataset is an improved version of KDD 99 dataset without any duplicate examples. The data set has 41 features. The used training dataset consists of 21 different attacks [8]. These attacks are further categorized into four different types: (1) denial of service attacks (DoS); (2) probing attacks (Probe); (3) remote to local attacks (R2L) and (4) user to root attacks (U2R). The number of examples of U2R and R2L attack classes is very less in the dataset. Many misuse detection algorithms failed to show an acceptable level of detection performance for these two attack categories [9, 10]. There is no ML algorithm that could be trained successfully on KDD dataset to perform acceptable level of misuse detection performance for U2R or R2L attack categories. The testing dataset has generous new attacks with signatures that are not correlated with similar attacks in the training dataset [11].

The effectiveness and accuracy of a ML classification system depends on the quality of input dataset. The quality of input depends upon the representative features of a dataset. The feature selection (FS) is a collection of well known dimensionality reduction techniques in the feature space. An FS technique selects a subset of relevant features and discards other irrelevant and redundant features. Adding irrelevant features to the dataset often confuse ML mechanism [12]. It is now common in the data mining to select relevant features in pre-processing step prior to actual learning phase due to the negative effect of irrelevant features on the ML system. The FS techniques have been extensibly investigated in the field of pattern recognition and ML for decades [13–15]. The investigation and research on FS algorithm is still continued due to improving ML performance, decreasing computational complexity and lowering storage requirement [16–18].

The objective of this paper is to select a minimal subset of the most relevant features of NSL-KDD dataset which will improve the overall accuracy of the IDS. The six subsets of features are selected using correlation based feature selection (CFS) method. A minimal subset of features is proposed to maintain the compactness and improve the performance of IDS. Random tree classifier is used for comparing the performance of seven reduced subset of features with all 41 features. The paper presents simulation results for only U2R attack class. We achieve better performance using only 12 selected features through the proposed approach as compared to others.

The rest of this paper is structured as follows: Sect. 2 describes feature selection techniques and related work. Section 3 describes the experimental setup. Section 4 presents results and discussions. Section 5 presents conclusions.

2 Feature Selection Techniques and Related Work

2.1 Feature Selection Techniques

The FS algorithms fall into two broad categories, the filter method or the wrapper method. The filter method is independent of the induction algorithm. The filter method produces most relevant subset of features based on the characteristics of the dataset. There are two manageable approaches namely sequential forward selection (SFS) and sequential backward elimination (SBE). In SFS, we start with no features and add them one at a time. In SBE, we start with all features and delete features one at a time. A pre-determined classifier is used to evaluate the selected subset of features in the wrapper method. The wrapper method is computationaly more expensive than the filter method [12–14, 20].

2.2 Related Work

The ten standard classification algorithms (Bayes Net, Naïve Bayes, J48, NBTree, Decision Table, JRip, OneR, multilayer perceptron, support vector machine, LBK) are empirically compared for 4 different types of attacks. The true positive rate of U2R class is reported in the range from 0.012 to 0.328. The true positive rate of R2L class is reported in the range from 0.001 to 0.107 [9]. A comparative study has been done using nine classifiers for four different types of attacks. The true positive rate of U2R class is reported in the range from 0.022 to 0.298. The true positive rate of R2L class is reported in the range from 0.001 to 0.096. The best results are also compared with three well known published results including winner of the KDD 99 intrusion detection competition [10]. There are discrepancies in the findings reported in the literature for U2R and R2L attack classes. The detection rate for U2R and R2L attack classes are far below from the acceptance level using standard training and testing dataset. The better results are reported in the literature using custom built datasets from standard training and testing dataset [19].

The relevancy and redundancy of a feature affects the performance of the classifier [20]. A couple of features selection algorithms have been evaluated to decide a quality subset of features for decision tree family of classifiers [21]. Enhanced support vector and decision function for features selection was proposed in [22]. Mukkamala et al. identified 17 and 12 features using Bayesian network and

CART classifier respectively [23]. They proposed a hybrid architecture involving ensemble and base classifiers. The hybrid features subset selection technique has been proposed using decision tree and SVM [24].

3 Experimental Setup

We used a laptop of Intel T2080, 1.73 GHz with 2 GB RAM. In this empirical study, WEKA 3.7.11 data mining tool was used with the heap size of 1048 MB [25]. We used default parameter setting of WEKA in this study.

3.1 Research Methodology

The objective of paper is to select a minimal subset of features to discriminate all the attacks occurring in the training dataset using CFS with six search methods. In each FS algorithm, the selected features are kept in the original training and testing data files and other features are removed permanently. Therefore, we have six sets of training and testing files obtained from six investigated FS algorithm. Some of the important features were commonly selected by all six FS algorithms. In the proposed approach, we picked these commonly selected features and built a new subset containing only 12 features. Random tree classifier is used for demonstrating the performance comparison between seven reduced datasets with all 41 features of dataset.

3.2 NSL-KDD 99 Dataset

In this paper we used NSL-KDD Cup99 dataset [7]. We considered KDDTrain+ 20 % and KDDTest+ full data sets. The total examples in training and testing dataset are 25192 and 22544 respectively. The total normal class examples in training and testing dataset are 12828 and 9712 respectively. The total examples of each attack under four categories in the training dataset are given in Table 1. The total examples of each attack in the testing dataset are presented in Table 2. The number of different attacks in training and testing dataset is 21 and 37 respectively. The KDDTest+ has some novel attacks that are not available in KDDTrain+ dataset. Two additional variants of the same dataset namely 5 classes and 2 classes are also available. Each dataset has same 41 features. These features can be categorized into four groups, namely basic features, content features, traffic features and same host features. The name of each feature is given in columns 2 and 11 of Table 3.

Table 1 Different attacks with number of examples in KDDTrain+

DOS	No.	Probe	No.	R2L	No.	U2R	No.
Neptune	8282	Satan	691	Guess_Password	10	Buffer_overflow	6
Teardrop	188	Nmap	301	Warezmaster	7	Loadmodule	1
Land	2	Portsweep	587	Warezclient	181	rootkit	4
Smurf	529	IPSweep	710	Multihop	2		
Pod	38		2289	ftpwrite	1		
Back	196			Imap	5		
				Spy	1		
				Phf	2		

Table 2 Different attacks with number of examples in KDDTest+

DOS	No.	Probe	No.	R2L	No.	U2R	No.
Neptune	4657	Satan	735	Guess_Password	1231	Buffer_overflow	20
Teardrop	12	Nmap	73	Warezmaster	944	Loadmodule	2
Land	7	Portsweep	157	Sendmail	14	rootkit	13
Smurf	665	IPSweep	141	Multihop	18	Sql Attack	2
Pod	41	Mscan	996	ftpwrite	3	Perl	2
back	359	Saint	319	Imap	1	Ps	15
mailbomb	293			Phf	2	Xterm	13
Processtable	685			Httptunnel	133		
worm	2			Xlock	9		
udpstrom	2			named	17		
Apache2	737			Xsnoop	4		
				Snmpgetattack	178		
				snmpguess	331		

Table 3 Feature subset selection techniques and selected features

Sr. no	Search method	Selected subset of features	#Features selected	Merit of subset	#Subsets formed
1	Best first	[2, 3, 4, 5, 6, 8.10, 12, 23, 25, 29, 30, 35, 36, 37, 38, 40]	17	0.725	680
2	Greedy stepwise	[2, 3, 4, 5, 6, 8, 10, 12, 23, 26, 29, 30, 35, 36, 37, 38, 40]	17	0.725	684
3	Genetic search	[2, 3, 4, 5, 6, 8, 12, 13, 22, 23, 25, 26, 27, 29, 30, 31, 32, 33, 36, 37, 38]	21	0.708	40
4	Scatter search V 1	[2, 3, 4, 5, 6, 8, 11, 12, 14, 23, 25, 29, 30, 35, 36, 37, 38, 40]	18	0.722	17866
5	Exhaustive search	[2, 3, 4, 5, 6, 8, 10, 12, 14, 23, 25, 29. 30, 31, 35, 36, 37, 38, 40]	19	0.721	860
6	Random search	[2, 3, 4, 5, 6, 8, 10, 11, 23, 25, 26, 27, 29, 30, 36, 37, 38]	17	0.726	896

3.3 Correlation Based Feature Selection Techniques

A well known correlation based feature selection (CFS) is used in this study. Features are said to be relevant if they are highly correlated with the predictive class and uncorrelated with each other [26]. The CFS is used to evaluate the merit of a subset of features consisting of k features by (1).

$$Ms = R_{FC} = \frac{Kr_{fc}}{\sqrt{K + k(K-1)r_{ff}}} \tag{1}$$

where
R_{FC} = correlation between the class and the features.
r_{fc} = average value of feature-class correlation.
r_{ff} = average value of features-feature correlation.

The value of correlation coefficient is a measure of merit of the selected subset that is considered as one of the performance metric to identify the best subset selection technique. The degree of correlation should be high between the feature and the class attribute and should be lowest between the features. As the number of features increases, degree of correlation between the class and the features increases. Because the new added features will be less correlated with the already selected features and may have well predominance over a higher correlation with the classes [26].

A search method is used to generate candidate feature subsets. The six search methods (Best First, Greedy-Stepwise, Genetic search, Scatter search, Random Search, Exhaustive search) with SBE approach are employed in this study. They are presented in the second column of Table 3.

3.4 Random Tree Classification Algorithm

In this paper, we employed random tree classifier for evaluating the performance of IDS on reduced subsets and all 41 features. The random tree is a decision tree based classification algorithm. The algorithm constructs a tree that considers a given number of random features at each node. It allows estimation of class probabilities and performs no pruning [12].

4 Results and Discussions

The second, third, fourth, fifth and sixth columns of Table 3 show the used search methods, selected subset of features, the number of selected features, merit of selected subset and the number of subsets evaluated respectively. The details of the selected features are shown in Table 4. The features are given in columns 2 and 11.

Table 4 Complete set of features and feature selection

Label	Feature	\multicolumn{6}{c}{Search methods}	Total
		1	2	3	4	5	6	
1	Duration							0
2	Protocol-type	✓	✓	✓	✓	✓	✓	6
3	Service	✓	✓	✓	✓	✓	✓	6
4	Flag	✓	✓	✓	✓	✓	✓	6
5	src_bytes	✓	✓	✓	✓	✓	✓	6
6	dst_bytes	✓	✓	✓	✓	✓	✓	6
7	Land							0
8	wrong_fragment	✓	✓	✓	✓	✓	✓	6
9	Urgent							0
10	Hot	✓	✓		✓	✓		4
11	num_failed_logins				✓	✓		2
12	logged_in	✓		✓	✓	✓	✓	5
13	num_compromised			✓				1
14	root_shell				✓	✓		2
15	su_attempted							0
16	num_root							0
17	num_file_creation							0
18	num_shells							0
19	num_access_files							0
20	num_outbound_cmds							0
21	is_host_login							0

Label	Feature	\multicolumn{6}{c}{Search methods}	Total
		1	2	3	4	5	6	
22	is_guest_login					✓		1
23	Count	✓	✓	✓	✓	✓	✓	6
24	srv_count							0
25	serror_rate	✓		✓	✓	✓	✓	5
26	srv_serror_rate		✓	✓		✓		3
27	rerror_rate			✓			✓	2
28	srv_rerror_rate							0
29	same_srv_rate	✓	✓	✓	✓	✓	✓	6
30	diff_srv_rate	✓	✓	✓	✓	✓	✓	6
31	srv_diff_host_rate			✓		✓		2
32	dst_host_count					✓		1
33	dst_host_srv_count				✓			1
34	dst_host_same_srv_rate							0
35	dst_host_diff_srv_rate				✓	✓	✓	4
36	dst_host_same_src_port_rate	✓	✓	✓	✓	✓	✓	6
37	dst_host_srv_diff_host_rate	✓	✓	✓	✓	✓	✓	6
38	dst_host_serror_rate	✓	✓	✓	✓	✓	✓	6
39	dst_host_srv_serror_rate							0
40	dst_host_rerror_rate				✓	✓	✓	4
41	dst_host_srv_rerror_rate							0

The labels assigned to the features are given in columns 1 and 10. The search methods are presented in columns 3 to 8 and 12 to 17 in Table 4. We present the total count of each feature i.e. how many times a feature is selected using six different search methods in columns 9 and 18. The random search and exhaustive search methods are more computationaly expensive than the best first and greedy-stepwise search methods.

A proposed set of features is chosen from the six selected subsets with highest count of 6 in columns 9 and 18 as shown in Table 4. This minimal set contains the features labeled as 2, 3, 4, 5, 6, 8, 23, 29, 30, 36, 37, 38 as in Table 4. These 12 features are commonly selected by all six investigated search methods using CFS. This subset does not contain any feature belonging to content category. The 15 features including 7 features of content category are not selected at all by any of the FS search method. This implies that these features are unimportant to detect all attacks presented in the training dataset. The third category of features contains the remaining 14 features.

We built a multiclass – hierarchical classifier model during training phase. The constructed model is tested by testing dataset. We saved the simulation results for all attack types available in the training data set. For brevity, we present, summary results of only U2R attack class for demonstrating the performance comparison on reduced data set and all 41 features. Four additional attacks of U2R class in the KDDTest+ dataset are Sqlattack, Perl, Sqlattack, ps, Xterm as shown in Table 2. These attacks are not detected by classifier. They are not considered in summary results. We present simulation results for all 41 features and for seven selected subset of features in Tables 5 and 6. Table 5 presents true positive rate (TPR), false positive rate (FPR), Recall, Precision, F-score and area under ROC curve (AUC).

On the basis of results in Table 5, the detection rate of load module attack is very poor in all experiments. The overall detection rate for U2R class is significantly decreased by load module attack. Table 6 presents the overall accuracy, root mean squared error (RMSE), time to build a model in the column second, third and eight respectively. Table 6 also presents the average of many standard metrics for three different attacks of U2R class. The average of TPR, FPR, F-Score and AUC are presented in the column fourth, fifth, sixth and seventh respectively in Table 6.

On the basis of Table 6, the performances of IDS never worsen using CFS with 6 search methods. The performance of classifier on the proposed subset is better than other subsets. The false alarm rates in all cases are negligibly small. The proposed subset of features is not only the smallest subset but also contributes in improving the overall accuracy and detection rate of U2R type of attacks. This subset containing only 12 feature is sufficient to distinguish all attacks available in the training data set.

The previous studies [9, 10, 23, 24] have used 5 class dataset. They have reported over all detection rates U2R of attack class. This paper used multi-class dataset and presented details of U2R class of attacks. The overall detection rate of U2R class is significantly decreased by very low detection rate for load module attack. The proposed approach outperforms the six other investigated FS methods.

Table 5 Summary results of random tree algorithm

Search method	Attack type	TPR	FPR	Recall	Precision	F-score	AUC
41 attributes	Buffer_overflow	0.1	0	0.1	1	0.18	0.55
	Loadmodule	0	0	0	0	0	0.5
	Rootkit	0.15	0.002	0.15	0.8	0.11	0.57
Best first	Buffer_overflow	0.38	0	0.38	1	0.631	0.72
	Loadmodule	0.002	0	0.002	0.007	0.003	0.50
	Rootkit	0.164	0.002	0.16	0.069	0.085	0.56
Greedy stepwise	Buffer_overflow	0.33	0	0.33	1	0.52	0.719
	Loadmodule	0.001	0	0.11	0.005	0.002	0.567
	Rootkit	0.15	0.002	0.15	0.59	0.085	0.50
Genetic search	Buffer_overflow	0.22	0	0.2	0.66	0.17	0.55
	Loadmodule	0	0	0	0	0	0.50
	Rootkit	0.35	0	0.35	0.42	0.49	0.73
Scatter search V 1	Buffer_overflow	0.22	0	0.63	0.54	0.39	0.6
	Loadmodule	0	0	0	0	0	0.5
	Rootkit	0.19	0	0.54	0.075	0.33	0.5
Exhaustive search	Buffer_overflow	0.19	0	0.1	1	0.34	0.55
	Loadmodule	0	0	0	0	0	0.50
	Rootkit	0.11	0	0.017	0.33	0.22	0.53
Random search	Buffer_overflow	0.2	0	0.2	1	0.33	0.60
	Loadmodule	0	0	0	0	0	0.50
	Rootkit	0.15	0.01	0.155	0.07	0.01	0.56
Proposed selected (12) attributes	Buffer_overflow	0.26	0	0.26	0.4	0.31	0.61
	Loadmodule	0.001	0	0.001	0.01	0.003	0.51
	Rootkit	0.39	0.002	0.39	0.54	0.48	0.75

Table 6 Average results of standard performance metrics

Search method	Accuracy	RMSE	TPR	FPR	F-score	AUC	Time
41 attributes	82.70	0.1254	0.09	6E-4	0.09	0.54	2.75
Best first	85.91	0.113	0.18	6E-4	0.23	0.59	1.00
Greedy stepwise	85.81	0.114	0.16	6E-4	0.23	0.59	1.26
Genetic search	83.49	0.122	0.19	0	0.20	0.59	1.48
Scatter search V 1	85.31	0.115	0.13	0	0.24	0.53	1.9
Exhaustive search	84.91	0.110	0.11	0	0.18	0.52	1.73
Random search	83.96	0.120	0.11	3E-4	0.11	0.55	1.31
Proposed selected (12) attributes	**85.93**	**0.110**	**0.22**	6E-4	**0.25**	**0.62**	1.33

5 Conclusion

In this paper, we proposed the feature selection for intrusion detection by correlation based feature selection utilizing six search methods. A subset of features is chosen from KDD dataset having 41 features. A proposed subset has only 12 features. Random tree classifier is used for comparing the performance. We demonstrated performance comparisons for seven different subsets of features including proposed subset and all existing 41 features. Empirical results reveal that overall accuracy and detection rate of U2R type of attacks have been improved for the proposed minimal subset as compared to other subsets.

References

1. Van der Geer J, et al.: Intrusion detection system: a review, the art of writing a scientific article. J. Sci. Commun. **163**, 51–9 2000; Managing Cyber Threats: Issues, Approaches, and Challenges, vol. 5. Springer (2006)
2. Han, J., Kamber, M., Pei, J.: Data mining, concepts and techniques. Southeast Asia Edition: (2006)
3. Tavallaee, M., Stakhanova, N., Ghorbani, A.A.: Toward credible evaluation of anomaly-based intrusion-detection methods. IEEE Trans. Syst. Man Cybern., Part C: Appl. Rev. **40**(5), 516–524 (2010)
4. Teodoro, G., Pedro, et al.: Anomaly-based network intrusion detection: techniques, systems and challenges. Comput. Secur. **28**(1), 18–28 (2009)
5. Witten, I.H., Frank, E., Hall, M.A.: Data Mining–Practical Machine Learning Tools and Techniques. Morgan Kaufmann (2011)
6. KDD Cup 1999. http://kdd.ics.uci.edu/databases/kddcup99/kddcup99.html, Accessed Oct 2014
7. Nsl-kdd data set for network-based intrusion detection systems. http://nsl.cs.unb.ca/KDD/NSL-KDD.html, Accessed March 2014
8. Revathi, S., Malathi, A.: A detailed analysis of KDD cup99 dataset for IDS. Int. J. Eng. Res. Technol. (IJERT) **2**(12) (2013)
9. Nguyen, H. Choi, D.: Application of Data Mining to Network Intrusion Detection: Classifier Selection Model. APNOMS 2008, LNCS 5297, pp. 399–408. Springer, Berlin (2008)
10. Sabhnani, M., Serpen, G.: Application of machine learning algorithms to kdd intrusion detection dataset within misuse detection context .In: MLMTA, pp. 209–215 (2003)
11. Sabhnani, M., Serpen, G.: Why machine learning algorithms fail in misuse detection on KDD intrusion detection data set. Intell. Data Anal. **8**(4), 403–415 (2004)
12. Chizi, B., Maimon, O.: Dimension reduction and feature selection. In: Data mining and knowledge discovery handbook, pp. 83–100. Springer, New York (2010)
13. Maaten, V., Laurens, J.P., Postma, E.O., Jaap, H., Herik, V.: Dimensionality reduction: a comparative review. J. Mach. Learn. Res. **10**(1–41), 66–71 (2009)
14. Liu, H., Motoda, H. (eds.): Computational methods of feature selection. CRC Press (2007)
15. Saeys, Y., Inza, I., Larrañaga, P.: A review of feature selection techniques in bioinformatics. Bioinformatics **23**(19), 2507–2517 (2007)
16. Jiliang, T., Alelyani, S., Liu, H.: Feature selection for classification: a review. Data classification: algorithms and applications. In: Aggarwal C. (ed.) CRC Data Mining and Knowledge Discovery Series. CRC Press, Chapman & Hall (2014)

17. Stańczyk, U.: Ranking of characteristic features in combined wrapper approaches to selection. Neural Comput. Appl. 1–16 (2015)
18. Wang, S.; Tang, J.; Liu, H.: Embedded Unsupervised Feature Selection (2015)
19. Engen, V. et al.: Exploring discrepancies in findings obtained with the KDD Cup'99 data set. Intell. Data Anal. **15**(2), 251–276 (2011)
20. Lei, Yu., Liu, H.: Efficient feature selection via analysis of relevance and redundancy. J. Mach. Learn. Res. **5**(2004), 1205–1224 (2004)
21. Piramuthu, S.: Evaluating feature selection methods for learning in data mining applications. Eur. J. Oper. Res. **156**, 483–494 (2004)
22. Zaman, S., Karray, F.: Features selection for intrusion detection systems based on support vector machines. In: Consumer Communications and Networking Conference, CCNC, pp. 1–8 (2009)
23. Peddabachigari, S., Abraham, A., Grosan, C., Thomas, J.: Modeling intrusion detection system using hybrid intelligent systems. J. Netw. Comput. Appl. **30**(1), 114–132 (2007)
24. Kermansaravi, Z., Jazayeriy, H., Fateri, S.: Intrusion detection system in computer networks using decision tree and svm algorithms. J. Adv. Comput. Res. **4**(3), 83–101 (2013)
25. Weka Data Mining Machine Learning Software. http://www.cs.waikato.ac.nz/ml/weka
26. Hall, M.A.: Correlation-based feature selection for machine learning, Thesis (1999)

CIOSOS: Combined Idiomatic-Ontology Based Sentiment Orientation System for Trust Reputation in E-commerce

Hasnae Rahimi and Hanan EL Bakkali

Abstract Due to the abundant amount of Customer's Reviews available in E-commerce platforms, Trust Reputation Systems remain reliable means to determine, circulate and restore the credibility and reputation of reviewers and their provided reviews. In fact before starting the process of Reputation score's calculation, we need to develop an accurate Sentiment orientation System able to extract opinion expressions, analyze them and determine the sentiment orientation of the Review and then classify it into positive, negative and objective. In this paper, we propose a novel semi-supervised approach which is a Combined Idiomatic-Ontology based Sentiment Orientation System (CIOSOS) that realizes a domain-dependent sentiment analysis of reviews. The main contribution of the system is to expand the general opinion lexicon SentiWordNet to a custom-made opinion lexicon (SentiWordNet++) with domain-dependent "opinion indicators" as well as "idiomatic expressions". The system relies also on a semi-supervised learning method that uses the general lexicon WordNet to identify synonyms or antonyms of the expanded terms and get their polarities from SentiWordNet and then store them in SentiWordNet++. The Sentiment polarity and the classification of the review provided by the CIOSOS is used as an input of our Reputation Algorithm proposed in previous papers in order to generate the Reputation score of the reviewer. We also provide an improvement in calculation method used to generate a "granular" reputation score of a feature or subfeature of the product.

Keywords Sentiment orientation · Ontology · SentiWordNet · Idiomatic expressions · Reputation score

H. Rahimi (✉) · H.E. Bakkali
Information Security Research Team (ISeRT), University Mohamed V Souissi,
ENSIAS, Rabat, Morocco
e-mail: hasnae.rahimi@gmail.com

H.E. Bakkali
e-mail: h.elbakkali@um5s.net.ma

© Springer International Publishing Switzerland 2015
Á. Herrero et al. (eds.), *International Joint Conference*, Advances in Intelligent
Systems and Computing 369, DOI 10.1007/978-3-319-19713-5_17

189

1 Introduction

The past decades have witnessed tremendous advances in the World Wide Web, the result of unprecedented trust reputation systems researches and innovative ways of interpreting and analysing sentiment in subjective data.

In fact, with the increasing availability of e-commerce platforms, customers share their experiences of using products and services. As a result, an abundant amount of subjective textual feedback are available on the web in the form of textual reviews, comments, opinions, discussions in forums and blog posts...etc.

Customers consider products' reputation scores references on which they rely to build their own product's reputation. Customers' reviews and products' reputation ratings are subject to dishonesty. As a result, we have to determine and study the trustworthiness of the provided review. However, we have to start by analysing the sentiment expressed in reviews in order to classify them. In opinion mining, sentiment analysis and classification of product reviews are a common problem and varieties of Sentiment Orientation Approaches SOA have been used to address the problem. Some approaches such as [1, 2, 3, 4] are based on lexical resources, others such as [5, 6] are based on the product's features ontology and papers such as [3, 7, 8] combine element of both. Product Ontology and lexicon base classifiers are unable to capture either implicit features of the product or opinion idiomatic expressions although it is common to find these expressions in product's reviews.

To overcome these limitations, we propose a novel semi-supervised approach which is a Combined Idiomatic-Ontology based Sentiment Orientation System (CIOSOS) that realizes a domain-dependent sentiment analysis of reviews. The main contribution of the system is to expand the general opinion lexicon SentiWordNet to a custom-made opinion lexicon (SentiWordNet++) with domain-dependent "opinion indicators" as well as "idiomatic expressions". The system relies also on a semi-supervised learning method that uses the general lexicon WordNet to identify synonyms or antonyms of the expanded terms. The Sentiment polarity and the classification of the review provided by the CIOSOS is used as an input of our Reputation Algorithm proposed in previous papers in order to generate the Reputation score of the reviewer. We also provide an improvement in calculation method used to generate a "granular" reputation score of a feature or subfeature of the product.

To the best of our knowledge, no existing opinion word lexicon or product's features ontology provide sufficient coverage of these "idiomatic expressions" and "indicators". The rest of the paper is organized as follows: Sect. 2 discusses related work. Section 3 describes the proposed approach. Section 4 concludes the paper.

2   Related Work

2.1 Overview of the Trust Reputation System Soliciting the Proposed System

In [9], we developed a Reputation Algorithm for Trust Reputation System which presents a new method to verify the trustworthiness of a user before accepting his rating and feedback in an e-commerce application. We proposed to add a new layer to the TRS which furnishes to each user who has given his opinion, prefabricated feedbacks to like or to dislike.

However in this work, we focus on the sentiment orientation system since the reputation algorithm has already been developed in previous work.

In the literature and in most research specialized in sentiment analysis or text mining systems, sentiment analysis systems or models can be broadly categorized as supervised such as [10], semi-supervised and unsupervised techniques such as [11] according to their way of maintenance.

Sentiment Orientation Approaches (SOA) can be also categorized as aspect-based approaches, lexicon-based approaches and ontology-based SOA according to the characteristics and methodology that they apply to determine the sentiment orientation of the review and classify it. Though, some approaches such as [8] and [3] combine elements of the previous categories.

2.2 Aspect-Based Sentiment Orientation Approach

Aspect-based SOA are systems that focus on the review's textual characteristics or aspects to extract and analyze in order determine the sentiment orientation of a review. Hence these characteristics or features can be found in the text of the review itself. Aspect-based SOA can be considered as the most used approaches in sentiment analysis. Either supervised or unsupervised, these approaches use different aspects of text as sources of features in order to get the opinion words which incarnate. The survey proposed in [10] presents several supervised learning algorithms using bag-of-words features common in text mining research, with best performance obtained using support vector machines SVM in combination with unigrams. Other results show that SVM exhibits also the best performance for sentiment classification.

In [12], they focus on opinion-mining using statistical techniques to determine the opinion of the customer review and analyze it word by word. After that they determine the polarity of the opinion text according to opinion expressed such as positive or negative moods of a reviewer. As a result, they calculate the frequency distribution and Bayesian probability of polarity for the entire product or features of product.

Classifying terms of a review into its grammatical roles, or parts of speech has also been explored in [13] and [14]. This part of speech, words string and root information are used with various combinations for performing classification on various data sets of consumer reviews.

2.3 Lexicon Based Sentiment Orientation Approach

Other researches such as [15] and [2] combine aspect-based and opinion lexicon-based approaches. In [2], authors use a distance based approach extract opinion words and phrases after extracting aspects. To calculate the polarity of each extracted opinion word WordNet was used. This work for instance provides a combination approach of the aspect-based approach and the lexicon-based one.

In fact, this research applied an opinion lexicon to assign a polarity to each opinion word of the review. Besides the aspect-based SOA, lexicon-based sentiment orientation approaches are proposed in many research papers. These techniques use opinion word dictionaries such as SentiWordNet and WordNET to determine opinion terms and phrases, their synonyms and antonyms.

Many papers such as [1, 2, 3, 4] use lexical resources and dictionaries like SentiWordNet and WordNet to build a data set of opinion words to be for instance both used for the machine learning training data and the part of speech tagging process.

Authors of [7] propose a trained classifier of Twitter messages. They propose to extend the feature vector of unigram model by the concepts extracted from DBpedia, the verb groups and the similar adjectives extracted from WordNet, they use SentiWordNet to extract opinion words and some useful domain specific features. They also built a dictionary for emotion icons, abbreviation and slang words in tweets which is useful before extending the tweets with different features.

Authors of [2] propose an aspect-based opinion mining system using an opinion lexicon which is WordNet. To determine the semantic orientation of the sentences a dictionary based technique of the unsupervised approach is adopted. To determine the opinion words and their synonyms and antonyms WordNet is used as a dictionary and proposed to classify the reviews as positive, negative and neutral for each feature. Negation is also handled in their proposed system.

2.4 Ontology-Based Sentiment Orientation Approach

Some other researches such as [5, 6, 8, 16] have been focused on the ontology of the product or the main topic of the review in order to determine the sentiment orientation. The work in [8] proposes a combined domain ontology based SOA where supervised learning technique is used to extract features and opinions from the movie reviews to enhance the existing sentiment classification tasks. The direct opinions on the feature level are identified to get the opinions of each feature.

However, it is hard for products' Ontology base classifiers to capture implicit features and idiomatic expressions or indicators referring to explicit features. However, it is obvious to find these expressions in product's reviews but they are usually not found in general products' features ontology. We propose to overcome this limitation by expanding a general product's ontology with domain-dependent idiomatic expressions and indicators related to the product's features referring to explicit features.

3 Proposed System: CIOSOS :Combined Idiomatic-Ontology Based Sentiment Orientation System

3.1 Overview of the CIOSOS Architecture

Our proposed system is a semi-supervised technique that aims to determine the sentiment orientation of customers' reviews and classify them in different categories according to their sentiment orientation (positive, negative, neutral). This approach is a combination of different elements of main sentiment orientation approaches namely aspect-based SOA, lexicon-based SOA and ontology-based SOA.

Our proposed architecture aims to extract main aspects of the review such as the product's features and their related opinion terms and expressions. For that purpose, we need a PoS tagger that tags these words and expressions. It is very important that the PoS tagging process provides a detailed and granular tagging output in order to have a granular classification and a granular reputation score. The system uses the PoS tagging output in order to extract opinion words and features and/or sub-features that they respectively describe (see Fig. 2). After that, the system verifies the coherence between the feature and the opinion word or phrase used to describe it using a semi-supervised learning method. Basing on the result of the test of coherence, we either consider the review incoherent or get the polarity of opinion words and apply the Reputation Algorithm in order to generate the reputation score of the feature or sub-feature of the product as well as the reputation score of the reviewer. We provide an improvement in calculation method used to generate the "all good granular reputation score" of the feature or sub-feature of the product reviewed. To get the polarity of the extracted opinion words, we're going to use an opinion lexicon SentiWordNet++ which is our extension of the existing Senti-WordNet. WordNet [17] is used to find synonyms or antonyms for the added terms and then get their polarities from SentiWordNet database.

Figure 1 gives the overview of the proposed architecture of the combined idiomatic-Ontology based Sentiment Orientation System (CIOSOS).

We generally classify opinion reviews in 3 categories: positive, negative and objective. The system performs this task in several steps as follows:

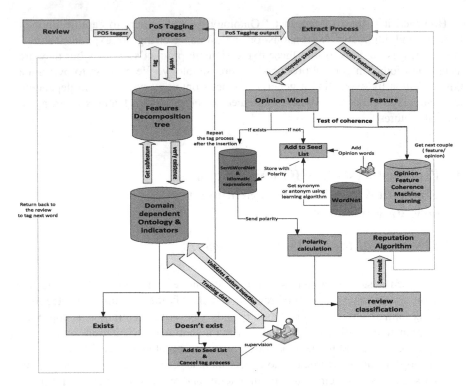

Fig. 1 Architecture combined idiomatic-ontology sentiment orientation system

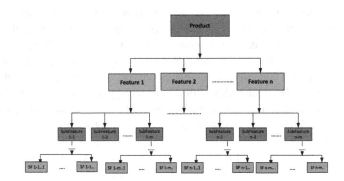

Fig. 2 Features/subfeatures decomposition tree

3.2 Data Collection

To determine the polarity of the sentences, based on aspects and lexicon resources, large numbers of reviews are collected from the e-commerce platform soliciting our Trust Reputation System. "Idiomatic expressions" are often used in the review and evoke implicitly a feature of the product. For instance, the following review contains explicit and implicit expressions and terms referring to features of the product: "The iPhone 6 Plus easily fits in the front pocket of my admittedly not-that-tight Levi's. It's not that much wider than an iPhone 5, and the added height doesn't seem like much of an issue. Plus, the new device is actually a bit thinner, and the rounded edges make it easier to slip into and out of your pocket". Before starting any process in the sentiment orientation approach we have to get the id of the product reviewed so we can use the right tables of the product's ontology database. Moreover, expressions such as "fits in the front pocket", "much wider than", "slip into and out of your pocket" are common expressions that can be written in a review to express satisfaction about the phone dimensions. As a result, we need to match these implicit or idiomatic expressions with the explicit feature "height". As a result, we have to feed the product's ontology database with implicit words and realize a mapping of the already existing explicit features and their "indicators" (implicit expressions).

Furthermore, we have to notice that some features can be divided to sub-features and the review can concern one of the sub-features of the product. For instance, "the flash of the camera of this phone is very bad". The subfeature is the flash and the feature is the camera and the product is a phone. The following architecture describes the decomposition tree of features and their characteristics (subfeatures).

3.3 PoS Tagging

After collecting the reviews, they are sent to the PoS tagging module where a PoS tagger tags all the words of the sentences to their appropriate part of speech tag. PoS tagging is an important phase of opinion mining, it is necessary to determine the features and opinion words from the reviews. Manual PoS tagging of the reviews take lots of time. Here, we will opt for a PoS tagger to tag all the words of reviews. the PoS tagging process needs the feature/sub-feature decomposition tree in order to index each feature with a main feature tag or a sub-feature one. furthermore, features belong to the product's ontology database and they can be either explicit or implicit such as indicators or idiomatic expressions.

```
For each Review(
-> identify features or expectable feature
for each feature (
->get into the feature decomposition tree related to the product ID and verify
-->IF the feature exists (
   if the feature is a main feature then
   Tag with #MFexp or #MFimp   // Main Feature either explicit or implicit
   ELSE IF it is a sub-feature
   tag with #SF_id_exp or #SF_id_imp   //either explicit sub-feature of a main
//feature mentioned with the ID or implicit one
      ELSE // means the feature doesn't exist in the Features Ontology
         //database. It was a text susceptible to be a feature
   --> Add the feature text to a seed list to be verified by a supervisor
   --> cancel the tag process. After the supervisor verifies, the tag process will be
      executed. ) ) )
```

The result of the PoS tagging process is that all of the interesting and useful words such as features, subfeatures, opinion words, expressions of negation, idiomatic expressions either concerning the product's features or opinion words are tagged using a PoS tagger. Then we have to move to the extraction process where we extract features of the reviewed product and the opinion words used to describe them. We start with the characteristics of the feature extraction.

3.4 Feature Extraction and Seed List Preparation

All the features and their sub-features are extracted from the reviews thanks to the PoS tag output. There are explicit and implicit features. Implicit features are idiomatic expressions stored in a different table in the ontology database and all of them matched to the explicit features by a foreign key referencing the primary key of explicit features table. Then we have to compare the extracted features from the review with the already existing ones in the database. The feature extraction is the posterior process succeeding the PoS tagging step. Opinion words are usually written close to features that they describe. The text expectable to be a feature can either exist in the ontology DB or not. If not, the tagging process will end and the feature text will be added to a seed list that a supervisor verifies in order to either add it as a feature in the DB or not (see Fig. 1). When the word is added within a 24 h as a maximum delay, the Tag process will be launched so as to generate the sentiment orientation of the review. Besides the features extraction, there is the opinion words extraction described in the following section.

3.5 Opinion Words Extraction and Seed List Preparation

All opinion words are extracted from the tagged output. Initially, the common opinion words along with their polarity are stored in the SentiWordNet opinion lexicon database. All the added opinion words, indicators or expressions are initially stored in the seed list. In fact, there are 2 types of expanded data: a single word or a composed expression). If the word is not found in the SentiWordNet DB, then the synonyms are determined with the help of the hierarchical structure of WordNet. We use WordNet for extracting the synonyms of verbs, adjectives and nouns, the verb hierarchies in which the verb synsets are arranged, the identical or similar adjectives (synset) and the concepts of nouns which are related by the relation is-a in WordNet. if any synonym matched, then the added opinion word is stored in SentiWordNet++ with its same polarity in the SentiWordNet DB and deleted from the seed list queue. If none of the synonym is matched, then the antonym is determined from the WordNet and the same process is repeated. If any antonym matched, then the added opinion word is stored in the SentiWordNet++ DB with its opposite polarity in the SentiWordNet DB. In this way the Senti-WordNet++ DB keeps on increasing. It grows every time whenever the synonyms or antonyms of the added expressions are found in WordNet. If it is a composed expression the same process is applied to each word and then the average polarity is calculated for the entire expression. If a word is not found, the process is stopped and the supervisor will handle this exception by assigning the right polarity to the specific word. The process will be repeated for the expression.

Subsequently, we use SentiWordNet++ for extracting the polarity of opinion expressions in the review related to each Feature or Subfeature so as to calculate the granular polarity which is the number of positive words divided by the number of negative ones incremented by one) and objectivity (the number of positive and negative words divided by the neutral ones incremented by one) as used in [7]. As a result, this method will generate a detailed and 'granular' Sentiment polarity by feature or Subfeature of the product (SP_F/SubF). Consequently, a single review can engender several sentiment polarities where each sentiment polarity SP is assigned to a specific feature or Subfeature: SP_F/SubF. The review can be classified in different categories (positive, negative or objective) by feature or subfeature of the product.

3.6 CIOSOS and the Trust Reputation System

In [9], we proposed a new architecture for Trust Reputation System TRS in e-commerce Context which includes review mining in order to calculate reputation scores. This architecture is based on an intelligent layer that proposes to each reviewer who has already given his recommendation, a collection of prefabricated reviews to like or to dislike. Then the proposed reputation algorithm calculates the

Reputation score of the reviewer according to his 'likes' and 'dislikes' and according to the trustworthiness of the proposed pre-fabricated reviews. Initially, we afford in a Knowledge database a list of pre-fabricated product's reviews from different categories with their sentiment polarity SP_F/SubF and their global Reputation score which is represented by a decimal in $[-10,10]$ range. In this paper, we propose to generate the Reputation score of each feature/subfeature of the product. We consider the Reputation score of the reviewer (RepScore_Rvwr) generated by the reputation algorithm a coefficient which is going to support the sentiment generated by the review. We add the product of the polarity and the coefficient and we divide the whole by the sum of the coefficients [9].

We then generate the Reputation score of the product By feature or Subfeature as follows:

$$Reputation\ Score\ of\ the\ product\ byfeature/Subfeature = \frac{\sum_{1st\ Rvwr}^{last\ Rvwr} \left(\frac{SP_F}{SubF}\right) * RepScore\ \_Rvwr}{\sum_{1st\ Rvwr}^{last\ Rvwr} RepScore\ \_Rvwr}$$

To generate the Global Reputation score of the product, we calculate the average of the reputation scores of the features and subfeatures of the product.

4 Conclusion and Future Work

In this paper, we propose a novel Combined Idiomatic-Ontology based Sentiment Orientation System (CIOSOS) that realizes a domain-dependent sentiment analysis of reviews. The system attempts to expand SentiWordNet with domain-dependent "opinion indicators" as well as "idiomatic expressions". The system relies on a semi-supervised learning method that uses WordNet to identify synonyms or antonyms of the expanded terms and get their polarities from SentiWordNet. We also propose to expand a general products' features ontology database with domain-dependent idiomatic expressions and indicators. We also provide an improvement in calculation method used to generate a "granular" reputation score of a feature or subfeature of the product.

In future work, we plan to provide experimental results of the CIOSOS functioning to extensively evaluate the effectiveness of our system. Furthermore, efforts would be done to make some enhancements in this technique in order to generate the strength of the sentiment in the review using for instance SentiStrengh database. We also plan to experimentally evaluate the entire Trust Reputation System including the CIOSOS to evaluate the effectiveness of the entire system in an e-commerce context.

References

1. Hamouda, A., Marei, M., Rohaim, M.: Building machine learning based senti-word lexicon for sentiment analysis. In: The Proceedings of the Journal of Advances in Information Technology, vol 2, issue 4, pp. 199–203. Nov 2011
2. Sharma, R., Nigam, S., Jain, R.: Mining of product reviews at aspect level. In: The Proceedings of the International Journal in Foundations of Computer Science & Technology (IJFCST), vol 4, issue 3, May 2014
3. Paramesha, K., Vishankar, V.: Optimization of cross domain sentiment analysis using SentiWordNet. In: The Proceedings of the International Journal in Foundations of Computer Science & Technology, vol 3, issue 5, p. 35. Sept 2013
4. khan, A., Baharudin, B.: Sentiment classification using sentence-level lexical based semantic orientation of online reviews. In: The Proceedings of the International journal of Computer Science Emerging Technology, vol. 2, issue 4, Aug 2011
5. Haider, S.Z.: An ontology based sentiment analysis a case study; Master Degree Project in Informatics
6. Kim, H-J., Song, M.: An ontology-based approach to sentiment classification of mixed opinions in online restaurant reviews. In: The Proceedings of the 5th International Conference, SocInfo 2013, pp. 95–108. Kyoto, Japan, November 25–27 2013
7. Hamdan, H., Béchet, F., Bellot, P.: Experiments with DBpedia, WordNet and SentiWordNet as resources for sentiment analysis in micro-blogging. In: The Proceedings of the Seventh International Workshop on Semantic Evaluation (SemEval 2013), pp. 455–459. Atlanta, Georgia, USA
8. Khin Phyu Phyu Shein: Ontology based combined approach for sentiment classification. In: The Proceedings of the 3rd International Conference on Communications and Information Technology, pp. 112–115
9. Rahimi, H., El Bakkali, H.: A new trust reputation system for E-commerce applications references. In: The Proceedings of the International Journal of Computer Science Issues (IJCSI) (2014)
10. Sharma, R., Nigam, S., Jain, R.: Supervised opinion mining techniques: a survey. In: The Proceedings of the International Journal in Foundations of Computer Science & Technology (IJFCST), vol. 4, issue 3, May 2014
11. Dasgupta, S., Vincent, N.G.: Mine the easy, classify the hard: a semi-supervised approach to automatic sentiment classification. In: The Proceedings of the 47th Annual Meeting of the ACL and the 4th IJCNLP of the AFNLP, pp. 701–709. Suntec, Singapore, 2–7 August 2009
12. Anwer, N., Rashid, A., Hassan, V.: Feature based opinion mining of online free format customer reviews using frequency distribution and bayesian statistics. In: The Proceedings of the Networked Computing and Advanced Information Management (NCM), 2010 Sixth International Conference, pp. 57–62. 16–18 August 2010
13. Kaushik, C., Mishra, A.: A scalable, lexicon based technique for sentiment analysis. In: The Proceedings of the International Journal in Foundations of Computer Science & Technology (IJFCST), Vol. 4, No. 5, September 2014
14. Manning, C.D.: Part-of-speech tagging from 97 % to 100 %: is it time for some linguistics?. In: The Proceedings of the 12th International Conference on Computational linguistics and Intelligent Text Processing, pp. 171–189
15. Hu, M., Liu, B.: Mining and summarizing customer reviews. In: The Proceedings of the Tenth ACM SIGKDD International Conference on Knowledge Discovery and Data Mining, pp. 168–177

16. Polpinij, J., Ghose, A.K.: An ontology-based sentiment classification methodology for online consumer reviews. In: the Proceedings of the 2008 IEEE/WIC/ACM International Conference on Web Intelligence and Intelligent Agent Technology, vol. 01, pp. 518–524
17. Miller, G.A., Beckwith, R., Fellbaum, C., Gross, D., Miller, K.: Introduction to WordNet: an on-line lexical database (Revised August 1993). In: The Proceedings of the International Journal of Lexicography, vol. 3, issue 4, pp. 235–244 (1990)

Neural Analysis of HTTP Traffic for Web Attack Detection

David Atienza, Álvaro Herrero and Emilio Corchado

Abstract Hypertext Transfer Protocol (HTTP) is the cornerstone for information exchanging over the World Wide Web by a huge variety of devices. It means that a massive amount of information travels over such protocol on a daily basis. Thus, it is an appealing target for attackers and the number of web attacks has increased over recent years. To deal with this matter, neural projection architectures are proposed in present work to analyze HTTP traffic and detect attacks over such protocol. By the advanced and intuitive visualization facilities obtained by neural models, the proposed solution allows providing an overview of HTTP traffic as well as identifying anomalous situations, responding to the challenges presented by volume, dynamics and diversity of that traffic. The applied dimensionality reduction based on Neural Networks, enables the most interesting projections of an HTTP traffic dataset to be extracted.

Keywords Intrusion detection · HTTP · Artificial neural networks · Exploratory projection pursuit

D. Atienza (✉) · Á. Herrero
Department of Civil Engineering, University of Burgos Spain C/Francisco de Vitoria s/n, 09006, Burgos, Spain
e-mail: dag0031@alu.ubu.es

Á. Herrero
e-mail: ahcosio@ubu.es

E. Corchado
Departamento de Informática y Automática, Universidad de Salamanca, Plaza de la Merced s/n, 37008 Salamanca, Spain
e-mail: escorchado@usal.es

© Springer International Publishing Switzerland 2015
Á. Herrero et al. (eds.), *International Joint Conference*, Advances in Intelligent Systems and Computing 369, DOI 10.1007/978-3-319-19713-5_18

1 Introduction

An attack or intrusion to a network would end up affecting any of the three computer security principles: availability, integrity and confidentiality, exploiting for example the Denial of Service, Modification and Destruction vulnerabilities [1]. The ever-changing nature of attack technologies and strategies is one of the most harmful issues of attacks and intrusions, increasing the difficulty of protecting computer systems. It means that new ways of attacking information systems and networks are being developed every single day.

Hypertext Transfer Protocol (HTTP) [2] is a stateless application-level protocol for distributed, collaborative, hypertext information systems. It was initially proposed for information exchanging over the World Wide Web. Nowadays, it is not only the usual way of exchanging information associated to web pages because new uses of such protocol are being proposed. HTTP users include household appliances, stereos, scales, firmware update scripts, command-line programs, mobile apps, and communication devices in a multitude of shapes and sizes [2]. On the other hand, common HTTP origin servers include home automation, units, configurable networking components, office machines, autonomous robots, news feeds, traffic cameras, ad selectors, and video-delivery platforms. This means that it is one of the protocols in the application layer of the TCP/IP stack [3] that is more frequently used and will still be in the coming future. Version 1.1 of such protocol was proposed in 1997 [3] and currently remains as the standard version.

The purpose of a web based attack is significantly different than other attacks related to information systems; in most traditional penetration testing exercises a network or host is the target of attack. Web based attacks focus on an application itself and functions on layer 7 of the Open Systems Interconnection [4]. As described for primitive web attacks, all web application attacks are comprised of at least one normal request or a modified request aimed at taking advantage of poor parameter checking or instruction spoofing [4]. Recent studies [5, 6] confirm that web based attacks continue to be on the rise so the automatic detection of such situations still is an open challenge.

Present study proposes a solution characterized by the use of unsupervised connectionist projection techniques providing a novel approach based on the visual analysis of the internal structure of the flow of HTTP data for the detection of web based attacks. Unsupervised learning is quite useful for identifying unknown or not previously faced attacks, known as 0-day attacks, based on the well-known generalization capability of the Artificial Neural Networks (ANNs).

The analysis of HTTP traffic has been approached from several different points of view up to now, but mainly from the machine learning perspective [7, 8]. Additionally, neural projection techniques have been previously applied to many different data, related to computer/network security [9–11]. Differentiating from those previous studies, present work addresses HTTP traffic to check whether the nature of such data allows neural visualization for the detection of attacks.

2   A Neural Approach for Visualization

This work proposes the application of projection models for the visualization of HTTP data. Visualization techniques have been applied to massive security datasets, such as those generated by network traffic [9], SQL code [10] or honeynets [11]. These techniques are considered a viable approach to information seeking, as humans are able to recognize different features and to detect anomalies by means of visual inspection. The underlying operational assumption of the proposed approach is mainly grounded in the ability to render the high-dimensional traffic data in a consistent yet low-dimensional representation. In most cases, security visualization tools have to deal with massive datasets with a high dimensionality, to obtain a low-dimensional space for presentation.

This problem of identifying patterns that exist across dimensional boundaries in high dimensional datasets can be solved by changing the spatial coordinates of data. However, an a priori decision as to which parameters will reveal most patterns requires prior knowledge of unknown patterns.

Projection methods project high-dimensional data points onto a lower dimensional space in order to identify "interesting" directions in terms of any specific index or projection. Having identified the most interesting projections, the data are then projected onto a lower dimensional subspace plotted in two or three dimensions, which makes it possible to examine the structure with the naked eye.

From the information security perspective, visualization techniques were previously proposed as *"visualizations that depict patterns in massive amounts of data, and methods for interacting with those visualizations can help analysts prepare for unforeseen events"* [12].

Due to the aforementioned reasons and based on previous successful applications [9–11], present study approaches the analysis of HTTP data from a visualization standpoint. That is MATLAB [13] implementations of some neural techniques, described in the following subsections, are applied for the analysis of such data. Additionally, Curvilinear Component Analysis (CCA) [14] has been applied to the data but it is not included in present paper for the sake of brevity as it does not visualize the dataset in a way that many different groups can be identified.

2.1 Principal Component Analysis

Principal Component Analysis (PCA) is a well-known statistical model, introduced in [15] and independently in [16], that describes the variation in a set of multivariate data in terms of a set of uncorrelated variables each, of which is a linear combination of the original variables. From a geometrical point of view, this goal mainly consists of a rotation of the axes of the original coordinate system to a new set of orthogonal axes that are ordered in terms of the amount of variance of the original data they account for.

PCA can be performed by means of ANNs or connectionist models such as [17] or [18]. It should be noted that even if we are able to characterize the data with a few variables, it does not follow that an interpretation will ensue.

2.2 Cooperative Maximum Likelihood Hebbian Learning

The Cooperative Maximum Likelihood Hebbian Learning (CMLHL) model [19] extends the Maximum Likelihood Hebbian Learning [20] model, which is based on Exploration Projection Pursuit. The statistical method of EPP was designed for solving the complex problem of identifying structure in high dimensional data by projecting it onto a lower dimensional subspace in which its structure is searched for by eye. To that end, an "index" must be defined to measure the varying degrees of interest associated with each projection. Subsequently, the data is transformed by maximizing the index and the associated interest. From a statistical point of view the most interesting directions are those that are as non-Gaussian as possible.

Considering an N-dimensional input vector (x), and an M-dimensional output vector (y), with W_{ij} being the weight (linking input j to output i), then CMLHL can be expressed as defined in Eqs. 1–4.

1. Feed-forward step:

$$y_i = \sum_{j=1}^{N} W_{ij}x_j, \forall i \tag{1}$$

2. Lateral activation passing:

$$y_i(t+1) = [y_i(t) + \tau(b - Ay)]^+ \tag{2}$$

3. Feedback step:

$$e_j = x_j - \sum_{i=1}^{M} W_{ij}y_i, \forall j \tag{3}$$

4. Weight change:

$$\Delta W_{ij} = \eta \cdot y_i \cdot sign(e_j)|e_j|^{p-1} \tag{4}$$

where: η is the learning rate, τ is the "strength" of the lateral connections, b the bias parameter, p a parameter related to the energy function [19–21] and A a symmetric matrix used to modify the response to the data [19]. The effect of this matrix is based on the relation between the distances separating the output neurons.

2.3 Self-Organizing Maps

The widely-used Self-Organizing Map (SOM) [22] was developed as a visualization tool for representing high dimensional data on a low dimensional display. It is also based on the use of unsupervised learning. However, it is a topology preserving mapping model rather than a projection architecture.

To mimic the biological brain maps, the SOM is composed of a discrete array of L nodes arranged on an N-dimensional lattice. These nodes are mapped into a D-dimensional data space while preserving their ordering. The dimensionality of the lattice (N) is normally smaller than that of the data, in order to perform the dimensionality reduction. The SOM can be viewed as a non-linear extension of PCA, where the global map manifold is a non-linear representation of the training data [23].

Typically, the array of nodes is one or two-dimensional, with all nodes connected to the N inputs by an N-dimensional weight vector. The self-organization process is commonly implemented as an iterative on-line algorithm, although a batch version also exists. An input vector is presented to the network and a winning node, whose weight vector W_C is the closest (in terms of Euclidean distance) to the input, is chosen, according to Eq. 5.

$$c = \arg \min_i (\|\mathbf{x} - W_i\|) \tag{5}$$

When this algorithm is sufficiently iterated, the map self-organizes to produce a topology-preserving mapping of the lattice of weight vectors to the input space based on the statistics of the training data.

3   Experiments and Results

As previously mentioned, many neural visualization models (see Sect. 2) have been applied to HTTP traffic to analyze its nature. Present section introduces the analyzed dataset as well as the main obtained results.

3.1 HTTP Dataset CSIC 2010

To check the validity of the proposed techniques, they have been confronted to a real-life publicly-available dataset, known as HTTP Dataset CSIC 2010 [24].

This dataset was automatically generated by creating traffic to an e-commerce web application. It contains several HTTP requests, labeled as normal or anomalous. Each HTTP request is defined by the following features: method, url, protocol, userAgent, pragma, cacheControl, accept, acceptEncoding, acceptCharset, accept-Language, host, connection, contentLength, contentType, cookie and payload.

The raw data were process, according to the following process:

1. The following features took a single possible value: protocol, userAgent, pragma, cacheControl, accept, acceptEncoding, acceptCharset, acceptLanguage, connection. As those variables do not provide any information to discriminate between normal and anomalous requests, they were deleted.
2. The cookie feature contains a session id for every request. As this kind of information is useless for applied models, it was removed too. Additionally, the URL feature was also removed from the dataset as it contains URLs from the e-commerce application, that cannot be used.
3. Duplicated HTTP requests were removed.
4. Categorical data were converted into numeric values, according to the following details:

   - method, host and contentType features have limited possible values. For example, possible values for method variable are: GET, POST, PUT. In this type of variable, we replaced each possible value with a different number.
   - contentLength is almost a completely numeric variable. Possible values for this variable are a number greater than 0 or null. Null values were replaced by 0.
   - payload variable contains character strings with different lengths and contents, so the possible values are nearly unlimited. Then, the payload value was replaced by its length.

Finally, the analyzed dataset is composed of 1,916 unique HTTP requests and 5 features for each one of them. Table 1 shows an example of an HTTP request on the final dataset.

3.2 Results

PCA Projection
Figure 1 shows the 2-principal component projection, obtained by applying PCA to the previously described data. Blue crosses and red circles represent normal and anomalous requests respectively. As can be seen in this Figure, normal and anomalous requests cannot be clearly differentiated from PCA projection. However, a deeper analysis has been performed, to extract some common characteristics of requests that are depicted in the same group, by knowing its position on this projection. In order to do that, Fig. 1 is divided into 5 areas. Regarding the

Table 1 Sample row from the final dataset

Method	Host	ContentLength	ContentType	Payload
1	0	221	1	12

method used on the HTTP request, there is a clear clustering in the PCA projection:

- Zones 1 and 2 contain requests that only use the GET method.
- Zones 3 and 4 contain requests that only use the POST method.
- Zone 5 contains requests that only use the PUT method.

The second principal component mainly takes into account the value of payload and contentLength. So, yellow and green zones contain the requests with the highest payload and contentLength values.

CMLHL Projection

Figure 2 shows the CMLHL projection of the analyzed data. **As it is shown in Fig. 2a, several groups can be clearly identified in CMLHL projection. An in-depth analysis of those groups reveals the following data:**

- **Zones 1 and 2** contain requests that use the POST and PUT methods.
- **Zone 3** contains requests that only use the GET method.
- **Zones 4 and 5** contain requests that only use the POST method.
- **Zone 6** contains requests that use the POST and GET methods.

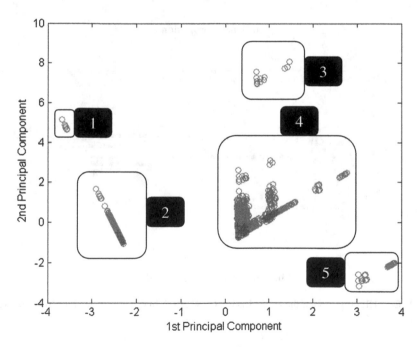

Fig. 1 PCA projection of the HTTP dataset CSIC 2010

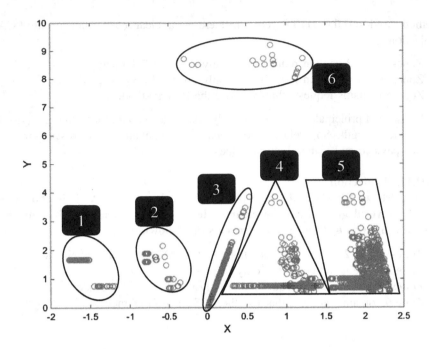

2.a. General projection.

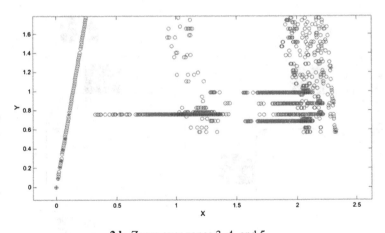

2.b. Zoom over zones 3, 4, and 5.

Fig. 2 CMLHL projection of the HTTP Dataset CSIC 2010. **2a** General projection. **2b** Zoom over zones 3, 4, and 5

As it happened in the case of PCA, CMLHL visualize the dataset in different groups. None of them contains only attack requests, but it can be seen (details on Fig. 2b) that all the attack requests are placed at the bottom-left side of groups 3, 4, and 5.

SOM Analysis

To study the SOM results, a 'specialized neuron' ratio has been calculated as performance criteria. It is defined as the number of neurons are specialized (in normal or anomalous data) and then, do not respond to samples of both normal and anomalous requests, divided by the total number of neurons in the given grid.

For the initial test, several experiments were conducted, with the following values for the SOM parameters:

- Size of the network: 4 × 4, 10 × 10, 20 × 20, and 30 × 30.
- Network topology: Grid and hexagonal.
- Distance criteria: 'dist', 'linkdist', 'mandist', and 'boxdist'.

Each combination of the above values was executed 10 times and the mean of the 'specialized neuron' ratio has been calculated, as shown in Table 2.

With the initial test, it was found that the best size for the SOM is 20 × 20 because their results (highest scores in the 'specialized neuron' ratio) are significantly better than those obtained for the other sizes. Additionally, Euclidean and Manhattan distance criteria provide slightly better results.

After obtaining an initial idea on the best parameter values, a second test was designed with the following parameter values:

- Size of the network: 15 × 15, 16 × 16, 17 × 17, 18 × 18, 19 × 19, 21 × 21, 22 × 22, 23 × 23, 24 × 24, and 25 × 25.
- Network topology: Grid and hexagonal.
- Distance criteria: Euclidean and Manhattan.

Table 2 'Specialized neuron' ratio for the different configurations in initial test

		Distance criteria			
Size	Topology	'dist'	'linkdist'	'mandist'	'boxdist'
4 × 4	Grid	0.64375	0.65000	0.66875	0.61875
4 × 4	Hex	0.64375	0.57500	0.65625	0.63750
10 × 10	Grid	0.66600	0.65800	0.64300	0.63600
10 × 10	Hex	0.65800	0.65800	0.66600	0.66400
20 × 20	Grid	0.72225	0.71725	0.72775	0.67125
20 × 20	Hex	0.73625	0.69850	0.74675	0.72075
30 × 30	Grid	0.66550	0.66250	0.65940	0.62170
30 × 30	Hex	0.65710	0.63760	0.67980	0.65780

As for the first test, each parameter combination is executed 10 times and the mean value is calculated. The performance of this second test is shown in Table 3.

As a result, it can be concluded that grid size close to 20×20 obtains good results. Manhattan distance works slightly better than Euclidean distance and hexagonal topology seems to be the best option. However, none of the experiments obtained a value of 1 in the 'specialized neuron' ratio.

Finally, for the SOM analysis, a representation of the best configuration found (21×21 size, hexagonal topology and Manhattan distance) is shown in Fig. 3. Each neuron is colored according to the type of HTTP requests it responds to:

- Gray: the neuron responds to no requests.
- Red: the neuron responds only to anomalous requests.
- Blue: the neuron responds only to normal requests.
- Orange: the neuron responds to both normal and anomalous requests.

Additionally, figures inside each neuron indicates the number of normal / anomalous HTTP requests the neuron responds to. As can be seen in Fig. 3, none of the neurons responds only to normal requests (blue color), and then normal request are mixed with anomalous ones (orange color).

Table 3 'Specialized neuron' ratio for the different configurations in second test

| | | Distance criteria | |
Size	Topology	*'dist'*	*'mandist'*
15×15	Grid	0.6951	0.7036
15×15	Hex	0.7120	0.7351
16×16	Grid	0.6973	0.6945
16×16	Hex	0.7105	0.7273
17×17	Grid	0.7118	0.7073
17×17	Hex	0.7142	0.7239
18×18	Grid	0.7188	0.7194
18×18	Hex	0.7204	0.7420
19×19	Grid	0.7249	0.7202
19×19	Hex	0.7288	0.7357
21×21	Grid	0.7195	0.7168
21×21	Hex	0.7231	0.7485
22×22	Grid	0.7207	0.7157
22×22	Hex	0.7283	0.7390
23×23	Grid	0.7130	0.7089
23×23	Hex	0.7161	0.7456
24×24	Grid	0.7122	0.7038
24×24	Hex	0.7097	0.7420
25×25	Grid	0.7034	0.7021
25×25	Hex	0.7014	0.7362

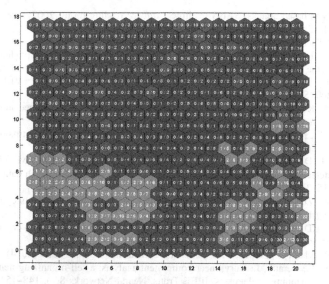

Fig. 3 SOM representation for the best configuration

4 Conclusions and Future Work

In present paper, several neural models have been proposed for the analysis of HTTP traffic, trying to detect web based attacks. As described in previous section, none of the applied models have been able to clearly differentiate normal from anomalous traffic. There are some kinds of data that can be certainly identified as normal or anomalous (see some groups in Sect. 3).

One of the main reasons for the low performance of neural visualization applied to the HTTP Dataset CSIC 2010 is that, the preprocessing process implied removing certain information that may support differentiating the different kinds of traffic. Thus, future work will be based on different ways of collecting and converting HTTP data for improving their neural visualization. More precisely, the original features of the dataset will be processed in a way that more information can be provided to the neural models.

References

1. Myerson, J.M.: Identifying enterprise network vulnerabilities. Int. J. Network Manage **12**(3), 135–144 (2002)
2. Fielding, R., Reschke, J.: Hypertext transfer protocol (HTTP/1.1): message syntax and routing. IETF RFC **7230** (2014)
3. Fielding, R., Gettys, J., Mogul, J., Frystyk, H., Berners-Lee, T.: Hypertext transfer protocol – HTTP/1.1. IETF RFC **2068** (1997)

 4. Crist, J.: Web based attacks. SANS institute - infosec reading room (2007)
 5. Ponemon Institute - Cost of Cyber Crime Study (2014)
 6. Kaspersky Security Bulletin 2014 (2014)
 7. Pastrana, S., Torrano-Gimenez, C., Nguyen, H., Orfila, A.: Anomalous web payload detection: evaluating the resilience of 1-grams based classifiers. In: Camacho, D., Braubach, L., Venticinque, S., Badica, C. (eds.) Intelligent Distributed Computing VIII, vol. 570, pp. 195–200. Springer International Publishing (2015)
 8. Choraś, M., Kozik, R.: Machine learning techniques applied to detect cyber attacks on web applications. Logic J. IGPL 23(1), 45–56 (2014)
 9. Corchado, E., Herrero, Á.: Neural visualization of network traffic data for intrusion detection. Appl. Soft. Comput. 11(2), 2042–2056 (2011)
10. Pinzón, C.I., De Paz, J.F., Herrero, Á., Corchado, E., Bajo, J., Corchado, J.M.: idMAS-SQL: intrusion detection based on MAS to detect and block SQL injection through data mining. Inf. Sci. 231, 15–31 (2013)
11. Herrero, Á., Zurutuza, U., Corchado, E.: A neural-visualization IDS for honeynet data. Int. J. Neural Syst. 22(2), 1–18 (2012)
12. D'Amico, A.D., Goodall, J.R., Tesone, D.R., Kopylec, J.K.: Visual discovery in computer network defense. IEEE Comput. Graphics Appl. 27(5), 20–27 (2007)
13. The MathWorks, Inc., Natick, Massachusetts, United States.: MATLAB (2014)
14. Demartines, P., Herault, J.: Curvilinear component analysis: a self-organizing neural network for nonlinear mapping of data sets. IEEE Trans. Neural Networks 8(1), 148–154 (1997)
15. Pearson, K.: On lines and planes of closest fit to systems of points in space. Phil. Mag. 2(6), 559–572 (1901)
16. Hotelling, H.: Analysis of a complex of statistical variables into principal components. J. Educ. Psychol. 24, 417–444 (1933)
17. Oja, E.: Principal components, minor components, and linear neural networks. Neural Networks 5(6), 927–935 (1992)
18. Fyfe, C.: A neural network for PCA and beyond. Neural Process. Lett. 6(1–2), 33–41 (1997)
19. Corchado, E., Fyfe, C.: Connectionist techniques for the identification and suppression of interfering underlying factors. Int. J. Pattern Recognit Artif Intell. 17(8), 1447–1466 (2003)
20. Corchado, E., MacDonald, D., Fyfe, C.: Maximum and minimum likelihood hebbian learning for exploratory projection pursuit. Data Min. Knowl. Disc. 8(3), 203–225 (2004)
21. Fyfe, C., Corchado, E.: Maximum likelihood hebbian rules. In: 10th European Symposium on Artificial Neural Networks (ESANN 2002), pp. 143–148 (2002)
22. Kohonen, T.: The self-organizing map. Proc. IEEE 78(9), 1464–1480 (1990)
23. Ritter, H., Martinetz, T., Schulten, K.: Neural Computation and Self-Organizing Maps; An Introduction. Addison-Wesley Longman Publishing Co., Inc., Chicago (1992)
24. HTTP DATASET CSIC 2010: http://www.isi.csic.es/dataset/

Analysis and Automation of Handwritten Word Level Script Recognition

M. Ravikumar, S. Manjunath and D.S. Guru

Abstract In this paper, a problem of automatic selection of feature sets and suitable classifiers for recognition of handwritten scripts at word level is addressed. The problem is brought out clearly with sufficient study by comparing state of the art techniques. Based on the analysis, three different models have been proposed in this paper. To accomplish the task, combination of various features and classifiers are tried out. The proposed work is on bi-script recognition. To conduct experimentation we have considered different features and classifiers which are recommended in the literature. The proposed work has been demonstrated for its effectiveness on our own dataset of reasonably large size. Experimental results reveal that, the proposed models perform better than the existing models.

Keywords Handwritten script identification · Word level · Feature combination · Classifier selection

1 Introduction

In country like India most of the handwritten documents are written in multiple scripts. Developing an OCR to such a multilingual document is a challenging issue as it has to recognize the scripts prior to recognize the characters. Recognizing

M. Ravikumar (✉) · D.S. Guru
Department of Studies in Computer Science, University of Mysore,
Mysore, Karnataka, India
e-mail: ravi2142@yahoo.co.in

D.S. Guru
e-mail: dsg@compsci.uni-mysore.ac.in

S. Manjunath
Department of Computer Science, Central University of Kerala,
Kasaragod, Kerala, India
e-mail: manju_uom@yahoo.co.in

© Springer International Publishing Switzerland 2015   213
Á. Herrero et al. (eds.), *International Joint Conference*, Advances in Intelligent
Systems and Computing 369, DOI 10.1007/978-3-319-19713-5_19

scripts can be done at three different levels: document level, line level and word level. Specific to Indic scripts, it is recommended to use word level script recognition as the words are written using different scripts rather than using same scripts for all words in a line or all words in a document.

Recognizing a script at word level in a document poses a challenging task, due to two reasons. The words written in the same script vary as they differ in visual appearance as well as it has inherent complexity of writing style of the writer. Obtaining features at word level which can discriminate a particular script is a tedious task. Therefore, choosing best feature extraction method(s) plays an important role. Once the discriminating features are extracted the next challenging issue is to select the appropriate classifier. For recognition problems, selection of classifiers is based on the performance of that particular classifier. Especially in case of script recognition in a multilingual document a single classifier alone may not work better for recognition of multiple scripts. In this case either selection of classifiers or fusion of classifiers has to be chosen [1]. The performance of classifier also depends on the features used, which indicates that features and classifiers are highly coupled. In this work we made an attempt to study on qualitative comparison of multiple classifiers to recognize scripts in a handwritten multilingual document using different feature extraction techniques. Also, we propose three different models to recommend automatically the features and classifiers to identify the scripts in a multilingual document.

The paper is organized as follows. In Sect. 2, a brief review on handwritten script recognition is presented. The proposed model for selection of features and classifiers for handwritten script identification is presented in Sect. 3. Experimental results are reported in Sect. 4 on our own data set and the obtained results are analyzed. In Sect. 5, the paper is concluded.

2 Related Work

Hochberg et al. [2], have proposed a method for script and language identification system for handwritten documents, in which, five connected component features such as relative Y centroid, relative X centroid, number of holes, sphericity and aspect ratio are extracted. Later, mean, standard deviation and skew of all the features are calculated and used LDA classifier for script identification.

Roy and Pal [3] and Roy et al. [4], have developed word level script identification for Indian postal automation using fractal based features, water reservoir based features, presence of small component and topological features. A neural network was recommended for classification purpose on Oriya and Roman scripts.

A word-level handwritten script identification technique has been proposed in [5]. A combination of shape and texture based features are used to identify the script of the handwritten word images written in Bangla, Devnagari, Malayalam, Telugu and Roman. Singh et al. [6] have made an attempt to study and compare different classifiers for script identification using different statistical test. They recommended to use, 39 distinctive features based on topological and convex hull for Devanari and Roman script and achieved a maximum of 99.87 % of accuracy for MLP among eight other classifiers.

Hangarge in his work [7] has proposed a Gaussian mixture model for word level script identification in which standard deviations of directional energy distributions are computed. Hangarge et al. [8] and Hangarge and Santhosh [9] have exploited directional discrete cosine transform (DDCT). LDA and KNN classifiers are used for classification purpose. Pardeshi et al. [10] have demonstrated the problem using eleven scripts with three classifiers viz., SVM, LDA and KNN with four feature extraction techniques through Radon Transform, Discrete Wavelet Transform (DWT), Discrete Cosine Transform (DCT) and Statistical Filters. They have observed that different combination works better to identify the scripts. It can be observed that the reported works are concentrated on directional features to tackle the problem. A detailed review on script identification from multi script environment is given in [1, 11]. A detailed review on various feature extraction methods used for handwritten script identification is explained in [12].

Table 1 gives an overview of the existing models for handwritten word level Indic script recognition. From Table 1 it can be understood that the researchers have recommended different features and classifiers with respect to particular script and have achieved good recognition accuracy. Unfortunately majority of the proposed algorithms are with fixed features or fixed classifiers for all the scripts and the results are demonstrated over a standard data set or their own created data set. In the literature we did not find any work on automation of selection of features and classifiers to recognize the scripts in multi lingual documents. We mean to say that given a document written using k scripts $(k > 1)$ which features and what classifiers should be selected are not addressed so far in the literature.

On the other hand, all the works reported in the literature assume that the texts in multilingual documents are almost horizontal. However, there are some applications such as forwarded notes, observation on a handwritten document, where we need to recognize the script of words which are skewed. Recognizing script in a skewed multilingual document remains a challenging and unanswered issue. Hence there is a great need for developing an algorithm which selects features and classifiers containing words with or without skew.

Table 1 A comparative analysis of different handwritten word level Indic script identification methods

References	Methodology		Types of Indic scripts classified	Best identification accuracy (%)
	Feature set	Classifier		
Roy et al. [3]	Topological and reservoir features	MLP	Bangla English	99.75
Roy and Pal [4]		Neural Network	Oriya, Roman	99.60
Sarkar et al. [5]	Topological and directional features	MLP	Bangla, Devanagri, English	99.29
Singh et al. [6]	Topological and convex hull based features		Devanagri, English	99.54
Hangarge [7]	Directional energy of words	GMM	Devanagri, Kannada, Telugu	98.7
Hangarge et al. [8]	DDCT	LDA	Devanagri, English, Kannada, Malayalam, Tamil, Telugu	96.95
		KNN		95.28
Hangrage and Santhosh [9]	2D DCT based features	LDA		99.70
Pardeshi et al. [10]	Radon transform, DWT, DCT, statistical filters	LDA	Bangla, Devanagari, Gujrati, Gurumukhi, Kannada, Malayalam, Oriya, Roman, Tamil, Telugu, and Urdu	97.46
		SVM		97.72
		KNN		92.60

3   Proposed Model

Let $W = \{w_1, w_2, w_3, \ldots, w_k\}$ be a set of words of k scripts, where k being the number of scripts in a multi lingual document and w_i be the number of words in the ith script, where $1 \leq i \leq k$. Let $F = \{f_1, f_2, f_3, \ldots, f_l\}$ be a set of feature extraction techniques preferred to be used. Let $C = \{c^1, c^2, c^3, \ldots, c^m\}$ be the set of m classifiers intended to be used in the development of script identification process.

Given W we apply feature extraction techniques and obtain individual feature set f for all k scripts. For each feature set f_j train all m classifiers in the set C and obtain the accuracy of script identification. It has to be noted that given W, f_j and C we get different accuracy. Select the classifier say C_j^k which has best accuracy to identify the scripts in W using feature set f_j. In order to estimate accuracy we need to train and test the classifier C^k and hence we recommend to, split the entire set of words of w_i into train and validation set. The training set will be used to train the classifiers and validation set will be used for testing part.

Similarly we carry out the experimentation for all l number of features with m number of classifiers and obtain a list of selected classifier $\{C_1^{m_1}(1..k), C_2^{m_2}(1..k),$

$C_3^{m_3}(1..k), \ldots, C_l^{m_p}(1..k)\}$ yielding highest accuracy for all l features. Then the classifier selected for the purpose of script identification in a multilingual document is given in Eq. (1). The same is shown in Fig. 1.

$$C_{f_j}^m = \max\{C_1^{m_1}(1..k), C_2^{k_2}(1..k), C_3^{m_3}(1..k), \ldots, C_m^{m_p}(1..k)\} \qquad (1)$$

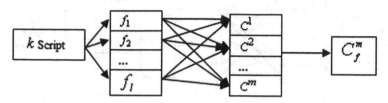

Fig. 1 Feature and classifier selection for script identification

This selection can be termed as feature based classifier selection as the selection of the classifier is based on individual feature set f_j.

As we have l number of features with m number of classifiers we can think of using fusion of features in order to recognize the script in an efficient way.

In case of feature level fusion, the feature extracted from l feature extraction techniques are combined using simple feature concatenation rule i.e., $F^+ = \{f_1, f_2, f_3, \ldots, f_l\}$ and fed to individual classifier and the classifier with highest recognition accuracy is selected as a method for script recognition. Let C_{F+}^m be the classifier selected for F^+, combined set of features as given in Eq. (2) and the same is diagrammatically shown in Fig. 2.

$$C_{F+}^m = \max\{C_{F+}^{m_1}, C_{F+}^{m_2}, C_{F+}^{m_3}, \ldots, C_{F+}^m\} \qquad (2)$$

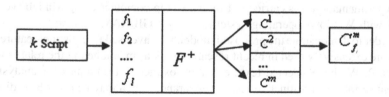

Fig. 2 Feature level fusion and classifier selection for script identification

In the third model we recommend to use combination of features and classifier based on classifier level fusion, where, the decisions of different classifiers are combined using majority voting technique. Let D be the decision of all m classifiers for F set of features including F^+ as given in Eq. (3).

$$D = \left\{ D_F^1(1..k), D_F^2(1..k), D_F^3(1..k), \ldots, D_F^m(1..k) \right\} \tag{3}$$

Then the script is recognized as one which has been selected by the majority of the classifiers. The same is shown in Fig. 3.

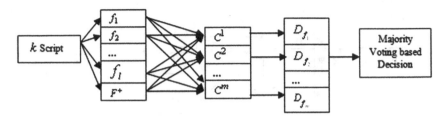

Fig. 3 Decision level fusion for script identification

4 Experimentation

For the purpose of experimentation, we created our own large dataset of six scripts i.e., Devanagari (D),English (E), Kannada (K), Malayalam (M), Tamil (Ta) and Telugu (Te) consisting of six thousand words (1000 words for each script) as there is no standard publically available dataset pertaining to handwritten words. In order to study the suitability of the proposed model we have created two data sets, one without skew i.e., the words are almost horizontal and another with skew where each word is skewed with ± 150. Experimentations are carried out on different Bi-script level i.e., document written using only two scripts. In order to conduct experimentation we split the data set into two parts training and testing. Two types of split are done with 50 and 60 % for training and remaining part is used for testing. The given training part is further divided into training set and validation set. The experimentation was carried out on a desktop computer using Matlab (version 2013), with Windows operating system having 2 GB RAM capacity.

In order to demonstrate the proposed model, we have used four distinct features set which are extensively used in recent research work as mentioned in the related work of Sect. 2. We have selected texture feature extraction techniques for analysis as different scripts have distinctive visual appearance which may be regards as distinct texture pattern. The feature set and their brief description are mentioned below.

4.1 Wavelet Decomposition (WD)

We performed single level wavelet decomposition on matrices of each channels of original gray scale input using Haar wavelet. It results into four coefficients matrix for each channel. They approximation matrix coefficients (cA) and three detail coefficient matrices called horizontal (cH), vertical (cV) and diagonal (cD) are obtained. To analyze the function of wavelet, we compute the reconstructed coefficients matrix for each of the details coefficients matrices of each channel. Then we extract features by calculating energy for each obtained reconstructed coefficients matrix. Since we obtained three reconstructed coefficients matrix for each channel, the number of features is 9.

4.2 Gabor Filter Response (GFR)

Gabor filters are band-pass filters which have both orientation- selective and frequency-selective properties and have optimal joint resolution in both spatial and frequency domains. Frequency and orientation representations of the Gabor filter are similar to those of the human visual system. Therefore, we exploited Gabor filter response to represent and discriminate texture of six classes. Gabor filter is a linear filter whose impulse response is defined by a harmonic function multiplied by a Gaussian function. Because of the multiplication convolution property, the Fourier transform of a Gabor filter's impulse response is the convolution of the Fourier of the harmonic function and the Fourier transform of the Gaussian function and it is given by Eq. (4)

$$g(x, y, \lambda, \theta, \psi, \sigma, \gamma) = \exp\left[\frac{-x'^2 + \gamma^2 y'^2}{2\sigma^2}\right] \cos\left[2\pi\frac{x'}{\lambda} + \psi\right] \qquad (4)$$

where, $x^1 = x \cos\theta + y \sin\theta$ and $y^1 = x \sin\theta + y \cos\theta$ and, λ represents the wavelength of the cosine factor, θ represents the orientation of the normal to the parallel stripes of a Gabor function, Ψ is the phase offset, σ (Sigma) is the Gaussian envelope and γ is the spatial aspect ratio specifying the ellipticity of the support of the Gabor function [13]. In this work we have used orientation θ (0°, 45°, 90°,135°) and σ (2, 4, 6, 8, 10) scale and hence we obtain 20 Gabor response and for each response we compute the energy. Hence, we obtain a vector of 20 features.

4.3 Local Binary Pattern Variance (LBPV)

Usually the high frequency texture regions will have higher variances and they contribute more to the discrimination of texture images. In LBPV, the variance of the local region is embedded into the LBP [14] histogram resulting in a LBPV histogram. Hence, the LBP histogram is obtained by adding the variance as a weight while calculation [15]. Although LBPV has the same feature dimensions as LBP, the use of LBPV adds additional contrast measures to the pattern histogram. LPRV produces 32 features for a given single image.

4.4 Directional Discrete Cosine Transform (DDCT)

In this method, first, 2D-DCT of a word image is computed as mentioned in [7]. Let the 2D-DCT coefficient matrix of a word image of size $N \times N$ be δ. Then, principal diagonal, upper $N - 2$ and lower $N - 2$ diagonals of δ is extracted before and after flipping δ, and further computed their standard deviations respectively. Along with this, features based on conventional DCT are also extracted. In this case, its coefficient matrix is divided into four zones and standard deviation of each zone is computed and feature vector of four features is obtained.

Before experimentation, the obtained features are reduced to lower dimension (10, empirically selected) using MCFS dimensionality reduction technique [16]. Dimensionality reduction technique is optional operation on the feature set. Once, the features are extracted then it has to be fed to classifier for learning the scripts. The classifiers that are used in this work are nearest neighbor (NN) and linear discriminant analysis (LDA) with NN. A brief description about NN and LDA is presented in the following sections.

4.5 Nearest Neighbor Classifier (NN)

NN classifier assign the label of any script class to the given known script based on the labels of its nearest neighbor. NN classifier is basically a function $fn(\partial, \ell, \Im, dist)$ where ∂ is the training set, ℓ is the label set, $dist$ is a distance measure and $\Im$ is the test script. Given a training set $\partial = (x_i, y_i)$ containing n training documents of a script NN classifier determines the distance between the test script to every training script and chose closest script to a test script. Query script is accepted as ith script if the label of the nearest script is 'i'.

4.6 Linear Discriminant Analysis (LDA)

LDA is a linear classifier and in a typical two class problem, it finds a suitable projection so that samples from different classes are well separated and samples within a particular class are minimally scattered. Then LDA determines a projection that maximizes the criteria function $J(V)$ or project the data points on line in the direction V which maximizes $J(V) = |\mu_1 - \mu_2| / - 0pt\sigma_1^2 + \sigma_2^2$, where μ_1 and μ_2 are the means σ_1^2 and σ_2^2 are the variance of two classes respectively. Given unknown sample is projected and compared with the training data using NN classifier. In case of NN classifier, Euclidean distance measure is used. In LDA we have empirically selected

12 dimensions for projection which has given best results for our own dataset. It should be noted that, we are not recommending using either of any above mentioned feature extraction techniques or classifiers or dimensionality techniques to use for script identification. Any possible techniques can be used on the proposed model.

4.7 Result Analysis

Using the above mentioned features and classifiers and using all the three proposed models we have obtained the results. The best results are tabulated in Tables 2 and 3. Tables 2 and 3 depict results for script recognition for words without skew and with skew respectively. The upper triangular region of the table gives accuracy of recognition, suitable feature and classifier for script recognition in case of 50 % training and 50 % testing and lower triangular region provides same information for 60 % training and 40 % testing of the data.

Supporting to our motivation, Tables 2 and 3 indicate that, common feature set and classifier for all bi-script recognition problems is not fare. The same may be true even in case of multiple script identification also. From upper triangular region of Table 2, we have three combinations {F$^+$+DL, WD + LDA, LBPV + LDA} which are selected automatically for recognition of bi-scripts. The probability of selection of F$^+$+DL is 0.60, WD + LDA is 0.33 and LBPV + LDA is 0.07 indicate that feature level fusion along with decision level fusion (F$^+$+DL) is better in most of the cases amongst all features and classifiers used in this work. From lower triangular region of Table 2, the three combinations selected are also same as upper triangular but with slight variation in the probability values. The probability of selection of F$^+$+DL is 0.40, WD + LDA is 0.27 and LBPV + LDA is 0.33 indicating once again (F$^+$+DL) is better in case of word level bi-script recognition without skew.

In case of Table 3, experimental results produce 4 combinations from upper triangular values {WD + LDA, F$^+$+DL, DDCT + LDA, LBPV + LDA} with probability of selection 0.47, 0.26, 0.20 and 0.07 respectively, giving top rank to WD + LDA. Under lower triangular values we have same 4 combinations with 0.13, 0.27, 0.27 and 0.33 giving top rank to LBPV + LDA. This creates confusion in selection of features and classifier in case of word level bi-script recognition with skew. Also, we expected Tables 2 and 3 to have a symmetric in terms of feature and classifier selection with change in number of training and testing samples. But it is observed that, there are few entries in the table (shaded entries) indicating that change in number of samples also have effect on selection of features and classifiers which poses the importance of selecting training samples for identification problems.

Table 2 Script identification accuracies for without skew, upper triangle indicates 50/50 and lower triangle indicates 60/40 % of training and testing

	E	D	K	M	Ta	Te
E	–	93.90 F^++DL	82.30 F^++DL	68.90 F^++DL	72.20 F^++DL	76.50 F^++DL
D	92.50 F^++DL	–	90.50 WD + LDA	96.00 WD + LDA	88.80 WD + LDA	95.70 F^++DL
K	83.00 F^++DL	88.25 WD + LDA	–	69.80 F^++DL	62.40 WD + LDA	76.90 F^++DL
M	72.50 F^++DL	96.00 WD + LDA	70.37 LBPV + LDA	–	67.50 LBPV + LDA	79.60 WD + LDA
Ta	71.62 LBPV + LDA	88.37 WD + LDA	58.12 WD + LDA	66.50 LBPV + LDA	–	74.60 F^++DL
Te	83.12 LBPV + LDA	95.62 F^++DL	80.50 LBPV + LDA	80.50 LBPV + LDA	72.50 F^++DL	–

Table 3 Script identification accuracies for with skew, upper triangle indicates 50/50 and lower triangle indicates 60/40 percentage of training and testing

	E	D	K	M	Ta	Te
E	–	89.80 WD + LDA	76.80 DDCT + LDA	66.10 WD + LDA	71.90 F$^+$+DL	62.90 F$^+$+DL
D	87.87 F$^+$+DL	–	74.50 WD + LDA	84.40 WD + LDA	86.30 WD + LDA	86.70 WD + LDA
K	75.20 LBPV + LDA	74.25 WD + LDA	–	76.90 DDCT + LDA	75.40 DDCT + LDA	70.20 LBPV + LDA
M	64.30 WD + LDA	89.25 LBPV + LDA/DL	76.30 LBPV + LDA	–	74.60 F$^+$+DL	68.20 F$^+$+DL
Ta	73.00 F$^+$+DL	86.87 F$^+$+DL	79.00 DDCT + LDA	80.50 DDCT + LDA	–	72.10 WD + LDA
Te	66.25 DDCT + LDA	87.62 F$^+$+DL	71.12 LBPV + LDA	69.37 LBPV + LDA	84.12 DDCT + LDA	–

5 Conclusion

In this paper, models for automatic selection of features and classifiers for recognizing handwritten script at word level are proposed. The models proposed are evaluated on feature extraction and classification techniques recommended for the purpose of handwritten script recognition. It is observed that in case of word level bi-script recognition, fusion of features and decision level fusion of classifier is the good choice. The proposed model can be used to solve a practical problem of automatic handwritten script recognition.

References

1. Singh, P.K., Sarkar, R., Nasipuri, M.: Offline script identification from multilingual Indic-script documents: a state-of-the art. Comput. Sci. Rev. (2014)
2. Hochberg, J., Bowers, K., Cannon, M., Kelly, P.: Script and language identification for handwritten document images. Int. J. Doc. Anal. Recogn. 2, 45–52 (1999)
3. Roy, K., Pal, U., Chaudri, B.B.: Neural network based word wise handwritten script identification system for Indian postal automation. In: Proceedings of International Conference on Intelligent Sensing and Information Processing, Chennai, pp. 581–586 (2005)
4. Roy, R., Pal, U.: Word-wise Handwritten script separation for Indian postal automation. In: Proceedings of IWFHR, La Baule, France (2006)
5. Singh, P.K., Mondal, A., Bhowmik, S., Sarkar, R., Nasipuri, M.: Word-level script identification from handwritten multi-script documents. In: Proceedings of the 3rd International Conference on Frontiers of Intelligent Computing: Theory and Applications (FICTA) (2014)
6. Singh, P.K., Sarkar, R., Das, N., Basu, S., Nasipuri, M.: Statistical comparison of classifiers for script identification from multi-script handwritten documents. Int. J. Pattern Recogn. 1(2), 152–171 (2014)
7. Hangarge, M.: Gaussian mixture model for handwritten script identification. In: Proceedings of International Conference on Emerging Trends in Electrical, Communication and Information Technologies, pp. 64–69 (2012)
8. Hangarge, M., Santosh, K.C., Pardeshi, R.: Directional discrete cosine transform for handwritten script identification. Int. Conf. Doc. Anal. Recogn. 1, 1–5 (2013)
9. Hangarge, M., Santhosh, K.C.: Word level handwritten script identification from multi-script documents. Recent Adv. Inf. Technol. 1–6 (2014)
10. Pardeshi, R., Chaudhuri, B.B., Hangarge, M., Santosh, K.C.: Automatic handwritten Indian scripts identification. In: 14th International Conference on Frontiers in Handwriting Recognition, pp. 375–380 (2014)
11. Ghosh, D., Dube, T., Shivaprasad, A.: Script recognition: a review. IEEE Trans. Pattern Anal. Mach. Intell. 32, 2142–2161 (2009)
12. Dalal, S., Malik, L.: A survey for feature extraction methods in handwritten script identification. Int. J. Simul. Syst. Sci. Technol. 10, 1–7 (2009)
13. Ma, H., Doermann D.: Word level script identification for scanned document images. In: SPIE Conference on Document Recognition and Retrieval, pp. 124–135 (2004)

14. Guo, Z., Zhang, L., Zhang, D.: Rotation invariant texture classification using LBP variance (LBPV) with global matching. Pattern Recogn. **43**, 706–719 (2010)
15. Ojala, T., Pietikainen, M., Maenpaa, T.: Multiresolution gray-scale and rotation invariant texture classification with local binary patterns. IEEE Trans. Pattern Anal. Mach. Intell. **24**(7), 971–987 (2002)
16. Cai, D., Zhang, C., He, X.: Unsupervised feature selection for multi-cluster data. In: Proceedings of 16th ACM SIGKDD International Conference on Knowledge Discovery and Data Mining, pp. 333–342 (2010)

Patterns Extraction Method for Anomaly Detection in HTTP Traffic

Rafał Kozik , Michał Choraś, Rafał Renk and Witold Hołubowicz

Abstract In this paper the new pattern extraction method for HTTP traffic anomaly detection is proposed. The method is based on innovative combination of (i) text segmentation technique—used to identify some common parts (tokens) of requests and (ii) statistical analysis—that captures the dynamic properties (variables) of data between tokens. In result, such approach allows to capture the structure of the message body received from the consecutive requests. Our experiments show that this technique allows for significant improvement of effectiveness when compared to other techniques that treat the message body as the whole. Another advantage is the fact that our tool does not need any prior knowledge about protocols and APIs that use HTTP as a transportation mean (e.g. RESTFull API, SOAP, etc.).

Keywords Anomaly detection · Pattern extraction · Application layer attacks · Web application security

1 Introduction

Hypertext Transfer Protocol (HTTP) is one of the most frequently used protocols of application layer in TCP/IP model. This is because of the fact that nowadays, the significant part of ICT solutions rely on web servers or web services and HTTP protocol is the reliable mean to enable communication between computers in distributed networks.

R. Kozik (✉) · M. Choraś · W. Hołubowicz
Institute of Telecommunications and Computer Science,
UTP University of Science and Technology, Bydgoszcz, Poland
e-mail: rafal.kozik@utp.edu.pl

M. Choraś
e-mail: michal.choras@utp.edu.pl

R. Renk · W. Hołubowicz
Adam Mickiewicz University, UAM, Poznan, Poland
e-mail: renk@amu.edu.pl

© Springer International Publishing Switzerland 2015   227
Á. Herrero et al. (eds.), *International Joint Conference*, Advances in Intelligent
Systems and Computing 369, DOI 10.1007/978-3-319-19713-5_20

Moreover, the number and importance of web sites, web services and web applications is constantly growing. It can be noticed that more and more sophisticated applications provided over the HTTP(S) protocol are currently developed to address wide range of users needs including entertainment, business, social activities, etc.

The growing popularity of publicly available web services is also a driving force for so called "web hacking" activities. According to Symantec [1] report the number of web attacks blocked per day has increased by 23 % in comparison to previous years.

Another factor, drawing the attention of hackers is the fact that HTTP is the most popular protocol allowed by Internet firewalls and since operating systems are constantly better and better protected, it is easier to focus on web-based applications which potentially have more security exploits.

The paper is structured as follows: in Sect. 2 we present general overview of the related methods and our major contributions. In Sect. 3 we demonstrate our solutions and the proposed method. The experimental setup and results are described in Sect. 4. Conclusions are given thereafter.

2 Related Work

On the market, there are plenty of solutions that aim at countering attacks targeting the application layer. Many of those solutions are signature-based. The Signature-based category of cyber attacks detection methods include Intrusion Prevention and Detection Systems (IDS and IPS) which use predefined set of signatures (often in form of regular expressions) in order to identify an attack. Commonly IPS and IDS are designed to increase the security level of computer networks through detection (in case of IDS) and detection and blocking (in case of IPS) of the network attacks.

There are also solutions called WAF (Web Application Firewall [2–4]). Commonly those solutions use white and black listing of requests that are sent from client to server. Some of them apply also signatures (regular expressions, patterns, etc.) in order to detect an attack. The patterns (or rules) are typically matched against content of requests (e.g. header or payload).

One of the most popular IDS/IPS software package, widely deployed worldwide, is Snort [5]. Since it is an open source project, its users are allowed to freely modify it as well as feed the Snort engine with rules obtained from different sources (e.g. not only form Snort homepage).

Commonly the signatures (in form of reactive rules) of an attack for the software like Snort are provided by experts form the cyber community. Typically, for deterministic attacks, it is fairly easy to develop patterns that will clearly identify particular attack. However, the problem of developing new signatures becomes more complicated when it comes to attacks patterns that can be obfuscated (e.g. SQL or XSS injection).

Another drawback of many WAF solutions is the fact that they often make some assumptions about the requests structure (e.g. [2, 3]). Different protocols that utilise HTTP as transportation exhibit different structures of a the payload. For example, the structure sent via plain HTML form will be different from GWT-RPC or SOAP call. In such cases some signatures will not match payload of different structure.

Some of the frequently used tools use static code analysis approaches in order to find the vulnerabilities that may be exploited by any cyber attack. Some examples of such tools include PhpMiner II [6], STRANGER [7], AMNESIA [8]. However, as it is stated in [9], the difficulty relates to the fact that many kinds of security vulnerabilities are hard to be found automatically (e.g. access control issues, authentication problems). Therefore, currently such tools are able to automatically find only relatively small fraction of application security flaws. Moreover, such solutions require the source code of server-side application, which can not be easily obtained for third party libraries or dedicated services.

In our previous work [10–12], we have introduced an innovative evolutionary algorithm for modelling genuine SQL queries generated by web-application in order to detect SQL injection attacks. In this paper, we focus on wider spectrum of attacks that can be targeted at the application layer. The list of top most critical risks related to web applications security, provided by OWASP (Open Web Application Security Project [13]) indicates "Injection" as the major vulnerability. The Injection flaws, such as SQL or XSS occur when improperly validated data containing malicious code is sent to an interpreter as the part of a command or query. The XSS flaws occur whenever an application takes untrusted data and sends it to a web browser without proper validation or escaping. The XSS allows attackers to execute scripts in the victim's browser which can hijack user sessions, deface web sites, or redirect the user to malicious sites.

Two main contributions of this paper are:

- a new and innovative approach to extract tokens (distinctive patterns) from consecutive requests,
- an effective and efficient statistical method for encoding data in between the tokens.

Our experiments show that this technique allows for significant improvement of effectiveness when compared to other techniques that treat the message body as the whole.

3 The Proposed Method Overview

The HTTP is a request-response plain-text protocol. Basically, the request is a type of method that can be called (executed) on a resource uniquely identified by URL address. There are different types of methods that (among others) allow to perform CRUD (Create, Read, Update, Delete) operations on a resource. Some of the methods are accompanied with request payload (e.g. POST, PUT). Different protocols that utilise HTTP for transportation exhibit different structures of the payload. For example, the structure sent via plain HTML form will be different from GWT-RPC or SOAP call.

In this approach, we take advantage of the request-response nature of the HTTP protocol and we use simple pre-classification approach shown in Fig. 1. Thus, instead of the whole packet analysis, we focus on the request payload in order to extract structure from the consecutive calls.

We represent the structure of the payload by means of tokens. The token of HTTP request is defined as the sequence of bytes that are common for all the requests sent to the same resource (phase-1 in Fig. 2). There could be several tokens identified for one request. Tokens allow to identify delimiters of these regions of the requests sequences that are likely to be related to data provided by client sending that request. Hence, this allows to identify possible points where malicious code can be injected. Once the tokens are identified, we describe the sequences between tokens using their statistical properties (phase-1 in Fig. 2).

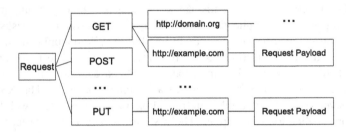

Fig. 1 HTTP requests pre-classification

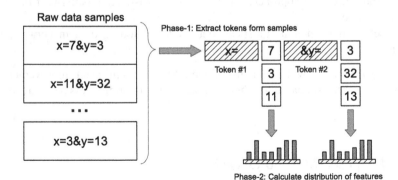

Fig. 2 Overview of the proposed algorithm

3.1 Request Tokenization

In order find a collection of tokens we have adapted LZW compression method (Lempel-Ziv-Welch [14, 15]). First, we calculate the LZW dictionary D, that transforms textual input to a set of natural numbers (see Eq. (1)).

$$D : word \rightarrow \{i : i \in N\} \tag{1}$$

The algorithm scans through the input set S for successively longer subsequence until it finds one that is not in the dictionary. If a given substring in not in the dictionary, then it is added and the whole procedure is repeated until the whole data set is processed. The detailed procedure is provided by Algorithm 1.

Data: Set of HTTP payloads S
Result: Dictionary D
s = empty string
while *there is still data to be read in S* **do**
  $ch \leftarrow$ read a character
  if $(s + ch) \in D$ **then**
    $s \leftarrow s+ch$;
  else
    $D \leftarrow D \cup (s + ch)$;
    $s \leftarrow ch$;
  end
end

Algorithm 1: Algorithm for establishing dictionary D.

The first advantage of this approach is that within a single scan of the data set, it is possible to gather manageable amount of candidates for tokens. The second advantage is that the dictionary can be used to compress the data in order to better utilise the server resources.

However, further processing of the dictionary is required in order to obtain the collection of tokens. First, we remove all the candidates that do not appear in all the samples that we used for structure extraction. In future, we plan to relax this constraint and use the majority rule instead. Afterwards, we need to remove those sequences that are sub-sequences of other ones.

Sometimes it may happen that there are not any tokens delimiting different subsequences of the payload. In that case we go directly to procedure described in Sect. 3.2.

3.2 Data Encoding

The data contained between tokens is described statistically using the method that follows the idea proposed by C. Kruegel in [16] (we use the same metric to measure dissimilarities between HTTP requests but we apply it to the different feature). Author used a character distribution model to describe the genuine traffic generated to web application. In that approach the Idealized Character Distribution (ICD) is obtained during the training phase from normal requests sent to web application. The IDC is calculated as the mean value of all character distributions. During the detection phase, the probability that the character distribution of a sequence is an actual sample drawn from its ICD is evaluated. For that purpose Chi-Square metric is used.

The equation used for computing the value of Chi-Square metric $D_{chisq}(Q)$ for a sequence Q is described by Eq. 2, where N indicates the length of a sequence Q (in our case 9), ICD the distribution established for all the samples, σ the standard deviation form the ICD, and $h()$ the distribution of the sequence that is being tested Q.

$$O_{chisq}(Q) = \sum_{n=0}^{N} \frac{1}{\sigma^2}[ICD_n - h(Q_n)]^2 \qquad (2)$$

In our case, we have used different features to calculate the distributions. Particularly, instead of the separate characters we have counted number of characters for which the decimal value in ASCII table belongs to the following ranges: $< 0, 31 >, < 32, 47 >, < 48, 57 >, < 58, 64 >, < 65, 90 >, < 91, 96 >, < 97, 122 >, < 123, 127 >< 128, 255 >$. It may look heuristically, but different ranges represent different type of symbols like numbers, quotes, letters or special characters. In result our histogram will have 9 bins. In the latter results section, we have demonstrated that this approach gives better results than typical 256 bin histograms.

4 Experiments and Results

For the experiments the CSIC'10 dataset [17] was used. It contains several thousands of HTTP protocol requests which are organised in form similar to the Apache Access Log. The dataset was developed at the Information Security Institute of CSIC (Spanish Research National Council) and it contains the generated traffic targeted to an e-Commerce web application. For convenience the data was split into anomalous, training, and normal sets. There are over 36000 normal and 25000 anomalous requests. The anomalous requests refer to a wide range of application layer attacks, such as: SQL injection, buffer overflow, information gathering, files disclosure, CRLF injection, XSS, server side include, and parameter tampering.

Moreover, the requests targeting hidden (or unavailable) resources are also considered as anomalies. Some examples classified to this group of anomalies include client requests for: configuration files, default files or session ID in URL (symptoms of http session take over attempt). What is more the requests, which parameters do not have appropriate format (e.g. telephone number composed of letters) are also considered anomalous. As authors of the dataset explained, such requests may not have a malicious intention but they do not follow the normal behaviour of the web application.

According to authors knowledge, there is no other publicly available dataset for web attack detection problem. The datasets like DARPA or KDD'99 are outdated and do not include many of the actual attacks.

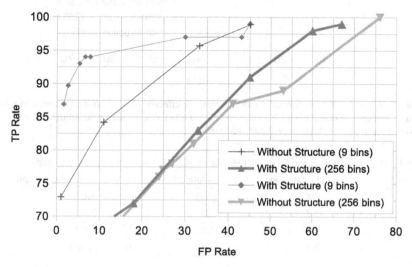

Fig. 3 ROC curves for Chi-Square metric comparing effectiveness of anomaly detection when structure for payload is extracted (With Structure) and otherwise (Without Structure). Experiment conducted for algorithm learned on 300 samples

In the first experiment, we investigated how proposed method will behave for different sizes of learning data. First, we have adapted classical 10-fold. For that approach the data obtained for learning and evaluation purposes is divided randomly into 10 parts (folds). One part (10 % of full dataset) is used for evaluation while the remaining 90 % is used for training (e.g. establishing model parameters). For each fold we deliberately picked only a subset of data to train the classifier in order to check the effectiveness of the proposed method. This way, we still have the same number of the testing samples (common baseline for comparison) even if we have used only a friction of available training data. The whole procedure is repeated 10 times, so each time different part is used for evaluation and different part of data set is used for the training. The result for all 10-folds are averaged to yield an overall error estimate. The whole 10-fold cross validation is repeated for different sizes of the training data, namely 1 %, 10 %, 20 %, and 100 %. Results are presented in Table 1.

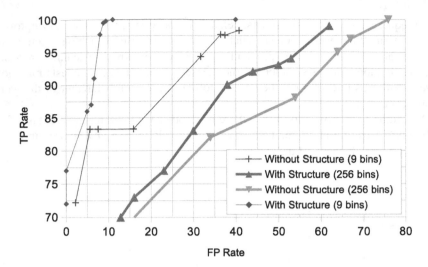

Fig. 4 ROC curves for Chi-Square metric comparing effectiveness of anomaly detection when structure for payload is extracted (With Structure) and otherwise (Without Structure). Experiment conducted for algorithm learned on 32000 samples

Table 1 True positive rate and false positive rate for different number of learning samples

TP rate [%]	FP rate [%]	Data set size	Number of samples
86.6	1.8	1 %	300
95.6	5.8	10 %	3000
96.9	6.8	20 %	6000
97.7	8.1	100 %	32400

In order to have the better overview of the effectiveness we have also plotted the ROC curves. The ROC curve for 300 learning samples is presented in Fig. 3, while the curve for 32400 samples is presented in Fig. 4. Additionally, we have compared our method when no structure from payload is extracted. In the performed experiments the Chi-Square distance was used. Moreover, we also have compared how our feature encoding methods can improve the overall effectiveness (please notice plots for 9 bin and 256 bin).

We have also tested the segmentation method on HTTP dumps for SOAP protocol. As it is shown in Fig. 5, for the consecutive SOAP calls (only one is shown on left side of the figure) our method extracted two distinctive tokens, which are listed on the right side of the figure and marked with dashed and solid lines respectively. It can be noticed that for the distribution we observe peak for the third feature, which are digits in the ASCII table.

5 Conclusions and Future Work

In this paper we have proposed the new method for HTTP Traffic Anomaly Detection. The method uses the information about the HTTP payload structure and we obtain the promising effectiveness and results. The proposed method combines compression-based technique (to identify some common parts of requests) and statistical analysis that allows to describe the dynamic properties of text in between tokens. Our experiments show that such technique allows for significant improvement of the effectiveness when compared to other techniques that treat the message body as the whole. Another advantage is the fact that our tool does not need any prior knowledge about protocols using HTTP as the transport. Our experiments show that the proposed method can be used to learn the payload structures also for other protocols like SOAP. For the future work we plan to evolve that method toward fully unsupervised learning. Currently, it uses only normal traffic (without attacks) to learn the structure and distribution of features for the extracted variables. For instance, we have not yet checked how the method will behave when the learning data will be contaminated with some attacks, which is quite likely to happen in the real production network environment.

Fig. 5 The segmentation method applied for the SOAP request

References

1. Symantec: 2014 Internet Security Threat Report, Volume 19. http://www.symantec.com/security_response/publications/threatreport.jsp (2014)
2. SCALP: Project homepage. http://code.google.com/p/apache-scalp/
3. PHPIDS: Project homepage. https://phpids.org/
4. OWASP Stinger: Project homepage. https://www.owasp.org/index.php/Category:OWASP_Stinger_Project
5. SNORT: Project homepage. http://www.snort.org/
6. Shar, L.K., Tan, H.B.K.: Predicting common web application vulnerabilities from input validation and sanitization code patterns. In: Proceedings of the 27th IEEE/ACM International Conference on Automated Software Engineering (ASE), pp. 310–313. IEEE (2012)
7. Yu, F., Muath, A., Tevfik, B.: Stranger: an automata based string analysis tool for PHP. Tools and algorithms for the construction and analysis of systems, pp. 154–157. Springer (2010)

8. CHalfond, W., Orso, A.: AMNESIA: analysis and monitoring for neutralizing SQL-injection attacks. In: Proceedings of the 20th IEEE/ACM International Conference on Automated Software Engineering, pp. 174–183 (2005)
9. Source Code Analysis Tools: Project homepage. https://www.owasp.org/index.php/Source_Code_Analysis_Tools
10. Choraś, M., Kozik, R., Puchalski, D.: Correlation approach for SQL injection attacks detection. In: Herrero, A., et al. (eds.) Advances in Intelligent and Soft Computing, vol. 189, pp. 177–186. Springer (2012)
11. Choraś, M., Kozik, R.: Real-time analysis of non-stationary and complex network related data for injection attempts detection. In: Proceedings of WSC17 Online Conference on Soft Computing in Industrial Applications, pp. 177–186 (2012)
12. Choraś, M., Kozik, R.: Evaluation of various techniques for SQL injection attack detection. In: Burduk, R. et al. (eds.) Proceedings of the 8th International Conference on Computer Recognition Systems (CORES 2013), Advances in Intelligent Systems and Computing, vol. 226, pp. 753–762. Springer (2013)
13. OWASP Top 10: The ten most critical web application security risks. http://www.snort.org/ (2013)
14. Welch, T.: A technique for high-performance data compression. IEEE Comput. **17**(69), 8–19 (1984)
15. Ziv, J., Lempel, A.: A universal algorithm for sequential data compression. IEEE Trans. Inf. Theory **23**, 337–343 (1977)
16. Kruegel, C., Toth, T., Kirda, E.: Service specific anomaly detection for network intrusion detection. In: Proceedings of ACM Symposium on Applied Computing, pp. 201–208 (2002)
17. Torrano-Gimnez, C., Prez-Villegas, A., lvarez, G.: The HTTP dataset CSIC 2010. http://users.aber.ac.uk/pds7/csic_dataset/csic2010http.html (2010)

Feature Selection for Hand-Shape Based Identification

Muhammad Hussain, Awabed Jibreen, Hatim Aboalsmah,
Hassan Madkour, George Bebis and Gholamreza Amayeh

Abstract The shape of a hand contains important information regarding the identity for a person. Hand based identification using high-order Zernike moments is a robust and powerful method. But the computation of high-order Zernike moments is very time-consuming. On the other hand, the number of high-order Zernike moments increases quadratically with order causing storage problem; all of them are not relevant and involve redundancy. To overcome this issue, the solution is to select the most discriminative features that are relevant and not redundant. There exists a lot of feature selection algorithms, different algorithms give good performance for different applications, and to choose the one that is effective for this problem is a matter of investigation. We examined a large number of state-of-the-art feature selection methods and found Fast Correlation-Based Filter (FCBF) and Sparse Bayesian Multinomial Logistic Regression (SBMLR) to be the best methods that are efficient and effective in reducing the dimension of the feature space significantly (by 62 %), i.e. the storage requirements and also slightly enhanced recognition rate (from 99.16 ± 0.44 to 99.42 ± 0.36).

Keywords Hand-based identification · Zernike moments · Feature selection · Pattern matching · Biometric technology

1 Introduction

Currently, there is an increased interest in biometric technology, which led to intensive research on fingerprint, face, iris, and hand recognition. A hand contains important information and can be used for identification (who the unknown subject is?)

M. Hussain (✉) · A. Jibreen · H. Aboalsmah · H. Madkour
Department of Computer Science, College of Computer and Information Sciences,
King Saud University, Riyadh, Saudi Arabia
e-mail: mhussain@ksu.edu.sa

G. Bebis · G. Amayeh
Department of Computer Science and Engineering, University of Nevada, Reno, USA

© Springer International Publishing Switzerland 2015
Á. Herrero et al. (eds.), *International Joint Conference*, Advances in Intelligent Systems and Computing 369, DOI 10.1007/978-3-319-19713-5_21

and verification (is the claimed identify of a subject correct?). Hand based identification has a large range of applications in both government and industry. Many researchers attempted to propose solutions for biometric-technology based on the hand-shape [1]. Guo et al. [2] proposed a contact free hand geometry-based identification system; its average identification rate is 96.23 %; it needs an infrared illumination device. Recently Sharma et al. [3], proposed a multimodal biometric system, which is based on hand-shape and hand geometry. In this method all processing has to be performed with respect to a reference point. Amayeh et al. [4] proposed a hand-based person identification and verification method in 2009, which is a simple and robust method and does not employ any constraint. After the acquisition of hand image and segmenting it into components (fingers and palms), this method extracts features using high-order Zernike moments from each component and then fuses this information to take the final decision. This is a peg-free hand-based identification approach; it is not affected by the motion of fingers or hand, and does not require landmark points' extraction [1]. The geometric information from each part of a hand is represented by high-order Zernike moments, which are invariant to rotation, translation and scaling, and lead to excellent recognition rate [4]. But these moments involve high computational cost. In addition, the number of these moments increases quadratically with their order and this number becomes very large causing the storage problem. To store the templates of a single subject, a huge amount of space is needed. Moreover, this number has an impact on the template matching efficiency and affects the recognition efficiency.

The solution of these problems is to select the most discriminatory features, which are relevant and not redundant. As such, to enhance the efficiency of the person identification system and to reduce the storage requirements, it is imperative to select and use the high-order Zernike moments with the highest discriminative power. There exist a large number of feature selection algorithms with their strengths and weaknesses. Which algorithm results in the best performance for the problem under consideration is a matter of investigation. In this study, we examined a large number of state-of-the-art algorithms for the selection of the most discriminative high-order Zernike moments [5]. We found only three of them suitable for selecting Zernike moments: Fast Correlation-Based Filter (FCBF) [6, 7], Sparse Bayesian Multinomial Logistic Regression (SBMLR) [7] and Spectrum Feature Selection Algorithm (Spectrum) [9]. We thoroughly explored them to find the most efficient and effective algorithms. Finally, we found that FCBF and SBMLR give the best performance. These algorithms help reduce the dimension of the feature space by 62 % and a slight enhancement in the recognition rate. In this paper, we focus only person identification problem. Our main contribution is to reduce the space required to store templates for hand based identification by selecting the most discriminative Zernike moments.

The rest of the paper is organised as follows. In Sect. 2, we give an overview of hand shape based recognition system. Section 3 discusses the feature selection methods. The results have been presented in Sect. 4 and finally Sect. 5 concludes the paper.

2 Hand Shape Based Identification

In this section, we present the hand shape based recognition system. Its flowchart is shown in Fig. 1.

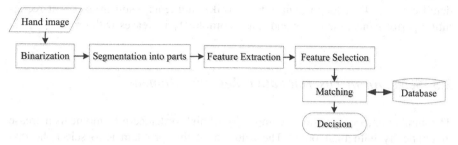

Fig. 1 Flowchart of the hand based Identification

2.1 Pre-processing

The image of a hand is acquired using a VGA resolution CCD camera and a flat lighting table. After image acquisition, it is converted into a binary form, and hand is separated from arm using segmentation. Finally, hand is segmented into palm and fingers; the detail can be found in [4]. After segmentation, the six hand parts that are used for recognition are little finger (F1), ring finger (F2), middle finger (F3), Index finger (F4), thumb finger (F5), and the Palm (P).

2.2 Feature Extraction

After extracting hand components, each component is described using high-order Zernike moments. These moments have high discriminative potential and effectively describe each component. Figure 2 shows the detail of feature extraction and feature selection.

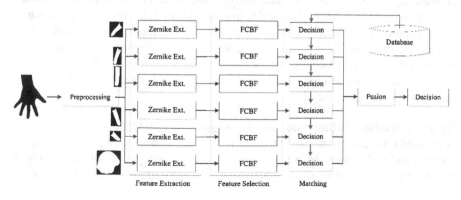

Fig. 2 Detail of feature extraction and feature selection

Zernike moment $Z_{n,m}$ depends on two parameters (n, the order and m, the repetition), its calculation involves a high computational cost [10]. The time complexity of computing a Zernike moment of order n of an image of size $M \times M$ is $O(n^2 M^2)$. The number of Zernike moments up to order n with repetition n is $(n/2 + 1)^2$ i.e. the number of high order Zernike moments increases quadratically with their order, for detail consult [4]. Though high-order Zernike moments result in excellent recognition performance, their time and space complexity increases with order.

2.3 Selection of Discriminative Zernike Moments

The number and computational complexity of high-order Zernike moments increase quadratically with their order. The solution of this problem is to select the discriminative Zernike moments. It is a well-known feature selection problem, and many powerful feature selection algorithms have been proposed during the last decade [5], which are mainly based on either filter model or wrapper model. Wrapper algorithms select features involving a certain classifier, which are optimized for that classifier; moreover, their time complexity is high. Filter algorithms discover the discriminant features optimizing some criterion based on their intrinsic properties and are usually fast [5]. Filter algorithms are mainly categorized as supervised (use labels of the instances) and unsupervised algorithms. In our case, labels are known, so for the selection of Zernike moments, we considered three state-of-the-art power filter based supervised algorithms: Fast Correlation-Based Filter (FCBF) [6, 7], Sparse Bayesian Multinomial Logistic Regression (SBMLR) [8] and Spectrum Feature Selection Algorithm (SPEC) [9]. In the following paragraphs, we give an overview of these algorithms.

Fast Correlation-Based Filter (FCBF)

It is a fast correlation-based filter algorithm, which is effective in selecting relevant features and removing redundant features for the classification of high dimensional data. It efficiently and significantly reduces the dimension of a feature space and improves classification accuracy. It operates in two phases: relevance analysis and redundancy analysis. The relevance analysis phase selects the features which are relevant to a target class. After selecting the relevant features, the redundancy analysis phase uses the concept of predominant correlation to discover redundancy among the selected relevant features and selects the predominant features. It gives better performance than both feature weighting (e.g. ReliefF) and subset search (e.g. CFS) algorithms for feature selection in terms of speed, dimensionality reduction and classification accuracy [6, 7].

Sparse Bayesian Multinomial Logistic Regression (SBMLR)

It is based on sparse multinomial logistic regression with Laplace prior that induces sparsity and makes the regularization parameters to be integrated out analytically. It is fully automatic and its space complexity scale only linearly with the number of model parameters. To determine the model parameters, it uses a simple but efficient training algorithm [8]. It is a kind of subset search algorithm.

Spectrum Feature Selection Algorithm (Spectrum)
In this algorithm, features are evaluated and ranked using a graph spectrum. A set of pair wise instance similarities S is represented as a graph G. Each feature that is consistent with the structure of the graph is assigned a value similar to those features that are close to it on the graph [9]. It is a feature waiting algorithm and ReliefF and Laplacian Score are special cases of this algorithm.

2.4 Pattern Matching

After extracting Zernike moments from hand component and selecting the most discriminative ones, Euclidean distance $d(Q, T_i)$ is computed between the query Q and all enrolment templates T_i in the database. The unknown person with query Q is identified with rank-one (having the smallest distance d) template. Similarly, a decision is made using each hand component, and the final decision is taken using decision based fusion with a majority vote.

3   Results and Discussion

For validation, we used the same dataset that was used in [4]. This dataset was collected from 99 subjects capturing 10 hand images from each subject. The total number of hand image samples is 990.

For evaluation, we randomly divided the 10 samples of each subject in the ratio n:10–n, denoted by (n, 10-n), where n samples were used as enrolment templates (for training) and the remaining samples (10-n) were used for testing. For experiments, we used $n = 6, 5, 4, 3$ and repeated the experiments 30 times for each n. To measure performance, we used commonly used measures: accuracy (the percentage of query images which are correctly identified) and Cumulative Match Characteristic (CMC) curve that is a plot of true match rate versus rank [11].

Tables 1, 2, 3, 4, 5, and 6 shows the identification rates based on F1, F2, F3, F4, F5 and P without and with feature selection as average percentage accuracies together with standard deviation (Acc ± Std) over 30 runs of the system with random selection of enrolment templates and query samples and the numbers of selected features by the four methods for different divisions. For F1 and the divisions (5, 5), (4, 6) and (3, 7), SBMLR gives the best results with only 38 features and FCBF gives the best accuracy for (6, 4) with 55 features in this case. The best identification accuracy for F2 is obtained with FCBF for all divisions, and the numbers of selected features are 50 and 47.

The best accuracy for F3 is also given by FCBF for all divisions with the numbers of selected features 68 and 51. For F4, SBMLR gives the best accuracy with 50 selected features in case of (4, 6) and (3, 7) divisions, whereas

Table 1 Identification rate based on F1 (little finger) only

Method	(6,4)		(5,5)		(4,6)		(3,7)	
	#F	Acc ± Std	#F	Acc ± Std	#F	Acc ± Std	#F	Acc ± Std
NoFS	121	96.57 ± 094	121	95.92 ± 0.66	121	94.70 ± 0.87	121	93.20 ± 1.15
SBMLR	38	96.73 ± 0.90	38	**95.91 ± 0.62**	38	**95.08 ± 0.93**	38	**93.32 ± 0.87**
FCBF	55	**96.99 ± 0.59**	38	92.87 ± 1.02	38	91.76 ± 0.72	38	90.01 ± 1.14
SPEC	55	94.89 ± 0.98	38	91.00 ± 1.03	38	89.14 ± 0.94	38	87.01 ± 1.29

Table 2 Identification rate based on F2 (ring finger) only

Method	(6,4)		(5,5)		(4,6)		(3,7)	
	#F	Acc ± Std	#F	Acc ± Std	#F	Acc ± Std	#F	Acc ± Std
NoFS	121	97.66 ± 0.67	121	97.14 ± 0.66	121	96.64 ± 0.61	121	95.64 ± 0.70
SBMLR	55	98.07 ± 0.54	47	95.79 ± 0.76	47	95.28 ± 0.70	47	93.80 ± 0.77
FCBF	50	**98.25 ± 0.70**	47	**97.85 ± 0.50**	47	**97.32 ± 0.58**	47	**96.17 ± 0.58**
SPEC	55	96.83 ± 0.86	47	95.53 ± 0.73	47	94.41 ± 0.81	47	92.76 ± 0.93

Table 3 Identification rate based on F3 (middel finger) only

Method	(6,4)		(5,5)		(4,6)		(3,7)	
	#F	Acc ± Std	#F	Acc ± Std	#F	Acc ± Std	#F	Acc ± Std
NoFS	121	98.04 ± 0.69	121	97.68 ± 0.63	121	97.01 ± 0.67	121	95.78 ± 0.67
SBMLR	50	97.78 ± 0.61	51	96.01 ± 0.68	51	95.29 ± 0.65	51	93.47 ± 0.86
FCBF	68	**98.72 ± 0.49**	51	**97.91 ± 0.56**	51	**97.65 ± 0.57**	51	**96.70 ± 0.87**
SPEC	60	97.04 ± 0.71	51	95.08 ± 0.89	51	94.22 ± 1.01	51	92.47 ± 0.82

Table 4 Identification rate based on F4 (index finger) only

Method	(6,4)		(5,5)		(4,6)		(3,7)	
	#F	Acc ± Std	#F	Acc ± Std	# F	Acc ± Std	#F	Acc ± Std
NoFS	121	98.47 ± 0.62	121	98.18 ± 0.61	121	97.79 ± 0.64	121	96.89 ± 0.61
SBMLR	50	98.87 ± 0.53	50	96.85 ± 0.59	50	**98.23 ± 0.53**	50	**97.57 ± 0.40**
FCBF	48	**99.02 ± 0.34**	50	**98.33 ± 0.41**	50	97.70 ± 0.5	50	97.40 ± 0.73
SPEC	50	97.31 ± 0.72	50	96.75 ± 0.89	50	96.16 ± 0.73	50	94.56 ± 0.93

Table 5 Identification rate based on F5 (thumb) only

Method	(6,4)		(5,5)		(4,6)		(3,7)	
	#F	Acc ± Std	#F	Acc ± Std	#F	Acc ± Std	#F	Acc ± Std
NoFS	121	90.84 ± 1.43	121	89.17 ± 1.69	121	87.52 ± 1.48	121	84.59 ± 1.14
SBMLR	58	91.68 ± 1.56	58	90.52 ± 1.10	58	**88.69 ± 1.06**	58	85.59 ± 1.20
FCBF	48	**91.60 ± 1.05**	75	**90.69 ± 1.13**	58	88.22 ± 0.99	58	**85.50 ± 1.18**
SPEC	50	88.86 ± 1.42	58	88.06 ± 1.10	58	85.66 ± 1.39	58	82.39 ± 1.30

Table 6 Identification rate based on P (palm) only

Method	(6,4)		(5,5)		(4,6)		(3,7)	
	#F	Acc ± Std	# F	Acc ± Std	#F	Acc ± Std	#F	Acc ± Std
NoFS	256	97.44 ± 0.85	256	96.72 ± 0.81	256	95.85 ± 1.08	256	94.05 ± 0.97
SBMLR	46	98.00 ± 0.74	46	97.50 ± 0.60	46	96.33 ± 0.65	46	**95.38 ± 0.77**
FCBF	46	**98.31 ± 0.60**	41	**98.43 ± 0.56**	46	**97.08 ± 059**	46	91.40 ± 1.05
SPEC	50	98.21 ± 0.46	46	98.18 ± 0.52	46	93.49 ± 0.69	46	95.07 ± 0.840

FCBC results in the best accuracy with 48 and 50 features in case of (6, 4) and (5, 5) divisions, respectively. For the thumb, the best accuracy is obtained using FCBF that selects 48, 75 and 58 features, respectively, for (6, 4), (5, 5) and (3, 7) divisions and SBMLR gives the best accuracy for (4, 6) selecting 58 features. For palm, again FCBF is the winner in case of (6, 4), (5, 5) and (4, 6) with selected features 46, 41 and 46, respectively, whereas in case of (3, 7), SBMLR gives the best accuracy with 46 features. The results discussed so far indicate that overall for all hand components, FCBF emerged out to be the winner for selecting the discriminative higher order Zernike moments.

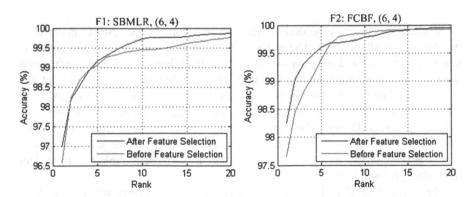

Fig. 3 CMC curves for F1(little finger) and F2(ring finger)

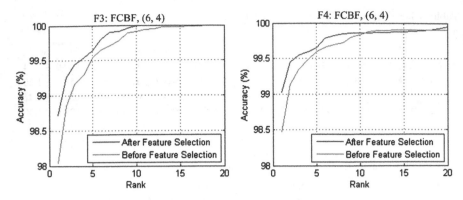

Fig. 4 CMC curves for F3 (middle finger) and F4 (index finger)

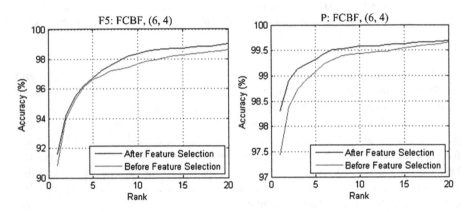

Fig. 5 CMC curves for F5 (thumb finger) and P (palm)

SBMLR occupies the second position in this competition. SPEC gives poor performance in all the cases. The reason why FCBF outperforms other algorithms is that FCBF concentrates not only on discovering the relevance but also in removing the redundancy. It also indicates that all high-order Zernike moments are not relevant, and a large number is redundant from the identification point of view.

Average CMC curves of the best cases for all hand components are shown in Figs. 3, 4, and 5. Overall, the curves corresponding to the systems with feature selection are above those related to the systems without feature selection. These curves indicate that accuracy rates increase with the increase in rank, i.e. the system has stable performance.

The decision level fusion with a majority vote was applied on the matching decisions based on the six hand components, the results are given in Table 7. For each division, fusion was done in two different ways: considering decisions of all components using one method, SBMLR or FCBF, and considering the decisions using SBMLR and FCBF (which result in the best accuracy) for different components. In case of (3, 7) and (4, 6), neither SBMLR nor FCBF gives the best accuracy when either method is used for all components. However, in case for (5, 5) and (6, 4), the best accuracy is obtained when SBMLR and FCBF, respectively, are used for all components. It indicates that as the number of enrolment templates decreases, only one feature selection is not enough for all components. In case of (3, 7), FCBF gives the best accuracy for F1, F4 and P, whereas SBMLR performs best for F2, F3 and F5, the fusion of decisions of F1, F4 and P using FCBF and those of F2, F4 and F5 using SBMLR gives the best results. Almost similar results are for (4, 6).

Table 7 shows that in case of 3 enrolment templates, 861 high order Zernike moments need to be stored for each template without feature selection but with feature selection only 300 of them will be stored. In this way, there will be a reduction of 62 % in the storage space.

Table 7 The results of decision level fusion with majority vote

(n, m)	F. S. Method	NoFS (861)	With F.S.	
		Acc $\pm$ std	#F	Acc $\pm$ std
(3, 7)	(1): SBMLR for all components	98.80 $\pm$ 0.43	290	98.77 $\pm$ 0.46
	(2): FCBF for all components		315	98.40 $\pm$ 0.67
	F1(1) + F2(2) + F3(2) + F4(1) + F5(2) + P(1)		300	**98.99 $\pm$ 0.58**
(4, 6)	(1): SBMLR for all components	98.75 $\pm$ 0.38	290	98.80 $\pm$ 0.46
	(2): FCBF for all components		315	98.73 $\pm$ 0.57
	F1(1) + F2(2) + F3(2) + F4(1) + F5(2) + P(2)		300	**99.03 $\pm$ 0.42**
(5, 5)	(1): SBMLR for all components	98.88 $\pm$ 0.39	290	**99.13 $\pm$ 0.38**
	(2): FCBF for all components		315	98.89 $\pm$ 0.45
	F1(1) + F2(2) + F3(2) + F4(2) + F5(2) + P(2)		298	99.10 $\pm$ 0.39
(6, 4)	(1): SBMLR for all components	99.16 $\pm$ 0.44	290	99.25 $\pm$ 0.33
	(2): FCBF for all components		315	**99.42 $\pm$ 0.36**

4 Conclusion

Hand based identification based on high-order Zernike moments is a robust method, but the computation of Zernike moments is time-consuming and their number increases with their order. To select the discriminative Zernike moments, we investigated a number of supervised filter methods. We found that SBMLR, FCBF and SPEC give the acceptable results. Further investigation revealed that only SBMLR and FCBF are the most suitable methods. If the number of enrolment templates is smaller (i.e. 3 or 4), then only one method does not give the best accuracy, in this case both SBMLR and FCBF for different parts result in the best accuracy. Feature selection reduces the number of high-order Zernike moments significantly. Only 300 high-order Zernike moments are discriminative out of 861 when the number of enrolment templates is 3 or 4, which significantly reduce the storage requirements.

Acknowledgment This project was supported by NSTIP strategic technologies programs, grant number 12-INF2582-02 in the Kingdom of Saudi Arabia.

References

1. Duta, N.: A survey of biometric technology based on hand shape. Pattern Recogn. **42**(11), 2797–2806 (2009)
2. Guo, J.M., Hsia, C.H., Liu, Y.F., Yu, J.C., Chu, M.H., Le, T.N.: Contact-free hand geometry-based identification system. Expert Syst. Appl. **39**(14), 11728–11736 (2012)
3. Sharma, S., Dubey, S.R., Singh, S.K., Saxena, R., Singh, R.K.: Identity verification using shape and geometry of human hands. Expert Syst. Appl. **42**(2), 821–832 (2015)
4. Amayeh, G., Bebis, G., Erol, A., Nicolescu, M.: Hand-based verification and identification using palm–finger segmentation and fusion. Comput. Vis. Image Underst. **113**(4), 477–501 (2009)
5. Han, Y., Yang, Y., Zhou, X Co-regularized ensemble for feature selection. In: Proceedings IJCAI-13, (2013)
6. Liu, H., Yu, L.: Feature selection for high-dimensional data: a fast correlation-based filter solution. In: Proceedings of the Twentieth International Conference on Machine Leaning (ICML-03), pp. 856–863, Washington, D.C. (2003)
7. Liu, H., Hussain, F., Tan, C.L., Dash, M.: Discretization: an enabling technique, data mining and knowledge discovery, vol. 6(4), pp. 393–423. Springer, Netherland (2002)
8. Cawley, G.C., Nicola L. C. Talbot, N.L.C., Girolami, M.: Sparse multinomial logistic regression via bayesian L1 regularisation. In: Proceedings of Neural Information Processing Systems (NIPS 2006), pp. 209–216 (2007)
9. Liu, H., Zhao, Z.,: Spectral feature selection for supervised and unsupervised learning. In: Proceedings of the 24th International Conference on Machine Learning (ICML 2007), pp. 1151–1157 (2007)
10. Khotanzad, A., Hong, Y.H.: Invariant image recognition by Zernike moments. IEEE Trans. Pattern Anal. Mach. Intell. **12**(5), 489–498 (1990)
11. Damer, N., Opel, A., Nouak, A.: CMC Curve Properties and Biometric Source Weighting in Multi-Biometric Score-level Fusion. Proc. FUSION **2014**, 1–6 (2014)

CISIS 2015: Infrastructure and Network Security

Designing and Modeling the Slow Next DoS Attack

Enrico Cambiaso, Gianluca Papaleo, Giovanni Chiola
and Maurizio Aiello

Abstract In the last years the Internet has become a primary tool for information dissemination, spreading itself on the entire world and becoming a necessary communication system. More recently, thanks to the advent of the Internet of Things paradigm, a wide range of objects (such as washing machines, thermostats, fridges) is able to communicate on the Internet. As a consequence of this large adoption, due to economic motivations, the Internet is often targeted by cyber-criminals. In this paper, we present a novel attack called Slow Next, targeting Internet services (IoT, cloud, mobile hosted, etc.). We analyze that the proposed menace is able to lead a Denial of Service on different categories of network protocols using a low amount of network bandwidth. Moreover, since connections behavior is legitimate, Slow Next is able to elude detection systems. The attack represents therefore a potential menace on the cybersecurity field.

Keywords Denial of service · Slow dos attack · Lbr dos attack · Network security · Internet of things

1 Introduction

The Internet is today the most important communication medium, connecting users around the world, through an infrastructure covering the entire globe. In the last years, several Internet-related phenomenons emerged, with the purpose of simplify

E. Cambiaso · G. Papaleo (✉) · M. Aiello
IEIIT-CNR, National Research Council, via De Marini 6, 16149 Genoa, Italy
e-mail: gianluca.papaleo@ieiit.cnr.it

E. Cambiaso
e-mail: enrico.cambiaso@ieiit.cnr.it

M. Aiello
e-mail: maurizio.aiello@ieiit.cnr.it

E. Cambiaso · G. Chiola
Università degli Studi di Genova, via Dodecaneso, 35, 16146 Genoa, Italy
e-mail: chiolag@acm.org

© Springer International Publishing Switzerland 2015
Á. Herrero et al. (eds.), *International Joint Conference*, Advances in Intelligent
Systems and Computing 369, DOI 10.1007/978-3-319-19713-5_22

and enhance users lives. Among these phenomenons, the advent of the Internet of Things (IoT) paradigm has introduced the concept of an inter-connection between highly heterogeneous networked entities and networks, accordingly to communication patterns such as human-to-human (H2H), human-to-thing (H2T), thing-to-thing (T2T), or thing-to-things (T2Ts) [1]. In particular, it consists in equipping an object of an Internet connection, thus increasing its potentialities and allowing users or other inter-connected objects to (even remotely) benefit of the object's functionalities. Because of its wide adoption and the spreading of always-connected devices, such as smart devices composing the Internet of Things (IoT), the Internet network has to be kept a safe place, protecting its users from malicious operations.

Among all the malicious operations executed every day on the global net, Denial of Service (DoS) attacks [2] are accomplished to deprive legitimate users of a particular service. Services of different nature are vulnerable to these threats. Under a DoS attack, targeted hosts (ranging from corporate servers to IoT devices) become unreachable, thus useless for their purpose. In particular, we are interested to DoS threats which use the network as the main tool of attack, thus excluding menaces such as exploit based DoS attacks.

The first generation of DoS attacks [3] is based on flooding the victim with a large amount of network packets, overwhelming its network resources. In the last years, DoS attacks evolved to a second generation of menaces known as Slow DoS Attacks (SDAs) [4]. Those threats make use of a low amount of bandwidth to reach the DoS, often affecting the application layer of the ISO/OSI model.

There are several categories of Slow DoS Attacks [4]. We are interested in particular to the attacks which directly affect the application layer of the victim, exploiting a particular server-side timeout. Indeed, some SDAs are focused on maintaining a connection with the victim established as long as possible. In order to do that, a server-side timeout has to be exploited.

Although there exist various attacks belonging to this category [4], the study of the field is still young, due to the novelty of the topic. In this paper we introduce an innovative menace belonging to this category of threats. The attack adopts indeed this behavior, establishing a large amount of connections with the application layer of the victim. As a consequence, the server would expect a Denial of Service, since it will not be able to serve legitimate clients.

This paper is organized as follows: in Sect. 2 we introduce Timeout Exploiting Slow DoS Attacks, appropriately categorizing them, also reporting related work relatively to the Slow DoS Attacks phenomenon. We then describe in detail the proposed attack in Sect. 3, analyzing different types of affected protocols. Then, we report the executed tests and obtained results in Sect. 4. Finally, in Sect. 5 we report the conclusions of the work.

2 Timeout Exploiting Slow DoS Attacks

In a previous work [5] we have made an analysis of network connection stream at the application layer.

If we consider the HTTP protocol, after a connection has been established, an HTTP request is sent from the client to the server. After reception, the request is interpreted by the server, in order to generate a response to send back to the client. Thus, two possible events could happen: (i) the connection is closed, or (ii) the connection is maintained alive, avoiding the client to re-establish the connection for a new request directed to the same server (*persistent connection*) [6].

Starting from rough network data (represented for instance as a PCAP packet capture file), in [5] we have extracted a list of connection streams. Then, each stream has been splitted, identifying characterizing parameters for network traffic representation. In particular, following parameters have been extracted from the server (see Fig. 1):

- Δ_{start}, identifying the time between the connection establishment and the begin of the request;
- Δ_{req}, identifying the time passed between the begin and the end of the request;
- Δ_{delay}, identifying the time passed between the end of a request and the start of the relative response;
- Δ_{resp}, identifying the time passed between the begin and the end of the response;
- Δ_{next}, identifying the time passed between the end of a response and the next request on the same stream.

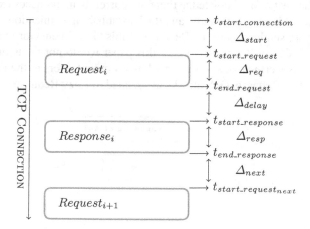

Fig. 1 TCP connection stream for an HTTP connection

Note that it's possible to extrapolate the parameters described above from connections speaking different TCP based protocols, such as HTTP/HTTPS, FTP, SSH, or SMTP. Also note that in this paper we extend the previous work [5], presenting the Δ_{start} parameter.

Analyzing previous works on the topic, [4] provides an accurate categorization of SDAs, built in function of the attacked resource. Nevertheless, the proposed work has to be adapted to completely and correctly include all the SDAs described above, exploiting a particular Δ-parameter. For simplicity, we'll call such threats as *Timeout Exploiting SDAs*.

Among the SDA threats described in [4], we could relate the Delayed Responses DoS category of attacks to an exploitation of the Δ_{delay} parameter. Indeed, a Delayed Responses DoS exhausts internal resources of the victim (i.e. CPU or memory) with legitimate requests, delaying the response sending. Analyzing instead attacks affecting the victim's network, in case a particular timeout is exploited, standing to [4], it may be one of the following: (i) request timeout or (ii) response timeout. If we consider the Δ-parameters exploitations described above, we could assert that the request timeout is directly related to the exploiting of the Δ_{req} parameter, while the response timeout can be associated with a Δ_{resp} exploitation.

Nevertheless, the taxonomy does not include attacks exploiting the Δ_{next} parameter or the Δ_{start} one. An extension is therefore needed. In particular, the proposed amendments to the taxonomy are reported in Fig. 2. For simplicity, we only report a subset of the SDAs categorization, only presenting Timeout Exploiting SDAs. This categorization extends the previous one, also including the Delayed Responses DoS category into the category of attacks against the victim's network. Moreover, we now introduce the Lazy Requests DoS attacks, relative to attacks exploiting the Δ_{start} parameter. In addition, we include the Intra-Requests DoS category of attacks, not covered in previous work, relating them to an exploitation of the Δ_{next} parameter. Although to the best of our knowledge there are currently no menaces exploiting the Δ_{start} parameter, it is easy to imagine an attack establishing connections and waiting some time before sending requests. Because of this "lazy" behavior, we categorize these menaces as Lazy Requests DoS. In this paper we do not focus on such category of threats. Nevertheless, since the exploited timeout refers to the time before a request start, Intra-Requests DoS attacks are related to Lazy Requests ones.

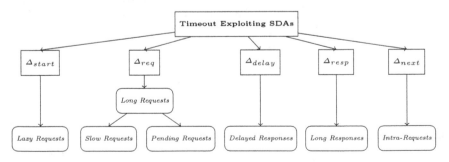

Fig. 2 Categorization of timeout exploiting slow DoS attacks

Considering the parameters described above, we are now interested to analyze the possible threats exploiting them.

In the arena of Slow DoS Attacks, the most known threat maybe is *Slowloris* [7], an attack that has also been executed during 2009 to accomplish a distributed DoS attack against the Iranian presidential elections. The attack works by establishing potentially endless requests with the victim. Similar menaces are *Slow HTTP POST* [4] and SlowReq [8] attacks. In the latter case, a minimum amount of network bandwidth is required to execute the attack.

Other attacks are focused instead on delaying the responses of the server. For instance, the *Apache Range Headers* attack [4] exploits the HTTP byte range parameter to force the server to replicate the requested resource, exhausting its RAM memory.

Other attacks such as *#hashDoS* [9] or *ReDoS* [10] are considered instead *meta attacks* [11], since they are not bounded to a specific implementation, but are limited to a high level definition of the attack. Those attacks exploits respectively hash tables and regular expression implementations, exhausting server's memory.

Instead, the *Slow Read* attack [12] simulates (at the transport layer) a tiny client-side reception buffer to slow down the responses of the server. Particularly, Slow Read tends to delay the response sending from the server, in terms of application layer payload. Moreover, the attack also tends to increase response sending duration. This threat is particularly interesting, since in this case attacker's behavior may be associated to a legitimate node really equipped with a tiny reception buffer.

Analyzing the timeouts scheme reported in Fig. 2, to the best of our knowledge, there are no threats exploiting the Δ_{next} parameter. Nevertheless, since this parameter is exploited by the attack introduced in this paper, we believe that the propose work represents an innovative contribution, providing a missing element to the field.

3 The Slow Next Attack

The *Slow Next* attack is an application layer DoS attack which establishes a particular amount of connections with the victim. In order to reach a DoS, that amount is lower limited to the maximum number of simultaneous connections managed by the attacked server, at application layer. Under these conditions, the buffer resources of the server would be saturated by the attacker and any new connection would't be served by the targeted system.

The attack makes use of the *persistent connection* feature implemented in protocols such as HTTP 1.1 [6] to maintain a connection opened after receiving a server-side response, obtained after the sending of a specific request. The attack exploits the Δ_{next} parameter described above. In particular, in order to exploit it, each connection would make a legitimate request to the server. Then, the server would produce a legitimate response, which is normally sent to the attacker. Typically, the attacker's aim would be to reduce both request and response size, thus reducing required attack bandwidth. After receiving the whole response, the malicious client would make use of a *Wait Timeout* [11] to delay the sending of another legitimate request to the server, through the established channel. In this way, the server would be forced to maintain the channel opened, thus maintaining the resource busy.

The attack is particularly effective, since the anomalous behavior can't be detected from a packet's payload analysis, due to the lawfulness of the exchanged messages.

Since the threat is not bounded to a specific protocol or packets payload, it can be defined as a meta attack [11]. Nevertheless, adopted attack payload may affect both performance (i.e. requesting an entire resource, or just a portion of it) and results (i.e. sending a random payload may cause a connection close).

For instance, relatively to the HTTP protocol, payloads reported in Code 1 and Code 2 may be involved in the message exchange during a Slow Next attack.

```
HEAD / HTTP/1.1\r\n
Host: [...]\r\n
Connection: Keep-Alive\r\n
\r\n
```
Code 1 HTTP Request Sample

```
HTTP/1.1 200 OK\r\n
[...]\r\n
Keep-Alive: timeout=5, max=100\r\n
Connection: Keep-Alive\r\n
Content-Type: text/html\r\n
\r\n
```
Code 2 HTTP Response Sample

From the behavioral point of view, this attack is particularly interesting, since (analyzing connections apart) it behaves legitimately. Indeed, each sent request and received response is compliant to the exploited protocol and it is sent/received in a legitimate way. Therefore, detection systems analyzing request/response inter-arrival times would not be able to detect an anomaly on the network. Indeed, in this case the exploited timeout is the one used to maintain a connection alive for the sending of an additional request on the same stream. Nevertheless, this behavior may also be related to a legitimate client waiting some seconds before requesting an additional resource to the server.

As reported in [13–16], the use of computational intelligence to design Intrusion Detection Systems based on neural networks, fuzzy systems, evolutionary computation, artificial intelligence systems, and soft computing can help to detect and mitigate these kind of threats. In [5, 17] we have applied computational based approaches to efficiently identify attacks belonging to such category.

3.1 Protocols Analysis

The Slow Next attack affects TCP based protocols implementing and supporting persistent connections [6] between client and server. We have analyzed some possible problems and precautions, studying how attack requirements vary by trying to face with these open issues.

Low Server-Side Timeout In terms of required resources, the proposed attack depends on the timeout used by the server relatively to the Δ_{next} parameter. In case of persistent connections, this timeout specifies the time the server will wait for a subsequent request before closing the connection. By setting this server timeout to a high value, performance problems may occur on an heavily loaded server. Indeed, by increasing the timeout, more server processes will be kept busy, thus not serving other idle clients. Conversely, a low value may lead to a high number of connection closures, hence reducing resources optimization.

Credentials Dependence Some network protocols are designed to provide access only to specific authenticated users. For instance, remote shell or file transfer protocols are often based on this model. As a consequence, an attacker willing to execute a Slow Next attack targeting such protocols may found some drawbacks. In particular, since the attack needs to send a legitimate request to the server and receive a legitimate response, client authentication is in this case needed.[1] Nevertheless, although the attack may be successful, in this case it is required to the attacker to authenticate to the server.

Particularly Affected Protocols In general, it is preferrable for the attacker to use less bandwidth possible. In order to do that, both request and response size should be minimized. In particular, considering the HTTP protocol, methods such as OPTIONS, HEAD, or TRACE may be used in conjunction to the protocol version 1.1, to reduce both request and response size by sending messages in accordance to the protocol. Instead, in case of the SMTP protocol, the NOOP command [18] may be used to send a null command to the server.

4 Executed Tests and Obtained Results

We have executed tests comparing the proposed menace to the currently available ones, on a real test environment in the institute LAN. In order to include a wide set of attacks, we have targeted the HTTP protocol which is affected from the tools we have in Sect. 2. We have excluded the Apache Range Header DoS attack, since it is nowadays mitigated, and during preliminary tests it resulted an uneffective attacking tool. We have also exluded meta attacks such as #hashDoS and ReDoS, since they

---

[1] Actually, before authenticating, some initial "unauthenticated" messages may be exchanged between client and server.

target application development implementation. Therefore, we have compared Slow Next to Slowloris, SlowReq, and Slow Read threats. In addition, we have included a legitimate situation.

Traffic sniffing operations have been accomplished on the targeted server. Each trial refers to a capture of $T_{capture} = 600\,$s of live network traffic. We have targeted an Apache2 web server, adopting a Wait Timeout of $T_{WT} = 60\,$s, which allows us to maintain the connections alive on most situations, since default Apache2 timeout is equal to $T = 300\,$s. Relatively to the Slow Next attack, as previously described, a different timeout is used, equal by default to $T_{keepalive} = 5\,$s. Therefore, only in this case, in order to avoid a server side connection closure, we have adopted a Wait Timeout equal to $T_{WT\_keepalive} = 4\,$s.

Payloads adopted during the attacks are default ones. Relatively to Slow Next, since tests refer to the HTTP protocol, HEAD requests are sent to the server. This choice allows us to reduce both request and response size.

Results obtained during the tests are shown in Table 1, where for each Δ parameter we have reported obtained average (μ) and standard deviation (σ).

We have observed that for all the attacks the required connections amount is successfully established (and maintained) with the server, thus making the primary goal of the threats accomplished. Relatively to the time needed to reach the DoS, all the attacks lead to similar results, since the DoS is reached almost instantly.

As shown in table, we have not reported traffic information relatively to the legitimate situation, since our purpose is to compare bandwidth consumption for the tested attacks, under reproducible conditions. Instead, due to its unpredictable behavior, these information would not be representative of a legitimate traffic. In particular, it's possible to notice that Slow Next appears to be the most bandwidth requiring threat. As previously described above, this is mainly caused by the reduced $T_{WT\_keepalive}$ value adopted by Apache2 web server. Although not in favor of Slow Next, our tests have been focused on the HTTP protocol since the other analyzed menaces are designed to target only such protocol.

Considering instead extrapolated Δ parameters, obtained Δ_{start} values are similar for all traffic conditions: indeed, the analyzed threats works by sending the first request (or part of the request) a few instants after a connection has been established. Concerning instead the Δ_{req} parameter, results show that Slow Next is undetectable using this metric, where other threats are. In particular, we have Δ_{req} or Δ_{resp} values equal to 0 when the request/response is composed by a single network packet. Analyzing other parameters, both Slowloris and SlowReq does not provide information retrieval, since they never send a complete request to the server. Hence, a detection of such menaces would be trivial [17].

Instead, relatively to the Slow Read attack, results show high Δ_{delay} values. In particular, once the request has been sent, a response message is sent from the server after some time (this time increases with the increase of load on the server). At this point, connections are maintained alive by the client/attacker by simulating a tiny reception buffer, at the transport layer. Therefore, although the anomaly found during the execution of is attack may not be addressed to a malicious node, we observe that it is possible to detect the attack through the Δ_{delay} parameter.

Table 1 Obtained results analyzing different situations on an Apache 2.2.22 web server for 600 s

Traffic nature			Legitimate	Slowloris	Slow req	Slow read	Slow Next
TRAFFIC	C→S	BYTES/S	–	2036.46	1721.76	6266.51	35226.16
		PACKETS	–	1376	1915	4989	27424
	S→C	BYTES/S	–	1166.73	1512.99	44980.39	68674.91
		PACKETS	–	1184	1701	4667	13976
Δ_{start}	μ		0.00912539	0.0275006	0.0271738	0.117814	0.000707241
	σ		0.0188789	0.160836	0.160704	0.562811	0.00283634
Δ_{req}	μ		0.0	424.287	549.215	0.0124508	0.0
	σ		0.0	132.624	0.932854	0.0880409	0.0
Δ_{delay}	μ		0.189399	–	–	28.3478	0.00077821
	σ		0.39023	–	–	65.3914	0.0219322
Δ_{resp}	μ		0.00181186	–	–	0.00291827	0.0
	σ		0.00906247	–	–	0.0044018	0.0
Δ_{next}	μ		1.44425	–	–	–	4.43899
	σ		1.30915	–	–	–	0.137005

Differently, results show that Slow Next results are considered legitimate for all the parameters: in case of the Δ_{next} metric, this value is different from legitimate one, but still in the same order. In particular, found values are compliant to our $T_{WT\_keepalive}$ choice. Nevertheless, although it would be simple to reduce $T_{WT\_keepalive}$ and increase σ to hinder Δ_{next} based detection approaches, the attack behavior is compliant to the protocol, and packet inspection operations would not be able to identify an anomaly.

5 Conclusions

In this paper we have introduced a novel Slow DoS Attack called Slow Next. The name of the attack derives from the Δ_{next} parameter it abuses, unexploited until now, accordingly to a previous analysis [5] of the typical TCP connection stream. In virtue of this novel exploitation, we believe that Slow Next represents an innovative contribution on the field. We have defined the menace as a meta attack, reporting some examples of practical implementations, and issues related to its deployment.

We have properly categorized the threat, presenting in detail the category of Timeout Exploiting Slow DoS Attacks, extending previous works on the topic [4, 5] and including Lazy Requests DoS attacks exploiting the Δ_{start} parameter.

We have analyzed attack behavior in terms of detectability, comparing the attack to other threats in the field. Our results show that the attack is more difficult to detect, due to its protocol and server compliant behavior. Nevertheless, we have analyzed how the attack ability to reach a DoS on a server may be reduced, due to possible server limits. This reasoning leads us to consider a possible mixed attack execution, combining Slow Next to other similar threats, with the aim of reducing detection. Further work will be focused on this strategy.

References

1. Chen, D., Chang, G., Sun, D., Jia, J., Wang, X.: Lightweight key management scheme to enhance the security of internet of things. Int. J. Wirel. Mob. Comput. 5(2), 191–198 (2012)
2. Gu, Q., Liu, P.: Denial of service attacks, Department of Computer Science Texas State UniversitySan Marcos School of Information Sciences and Technology Pennsylvania State University Denial of Service Attacks Outline, pp. 1–28 (2007)
3. Kumar, S., Singh, M., Sachdeva, M., Kumar, K.: Flooding based DDoS attacks and their influence on web services. (IJCSIT) Int. J. Comput. Sci. Inf. Technol. 2(3), 1131–1136 (2011)
4. Cambiaso, E., Papaleo, G., Chiola, G., Aiello, M.: Slow DoS attacks: definition and categorisation. Int. J. Trust Manag. Comput. Commun.—In press article (2013)
5. Aiello, M., Cambiaso, E., Scaglione, S., Papaleo, G.: A similarity based approach for application DoS attacks detection. In: The Eighteenth IEEE Symposium on Computers and Communications (2013)
6. Fielding, R., Gettys, J., Mogul, J., Frystyk, H., Masinter, L., Leach, P., Berners-Lee, T.: RFC 2616, Hypertext transfer protocol—HTTP/1.1. http://www.rfc.net/rfc2616.html

7. Giralte, L.C., Conde, C., de Diego, I.M., Cabello, E.: Detecting denial of service by modelling web-server behaviour. Comput. & Electr. Eng. (2012)
8. Aiello, M., Papaleo, G., Cambiaso, E.: SlowReq: a weapon for cyberwarfare operations. Characteristics, limits, performance, remediations. In: International Joint Conference SOCO'13-CISIS'13-ICEUTE'13, pp. 537–546 (2013)
9. Siriwardena, P.: Security by design. In: Advanced API Security, pp. 11–31. Springer (2014)
10. Jain, A., Chhabra, G.S.: Anti-forensics techniques: an analytical review. In: 2014 Seventh International Conference on Contemporary Computing (IC3), pp. 412–418 (2014)
11. Cambiaso, M.A.E., Papaleo, G.: Taxonomy of slow dos attacks to web applications. In: Recent Trends in Computer Networks and Distributed Systems Security, pp. 195–204. Springer, Heidelberg (2012)
12. Park, J., Iwai, K., Tanaka, H., Kurokawa, T.: Analysis of slow read DoS attack.In: 2014 International Symposium on Information Theory and its Applications (ISITA), pp. 60–64 (2014)
13. Corchado, E., Herrero: Neural visualization of network traffic data for intrusion detection. Appl. Soft Comput. **11**(2), 2042–2056 (2011)
14. Herrero, Navarro, M., Corchado, E., Julin, V.: RT-MOVICAB-IDS: addressing real-time intrusion detection. Future Gener. Comput. Syst. **29**(1), 250–261 (2013)
15. Kozik, R., Chora, M., Renk, R., Houbowicz, W.: Modelling HTTP requests with regular expressions for detection of cyber attacks targeted at web applications. In: International Joint Conference SOCO14-CISIS14-ICEUTE14, pp. 527–535 (2014)
16. Wu, S.X., Banzhaf, W.: The use of computational intelligence in intrusion detection systems: a review. Appl. Soft Comput. **10**(1), 1–35 (2010)
17. Aiello, M., Cambiaso, E., Mongelli, M., Papaleo, G.: An on-line intrusion detection approach to identify low-rate DoS attacks. In: 2014 International Carnahan Conference on Security Technology (ICCST), pp. 1–6 (2014)
18. Klensin, J.: RFC 2821: simple mail transfer protocol. http://tools.ietf.org/rfc/rfc2821

Non-Interactive Authentication and Confidential Information Exchange for Mobile Environments

Francisco Martín-Fernández, Pino Caballero-Gil
and Cándido Caballero-Gil

Abstract The growing number of devices that can connect to the Internet has given rise to a new concept that is having much impact nowadays, the Internet of Things. Thus, it is necessary to devise innovative security schemes to become accustomed to this new dimension of the Internet, where everything is connected to everything. This paper describes a new scheme for authentication and exchange of confidential information in the non-secure environment of the Internet of Things. The proposal is based on the concept of non-interactive zero-knowledge proofs, allowing that in a single communication, relevant data may be inferred for verifying the legitimacy of network nodes, and for sharing a session key. The proposal has been developed for the platforms built on the Android Open Source Project so it can be used both in smartphones and wearable devices. This paper provides a full description of the design, implementation and analysis of the proposed scheme. It also includes a comparison to similar schemes, which has revealed promising results.

Keywords Authentication · Security · Internet of things

1 Introduction

The number of devices connected to the Internet is growing exponentially on a daily basis. Nowadays it is quite common for any electronic device to communicate with other smart electronic devices. This phenomenon is known as the Internet of Things (IoT) and arises from the need to monitor and interconnect all electronic devices that

F. Martín-Fernández · P. Caballero-Gil (✉) · C. Caballero-Gil
Department of Computer Engineering, University of La Laguna,
38271 La Laguna, Tenerife, Spain
e-mail: pcaballe@ull.edu.es

F. Martín-Fernández
e-mail: francisco.martin.07@ull.edu.es

C. Caballero-Gil
e-mail: ccabgil@ull.edu.es

© Springer International Publishing Switzerland 2015
Á. Herrero et al. (eds.), *International Joint Conference*, Advances in Intelligent
Systems and Computing 369, DOI 10.1007/978-3-319-19713-5_23

are useful to humans [2]. In this novel situation, new challenges related to wireless security appear since wireless communication is the usual communication between hyperconnected devices. Thus, new lightweight cryptographic algorithms are necessary, so much research in this field is being done [3]. With the advent of increasingly powerful technology, which is reduced in size and weight, the cryptographic schemes used in wireless communications have been changing quickly in the last years. With the emergence of this new paradigm of interconnected objects, where the physical dimension mimics the logical dimension, it is necessary to encode more than 4.9 Billion objects in 2015 [15].

As aforementioned, a key aspect to consider is the way of communication between these objects because due to the mobile nature and the small size of many of these devices, this communication is usually wireless. In addition, the usual features of these objects cause that the wireless communication between them is established in the form of Mobile Ad-hoc NETworks (MANETs), which are networks composed of mobile devices, wirelessly connected, and generally characterized by properties of autoconfiguration. Each device that is part of a MANET has freedom to move, what implies that the link conditions between different devices change dynamically and that each node acts as a router for other nodes communications. Another important aspect of these networks is that, in general, they can operate independently or be connected to the Internet. This latter possibility is very useful in situations in which some devices do not have a direct Internet connection.

As regards security, in MANETs there is a number of different types of threats that can affect their use. In the literature on this issue, many of the proposals to protect communications in MANETs [10] are based either on secret-key cryptography [1] or on public-key cryptography [23]. The security of many symmetric schemes is strong, but their major drawback is the difficulty of the distribution of secret keys shared between pairs of nodes, what requires a secure channel. In an environment like MANETs applied to the Internet of Things, the assumption regarding the existence of a completely secure channel to transmit symmetric keys is a utopia. In addition, if the MANET is large and only based on symmetric cryptography, the number of secret keys that would be required would be very high. That is why, asymmetric cryptography was born in order to solve the problem of secure secret key distribution.

In particular, node authentication is here performed using an approach based on the idea of Zero-Knowledge Proof (ZKP) [13], which defines a method to prove the knowledge of certain piece of information, without revealing anything about it. Typical ZKPs are based on several challenges and responses, involving a successive exchange of messages, what implies the need of having a stable and continuous connection between nodes [7]. However, this assumption is unusual in a volatile environment like IoT, where sometimes devices move at a high speed, e.g. vehicles in Vehicular Ad-hoc NETworks (VANETs). In these cases, a massive exchange of messages to run a typical ZKP can be unfeasible due to possible connection failures during the protocol. In order to deal with this problem, the idea of Non-Interactive ZKP (NIZKP) has emerged in the related literature [18]. In a NIZKP, all the challenges of a typical ZKP are condensed into a single package sent in a few messages. This leads to the minimization of the time necessary for the exchange of messages.

This paper is divided into several sections. Section 2 describes the proposed non-interactive authentication scheme in detail. Section 3 provides some results of an implementation of the proposal. Section 4 includes a brief comparison between the proposal and other similar schemes. Finally, some conclusions and open problems are mentioned in Sect. 5.

2 Non-Interactive Authentication

Non-interactive authentication is characterized by the requirement of fewer messages than interactive authentication, which is fundamental for volatile and mobile environments, like the ones on the Internet of Things. For instance, it is of vital importance in intelligent vehicular networks, where nodes move at high speed and barely have a few milliseconds for communication. One of the most important factors of both interactive and non-interactive ZKP is the choice of the mathematical problem used as basis. In this paper, the chosen problem is the graph isomorphism problem. An isomorphism between two graphs is a bijection that preserves the adjacency relationship, i.e. any two vertices of a graph are adjacent if and only if so are their images in the other graph. The graph isomorphism problem consists in determining whether two graphs are isomorphic or not. This problem has been used in cryptography [16, 17] since an efficient algorithm to solve it in general is yet unknown. In particular, the determination of whether two graphs with the same number v of vertices and the same number e of edges are isomorphic or not involves a brute force attack because it requires checking whether some of the $v!$ possible bijections preserve adjacency. In general, the graph isomorphism problem is one of a few problems in computational complexity theory belonging to NP, but it is not known whether it belongs to P or to NP-complete subset [14]. Therefore, this paper proposes the use of difficult graphs, such as, for instance, strongly regular graphs, Hadamard matrices or projective planes [22].

The proposed scheme is based on a variant of NIZKP that requires just a single message to verify the knowledge, and can be adapted to a required security level so that the greater number of different challenges considered in the scheme, the better security level for the verifier. Specifically, the parameters of the proposal are shown in Table 1.

According to the proposal, each node has to send a message to authenticate itself as a legitimate network node. That message is composed of several commitments defined by isomorphic graphs generated from an original graph known by all legitimate nodes. For example, the original graph might be a graph where the nodes represent all the users of the network.

As regards the message, this is divided into $n+1$ segments that are all encrypted with different keys, except the first segment that is not encrypted. Thus, a legitimate network user can authenticate itself to join a communication session with another node if this latter node is able to decrypt all the segments of the message broadcast by the first one, so that it can reach the last segment, which contains the secret to

Table 1 Proposal parameters

Notation	Meaning
G	Graph known by all legitimate nodes, on which they know how to solve a hard problem
Sol_G	Solution to the hard problem in G
Cha_i	$i-th$ Challenge proposed by the verifier
G_i	$i-th$ isomorphic Graph used as a commitment
Iso_i	Isomorphism between G and G_i
Res_i	$i-th$ Response corresponding to the Challenge Cha_i on the Graph G_i
$h(\cdot)$	Hash function
$LSB(\cdot)$	Least Significant Bit of an input string
$E_{k_i}(\cdot)$	Symmetric Encryption with key k_i
$Subkey$	Secret Subkey to reveal

share. The encryption key of each segment depends on the previous segment. Thus, although someone wants to decipher only the last segment, this is impossible because that would require the decryption of all previous segments. The security level of the scheme depends on the number of segments of the message, which represent different challenges. The greater the number of segments, the more complex it is to reach the last segment and to obtain the confidential information. That is to say, the scheme can be used to share subkeys in order to establish a common session key. In this way, after a two-way authentication process using the same procedure, based on the idea of the Diffie-Hellman scheme [9], both nodes will know the shared session key obtained with both exchanged subkeys (see Fig. 1).

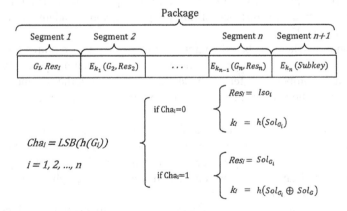

Fig. 1 Components of sent messages

Each segment contains an isomorphic copy of the original graph. A one-way hash function known by all legitimate network nodes must be chosen so that this hash function can be used to define the challenge that the receiver must solve on each isomorphic graph. Furthermore, the hash function is used to define the encryption key for each message segment.

The operations that the receiver must perform on the message are:

1. Process the first segment of the message.
2. Compute, by using the hash function, the challenge that matches the information included in the segment.
3. Check whether the response corresponds to the challenge and isomorphic graph or not.
4. From the challenge, compute the key to decrypt the next segment.
5. Apply steps 2 to 4 until the last segment, which once deciphered contains the information needed to establish the shared secret.

If the used hash function, problem and graph are adequate, the probability that $Cha_i = 0$ is $1/2$. Thus, the probability that a legitimate node knows the key k_1 is $1/2$, that it knows the two keys k_1 and k_2 is $1/2^2$,..., and that it knows the n keys k_1, k_2, ..., k_n is $1/2^n$.

The challenges have been chosen as in the well-known ZKP based on isomorphic graphs. However, in the NIZKP here proposed, the challenges are defined from the result of a Boolean output of a hash function defined through the LSB applied on each committed isomorphic graph. Thus, for each challenge, the response is defined as follows:

- If *Challenge* $= 0$, the response is the isomorphism.
- If *Challenge* $= 1$, the response is the solution to the problem in the isomorphic graph.

After running the algorithm, there is access to the last segment of the message through its decryption with the key returned by the previous segment. The last segment allows the obtaining of the contribution of the sender node to the session key shared with each potential partner or any other sensitive information.

All legitimate network users know both the original graph and a secret key for this graph, which is a solution to a hard problem in this graph. The scheme proposed as a solution, a Hamiltonian Cycle, because the Hamiltonian Cycle Problem for arbitrary graphs can be NP-complete, that will be the kind of problems that the schema use. Hamiltonian Cycle is a cycle in a graph that visits each vertex only once.

The Hamiltonian Cycle Problem consists in determining whether there is a path in the graph that visits each vertex exactly once. This problem is often considered to be NP-complete, but there are some particular graphs for which the problem is polynomial or even linear. Because of this, in this paper, if the Hamiltonian Cycle Problem is used, we suggest the use of non-planar graphs. A graph is planar if it can be drawn in a plane without graph edges crossing.

Another choice for the implementation is the hash function. In particular, the hash function used to compute the challenges and encryption keys of each message segment has been the new standard hash function SHA-3 [4, 5] because of its adequate features.

As for the symmetric encryption used to encrypt the segments of the message, in the implementation we have applied the stream cipher used in the fourth generation of mobile communications (4G LTE) [12], known as SNOW 3G [11], because one of its advantages is a linear computational complexity both in encryption and decryption, what ensures efficiency and speed of encryption and decryption processes.

3 Implementation and Evaluation

One of the most fundamental aspects of any proposed authentication scheme is whether it has been tested in real environments or not. In order to perform a valid test, its implementation in devices that are widely used is necessary. The scheme described here is intended for its use in areas related to the Internet of Things. Therefore, this scheme has been implemented in this work for different Androids platforms (Android, Android Wear, Android TV, Android Auto, etc.), which are examples of the proliferation of devices in this new dimension of the Internet. The evaluation was only considered on Android smartphones and other Android devices because this operating system is the most widespread in mobile devices. Consequently, it will be possible to immediately implement the scheme in order to reach a large mass of users, without acquiring new devices. Indeed, Android is the most popular smartphone operating system, with over 70 % market share worldwide [20]. Android Wear is the operating system based on Android for wearable devices of the same company. Android Wear has more than 90 % of the market share in the devices of its kind. The Android ecosystem is growing rapidly, and there are many new Android operating systems in environments as diverse as smart TVs, vehicles, etc. Therefore, all the results presented here are the result of the implementation of the scheme in these platforms belonging to Android Open Source Project. The source code is open source under a Git repository on GitHub Platform [21].

An important part of the analysis of the scheme implementation has been the measure of package sizes, and the conclusion is that very satisfactory sizes have been achieved. The message size is given by the dimension of the graph used to represent the network. The more nodes a graph has got, the more space it is needed, which involves that the segments will be larger, resulting thus, in larger package sizes. We have analysed which sizes per segment are recommended depending on the number of nodes in the graph, which are shown in Table 2.

Regarding the package generation time, we have obtained acceptable results. The computational time required for the devices to create the packages to be sent depends on the number of segments that are defined in the package. Thus, the larger the number of segments, the higher its security, and the more time the microprocessor of the device requires for generating the package. For the tests, LG Nexus smartphones

Table 2 Segment size based on the number of nodes

Number of nodes	Segment size
1	17 bytes
5	49 bytes
10	161 bytes
15	305 bytes
50	2721 bytes (2.7 Kb)
75	5921 bytes (5.9 Kb)
100	10321 bytes (10.3 Kb)
150	22849 bytes (22.8 Kb)
200	40273 bytes (40.2 Kb)

have been used because these are the devices that Google recommends for Android developers. The specific model that has been used here is the LG Nexus 5. Considering all these features, the programming language that has been used is Java because it is the most widely used platform for programming in the Android Open Source Project. The time required for segment generation has been studied depending on the number of nodes in the graph, and the results are shown in Table 3.

Table 3 Segment generation time depending on the number of nodes

Number of nodes	Segment time generation
1	4 ms
5	5 ms
10	6 ms
15	7 ms
50	50 ms
75	210 ms
100	570 ms
150	2806 ms (~3 s)
200	9305 ms (~9 s)

It was necessary to measure the time required for a receiver to decrypt and process a package. In particular, we analysed the time it would take a mobile device to decipher the package received from other mobile devices. These package-processing times are really short, as evidenced by the processing times per segment shown in Table 4.

The conclusion is that for large networks with more than 100 nodes, the use of clusters [8] is advised so that the protocol can be applied with fewer nodes.

Table 4 Segment processing time depending on the number of nodes

Number of nodes	Segment processing time
1	1 ms
5	1.5 ms
10	2 ms
15	3 ms
50	8 ms
75	20 ms
100	32 ms
150	80 ms
200	170 ms

4 Comparative Analysis

In recent related literature, it is not easy to find novel schemes proposed for efficient authentication of mobile devices to be used specifically in the Internet of Things. In order to provide a brief comparative analysis of the proposal, a representative lightweight authentication scheme [19] used in the Internet of Things was implemented and analysed.

Since the ZKP authentication algorithm proposed in [19] is also based on isomorphic graphs, it was chosen for the comparative analysis. Thus, it was then implemented and evaluated. Such a mechanism allows authentication with varying confidence and security levels. That work describes its implementation on conventional computers (with different configurations), so that the starting conditions are more powerful than when using mobile devices as in this paper, mainly because of the efficiency of the used programming language. On the one hand, the scheme proposed in this paper uses Java as a programming language, so that the implementation is as cross-platform as possible, and thus it can be applied to many current mobile devices. On the other hand, the scheme proposed in [19] uses random graphs of 41 nodes so in order to allow the comparison, the scheme proposed in this paper has also been implemented with graphs of 41 nodes. The comparison results are shown in Table 5, where the three different hardware configurations described in the paper are shown.

Table 5 Comparison between schemes

		Conf1. [19]	Conf2. [19]	Conf3. [19]	Our scheme
10 Challenges	Time (ms)	469	484	1522	**454**
	Size (bytes)	**4045**	4045	4045	17826

Regarding the results shown in Table 5, a remarkable consideration is the fact that the scheme proposed in [19] uses an interactive ZKP with 6 exchanges of messages, while the proposal of this paper is based on a NIZKP, in which only one single message is necessary.

After reviewing relevant bibliography, we have chosen a scheme that implements an authenticated Diffie-Hellman protocol for the aforementioned comparison. This scheme is called Password-Authenticated Key (PAK) Diffie-Hellman exchange [6] and it suggests adding mutual authentication based on a memorisable password to the basic unauthenticated Diffie-Hellman key exchange. For the comparison, we have implemented PAK scheme for Android Open Source Project (AOSP) devices. The results of the comparison are shown in Table 6. As a conclusion, it may be said that in general the scheme proposed in this paper has a similar performance to the PAK scheme, and even in some cases the proposed scheme slightly improves the results of the PAK scheme.

Table 6 PAK scheme versus proposed scheme

PAK scheme	Our scheme	
Time (ms)	Challenges	Time (ms)
197	3	86
	4	112
	5	153
	6	176
	7	195
	8	221

5 Conclusions

Nowadays, most electronic devices can connect to a network. This feature allows almost anything to connect to any other thing, what is known as the Internet of Things. This proliferation of connected devices is so popular that, to secure communications, it is necessary to adapt existing cryptographic algorithms to make them more lightweight, due to the special requirements of most of these devices. In this paper, a novel authentication scheme for the IoT, based on the idea of non-interactive zero-knowledge proofs, has been presented. The scheme proposed here is based just on a single authenticated message that has to be sent in order to be able to exchange confidential information with another network user in an insecure network environment, such as the IoT. The proposed scheme can also be used to share confidential information based on the idea of Diffie-Hellman protocol. In addition to the design of the proposal, this paper also includes conclusions obtained from the implementation and analysis of the scheme, showing promising results of a preliminary comparison with another similar proposal. The described development was performed on the Android platform under the seal of Android Open Source Project, to take advantage of its market share in the emerging Internet of Things. Finally, it must be said that this paper is part of some work in progress. Therefore, a more complete performance analysis is being done through the implementation in more varied devices, and a more thorough comparative analysis with different schemes is being performed.

Acknowledgments Research supported under TIN2011-25452, IPT-2012-0585-370000, BES-2012-051817, and RTC-2014-1648-8.

References

1. Agren, M.: On some symmetric lightweight cryptographic designs. Diss. Lund University (2012)
2. Atzori, L., Iera, A., Morabito, G.: The internet of things: a survey. Comput. Netw. **54**(15), 2787–2805 (2010)
3. Bansod, G., Raval, N., Pisharoty, N.: Implementation of a new lightweight encryption design for embedded security. IEEE Trans. Inf. Forensics Secur. **10**(1) (2015)
4. Bertoni, G., Daemen, J., Peeters, M., Van Assche, G., Keccak sponge function family main document. Updated submission to NIST, Round 2, version 2.1 (2010)
5. Bertoni, G., Daemen, J., Peeters, M., Van Assche, G.: The Keccak SHA-3 submission. http://keccak.noekeon.org/Keccak-submission-3.pdf
6. Brusilovsky, A., Faynberg, I., Zeltsan, Z., Patel, S.: Password-Authenticated Key (PAK) Diffie-Hellman Exchange. RFC 5683 (2010)
7. Caballero-Gil, P., Caballero-Gil, C., Molina-Gil, J., Hernández-Goya, C.: Self-organized authentication architecture for mobile Ad-hoc networks. In: IEEE International Symposium on Modeling and Optimization in Mobile, Ad Hoc, and Wireless Networks and Workshops WiOPT, pp. 217–224 (2008)
8. Caballero-Gil, C., Caballero-Gil, P., Molina-Gil, J.: Knowledge management using clusters in VANETs-description, simulation and analysis. In: International Conference on Knowledge Management and Information Sharing, pp. 170–175 (2010)
9. Diffie, W., Hellman, M.E.: New directions in cryptography. IEEE Trans. Inf. Theory **22**, 644–654 (1976)
10. Eisenbarth, T., Kumar, S., Paar, C., Poschmann, A., Uhsadel, L.: A survey of lightweight-cryptography implementations. IEEE Des. Test Comput. **4**(6), 522–533 (2007)
11. Ekdahl, P., Johansson, T.: A new version of the stream cipher SNOW. Selected Areas in Cryptography. Lecture Notes in Computer Science, vol. 2595, pp. 37–46 (2003)
12. ETSI/SAGE: Specification of the 3GPP confidentiality and integrity algorithms UEA2 and UIA2. Document 2, SNOW 3G Specification, version 1.1 (2005)
13. Feige, U., Fiat, A., Shamir, A.: Zero-knowledge proofs of identity. J. Cryptol. **1**(2), 77–94 (1988)
14. Garey, M.R., Johnson, D.S.: Computers and intractability: a guide the theory of NP-completeness. Freeman and Co., (1979)
15. Gartner, Analysts to explore the disruptive impact of IoT on business. In: Gartner Symposium/ITxpo, Spain (2014)
16. Goldreich, O., Micali, S., Wigderson, A.: Proofs that yield nothing but their validity or all languages in NP have zero-knowledge proof systems. J. ACM **38**(3), 690–728 (1991)
17. Goldwasser, S., Micali, S., Rackoff, C.: The knowledge complexity of interactive proof systems. SIAM J. Comput. **18**(1), 186–208 (1989)
18. Groth, J.: Short non-interactive zero-knowledge proofs. In: Advances in Cryptology—ASIACRYPT, pp. 341–358 (2010)
19. Grzonkowski, S., Zaremba, W., Zaremba, M., McDaniel, B.: Extending web applications with a lightweight zero knowledge proof authentication. In: ACM International Conference on Soft Computing as Transdisciplinary Science and Technology, pp. 65–70 (2008)

20. Kantar, Kantar Worldpanel ComTech. http://www.kantarworldpanel.com/global/smartphone-os-market-share/ (2015)
21. Martin-Fernandez, F.: Source code of the proposed authenticated scheme. https://github.com/pacomf/ASD (2014)
22. Schweitzer, P.: Problems of unknown complexity. Graph isomorphism and Ramsey theoretic numbers. Diss. University of Saarlandes (2009)
23. Toorani, M., Beheshti, A.: LPKI—a lightweight public key infrastructure for the mobile environments. In: IEEE Singapore International Conference on Communication Systems, pp. 162–165 (2008)

Spatio-Temporal Spreading of Mobile Malware: A Model Based on Difference Equations

A. Martín del Rey, A. Hernández Encinas, J. Martín Vaquero,
A. Queiruga Dios and G. Rodríguez Sánchez

Abstract In this work a novel mathematical model to simulate the spatio-temporal spreading of mobile malware is introduced. It is a compartmental model where the mobile devices are grouped into two classes: susceptibles and infected devices, and the malware spreads via bluetooth. There are few models dealing with the spreading of mobile malware using bluetooth connections and all of them only study the temporal evolution. Due to the main characteristics of bluetooth it is of interest to simulate not only the temporal evolution but also the spatial spreading, and consequently, this is the main goal of this work. In our model the global dynamic is governed by means of a system of difference equations and the transmission vector is defined by the bluetooth connections. Explicit conditions for spreading are given in terms of the number of susceptible individuals at a particular time step. These could serve as a basis for design control strategies.

Keywords Malware · Epidemic spreading · Difference equations · Spatio-temporal dynamics

A. Martín del Rey (✉) · A. Hernández Encinas · J. Martín Vaquero ·
A. Queiruga Dios · G. Rodríguez Sánchez
Department of Applied Mathematics, University of Salamanca, Salamanca, Spain
e-mail: delrey@usal.es

A. Hernández Encinas
e-mail: ascen@usal.es

J. Martín Vaquero
e-mail: jesmarva@usal.es

A. Queiruga Dios
e-mail: queirugadios@usal.es

G. Rodríguez Sánchez
e-mail: gerardo@usal.es

© Springer International Publishing Switzerland 2015
Á. Herrero et al. (eds.), *International Joint Conference*, Advances in Intelligent
Systems and Computing 369, DOI 10.1007/978-3-319-19713-5_24

1 Introduction

Smartphones are mobile phones with extended capabilities that make them very similar to handheld personal computers and tablets. The uses of smartphones are varied: calling, SMS, instant messaging, web browsing, social networking, ubiquitous access, etc., and in a near future they will play an important role in the Internet of Things. Smartphone use has greatly expanded since the beginning of the 21st century. In this sense, the information technology research and advisory company IMS Research expects smartphones to reach 1 billion in annual sales in 2016 (half the mobile device market).

Smartphones are used both within the corporate world and particular consumers. The companies work in the so-called "open space", where the employees are connected wherever they work: in the office, at home, or on the road. As mobile devices are intrinsically less secure, the security of smartphone is emerging as a key task, being the most important threat that one due to the action of malware.

Malware can be defined as an undesirable and dangerous software that gets installed on the mobile device often through stealth or deception. The main threats of mobile malware are information leakage (there are sensible information stored in the smartphones: IMSI (International Mobile Subscriber Identity), IMEI (International Mobile system Equipment Identity), contacts, GPS position, etc.), monetary loss (some types of mobile malware causes financial loss by sending expensive calls or SMS without the user's authorization), and restricted device usage (the legitimate users cannot use their smartphones due to changing background images or invisible texts on the devices).

Consequently, it is very important to design computational tools, based on mathematical models, that allows us to predict the behavior of a malware spreading in a smartphone network. These mathematical models are compartmental models where the devices are classified into different classes or compartments: susceptible, infected, recovered, etc. Although there exist several mathematical models to study the spreading of malware in a computer network (see, for example, [4, 8, 16, 19] and references therein) there are so few dedicated to the study of mobile phone malware.

In a similar way to which it is done with computer networks, the great majority of these works are based on systems of ordinary differential equations (see, for example [2, 13, 17]). In a more precise way, a general model is proposed in [12] to study the propagation of malware in cell phones networks; in [5, 18] two mathematical models to characterize the propagation of bluetooth worms are introduced; a model called *probabilistic queuing* is introduced in [9] which is based on a modification of Kephart-White model for mobile environments; the mathematical simulation the spreading of wireless worms between mobile devices is given in [14]; Also a mathematical model based on a system of differential equations is shown in [15] in order to model the botnet interactions.

All these models are based on the use of differential equations; this is a consequence of fact that the traditional use of continuous mathematical tools in epidemic modeling (infectious diseases on humans or animals, and malware on personal computers). Other models based on different mathematical primitives have also been appeared in the scientific literature: in [10] a stochastic model based on semi-Markov processes is presented, in [6, 11] both models based on cellular automata are proposed, and in [7] a model using recurrence relations is introduced.

In these models, the temporal evolution of the number of the different classes of mobile devices is simulated and analyzed but, unfortunately, to the best of our knowledge any model has been proposed simulating the spatio-temporal evolution of the compartments. This is precisely the aim of this work: to design a novel mathematical model to simulate the behavior of spatio-temporal spreading of mobile malware based on a system of difference equations. This is a SIS model where susceptible and infected devices are considered, and the transmission vector is given by bluetooth connections.

The rest of the paper is organized as follows: In Sect. 2 the description of the model to simulate malware spreading and some illustrative simulations are introduced; The analysis of the model is done in Sect. 3, and, finally, the conclusions and further work are shown in Sect. 4.

2 Description of the Model for Mobile Malware Propagation

2.1 Main Assumptions

In the mathematical model introduced in this work to study the spatio-temporal spreading of mobile malware, the following assumptions are done related to the environmental situation, the characteristics of the devices and the mobile malware spreading:

(1) Bluetooth technology stands for the transmission vector of the mobile malware.
(2) The mobile devices are located in an enclosed area which is represented as a series of tessellated identical square shapes. This is a finite area and, consequently, there are a finite number of square grid cells: $\{(1, 1), \dots, (n, m)\}$.
(3) The neighborhood of each cell (i, j) is denoted by $\mathcal{V}_{ij}$ and stands for the four existing nearest cells located around in (see Fig. 1):

$$\mathcal{V}_{ij} = \{(i - 1, j - 1), (i, j - 1), (i, j + 1), (i + 1, j)\}. \tag{1}$$

Fig. 1 Von Neumann neighborhood

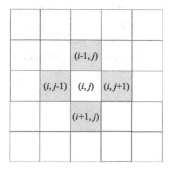

Due to the finiteness of the area, null boundary conditions must be considered, that is:

$$\mathcal{V}_{1j} = \{(1, j-1), (1, j), (1, j+1), (2, j)\}, \tag{2}$$
$$\mathcal{V}_{nj} = \{(n-1, j), (n, j-1), (n, j), (n, j+1)\}, \tag{3}$$
$$\mathcal{V}_{i1} = \{(i-1, 1), (i, 1), (i+1, 1), (i, 2)\}, \tag{4}$$
$$\mathcal{V}_{im} = \{(i-1, m), (i, m-1), (i, m), (i+1, m)\}, \tag{5}$$
$$\mathcal{V}_{11} = \{(1, 1), (1, 2), (2, 1)\}, \tag{6}$$
$$\mathcal{V}_{n1} = \{(n-1, 1), (n, 1), (n, 2)\}, \tag{7}$$
$$\mathcal{V}_{1m} = \{(1, m-1), (1, m), (2, m)\}, \tag{8}$$
$$\mathcal{V}_{nm} = \{(n-1, m), (n, m-1), (n, m)\}. \tag{9}$$

(4) The total number of mobile devices remains constant over the time: N. Moreover if N_{ij}^t stands for the number of devices moving through the cell (i, j) at time t then:

$$N = \sum_{\substack{1 \le i \le n \\ 1 \le j \le m}} N_{ij}^t, \tag{10}$$

for every t.

(5) The bluetooth devices are grouped in two classes: susceptible devices and infected devices. The number of susceptibles (*resp.* infected) devices located in the cell (i, j) at time t is denoted by S_{ij}^t (*resp.* I_{ij}^t) such that $N_{ij}^t = S_{ij}^t + I_{ij}^t$ for every t. Furthermore, we will set

$$S^t = \sum_{\substack{1 \le i \le n \\ 1 \le j \le m}} S_{ij}^t, \quad I^t = \sum_{\substack{1 \le i \le n \\ 1 \le j \le m}} I_{ij}^t. \tag{11}$$

(6) The mobile devices can move freely in the area taking into account the following: (1) during each step of time the movement is limited to the cell where the device is located, and (2) the mobile devices can move from the main cell to a neighbor cell when the time step changes.

(7) It is supposed that the percentage of mobile devices that move from one cell to a neighbor cell depends on the number of mobile devices located in the main cell so that the greater the number of devices in the cell, the greater the percentage of devices that moves to a neighbor cell. In this sense, this percentage will be defined by the parameter

$$0 \le m_{ij} = \frac{N_{ij}^t}{N^t} \le 1. \tag{12}$$

(8) Finally, it is suppose that the total number of devices that move from a cell to a neighbor depends on the number of devices located at the neighbor cell. Consequently, the percentage of devices that migrate from the cell (i,j) to the neighbor cell (u, v) is defined by the parameter $N_{uv}^t/\tilde{N}_{ij}^t$, where $\tilde{N}_{ij}^t$ is the total number of devices located in the neighborhood of (i,j) at time t:

$$\tilde{N}_{ij}^t = \sum_{(u,v)\in\mathcal{V}_{ij}} N_{uv}^t. \tag{13}$$

2.2 Dynamic of the Model

The dynamic of the model is as follows: during every step of time t the mobile devices located in the cell (i,j) interact with each other (being infected -or not- or recovered -or not-) and subsequently some of them move to a neighbor cell, locating there at step of time $t + 1$.

The susceptible devices of the cell (i,j) are infected during the step of time t at infection rate $0 \le a_{ij} \le 1$, which depends on the characteristics of the cell (level of noise present in the cell hindering the bluetooth connections, etc.).

On the other hand, the infected devices of (i,j) recovered at recovery rate $0 \le b \le 1$.

Consequently the number of susceptible and infected devices located in the cell (i,j) at time step $t + 1$ is given by the following transition difference equations:

$$S_{ij}^{t+1} = \left(1 - m_{ij}^t\right) S_{ij}^t - a_{ij}\left(1 - m_{ij}^t\right) S_{ij}^t I_{ij}^t + b\left(1 - m_{ij}^t\right) I_{ij}^t$$
$$+ \sum_{(u,v)\in\mathcal{V}_{ij}} \frac{N_{ij}^t}{\tilde{N}_{uv}^t} m_{uv}^t \left(S_{uv}^t + bI_{uv}^t - a_{uv}S_{uv}^t I_{uv}^t\right), \tag{14}$$

$$I_{ij}^{t+1} = \left(1 - b - m_{ij}^t + bm_{ij}^t\right) I_{ij}^t + a_{ij}\left(1 - m_{ij}^t\right) S_{ij}^t I_{ij}^t$$
$$+ \sum_{(u,v)\in\mathcal{V}_{ij}} \frac{N_{ij}^t}{\tilde{N}_{uv}^t} m_{uv}^t \left((1 - b) I_{uv}^t + a_{uv}S_{uv}^t I_{uv}^t\right). \tag{15}$$

The meaning of each term of these equations is shown in Table 1.

2.3 Some Illustrative Simulations

In what follows we illustrate the model introduced in the last section with some examples. In all of them it is supposed that the area is tessellated into 20×20 identical square cells and $0 \le t \le 500$.

In the first group of simulations we show the temporal evolution of the two classes of mobile devices when $b = 0.15$, $S_{ij}^0 = 10$, and

$$I_{ij}^t = \begin{cases} 1, \text{ if } (i,j) = (10, 10) \\ 0, \text{ otherwise} \end{cases} \tag{16}$$

for $1 \leq i,j \leq 20$. In Fig. 2 the evolution of the number of susceptible and infected devices for different values of the infection rate is shown. Note that in these simulations it is supposed that the infection rate a_{ij} is constant for every cell (i,j).

In the second group of simulations the spatio-temporal evolution of the susceptible and infected mobile devices is introduced. The parameters used are the following: $a_{ij} = 0.03$ for every (i,j), $b = 0.15$, $S_{ij}^0 = 20$ and $0 \leq I_{ij}^t \leq 5$ is randomly chosen for $1 \leq i,j \leq 20$. In Fig. 3 the distribution of the percentage of susceptible and infected devices between the cells is shown at $t = 0, 5, 10$ and 15.

Table 1 Description of the terms appearing in transition difference equation

Term of the equation	Description
$\left(1 - m_{ij}^t S_{ij}^t\right)$	Susceptible devices of (i,j) at t that have not been infected and remain in (i,j) at $t+1$
$a_{ij}\left(1 - m_{ij}^t\right) S_{ij}^t I_{ij}^t$	Susceptible devices of (i,j) at t that have been infected and remain in (i,j) at $t+1$
$b\left(1 - m_{ij}^t\right) I_{ij}^t$	Infected devices of (i,j) at t that have been recovered and remain in (i,j) at $t+1$
$\sum_{(u,v) \in \mathcal{V}_{ij}} \frac{N_{ij}^t}{\bar{N}_{uv}^t} m_{uv}^t \left(S_{uv}^t - a_{uv} S_{uv}^t I_{uv}^t\right)$	Susceptible devices of $\mathcal{V}_{ij}$ at t that have not been infected and move to (i,j) at $t+1$
$\sum_{(u,v) \in \mathcal{V}_{ij}} \frac{N_{ij}^t}{\bar{N}_{uv}^t} m_{uv}^t b I_{uv}^t$	Infected devices of $\mathcal{V}_{ij}$ at t that have been recovered and move to (i,j) at $t+1$
$\left(1 - b - m_{ij}^t + bm_{ij}^t\right) I_{ij}^t$	Infected devices of (i,j) at t that have not been recovered and remain in (i,j) at $t+1$
$a_{ij}\left(1 - m_{ij}^t\right) S_{ij}^t I_{ij}^t$	Susceptible devices of (i,j) at t that have been infected and remain in (i,j) at $t+1$
$\sum_{(u,v) \in \mathcal{V}_{ij}} \frac{N_{ij}^t}{\bar{N}_{uv}^t} m_{uv}^t (1 - b) I_{uv}^t$	Infected devices of $\mathcal{V}_{ij}$ at t that have not been recovered and move to (i,j) at $t+1$
$\sum_{(u,v) \in \mathcal{V}_{ij}} \frac{N_{ij}^t}{\bar{N}_{uv}^t} m_{uv}^t a_{uv} S_{uv}^t I_{uv}^t$	Susceptible devices of $\mathcal{V}_{ij}$ at t that have been infected and move to (i,j) at $t+1$

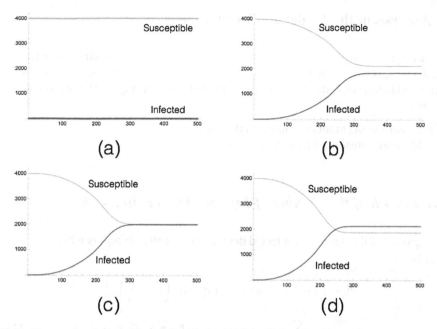

Fig. 2 Temporal evolution of the number of susceptible and infected mobile devices: (a) $a_{ij} = 0.01$; (b) $a_{ij} = 0.028$; (c) $a_{ij} = 0.03$; (d) $a_{ij} = 0.032$

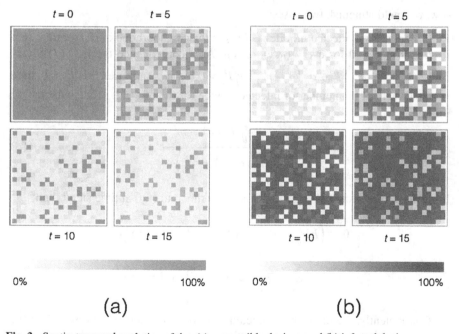

Fig. 3 Spatio-temporal evolution of the: (a) susceptible devices, and (b) infected devices

3 Analysis of the Model and Discussion

In any epidemiological model is necessary to study under what circumstances the epidemic spreads. Taking into account the particular characteristics of the model proposed in this work, the following two conditions will be required for such behavior occurs:

(1) Growth of the number of infected devices.
(2) Malware spreading between the cells.

3.1 Growth of the Number of Infected Devices in a Cell

The growth of the number of infected devices in the cell (i,j) occurs when $I_{ij}^{t+1} > I_{ij}^t$, that is:

$$I_{ij}^t < \left(1 - b - m_{ij}^t + bm_{ij}^t\right) I_{ij}^t + a_{ij}\left(1 - m_{ij}^t\right) S_{ij}^t I_{ij}^t$$
$$+ \sum_{(u,v)\in \mathcal{V}_{ij}} \frac{N_{ij}^t}{\tilde{N}_{uv}^t} m_{uv}^t \left((1 - b) I_{uv}^t + a_{uv} S_{uv}^t I_{uv}^t\right). \qquad (17)$$

Now, we can distinguish two cases:

(a) If there are not infected mobile devices in the neighborhood of (i,j) at time step t, then $I_{uv}^t = 0$ for every $(u,v) \in \mathcal{V}_{ij}$. As a consequence the Eq. (17) yields $\left(bm_{ij}^t - b - m_{ij}^t\right) I_{ij}^t + a_{ij}\left(1 - m_{ij}^t\right) S_{ij}^t I_{ij}^t > 0$. As it is supposed that $I_{ij}^t > 0$, then the growth of the number of infected devices of the cell (i,j) occurs when:

$$S_{ij}^t > \frac{m_{ij}^t + b\left(1 - m_{ij}^t\right)}{a_{ij}\left(1 - m_{ij}^t\right)}. \qquad (18)$$

(b) If there are infected devices in the neighborhood of (i,j) at time step t then:

$$0 < \left(bm_{ij}^t - b - m_{ij}^t\right) I_{ij}^t + a_{ij}\left(1 - m_{ij}^t\right) S_{ij}^t I_{ij}^t$$
$$+ \sum_{(u,v)\in \mathcal{V}_{ij}} \frac{N_{ij}^t}{\tilde{N}_{uv}^t} m_{uv}^t \left((1 - b) I_{uv}^t + a_{uv} S_{uv}^t I_{uv}^t\right). \qquad (19)$$

Consequently, the number of infected devices in the cell (i,j) grows when the number of new infected mobile devices in (i,j) is greater than the number of infected devices that disappear from (i,j).

3.2 Malware Spreading Between Cells

Suppose that $I_{ij}^t = 0$ at a particular time step t, and there are infected mobile devices located in the neighbor cells of (i,j) at t. Then the malware spreading to the cell (i,j) occurs when $I_{ij}^{t+1} > 0$, that is, when the following equation holds:

$$\sum_{(u,v)\in\mathcal{V}_{ij}} \frac{N_{ij}^t}{\tilde{N}_{uv}^t} m_{uv}^t \left((1-b)I_{uv}^t + a_{uv}S_{uv}^t I_{uv}^t\right) > 0. \tag{20}$$

Moreover, this condition is satisfied if there exists a neighbor cell (u,v) such that:

$$\frac{N_{ij}^t}{\tilde{N}_{uv}^t} m_{uv}^t \left((1-b)I_{uv}^t + a_{uv}S_{uv}^t I_{uv}^t\right) > 0. \tag{21}$$

Then a simple computation shows that the malware spreading occurs if there exists a neighbor cell of (i,j), (u,v), such that

$$I_{uv}^t > 0, \tilde{N}_{uv}^t > 0, N_{ij}^t > 0, m_{uv}^t > 0, S_{uv}^t > \frac{b-1}{a_{uv}}. \tag{22}$$

4 Conclusions

In this work a novel mathematical model to simulate the spatio-temporal spreading of mobile malware via bluetooth connections is proposed. The mobile devices are located in an enclosed area which can be tessellated into identical square cells whose dimensions depend on the connection range of the bluetooth devices.

The number of the mobile devices moving between the cells remains constant over the time and they can be classified into two compartments: susceptibles and infected devices. As a consequence the model proposed is a compartmental SIS model and its dynamic is governed by a system of difference equations.

The variables involved in the system represent the number of susceptible and infected mobile devices located in a cell at a particular time step t: S_{ij}^t and I_{ij}^t, respectively. As a consequence simulations of the spatio-temporal spreading of the mobile malware are obtained.

The analysis of the model allows us to state the following:

(1) If there are not infected devices in the neighborhood of a particular cell (i,j), the number of infected devices located in (i,j) increases when

$$S_{ij}^t > \frac{m_{ij}^t + b\left(1 - m_{ij}^t\right)}{a_{ij}\left(1 - m_{ij}^t\right)}, \tag{23}$$

where a_{ij} is the transmission rate corresponding to the cell (i,j), b is the recovery rate, and m_{ij}^t is the fraction of mobile devices that leave the cell (i,j).

(2) If there exist infected devices in the neighborhood of a cell, the number of infected devices in such cell increases when the number of new infected devices of the cell is greater than the number of infected devices that disappear from the cell.

(3) A necessary condition for malware spreading to a particular cell (i,j) is that

$$S_{uv}^t > \frac{b-1}{a_{uv}}, \tag{24}$$

where (u, v) is a neighbor cell of (i,j).

Future work aimed at designing and analyzing a general model where several compartments will be considered (susceptible, exposed, infected, recovered, quarantined, etc.) where different tessellated shapes (triangular shapes, square shapes and hexagonal shapes) and different neighborhoods (Von Neumann, Moore, etc.) must be considered and studied. Furthermore, future work could deal with more than one malware or the successive infection with different specimens. Finally it is also of interest to study the usefulness of this model in other epidemiological models such as [1, 3].

Acknowledgments This work has been supported by Ministerio de Economía y Competitividad (Spain), and by Junta de Castilla y León (Spain).

References

1. Allen, L., Burgin, A.M.: Comparison of deterministic and stochastic SIS and SIR models in discrete time. Math. Biosci. **163**, 1–33 (2000)
2. Cheng, S.M., Ao, W.C., Chen, P.Y., Chen, K.C.: On modeling malware propagation in generalized social networks. IEEE Commun. Lett. **15**, 25–27 (2011)
3. Ghoshal, G., Sander, L.M., Sokolov, I.M.: SIS epidemics with household structure: the self-consistent field method. Math. Biscay. **190**, 71–85 (2004)
4. Gleissne, W.: A mathematical theory for the spread of computer viruses. Comput. Secur. **8**, 35–41 (1989)
5. Jackson, J.T., Creese, S.: Virus propagation in heterogeneous bluetooth networks with human behaviors. IEEE Trans. Depend. Secur. **9**, 930–943 (2012)
6. Martín del Rey, A., Rodríguez Sánchez, G.: A CA model for mobile malware spreading based on bluetooth connections. In: Herrero, A., Baruque, B., Klett, F., Abraham, A., Snasel, V., Carvalho, A.C.P.L.F., Bringas, P.G., Zelinka, I., Quintian, H., Corchado, E. (eds.) CISIS 2013. Adv. Intel. Syst. Comput. vol. 234, pp. 619–629. Springer, Heidelberg (2014)
7. Merler, S., Jurman, G.: A combinatorial model of malware diffusion via bluetooth connections. PLos ONE **8**, art. no. e59468 (2013)

8. Mishra, B.K., Saini, D.: Mathematical models on computer viruses. Appl. Math. Comput. **187**, 929–936 (2007)
9. Mickens, J.W., Noble, B.D.: Modeling epidemic spreading in mobile environments. In: The 4th ACM Workshop on Wireless Security, pp. 77–86. ACM (2005)
10. Peng, S., Wu, M., Wang, G., Yu, S.: Propagation model of smartphone worms based on semi-Markov process and social relationship graph. Comput. Secur. **44**, 92–103 (2014)
11. Peng, S., Wang, G., Yu, S.: Modeling the dynamics of worm propagation using two-dimensional cellular automata in smartphones. J. Comput. Syst. Sci. **79**, 586–595 (2013)
12. Ramachandran, K., Sikdar, B.: On the stability of the malware free equilibrium in cell phones networks with spatial dynamics. In: The 2007 IEEE International Conference on Communications, pp. 6169–6174. IEEE Press (2007)
13. Ramachandran, K., Sikdar, B.: Modeling malware propagation in networks of smart cell phones with spatial dynamics. In: The 26th IEEE International Conference on Computer Communications, pp. 2516–2520. IEEE Press (2007)
14. Rhodes, C.J., Nekovee, M.: The opportunistic transmission of wireless worms between mobile devices. Phys. A **387**, 6837–6844 (2008)
15. Song, L.P., Jin, Z., Sun, G.Q.: Modeling and analyzing of botnet interactions. Phys. A **390**, 347–358 (2011)
16. Toutonji, O.A., Yoo, S.M., Park, M.: Stability analysis of VEISV propagation modeling for network worm attack. Appl. Math. Model. **36**, 2751–2761 (2012)
17. Wei, X., Zhao-hui, L., Zeng-qiang, C., Zhu-zhi, Y.: Commwarrior worm propagation model for smart phone networks. J. China U Posts Telecommun. **15**, 60–66 (2008)
18. Yan, G., Eidenbenz, S.: Modeling propagation dynamics of bluetooth worms. In: The 27th International Conference on Distributed Computing Systems, pp. 42–51. IEEE Press (2007)
19. Zhu, Q., Yang, X., Ren, J.: Modeling and analysis of the spread of computer virus. Commun. Nonlinear Sci. **17**, 5117–5124 (2012)

Privacy-Preserving Protocol for Smart Space Services with Negotiable Comfort Preferences

Adam Wójtowicz and Daniel Wilusz

Abstract In this paper an architecture and a protocol for secure and privacy-preserving service usage in smart spaces are presented. Comfort preferences that are defined and stored on end-users mobile devices are subject to negotiations. The approach relies on a single trusted party operating as a public service in the "security infrastructure as a service" model. It is designed to assure resilience against attacks on users privacy from the side of service providers (SPs), user's attempts to use the services in an unauthorized manner, user's attempts to violate payment policies, as well as payment authority's attacks on privacy of users' payment patterns. From the SP perspective, the proposed approach minimizes the risk of unauthorized service usage, or denial of service attacks. All players benefit from the fast and secure micropayments allowing for pay-per-use model implementation, which is natural in ubiquitous services usage scenarios.

Keywords User privacy · Smart spaces · Communication protocol

1 Importance of Privacy in Smart Spaces

Smart spaces (sometimes referred to as Ambient Intelligence or Intelligent Environments) use ubiquitous and pervasive technology to enrich the environment. Combining the definitions provided in [1, 2], it can be said that smart spaces are digital environments that proactively but sensibly support people in their daily lives in a non-intrusive way with respect to their privacy. Smart spaces support living or processes in such application areas as: smart houses, healthcare, public transport,

A. Wójtowicz (✉) · D. Wilusz
Department of Information Technology, Poznań University of Economics,
Poznań, Poland
e-mail: awojtow@kti.ue.poznan.pl

D. Wilusz
e-mail: wilusz@kti.ue.poznan.pl

© Springer International Publishing Switzerland 2015   285
Á. Herrero et al. (eds.), *International Joint Conference*, Advances in Intelligent
Systems and Computing 369, DOI 10.1007/978-3-319-19713-5_25

education services, emergency services and production [1]. Smart spaces seem to provide users with various kinds of benefits, however privacy issues need to be taken into account.

General classification of privacy borders has been proposed in [3, 4]; exceeding these borders violates one's privacy. First, natural borders were distinguished. All barriers may be included here that naturally protect one's privacy, such as walls, curtains, darkness, envelopes or even facial expressions. Social borders constitute the next set of privacy protection factors. These factors rely on the trust in social roles such as family members, clerks, who will keep information secret, protect confidentiality and meet due diligence when processing private information. The third sort of privacy borders are special and temporal ones. These borders relay on the assumption that parts of people's lives are separated (e.g. professional and family life) and events from the past should not influence present perception of a person (e.g. one's behavior in high school should not influence their professional occupation). The last but not least type of privacy borders is ephemeral and transitional effects. These borders assume that single or spontaneous statements or behaviors will be quickly forgotten and will not affect personal life in the long term.

Smart spaces based on pervasive computing and intelligent systems pose a risk to the above described borders. Smart spaces, by monitoring a user position, behavior and habits, may breach the natural privacy borders. Moreover, data collected by intelligent systems, if used for purposes other than well-being of a user, could create a risk of exceeding social borders. Protection of social borders is especially important when the data are processed by a third party. Any institution processing Personally Identifiable Information (PII) needs to be fully trusted or, preferably, the privacy of data processing needs to be assured by system design. Archival data stored by intelligent systems could pose the risk for spatial and temporal borders if recording events from professional, social and family life. Finally, the archival data stored by smart spaces could reveal some spontaneous actions a user prefers to be forgotten. In conclusion, the risk of privacy breach by smart spaces is a security issue, which needs to be dealt with [3, 4].

Smart spaces can pose a threat to some of OECD privacy framework principles. The first principle prone to violation is Collection Limitation Principle. It states that any collection of private information should be limited and accessed only in lawful way with the consent of data subject. The next principle is Purpose Specification which assures clear purposes of data collection known to data subject. Finally, the Use Limitation Principle prevents revealing any data without the consent of data subject or authority of law [5, 6]. According to European Commission survey, although the majority of Europeans have accepted the disclosure of PII as a part of modern life, they are still concerned about their privacy [7]. On the one hand, privacy solutions have to comply with ethical and legal requirements, but on the other hand, practical, legal and technical solutions should be proposed to foster implementation of profitable business models [8].

Users preferences regarding comfort in smart spaces are a specific case of privacy sensitive data. They carry information on users' spatial, temporal, financial, behavioral usage patterns that, if disclosed, could draw a detailed picture of users' life in public spaces. Protection of private data is especially significant during comfort negotiation in scenarios where many non-trusted heterogeneous smart spaces, service consumers and payment processes are employed. The above mentioned circumstances motivate the authors to propose a universal model for "security infrastructure as a service" in the area of negotiated comfort in smart spaces.

The reminder of this paper is organized as follows. Section 2 describes the state-of-the-art research on privacy protection in smart spaces. The original architecture and protocol for privacy-preserving and secure service usage in smart spaces is presented in Sect. 3. The advantages of the proposed solution are discussed in Sect. 4. Finally, Sect. 5 concludes the paper.

2 Related Work

2.1 Approaches to Privacy Protection

The basic approaches to privacy protection employ access control policies that specify the conditions of information release. Plenty of access control methods have been proposed. Role Based Access Control (RBAC) is commonly used in information systems. However, it is rather suited for structured systems and requires fixed context. As smart spaces are characterized by continuously changing context, RBAC is not sufficient to ensure high level privacy. On the other hand, Discretionary Access Control (DAC) method allows for expressing data access conditions under continuously changing context. DAC conditions can be based on spatio-temporal, social or even physiological context [8]. However, the access control policies do not assure privacy protection on desired level. Moreover, even if access control is carefully defined, still private information may leak, as eavesdropper can infer personal information by observing properties of smart space [9]. Another weakness of general access control methods is their rigidity.

Access control extended by obfuscation of private data seems to be a flexible solution for privacy protection in smart spaces. Systems utilizing semantic obfuscation allow the information owner to specify the conditions under which context information is disclosed. Moreover, user can define the detail level, depending on context, at which disclosure of private information is allowed [10]. The serious issue of data obfuscation is possible conflict of interests with service providers, when obfuscated data quality does not meet the provider's requirements [8]. To solve this problem, the quality of context negotiation methods have been proposed [11]. Although quality of context negotiation methods allow for a trade-off between privacy protection and service requirements, it is still possible that the service, to operate, requires data of too high accuracy to protect user privacy [8].

As the access control and obfuscation methods prevent data release, the alternative approach is realized by anonymity solutions, which aim at identity protection. One of the anonymization methods is k-anonymity, which is based on quasi-identifiers [9]. The use of quasi-identifiers ensures that each record can be linked to at least k users [8]. To achieve k-anonymity, two data protection techniques can be used: generalization and suppression. These techniques enable reliability of the information as well as privacy protection [12]. Generalization is based on replacing a value by a less specific one, which still is close to the original value. Suppression removes information carrying values [13]. Although k-anonymity seems to reliably protect user privacy, there are some issues to be examined. First, the k-anonymized data are generalized, and this decrease in accuracy reduces the performance of smart space systems. Second, as the result of the release of multiple data describing the same or overlapping sets of users, the composition or intersection attacks may be performed in order to reduce the anonymizing set [9].

Differential privacy is another privacy-preserving method, which may be applied to smart spaces. The aim of differential privacy is to ensure that receiving any result is independent on the presence or the absence of any particular entity [9]. In this way differential privacy allows provision of reliable statistical information on a set of users while protecting privacy of individuals. As the differential privacy results from data release mechanism, not the way users interact, it is resistant to data analysis or linkage attacks [14]. Unfortunately, the application of differential privacy in smart spaces could turn out to be unpractical. Since the differential privacy assures that the change in data of individual user should not influence the changes in probability of getting particular total outcome, adjusting the properties of smart space to any single user becomes impossible [9].

The analysis of state-of-the-art approaches to protection of smart space user data reveals that a robust as well as practical solution is still required. Such a solution should comply with privacy principles elaborated by OECD [6]. Moreover, crossing the privacy borders by any actor of any smart space subsystem should be prevented "by design", i.e. with the use of proper system architecture and adequate communication protocols.

2.2 Privacy in Smart Spaces

On the one hand, smart spaces facilitate daily life, however a number of private data needs to be revealed by the users. The privacy concerns led to research aimed at user data protection. A concept utilizing Security Assertion Markup Language (SAML) and Extensible Access Control Markup Language (XACML) was proposed in [15]. The concept allows an owner of home automation system to define user specific privacy policy. However, in this work the party playing the role of service provider is a natural person who shares private data for different purposes e.g. energy consumption statistics. In most cases, the natural persons play role of end-users and consume services provided by publicly available smart spaces.

The solution presented in [16] relies on privacy policy enforcement. The privacy enforcement authority has been proposed to certify and monitor privacy policies of service providers. However, such solution does not protect users privacy comprehensively, since in the case of comfort negotiation in smart spaces the service providers should not know the identity and even preferences of a particular user.

Another approach to privacy protection is based on the use of pseudonyms. The mechanism for pseudonyms generation for specific group members, where user identifies herself as a group member, is presented in [17]. However, for comfort negotiation in smart spaces, this solution is insufficient, as negotiating members need to be recognizable.

As the above presented solutions are not sufficient for user privacy protection in smart spaces, the new solution has been proposed. The original model of the architecture and the protocol for secure and privacy-preserving comfort negotiation and service usage in smart spaces is described in next section.

3   The Architecture and the Protocol

3.1   The Architecture

In the Fig. 1 the proposed architecture is depicted as well as sequential data flow for a single request from a single user, constituting the proposed protocol. CP denotes Comfort Preferences, USSS – Ubiquitous Services of Smart Spaces, AAS – Authentication and Authorization Services, CNS – Comfort Negotiation Services, CNR – Comfort Negotiation Results, PP – payment policy.

The model relies on a single trusted side, namely AAS, responsible for context handling, authentication (identity provider functionality), pseudonym generation and requests anonymization, and obtaining accounting data. The AAS constitutes a middleware between the mobile user and negotiation and payment services, while the user can still interact directly and efficiently with smart space services. The AAS operates as a public service in the model "security infrastructure as a service", and it is accessible to mobile end-users, smart space SPs (called USSS), Payment SPs, and negotiation brokers (CNS).

It is assumed that besides AAS, and partially CNSs, all the other actors or software modules are untrusted. Despite this fact, the architecture is designed to assure resilience against attacks on users privacy (confidentiality of user's both identity and CP data) from the side of USSS, user's attempts to use the services in unauthorized manner, user's attempts to violate payment policies, as well as payment authority's attacks on users privacy (confidentiality of the payment patterns).

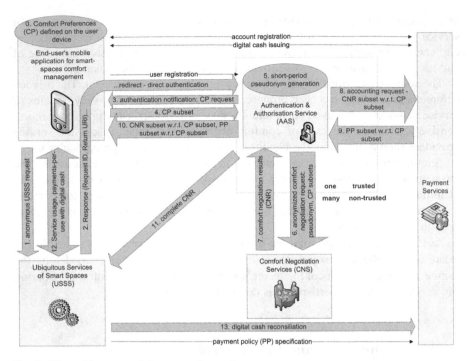

Fig. 1 The architecture and the protocol overview

CNSs have to be trusted w.r.t. negotiation process itself (obviously the model is not resilient to negotiation result forgery or unreliable negotiation process), but have not to be trusted w.r.t. attacks on users privacy (confidentiality of user's both identity and CPs).

3.2 Protocol Prerequisites

The required initial steps of the proposed protocol include:

- Payment Policy specification
  A manager of USSS defines payment policy reflecting the assumed business model and the service functionality. PP contains data related to payment models (pay-per-use, subscriptions), role priorities, pricing, discounts, time- and space-specific conditions, usage constraints, etc.
- User registration
  User registers herself in the Authentication and Authorization Service. Her PIIs are encrypted and do not leave the AAS. This step is mandatory for non-repudiation assurance.

- Defining Comfort Preferences (CP) on the user device
  User defines her CP and stores them encrypted on the device. Disclosure of these data will concern only subset required for the intended usage session, and only to the trusted AAS, and not to the USSS.
  CP are expressed as sets composed of statements (e.g. `temp >=15`, `hum <=80 %`), rules (e.g. `temp >=25 => hum <=70 %`), and arithmetic expressions (e.g. `hum < -13*ATAN (temp/5) + 80`), where temp denotes temperature [°C], and hum denotes humidity [%].
- Payment account registration and digital cash issuing
  After a user registers her account, payment authority issues digital cash as a result of user request. User "buys" digital cash for real money. The architecture of semi-off-line, anonymous micropayment system for Internet of Things, proposed by the one of the authors of this paper is utilized to provide payment services [18, 19]. This solution utilizes the concept of hash chain to make micropayments effective. Each electronic coin consists of hash chain with the last hash digitally signed by a bank. Such solution enables micropayments as the last hash validates the remaining ones. Moreover, the blind signature scheme is used to ensure anonymity of a payment. As a bank blindly signs the last hash in the chain, the identity of a user cannot be revealed during payment. Cryptographic formalisms and protocols of this solution are described in details in [18].

3.3 The Protocol – Main Steps

1. Anonymous service request
   The user requests the service of the intelligent building or other smart space. She is not disclosing to the USSS what her comfort preferences are.
2. Response (Request ID, Return URI); Redirect to AAS for direct user authentication in AAS.
   USSS responds with Request ID for further reference and Redirect URI for further callback from AAS (c.f. step 11). The user's client application is redirected to the AAS for direct authentication with the credentials defined in the initial User registration phase (mentioned in the previous section).
3. Authentication notification; CP request
   AAS confirms authentication (if successful) and requests CP from a user.
4. CP subset
   User client sends CP subset relevant to the planned service usage scope.
5. Short-period pseudonym generation
   Short-period pseudonym is generated for user to be used in AAS and CNS in order to hide user identity. If a user uses another service in next request, a new pseudonym will be generated for her, in order to assure unlinkability of the usage and preferences data in the CNS. Furthermore, even if the service used and CP do not change, for each negotiation session a new pseudonym is

generated for a user. Mapping of a user identity data to all used pseudonyms and to Request IDs (c.f. step 2) is persistently stored within the AAS.

6. Anonymous comfort negotiation request; Pseudonym, CP subsets

   AAS sends anonymous request for negotiate desired comfort parameters within smart space to the CNS. The request contains a set composed of statements, rules, and arithmetic expressions.

   CNS collects many such requests sent from AAS on behalf of different users and performs the negotiations by finding the set of comfort values compliant to all constraints of all users. Specific algorithms applied for negotiation are beyond the scope of this paper. However, it has to be noted, that for the performance optimization it is assumed that CNS employs incremental approach to negotiations algorithm, i.e. most frequently any new CP subset does not require renegotiating the whole dataset.

7. CNR – Comfort negotiation results

   The complete negotiations results are sent back to the AAS. Knowing the complete negotiations results AAS intersects them with user's CP. The intersection represents actual comfort settings limited to user's area of interest, that are as close to the user's CPs as it is possible, and is denoted as CNR subset w.r.t. CP subset. In the alternative (advanced) version of the protocol, before step 7, CNS exchanges messages with Payment Services in order to get to know payment policies for payment-specific negotiations (having payment-dependent priorities).

8. Accounting request; CNR subset w.r.t. CP subset

   AAS requests service accounting from Payment Service, i.e. payment plan for a given user, based on negotiated comfort settings. The anonymized request containing CNR subset w.r.t. CP subset is sent to Payment Service.

9. PP subset w.r.t. CP subset

   PP subset, i.e. payment plan that prices user's negotiated comfort and is compliant with user's CP, is sent back to AAS.

10. CNR subset w.r.t. CP subset; PP subset w.r.t. CP subset

   AAS sends CNR generated for a given user and PP generated for a given user to the user. Then, payment conditions are reviewed by the user and are subject to her acceptance.

11. Complete CNR

   As the complete CNRs contain final comfort values and do not contain original preferences of individual users, they can be sent to the USSS. Knowing the CNR, the USSS can apply the negotiated comfort values to the services. It enables service usage.

12. Service usage; payment-per-use with digital cash

   Services are used with the desired comfort settings. Payments for the service usage are continuously made within the "pay-per-use" model with off-line digital micropayments.

13. Digital cash reconciliation

After a longer usage period, the digital cash is subject to reconciliation directly between USSS provider and Payment Service. USSS provider obtains real money.

4 Discussion

The proposed architecture and protocol provide a number of advantages for both end-users and service providers of smart space services with negotiable comfort settings. First, it provides an end-user with anonymity while she uses smart spaces services. Service provider does not know the user identity directly, nor user comfort preferences, thus the identity cannot be deducted indirectly based on analysis of the preferences and other metadata. The unlinkability feature of the user's service requests, usage patterns and preferences is assured by applying short-term pseudonyms issued by the AAS.

Additional level of privacy protection is achieved with the storage model: comfort preferences are defined, stored and maintained locally, on the mobile device, as opposed to the cloud. They leave the device only when are sent to the trusted middleware, if the user decides so, and only in the minimal subsets that are necessary for the given service request. This privacy-enhancing approach does not limit user's functionality i.e., user still has a full flexibility regarding comfort modelling statements and rules.

From the service provider perspective, the proposed architecture and the protocol, specifically the AAS middleware, minimize the risk of the service unauthorized usage, or denial of service attacks, which normally are more probable in the anonymity preserving systems. Non-repudiation is assured for the cases of the reported security incidents. In such cases real-world identity data stored in the AAS are mapped to the history of short-term pseudonyms issued for the user, and later can be used to deanonymize activities logged at the SP side, if AAS and SP cooperate.

Another advantage of the proposed approach seen from the SP perspective is the ability to define flexible usage and payment policies that can be efficiently enforced within the framework. Moreover, SPs security level seen from both technological and legal perspective is increased: authentication handling, payments handling and PII storage is "outsourced" to the highly specialized infrastructure.

Both end-user and SPs benefit from the fast and secure micropayments allowing for pay-per-use model implementation, which is natural in ubiquitous services usage scenarios. Due to cryptographic solution employed, the transaction are "semi-offline", which means that user's connection with payment services is not needed during the service usage. Obviously, the user is required to have the connections with the SP and AAS, and SP is required to have connections with AAS and Payment Services.

As a part of the future work the module that protects the user anonymity against the statistical attacks based on the analysis of the service usage patterns, including user spatial and temporal locations and path analysis is planned to be designed and implemented.

5 Conclusions

In the public and massively multiuser smart space ecosystems there are a number of actors: service providers, users, payment brokers, and comfort negotiation brokers, that not necessarily trust one another regarding sensitive information confidentiality or unauthorized service usage. Providing a trusted middleware operating as a public service in the "security infrastructure as a service" model allows for data flow between the actors by removing security and privacy concerns from each side.

The presented architecture and protocol rely on a trusted party responsible for context handling, authentication, pseudonym generation, requests anonymization, and obtaining accounting data. It constitutes a middleware between the mobile user and negotiation and payment services, while the user can still interact directly and efficiently with smart space services.

On the other hand, the presented model, contrary to highly centralized approaches, does not violate the assumptions of the openness. The functionalities of smart spaces, payment or negotiations are not coupled together in a single service, thus competition between different providers can leverage services quality. The proposed architecture and protocol are suitable especially for public smart spaces intended for simultaneous usage of large number of users in a non-incidental time period, e.g. public transport, classrooms, or hospital spaces.

Acknowledgments This research work has been supported by the GOLIATH project jointly funded by the Poland NCBR and Luxembourg FNR Lead Agency agreement, under NCBR grant number POLLUX-II/1/2014 and Luxembourg National Research Fund grant number INTER/POLLUX/13/6335765.

References

1. Augusto, J.C.: Ambient intelligence: the confluence of ubiquitous/pervasive computing and artificial intelligence. In: Alfons Dr., Schuster J. (eds.) Intelligent Computing Everywhere, pp. 213–234. Springer, Heidelberg (2007)
2. Brelsford, C.: Other perspectives on ambient intelligence. In: Password, issue 23, Philips Research, Eindhoven (2005)
3. Bohn, J., Coroamă, V., Langheinrich, M., Matern, F., Rohs, M.: Living in a World of smart everyday objects—social, economic, and ethical implications. In: Journal of Human and Ecological Risk Assessment, vol. 10, issue 5, pp. 763–786. Taylor & Francis (2004)
4. Marx, G.T.: Murky conceptual waters: the public and the private. In: Ethics and Information Technology, vol. 3, issue 3, pp. 157–169. Springer, Heidelberg (2001)

5. Čas, J.: Ubiquitous computing, privacy and data protection: options and limitations to reconcile the unprecedented contradictions. In: Gutwirth, S., Poullet, Y., de Hert, P., Leenes, R. (eds.) Computers, Privacy and Data Protection: an Element of Choice, pp. 139–169. Springer, Heidelberg (2011)
6. Organization for Economic Co-operation and Development: The OECD Privacy Framework, http://www.oecd.org/sti/ieconomy/oecd_privacy_framework.pdf (2013)
7. European Commission: Attitudes on Data Protection and Electronic Identity in the European Union http://ec.europa.eu/public_opinion/archives/ebs/ebs_359_en.pdf (2011)
8. Bettini, C., Riboni, D.: Privacy protection in pervasive systems: state of the art and technical challenges. In: Pervasive and Mobile Computing (in press). Elsevier (2014)
9. Chau, J.C., Little, T.D.C.: Challenges in retaining privacy in smart spaces. In: Procedia Computer Science, vol. 19, pp. 556–564. Elsevier, New York (2013)
10. Wishart, R., Henricksen, K., Indulska, J.: Context Privacy and obfuscation supported by dynamic context source discovery and processing in a context management system. In: Ubiquitous Intelligence and Computing. LNCS, vol. 4611, pp. 929–940. Springer, Heidelberg (2007)
11. Sheikh, K., Wagdam, M., van Sinderen, M: Quality-of-context and its use for protecting privacy in context aware systems. In: Journal of Software, vol. 3, pp. 83–93. Academy Publisher, Finland (2008)
12. Samarati, P.: k-Anonymity. In: Encyclopedia of Cryptography and Security, pp. 663–666. Springer, Heidelberg (2011)
13. Sweeney, L.: Achieving k-anonymity privacy protection using generalization and suppression. In: International Journal on Uncertainty, Fuzziness and Knowledge-based Systems, vol. 10 Issue 5, pp. 571–588. World Scientific Publishing, New Jersey (2011)
14. Dwork, C.: Differential privacy. In: Encyclopedia of Cryptography and Security, pp. 338–340. Springer, Heidelberg (2011)
15. Jung, M., Kienesberger, G., Granzer, W., Unger, M., Kastner, W.: Privacy enabled web service access control using SAML and XACML for home automation gateways. In: IEEE Internet Technology and Secured Transactions (ICITST), pp. 584–591 (2011)
16. Oyomno, W., Jäppinen, P., Kerttula, E.: Privacy preservation for personalised services in smart spaces. In: IEEE Internet Communications (BCFIC Riga), pp. 181–189 (2011)
17. Wakeman, I., Chalmers, D., Fry, M.: Reconciling privacy and security in pervasive computing. In: Proceedings of the 5th International Workshop on Middleware for Pervasive and Ad-Hoc Computing: Held at the ACM/IFIP/USENIX (MPAC '07), 8th International Middleware Conference, pp. 7–12. ACM (2007)
18. Wilusz, D., Rykowski, J.: The architecture of coupon-based, semi-off-line, anonymous micropayment system for internet of things. In: Camarinha-Matos, L.M., Tomic, S., Graça, P. (Eds.) Technological Innovation for the Internet of Things. AICT, vol. 394, pp. 125–132. Springer, Heidelberg (2013)
19. Wilusz, D., Rykowski, J.: Requirements and general architecture of a payment system for the future internet. In: Informatyka Ekonomiczna (Business Informatics), vol. 2, pp. 91–103. Wydawnictwo Uniwersytetu Ekonomicznego we Wrocławiu, Wrocław (2013)

An Intermediate Approach to Spritz and RC4

Rafael Álvarez and Antonio Zamora

Abstract Spritz is a recently announced sponge-based stream cipher intended to be a drop-in replacement for RC4. It is more secure, more complex and more versatile than RC4 but, unfortunately, it cannot compete with it in terms of performance. In this paper we analyze the modification of RC4 to include some of the concepts from Spritz in order to make a hybrid that is faster than Spritz and more secure than RC4, which has been attacked recently, especially in terms of short term biases in its output. We analyze the performance and statistical randomness of this hybrid, that we call RC4itz, and compare it to AES in output feedback mode, RC4 and Spritz. RC4itz presents good characteristics and is an interesting tradeoff between the security of Spritz and the speed of RC4.

Keywords Stream cipher · PRNG · Cryptography · Spritz · RC4 · RC4itz

1 Introduction

The RC4 stream cipher algorithm (see [13]) is one of the most widely used software stream ciphers since it is part of the Secure Sockets Layer (SSL) and Wired Equivalent Privacy (WEP) protocols, among others. Its internal state consists of 256 byte array functioning as an 8×8 bit substitution box (S-Box) that dynamically changes as data is encrypted. Recently, there have been several attacks on RC4 with varying degrees of usefulness and success (see [7, 9, 12, 15]), therefore, the cryptographic community has lost its confidence on RC4.

Spritz (see [14]) is a sponge-based cryptographic function that can be employed as a stream cipher as well as a hash function (among other applications). It is a nontrivial evolution of the well-known RC4 stream cipher, incorporating the sponge

R. Álvarez (✉) · A. Zamora
Department of Computer Science and Artificial Intelligence (DCCIA), University of Alicante (Campus de San Vicente), Ap. 99, E-03080 Alicante, Spain
e-mail: ralvarez@dccia.ua.es

A. Zamora
e-mail: zamora@dccia.ua.es

© Springer International Publishing Switzerland 2015
Á. Herrero et al. (eds.), *International Joint Conference*, Advances in Intelligent Systems and Computing 369, DOI 10.1007/978-3-319-19713-5_26

paradigm over a permutation of N values and extending the internal state of RC4 with more registers. It is meant as a drop-in replacement for RC4.

Although the Spritz authors have taken multiple steps to improve the security of RC4, including exhaustive statistical analysis of many possible candidates for each function in Spritz, this has also brought an increase in complexity and computational cost. For this reason, there are some situations where the slower performance of Spritz might make RC4 still desirable over its more secure counterpart.

Therefore, it is interesting to find an intermediate approach between Spritz and RC4, incorporating certain modifications inspired by Spritz to RC4, obtaining a hybrid algorithm that can be faster than Spritz and with better statistical characteristics than RC4.

In this paper, we analyze the performance of Spritz against RC4, including AES as a benchmark since it is the current encryption standard. We also propose and analyze, in terms of performance and security, a hybrid algorithm that incorporate the output/update function of Spritz (that was chosen after exhaustive search by the Spritz authors, see [14]) to the RC4 structure, together with forced output skipping during the key schedule. Since it is something between both, RC4itz appears to be a suitable name for this algorithm.

This approach becomes a tradeoff between the high complexity and security but slower performance of Spritz and the higher performance and simplicity but questionable security of RC4. It can be highly beneficial in real time applications where data decryption must be performed online for further inspection like in intrusion detection and prevention systems or malware/threat identification systems (honeynets and honey-pots), etc.

The RC4itz hybrid presents interesting properties in terms of performance and security and is described in detail in Sect. 2; also, a performance, randomness and security analysis and comparison to AES, RC4 and Spritz is provided in Sect. 3; some concluding remarks are given afterwards.

2 Description

We have implemented and tested several algorithms; besides RC4 (see [13]), Spritz (see [14]) and the hybrid discussed in this paper (RC4itz), we also include AES (see [5]) in ouput-feedback mode (OFB).

2.1 AES

The Advanced Encryption Standard (AES) is a staple in current cryptography. Although natively slower than some stream ciphers (like RC4, for example) it has the advantage of being hardware accelerated in many modern processors. This makes it extremely fast, with speeds exceeding 1 GB/s on many consumer machines, and an excellent building block for many other types of cryptosystems that attempt to incorporate its speed and security into their designs.

When used in some operation modes, like OFB or counter mode (CTR), AES behaves like a stream cipher and is therefore relevant in this comparison.

2.2 RC4

Standard RC4 is a very well known and simple algorithm. Its state consists only of the contents of a 256 byte array functioning as a bijective 8×8 bit S-Box, S, and 8 bit registers i and j. It is simple enough to be memorized and can be incredibly fast with some careful optimizations. It basically consists of a key schedule function and an update/output function. Although RC4 can be specified so that the internal state, S, is a permutation of the numbers 0 to $N - 1$, we analyze here the most common version where $N = 256$.

```
func RC4KSA(K[]) {
    for i = 0 to 255
        S[i] = i

    j = 0
    for i = 0 to 255
        j += S[i] + K[i mod len(K)]
        Swap(S[i],S[j])
}

func RC4PRG() {
    i++
    j += S[i]
    Swap(S[i],S[j])
    return S[S[i] + S[j]]
}
```

(Pseudo-code for RC4.)

2.3 Spritz

As can be observed in the following pseudo-code, Spritz (see [14]) is much more complex than RC4. It is based on a sponge-like (see [3]) function design and expands on the internal state of RC4 with additional 8 bit registers: i, j, k, z, and a; besides the S-Box, S.

Although it is meant as a drop-in replacement for RC4, due to its sponge interface, Spritz allows many applications beyond that of a traditional stream cipher, including hash functions, authenticated encryption, etc.

In the case of a stream cipher application, the standard interface would be to call *InitializeState()*, then *Absorb()* the key and, finally, call *Squeeze()* to obtain the keystream data required to encrypt the current message block.

Please note that, as in RC4, we are only considering the version of Spritz where $N = 256$. The authors refer to this variant as Spritz-XOR, since N is a power of 2, allowing for more efficient XOR operations during encryption, rather than the addition/subtraction required when N is an arbitrary integer. For further details regarding Spritz, refer to the specifications paper [14].

```
func InitializeState() {
    i = j = k = z = a = 0
    w = 1
    for v = 0 to 255
        S[v] = v
}


func Absorb(I[]) {
    for v = 0 to len(I) - 1
        AbsorbByte(I[v])
}


func AbsorbByte(b) {
    AbsorbNibble(Low(b))
    AbsorbNibble(High(b))
}
func AbsorbNibble(x) {
    if a = 128
        Shuffle()

    Swap(S[a],S[[N/2] + x])
    a = a + 1
}


func AbsorbStop() {
    if a = 128
        Shuffle()
    a++
}


func Shuffle() {
    Whip(512)
```

```
        Crush()
        Whip(512)
        Crush()
        Whip(512)
        a = 0
}


func Whip(r) {
    for v = 0 to r - 1
        Update()

do w++
    until GCD(w, N) = 1
}

func Crush() {
    for v = 0 to 127
        if S[v] > S[N - 1 - v]
            Swap(S[v],S[N - 1 - v])
}

func Squeeze(r) {
    if a > 0
        Shuffle()

    P = new array(r)
    for v = 0 to r - 1
        P[v] = Drip()

    return P
}

func Drip() {
    if a > 0
        Shuffle()

    Update()
    return Output()
}

func Update() {
    i += w
    j = k + S[j + S[i]]
    k = i + k + S[j]
    Swap(S[i],S[j])
```

```
}

func Output() {
    z = S[j + S[i + S[z + k]]]
    return z
}
```

(Pseudo-code for Spritz for $N = 256$.)

2.4 RC4itz

The hybrid discussed in this paper introduces several modifications from Spritz to RC4. Basically, we modify the key scheduling algorithm and the update/output functions to be similar to those in Spritz while expanding the internal state to registers i, j, k and z. We also introduce output skipping (1024 bytes) after the key schedule to further improve the randomness characteristics of the output sequence.

This modifications produce an algorithm that is between RC4 and Spritz, obtaining performance and security characteristics within those of both algorithms (see Sect. 3).

```
func RC4itzKSA(K[]) {
    for i = 0 to 255
        S[i] = i

    j = 0
    for i = 0 to 255
        j = k + S[j+S[i]] + K[i mod len(K)]
        k += i + S[j]
        Swap(S[i],S[j])

    SkipOutput(1024)
}

func RC4itzPRG() {
    i++
    j = k + S[j + S[i]] + K[i mod len(K)]
    k += i + S[j]
    Swap[S[i],S[j]]
    z = S[j + S[i + S[z + k]]]
    return z
}
```

```
func SkipOutput(r) {
    for v = 0 to r - 1
        i++
        j = k + S[j + S[i]]
        k += i + S[j]
        Swap[S[i],S[j]]
        z = S[j + S[i + S[z + k]]]
}
```

(Pseudo-code for the modified key schedule and output/update functions in RC4itz.)

3 Results

3.1 Performance

All implementations have been done in the Go programming language (see [8]), without special instruction sets, hardware acceleration nor other advanced optimizations with the aim of making the comparison as fair as possible. The testing platform has been a MacBook Pro with a 2.6 GHz Intel Core i5 and 16 GB of RAM, running Mac OS X Yosemite (version 10.10.1).

As shown in Table 1, Spritz is more complex than RC4 and is, therefore, much slower; it is similar in speed as non-accelerated AES. Spritz is also the slowest algorithm in terms of key setup.

Regarding the performance of the RC4itz hybrid, it lies between RC4 and Spritz in key setup and encryption performance. The key setup is slower than RC4 since it performs output skipping, but still faster than Spritz. The same happens with the encryption performance, the additional operations and state registers make RC4itz a little slower than RC4, but much faster than Spritz.

Table 1 Performance results

	Key setup (ns)	Encryption (MB/s)
AES	22.839	69.59
RC4	13.228	234.73
Spritz	27.147	54.99
RC4itz	20.512	107.14

3.2 Statistical Analysis

We have performed statistical analysis of the different algorithms with three different sets of tests: a simple custom battery of statistical tests, RandTest (that is available on GitHub, see [1]), that includes autocorrelation and linear complexity (Berlekamp-Masey algorithm, see [2, 11]) among other tests; PractRand (version 0.92, see [6]) that is a very modern and comprehensive battery of tests, as well as very efficient, allowing to test long sequences; and the Dieharder suite (see [4]) that is an evolution of the original Diehard test battery (see [10]).

3.2.1 RandTest

This randomness test suite checks for bit frequency (Frequency), bit pair frequency (Serial), 8 bit and 16 bit pattern frequency (Poker 8 and Poker 16 respectively), runs (contiguous set bits) and gaps (contiguous unset bits) in the sequence (Runs), autocorrelation, and linear complexity.

We have tested sequences of 128,000 bits in length for each algorithm. As shown in Table 2, all algorithms (including the RC4itz hybrid) pass the tests. A test is passed if its value is below the correction value for that test, except for linear complexity that is expected to be equal to half the length of the sequence (64,000 in this case).

3.2.2 PractRand

The setup employed uses a 1 GB keystream sequence from each algorithm, performing 99 different tests on each sequence. Most of the tests are passed as *normal* by all algorithms but we include in Table 3 the results for a specific test where there are some differences: AES result is considered *suspicious* by PractRand, while RC4 and RC4itz results are considered *unusual*; on the contrary, Spritz result is considered as *normal*.

Table 2 RandTest results

Test	Correction	AES	RC4	Spritz	RC4itz
Frequency	2.7060	0.1403	0.0061	0.0475	0.4351
Serial	4.6050	0.1547	2.5355	3.7995	1.7114
Poker 8	284.3	243.8	236.0	277.2	258.0
Poker 16	65999	65515	65498	65564	65973
Runs	30.813	15.551	29.066	26.653	16.958
Autocorr.	1.2820	0.7989	0.8061	0.7961	0.7938
Linear Comp.	64000	64000	64000	64000	64000

Please note that none of the algorithms fail this test per se, they are just on the outer edge of the valid interval. The rest of the tests are considered normal for all algorithms and, therefore, omitted here for the sake of brevity.

We can conclude that RC4itz is similar to the other algorithms under PractRand.

Table 3 Significant PractRand results

Test	AES	RC4	Spritz	RC4itz
FPF-14+6/16:cross	p = 3.2e-5	p = 1.3e-3	p = 0.143	p = 8.4e-5

3.2.3 Dieharder

The original Diehard battery of statistical tests was developed by George Marsaglia in 1995 (see [10]), while the Dieharder (see [4]) suite of tests is an improved version of diehard that includes tests from the Statistical Test Suite (STS) developed by the National Institute for Standards and Technology (NIST).

As before, all algorithms pass all tests in the battery and only a few are marked as *weak* (the test is passed but on the outer edge of the valid interval); these special cases are shown in Table 4. Each of AES, RC4 and Spritz obtain a *weak* result on a single different test out of the whole battery, while RC4itz achieves a *passed* status on all tests; this is within the expected tolerance so we can conclude that all four algorithms are acceptable pseudo-random number generators from the perspective of the Dieharder suite of tests.

3.3 Security Considerations

There have been a lot of recent attacks on RC4 (see [9, 12, 15]) and even some biases found in reduced variants of Spritz (see [16]), although these do not affect the full $N = 256$ version of Spritz. The authors of Spritz (see [14]) describe very well many of the problems with RC4, and how they try to solve them in Spritz.

Table 4 Significant Dieharder results

Test	AES	RC4	Spritz	RC4itz
sts-serial, ntup=7	weak	passed	passed	passed
rgb-lagged-sum, ntup=14	passed	weak	passed	passed
dab-filtree2, ntup=0	passed	passed	weak	passed

The hybrid described here is an attempt to improve RC4 security while being faster than the complete sponge-design of Spritz, with a structure much closer to that of RC4.

In order to improve security regarding the short term biases of RC4, RC4itz skips 1024 bytes of output after key setup; some authors state that skipping 256 bytes is enough to eliminate these biases [7], but 1024 is still within reasonable performance (see Sect. 3.1) for the key schedule function.

Also, the modified update/output function should improve RC4 in long term biases, since it has been extensively tested as part of the Spritz design.

RC4itz is, of course, a certain tradeoff between the security and sponge design of Spritz and the simplicity and performance of RC4; and should undergo further external cryptanalysis before its use in production, like any other new cryptographic primitive.

4 Conclusions

We have described an intermediate approach between RC4 and Spritz in terms of performance and security in the form of a hybrid algorithm that we call RC4itz.

It is based on RC4 but incorporates modifications in the key schedule and update/output functions as well as output skipping to prevent the short term randomness deficiencies associated with RC4. Although not as secure as Spritz, it can be useful where the additional applications provided by the sponge-based design are not needed or more performance is desired.

In terms of performance, RC4itz lies precisely between the two: it is faster than Spritz but slower than the very simple design of RC4. In terms of randomness, it presents similar statistical results to the other algorithms analyzed in this paper (AES, RC4 and Spritz) and should benefit from the extensive bias analysis performed during the design of Spritz by the usage of a very similar update/output function, despite the fact that the structure is not sponge-based and much closer to that of RC4.

As future work, optimized implementation analysis as well as further cryptanalysis for RC4itz are planned.

Acknowledgments Research partially supported by the Spanish MINECO under Project Grant TIN2011-25452 (TUERI).

References

1. Álvarez, R.: RandTest. http://github.com/rias/randtest (2004)
2. Berlekamp, E.R.: Algebraic Coding Theory. McGraw Hill, New York (1968)
3. Bertoni, G., Daemen, J., Peeters, M., Van Assche, G.: Cryptographic sponge functions. http://sponge.noekeon.org/ (2011)

4. Brown, R.G.: Dieharder: a random number test suite. http://www.phy.duke.edu/rgb/General/dieharder.php (2003)
5. Daemen, J., Rijmen, V.: AES proposal: Rijndael. In: First Advanced Encryption Standard (AES) Conference (1999)
6. Doty-Humphrey, C.: PractRand. http://pracrand.sourceforge.net/ (2014)
7. Fluhrer, S.R., Mantin, I., Shamir, A.: Weaknesses in the key scheduling algorithm of RC4. Selected areas in cryptography, Lecture Notes in Computer Science, vol. 2259, pp. 1–24 (2001)
8. The Go Programming Language. http://www.golang.org
9. Klein, A.: Attacks on the RC4 stream cipher. Des. Codes Crypt. **48–3**, 269–286 (2008)
10. Marsaglia, G.: The marsaglia random number CDROM including the Diehard battery of tests of randomness. http://www.stat.fsu.edu/pub/diehard/ (1995)
11. Masey, J.L.: Shift-register synthesis and BCH decoding. IEEE Trans. Inf. Theory **15**, 122–127 (1969)
12. Paul, G., Maitra, S.: RC4 Stream Cipher and its Variants. CRC Press, Boca Raton (2012)
13. Rivest, R.L.: The RC4 Encryption Algorithm. RSA Data Security Inc., Redwood City (1992)
14. Rivest, R.L., Schuldt, J.: Spritz—a spongy RC4-like stream cipher and hash function. http://people.csail.mit.edu/rivest/pubs/RS14.pdf Presented at CRYPTO 2014 Rump Session (2014)
15. Sengupta, S., Maitra, S., Paul, G., Sarkar, S.: RC4: (Non-) random words from (non-) random permutations. IACR Crypt. ePrint Arch. **2011**, 448 (2011)
16. Zoltak, B.: Statistical weakness in Spritz against VMPC-R: in search for the RC4 replacement. http://eprint.iacr.org/2014/985.pdf IACR e-print (2014)

An Efficient Identity-Based Proxy Signature Scheme in the Standard Model with Tight Reduction

Xiaoming Hu, Hong Lu, Huajie Xu, Jian Wang and Yinchun Yang

Abstract Identity-based proxy signature (IDPS) is a special signature. It has many applications, such as distribution networks, mobile communication, etc. Numerous IDPS schemes have been proposed. However, most existing IDPS schemes suffer the following shortcoming: loose security reduction. This problem is very important because loose security reduction weakens the security of IDPS and makes IDPS more vulnerable to attack. In this study, based on Kang et al.'s proof technical, we propose a new identity-based proxy signature scheme with a detailed security proof in the standard model. We also reduce the security of our scheme to the hardness assumption of computational Diffie-Hellman. In order to present the advance of our scheme, we make a theory comparison between our scheme with other identity-based proxy signature schemes in terms of security reduction and computational cost. The comparison shows that our scheme has tightest security reduction. What's more, our scheme needs less computational cost that is almost half of other schemes.

Keywords Information security · Identity-based signature · Identity-based proxy signature · Tight reduction · CDH

1 Instruction

In 1984, Shamir [1] first proposed the concept of identity-based cryptography that can simplify the key management process in certificate-based public key settings. IDPS scheme is a special identity-based signature (IBS) scheme [2–4], which can

X. Hu (✉) · H. Lu · J. Wang · Y. Yang
College of Computer and Information Engineering,
Shanghai Second Polytechnic University, Shanghai 201209, China
e-mail: xmhu@sspu.edu.cn

H. Xu
School of Computer and Electronic Information, Guangxi University,
Nanning 530004, China

© Springer International Publishing Switzerland 2015
Á. Herrero et al. (eds.), *International Joint Conference*, Advances in Intelligent
Systems and Computing 369, DOI 10.1007/978-3-319-19713-5_27

309

delegate an original signer's signing right to a proxy signer who can sign instead of the original signer. The property of IDPS is very useful in many situations, such as distribution networks, grid computing, mobile agent applications and mobile communications. Therefore, IDPS has aroused many experts and a lot of work has been done on IDPS schemes.

The first IDPS scheme was proposed by Zhang and Kim [5] in 2003. But they didn't give a formal security analysis. In 2005, Xu et al. [6] presented a security model for identity-based proxy signature scheme. Since then, many identity-based proxy signature schemes have been proposed, such as [7, 8]. However, [7]'s scheme and [8]'s scheme are both only secure in the random oracle model [9] which has a weak security level. Canetti et al. [10] pointed out that there existed a scheme that was secure in the random oracle model but was insecure in real situation. Therefore, constructing identity-based proxy signature secure in the standard model is more practical. In 2010, Cao and Cao [11] proposed an identity-based proxy signature scheme that was secure in the standard model. But their scheme was insecure because it could not resist the attack of the original signer. In 2013, Gu et al. [12] presented a frame and a detailed security model for identity-based proxy signature scheme. However, all above schemes have the drawback, i.e., efficiency of the security reduction is very low. In this study, we propose a new identity-based proxy signature scheme. We also address the detailed security proof with tight reduction in the standard model. What's more, we make a comparison on computational efficiency and reduction efficiency with other identity-based proxy signature schemes. Next, we give a summary on our scheme with other identity-based proxy signature schemes in terms of four features: security model (what is the security model that is used in these schemes?), formal proof (whether a formal security proof is provided?), pairing cost (how many pairing operations are used?), security reduction (what is the level of security reduction?). From the Table 1, we can see that our scheme obtains better performance than other identity-based proxy signature schemes.

Table 1 A summary of identity-based proxy signature schemes

-scheme	-security model	-formal proof	-pairing cost	-security reduction
[3]	No	No	3	No
[6]	Random oracle	Yes	5	Loose
[7]	Random orcale	Yes	4	Loose
[8]	Random orcale	No	3	No
[11]	Standard model	Yes	6	Loose
[12]	Standard model	Yes	6	Loose
Our	Standard model	Yes	3	Tight

2 Preliminaries

2.1 Bilinear Map

Let G_1 be a cyclic additive group which is generated by g and has a prime order p. G_2 is a cyclic multiplicative group which has the same order p. Define e: $G_1 \times G_1 \rightarrow G_2$ to be a bilinear map if it holds:

– Bilinearity: for all $g, h \in G_1$ and all $a, b \in Z_p$, $e(g^a, h^b) = e(g, h)^{ab}$.

– Non-degeneracy: there exists a $g \in G_1$, such that $e(g, g) \neq 1$.

– Computability: for $g, h \in G_1$, there exists an efficient algorithm to compute $e(g, h)$.

2.2 Computational Diffie-Hellman (CDH) Problem and CDH Assumption

Let $g \in G_1$ be a generator. Given (g, g^a, g^b) with unknowing a and b, the CDH problem is to compute g^{ab}.

The CDH assumption holds (ϵ, t) if no algorithm can solve the CDH problem in G_1 with running time at most t and probability at least ϵ.

2.3 Identity-Based Signature Scheme

An IBS scheme includes four parts: Setup, Extract, Sign and Verify.

Setup: Given a system security parameter k, the algorithm generates the system public parameters *params* and the master key t.

Extract: Given *params*, t and an identity *ID*, the algorithm produces a private key d_{ID} on *ID*.

Sign: Given *params*, the private key d_{ID} on *ID* and a message m that is signed, the algorithm constructs a signature σ on m.

Verify: Given *params* and a signature σ on m with identity *ID*, the algorithm outputs *true* if it is a valid signature. Otherwise, it outputs *false*.

2.4 Identity-Based Proxy Signature Scheme

An IDPS scheme consists of five algorithms: PSetup, PExtract, Delegate, Proxysign, PVerify.

PSetup: Given a security parameter, the algorithm generates the system public parameters *params* and the master key t.

PExtract: Given *params*, t and an identity *ID*, the algorithm produces a private key d_{ID} on *ID*. Using the method, the original signer and the proxy signer get their private keys d_{ID_o} and d_{ID_p}, respectively.

Delegate: Given *params*, the private key d_{ID_o} of the original signer and a warrant w, the algorithm generates a delegation R on w.

Proxysign: Given *params*, the delegation R on w, the private key d_{ID_p} of the proxy signer and a message that is signed, the algorithm generates a proxy signature σ.

PVerify: Given *params* and a proxy signature σ on m with warrant w, the algorithm outputs *true* if it is a valid signature. Otherwise, it outputs *false*.

3 Our Identity-Based Signature Scheme

In this section, based on Tian et al.'s strong designed verifier signature scheme [13], we present an efficient identity-based signature scheme. Then we give a detailed security analysis of the proposed scheme.

3.1 Construction of the Identity-Based Signature Scheme

Setup: Let $H: \{0, 1\}^* \to \{0, 1\}^n$ be a hash function and g is a generator of G_1. Define a n-length vector $U = (u_i)$, where $u_i \in Z_p^*$ and $1 \leq i \leq n$. Pick randomly $t \in Z_p^*$, $g_2 \in G_1$ and set $g_1 = g^t$. Publish the system public parameters *params* = $\{G_1, G_2, e, H, U, g, g_1, g_2\}$.

Extract: Let $ID \in \{0, 1\}^*$ be the identity. PKG computes the private key for *ID* as follows.

- Choose randomly $r_1 \in Z_p^*$ and compute $I_{ID} = H(ID||r_1)$. Define $I_{ID[i]}$ to denote the ith bit of I_{ID} and $I_{(ID)}$ to denote the set of all i with $I_{ID[i]} = 1$, i.e., $I_{(ID)} = \{i | I_{ID[i]} = 1, 1 \leq i \leq n\}$.
- Compute $d_{ID,1} = (\prod_{i \in I_{(ID)}} u_i)^t$, $d_{ID,2} = r_1$. Then, $(d_{ID,1}, d_{ID,2})$ is the private key for *ID*.

Sign: Let $m \in \{0, 1\}^*$ be the message to be signed. The signer with the private key $(d_{ID,1}, d_{ID,2})$ on *ID* computes the signature as follows.

- Choose randomly $r_2 \in Z_p^*$ and compute $\sigma_1 = g^{r_2}$, $\sigma_2 = d_{ID,1} \cdot g_2^{r_2 \cdot H(\sigma_1 || m)}$.
- Then, $(\sigma_1, \sigma_2, \sigma_3)$ is the signature on m with identity *ID*.

Verify: Given $(\sigma_1, \sigma_2, \sigma_3)$ on m with identity *ID*, the receiver computes $I_{ID} = H(ID||r_1)$ and checks if $e(\sigma_2, g) = e(\prod_{i \in I_{(ID)}} u_i, g_1)e(g_2, \sigma_1^{H(\sigma_1||m)})$ holds.

If the above equation holds, the receiver accepts the signature. Otherwise, he rejects it. The above scheme is correct because

$$e(\sigma_2, g) = e(d_{ID,1} \cdot g_2^{r_2 \cdot H(\sigma_1 || m)}, g)$$

$$= e((\prod_{i \in I_{(ID)}} u_i)^t \cdot g_2^{r_2 \cdot H(\sigma_1 || m)}, g)$$

$$= e((\prod_{i \in I_{(ID)}} u_i)^t, g^t) \cdot e(g^{r_2 \cdot H(\sigma_1 || m)}, g_2)$$

$$= e((\prod_{i \in I_{(ID)}} u_i)^t, g_1) \cdot e(\sigma_1^{H(\sigma_1 || m)}, g_2)$$

3.2 Security Analysis of Our Scheme

We adopt the similar proof method as Tian et al. [13] and Kang et al. [14] to analyze the security of our identity-based signature scheme.

Theorem 1 *Assume that the CDH problem is hard with (t', ϵ') in G_1, the proposed identity-based signature scheme is secure with (t, ϵ, q_e, q_s), where*

$$\epsilon' \geq \frac{1}{2}\epsilon(1 - adv_{hash-c}), t' = t + (n + 5 + q_e + 2q_s)C_E + (n + q_s + 2)C_M.$$

C_E denotes the amount of time to run one exponentiation operation in G_1, C_M denotes the amount of time to run one multiplication operation in G_1, adv_{hash-c} denotes the probability to occur hash collision, q^e denotes the maximal times to make extracting query and q_s denotes the maximal times to make signing query.

Proof Assume that (g, g^a, g^b) is an instance of the CDH problem which is given to the simulator (challenger) B. If the adversary A can break our identity-based signature scheme, then B can use A to construct another algorithm to solve the given CDH problem. In order to solve the problem, B and A perform the following process.
Setup: B picks randomly $\alpha_i, \beta_i, \gamma \in Z_p^*, 1 \leq i \leq n$. Set $g_1 = g^b, g_2 = g^\gamma$ and $U = (u_i)$, $u_i = g^{a(-1)^{\alpha_i}} \cdot g^{\beta_i}$ $(1 \leq i \leq n)$. Define two functions

$$F(I_{ID}) = \sum_{i \in I_{(ID)}} (-1)^{\alpha_i} mod p, J(I_{ID}) = \sum_{i \in I_{(ID)}} \beta_i mod p,$$

where $I_{(ID)}$ denotes the set of all i with $I_{ID[i]} = 1$ and $I_{ID[i]}$ denotes the ith bit of I_{ID}. Then, b is the master key which is unknowing by B. B sends the system parameters $\{G_1, G_2, e, H, U, g, g_1, g_2\}$ to A.
Query: In this stage, A is allowed to make two kinds of queries for polynomial times: Extract query ($O_{ext}()$) and Sign query ($O_{sig}()$). B answers as follows.

– $O_{ext}()$: Given an identity ID_i from the adversary A, B chooses a random element $r_{i,1} \in Z_p^*$ and computes $I_{ID_i} = H(ID_i||r_{i,1})$. If $F(I_{ID_i}) \neq 0$, B aborts the game. If $F(I_{ID_i}) = 0$, B computes $d_{ID_{i,1}} = g^{bJ(I_{ID_i})}$, $d_{ID_{i,2}} = r_{i,1}$ and returns $(d_{ID_{i,1}}, d_{ID_{i,2}})$ to A. This is a valid private key for identity ID, because

$$d_{ID_{i,1}} = g^{bJ(I_{ID_i})} = g^{abF(I_{ID_i})} \cdot g^{bJ(I_{ID_i})}$$
$$= (g^{a \cdot \Sigma_{i \in I_{(ID_i)}} (-1)^{\alpha_i}} \cdot g^{a \cdot \Sigma_{i \in I_{(ID_i)}} \beta_i})^b$$
$$= (\prod_{i \in I_{(ID_i)}} u_i)^b.$$

– $O_{sig}()$: Given an identity ID_i and a message m_i from the adversary A, B products a signature for m_i on ID_i as follows. B picks two random elements $r_{i,1}, r_{i,2} \in Z_p^*$ and computes $I_{ID_i} = H(ID_i||r_{i,1})$. If $F(I_{ID_i}) \neq 0$, B aborts the game. If $F(I_{ID_i}) = 0$, B uses $O_{ext}()$ to compute the private key $(d_{ID_{i,1}}, d_{ID_{i,2}})$ for ID_i. Then, B uses the **Sign** algorithm to compute the identity-based signature $(\sigma_1, \sigma_2, r_1)$ for m_i. B returns it to A.

Forgery: If B doesn't abort in the above simulation, then finally A outputs a forged identity-based signature $(\sigma_1^*, \sigma_2^*, r_1^*)$ on a message m^* with an identity ID^*. Because $(\sigma_1^*, \sigma_2^*, r_1^*)$ is a valid signature, it satisfies the verification equation,

$$e(\sigma_2^*, g) = e(\prod_{i \in I_{(ID^*)}} u_i, g_1) \cdot e(g_2, \sigma_1^{*H(\sigma_1^*||m^*)}). \tag{1}$$

After B gets the signature $(\sigma_1^*, \sigma_2^*, r_1^*)$, B computes $I_{ID^*} = H(ID^*||r_1^*)$. If $F(I_{ID^*}) = 0$, B aborts the game. If $F(I_{ID^*}) \neq 0$, B uses the equation (1) to compute

$$g^{ab} = \frac{\sigma_1^*}{\sigma_2^{*\gamma H(\sigma_2^*||m^*)} \cdot g^{bJ(I_{ID^*})}}.$$

Thus, B solves the given CDH problem.

Next, we analyze the success probability of B, i.e., B doesn't abort in the above simulation. Define the following events:

– E1: B doesn't abort in $O_{ext}()$, i.e., $F(I_{ID_i}) = 0$ for all $1 \leq i \leq q_e$.
– E2: B doesn't abort in $O_{sig}()$, i.e., $F(I_{ID_i}) = 0$ for all $1 \leq i \leq q_s$.
– E3: $F(I_{ID_i}) \neq 0$ in forgery stage.

According to the analysis and experiment presented by Kang et al. [14], B needs to select 40 times averagely to find a value $r_{i,1}$ that satisfies $F(I_{ID_i}) = 0$ with $I_{ID_i} = H(ID_i||r_{i,1})$ for 512-bit security level, i.e., $n = 1024$ bit. On the other hand, the analysis of Tian et al. [13] and Kang et al. [14] shows that the probability of $F(I_{ID_i}) \neq$

0 is at most $\frac{1}{2}(1 - adv_{hash-c})$, where adv_{hash-c} is the probability of finding a hash collision on any two inputs. Therefore, $\epsilon' = \Pr[\text{E1} \wedge \text{E1} \wedge \text{E3}] \geq (1 - adv_{hash-c})$.

The time complexity is decided by the key computational operations in simulation process. It needs $(n + 5 + q_e + q_s)$ exponentiation operations in G_1 and $(n + q_s + 2)$ multiplication operations in G_1. Therefore, $t' = t + (n + 5 + q_e + 2q_s)C_E + (n + q_s + 2)C_M$, where C_E and C_M denote one time exponentiation operation and one multiplication operation in G_1, respectively.

4 Our Identity-Based Proxy Signature Scheme

4.1 Construction of Our Identity-Based Proxy Signature Scheme

PSetup: This algorithm is the same as Setup of our above identity-based signature scheme. The system public parameters are $\{G_1, G_2, e, H, U, g, g_1, g_2\}$.

PExtract: The algorithm is also the same as **Extract** of our above scheme. Using the algorithm, the original signer and the proxy signer obtain the public-private key pairs $(ID_o, (d_{ID_{o,1}}, d_{ID_{o,2}}))$ and $(ID_p, (d_{ID_{p,1}}, d_{ID_{p,2}}))$, respectively, where

$$d_{ID_{o,1}} = (\prod_{i \in I_{(ID_o)}} u_i)^t, d_{ID_{o,2}} = r_{o,1}, d_{ID_{p,1}} = (\prod_{i \in I_{(ID_p)}} u_i)^t, d_{ID_{p,2}} = r_{p,1},$$

and $r_{o,1}, r_{p,1}$ are two random elements from Z_p^*, respectively.

Delegate: The original signer creates a warrant $w \in \{0, 1\}^*$ that includes the identities of the original signer and the proxy signer, the date and the message type that is signed. Next, the original signer does as follows.

– Choose randomly $r_2 \in Z_p^*$, and compute $R_1 = g^{r_2}, R_2 = d_{ID_{o,1}} \cdot g_2^{r_2 H(w)}$.
– Then, (R_1, R_2, r_1) is the delegation on w and it is sent to the proxy signer.

Proxysign: Given a message $m \in \{0, 1\}^*$ that is signed, the proxy signer chooses a random element $r_3 \in Z_p^*$ and computes $\sigma_1 = R_2 \cdot d_{ID_{p,1}} \cdot g_2^{r_3 H(w||m)}, \sigma_3 = g^{r_3}$. Set $\sigma_2 = R_1, \sigma_4 = r_{o,1}, \sigma_5 = r_{p,1}$. Then, $(\sigma_1, \sigma_2, \sigma_3, \sigma_4, \sigma_5)$ is the proxy signature on m.

PVerify: Given a signature $(\sigma_1, \sigma_2, \sigma_3, \sigma_4, \sigma_5)$ on m with w, the signature receiver computes $I_{ID_o} = H(ID_o||\sigma_4), I_{ID_p} = H(ID_p||\sigma_5)$ and checks that

$$e(\sigma_1, g) = e(\prod_{i \in I_{(ID_o)}} u_i \cdot \prod_{i \in I_{(ID_p)}} u_i, g_1) \cdot e(g_2, \sigma_2^{H(w)} \cdot \sigma_3^{H(w||m)}).$$

Correctness: the above signature is correct because

$$e(\sigma_1, g) = e(R_2 \cdot d_{ID_{p,1}} \cdot g_2^{r_3 H(w\|m)}, g)$$

$$= e(d_{ID_{o,1}} \cdot g_2^{r_2 H(w)} \cdot d_{ID_{p,1}} \cdot g_2^{r_3 H(w\|m)}, g)$$

$$= e(d_{ID_{o,1}} \cdot d_{ID_{p,1}}, g) \cdot e(g_2^{r_3 H(w\|m)} \cdot g_2^{r_2 H(w)}, g)$$

$$= e(\prod_{i \in I_{(ID_o)}} u_i \cdot \prod_{i \in I_{(ID_p)}} u_i, g^t) \cdot e(g_2, g^{r_2 H(w)} \cdot g^{r_3 H(w\|m)})$$

$$= e(\prod_{i \in I_{(ID_o)}} u_i \cdot \prod_{i \in I_{(ID_p)}} u_i, g_1) \cdot e(g_2, \sigma_2^{H(w)} \cdot \sigma_3^{H(w\|m)}).$$

4.2 Security Analysis of Our Scheme

Theorem 2 *If there exists an adversary A1 that can* $(t, \epsilon, q_e, q_w, q_s)$ *break our identity-based proxy signature scheme, there exists another algorithm B1 that can use A1 to* (t', ϵ') *solve the CDH problem in* G_1*, where*

$$\epsilon' \geq \frac{1}{4}\epsilon(1 - adv_{hash-c}),$$

$$t' = t + (n + 7 + q_e + 2q_w + 5q_s)C_E + (n + 2 + q_w + 3q_s)C_M,$$

and CE denotes one time exponentiation operation in G_1*,* C_M *denotes one multiplication operation in* G_1*,* adv_{hash-c} *denotes the probability of a hash collision.* q_e *is the maximal number of extracting query,* q_w *is the maximal number of delegating query and* q_s *the maximal number of proxy signing query.*

Proof The security is similar to that of our above identity-based signature scheme. Assume that (g, g^a, g^b) is an instance of the CDH problem that is given to the simulator (challenger) B1.

Setup: B1 sets the system parameters *params* $= \{G_1, G_2, e, H, U, g, g_1, g_2\}$ as *Setup* of our above scheme and b is the master key. B1 sends *params* to A1.

Query: A1 is allowed to make three kinds of queries: Extract query ($O_{pext}()$), Delegate query ($O_{pdel}()$) and Sign query ($O_{psig}()$).

- $O_{pext}()$: The oracle works as $O_{ext}()$ of our above scheme.
- $O_{pdel}()$: Given a warrant $w \in \{0,1\}^*$ and an identity of the original signer ID_o, B1 first uses $O_{pext}()$ to generate a private key for ID_o. Then, B1 chooses randomly $r_2 \in Z_p^*$ and computes $R_1 = g^{r_2}$, $R2 = d_{ID_{o,1}} \cdot g_2^{r_2 H(w)}$. B1 returns (R_1, R_2, r_1) as the delegation on w with ID_o to A1.
- $O_{psig}()$: Given $m \in \{0,1\}^*$, a warrant w that includes the identities of the original signer ID_o and the proxy signer ID_p. B1 first uses $O_{pdel}()$ to generate a delegation

(R_1, R_2, r_1) on w with ID_o. Then, B1 uses **Proxysign** to produce a signature $(\sigma_1, \sigma_2, \sigma_3, \sigma_4, \sigma_5)$ on m and returns it to A1.

Forgery: A1 finally outputs a forged identity-based proxy signature $(\sigma_1^*, \sigma_2^*, \sigma_3^*, \sigma_4^*, \sigma_5^*)$ on a message m^* and a warrant w^* with corresponding identities: ID_o^* for the original signer and ID_p^* for the proxy signer. Because $(\sigma_1^*, \sigma_2^*, \sigma_3^*, \sigma_4^*, \sigma_5^*)$ is a valid signature, it satisfies

$$e(\sigma_1^*, g) = e(\prod_{i \in I_{(ID_o^*)}} u_i \cdot \prod_{i \in I_{(ID_p^*)}} u_i, g_1) \cdot e(g_2, \sigma_2^{*H(w)} \cdot \sigma_3^{*H(w^*||m^*)}).$$

where $I_{ID_o^*} = H(ID_o^*||\sigma_4^*)$, $I_{ID_p^*} = H(ID_p^*||\sigma_5^*)$. If $F(I_{ID_o^*}) = 0$ and $F(I_{ID_0^*}) = 0$, then B1 aborts the game. Otherwise, If $F(I_{ID_o^*}) = 0$ and $F(I_{ID_p^*}) \neq 0$, then B1 computes

$$g^{ab} = (\frac{\sigma_1^*}{\sigma_2^{*\gamma H(\sigma_2^*||m^*)} \cdot \sigma_3^{*\gamma H(w^*||m^*)} \cdot g^{bJ(I_{ID_o^*})} \cdot g^{bJ(I_{ID_p^*})}})^{F(I_{ID_p^*})^{-1}}.$$

If $F(I_{ID_o^*}) \neq 0$ and $F(I_{ID_p^*}) = 0$, then B1 computes

$$g^{ab} = (\frac{\sigma_1^*}{\sigma_2^{*\gamma H(\sigma_2^*||m^*)} \cdot \sigma_3^{*\gamma H(w^*||m^*)} \cdot g^{bJ(I_{ID_o^*})} \cdot g^{bJ(I_{ID_p^*})}})^{F(I_{ID_o^*})^{-1}}.$$

If $F(I_{ID_o^*}) \neq 0$ and $F(I_{ID_p^*}) \neq 0$, then B1 computes

$$g^{ab} = (\frac{\sigma_1^*}{\sigma_2^{*\gamma H(\sigma_2^*||m^*)} \cdot \sigma_3^{*\gamma H(w^*||m^*)} \cdot g^{bJ(I_{ID_o^*})} \cdot g^{bJ(I_{ID_p^*})}})^{(F(I_{ID_o^*})+F(I_{ID_p^*}))^{-1}}.$$

According to the analysis of Kang et al. [14], the probability of $F(I_{ID_o^*}) = 0$ is at most $\frac{1}{2}$. Therefore, the success probability of B1 that solves the CDH problem is $\epsilon' \geq \frac{1}{4}\epsilon(1 - adv_{hash-c})$. Analysis of the time complexity is similar to Theorem 1, finally we can get $t' = t + (n + 7 + q_e + 2q_w + 5q_s)C_E + (n + 2 + q_w + 3q_s)C_M$.

4.3 Efficiency Analysis

In order to obtain a meaningful and fair result, our compare our identity-based proxy signature scheme with other identity-based proxy signature schemes which are secure in the standard model (including [11, 12]) in terms of the complexity of security reduction and computational cost. q_{pw} denotes the maximal number of

Table 2 Comparisons of security and pair cost with other schemes

Scheme	Security reduction	Pair cost
Cao [11]	$\dfrac{\epsilon}{192(q_e+q_w+q_{pw}+q_s)(q_{pw}+q_s)q_s(n+1)}$ (1)	$6C_P$
Gu [12]	$\dfrac{\epsilon(q-q_e)^2(q-q_w)^2}{q^5}$ (2)	$6C_P$
Our	$\dfrac{\epsilon(1-adv_{hash-c})}{4}$	$3C_P$

proxy signing key query. C_P denotes one pairing operation in G_2. Pair Cost denotes the total pairing costs in whole scheme.

We only consider the most consuming operations, i.e., pairing computation in G_2. From the Table 2, it can find that the security reduction of our identity-based proxy signature scheme is very tight with a constant probability ($\approx \frac{1}{4}\epsilon$) while the security of other both schemes degrades by a factor (1) in [11] and (2) in [12]. What's more, our scheme only needs three pairing operations which almost are half of other both schemes' costs. So, our scheme has tighter security reduction and lower computational cost than Cao-Cao's scheme and Gu et al.'s scheme.

Remark On the efficiency analysis, we make a comparison between our scheme and other schemes by using a most usual method. In other words, we provide the theory analysis of the computational efficiency that counts the main computational operation (i.e., pairing operation) in G2. However, one can implement our identity-based proxy signature scheme by using some standard cryptographic libraries, such as MIRACLE library from Miracle Shamus Software and Ltd., Crypto++ library, etc. If one adopts the same experiment platform for all schemes and chooses the same system parameters, he will get the same result as theory analysis.

Because cryptography is largely demanding and is also complex, applying Computational Intelligence (CI) into the field of cryptography comes nature. Many CI methods [15], such as Artificial Neural Networks (ANNs), Evolutionary Computation (EC), Genetic Algorithms (GAs), Ant Colony Optimization (ACO), etc., have been used to design or analyze the cryptography. For example, Sardha Wijesoma et al. [16] adopted the genetic algorithm of CI to perform the on-line signature verification that yields good results with moderate complexity. Due to the effectiveness of CI in handling hard problems in cryptography, we will consider to improve the performance of our identity-based proxy signature scheme by CI methods in the future.

5 Conclusions

Identity-based proxy signature has attracted the interest of many experts. Much work has been done on identity-based proxy signature schemes. High efficiency and high security are both key properties of a good identity-based proxy signature. However, designing an identity-based proxy signature scheme with high efficiency and high security is always a problem worth of study. In this study, we propose an identity-based proxy signature scheme that holds high security and high efficiency. As far as our knowledge, our identity-based proxy signature is better than other similar schemes in terms of security and efficiency. The next step is to make an experimental analysis.

Acknowledgments This work is supported by the Innovation Program of Shanghai Municipal Education Commission (No.14ZZ167), the National Natural Science Foundation of China (No.61103213), the Guangxi Natural Science Foundation (No.2014GXNSFAA11838-2) and the Construct Program of the Key Discipline in SSPU (fourth): Software Engineering (No. XXKZD1301).

References

1. Shamir, A.: Identity-Based cryptosystems and signature schemes. In: Blakely, G.R., Chaum, D. (eds.) CRYPTO 1984. LNCS, vol. 196, pp. 47–53. Springer, Heidelberg (1985)
2. Liu, H., Wang, S., Liang, M.: New construction of efficient hierarchical identity based signature in the standard model. Int. J. Secur. Appl. **7**(4), 211–222 (2013)
3. Zhang, L., Hu, Y., Wu, Q.: New identity-based short signature without random oracles. Procedia Eng. **15**, 3445–3449 (2011)
4. Tsai, T., Tseng, Y., Wu, T.: Provably secure revocable id-based signature in the standard model. Secur. Commun. Netw. **6**(10), 1250–1260 (2013)
5. Zhang, F., Kim, K.: Efficient ID-based blind signature and proxy signature from bilinear pairings. ACISP'03. LNCS, vol. 2727, pp. 312–323. Springer, Heidelberg (2003)
6. Xu, J., Zhang, Z., Feng, D.: ID-based proxy signature using bilinear pairings. In: Chen, G., Pan, Yi, Guo, M., Lu, J. (eds.) ISPA-WS 2005. LNCS, vol. 3759, pp. 359–367. Springer, Heidelberg (2005)
7. Singh, H., Verma, G.: ID-based proxy signature scheme with message recovery. J. Syst. Softw. **85**, 209–214 (2012)
8. Zhao, Y.: Identity-based proxy signature protocols from pairing. In: PMINES 2011, pp. 397–400. IEEE, Los Alamitos (2011)
9. Bellare, M., Rogaway, P.: Rondom oracles are practical: a paradigm for designing efficient protocols. In: Advances in the 1st ACM Conference on Computer and Communication Security, ACM, pp. 62–73. USA (1993)
10. Canetti, R., Goldreich, P., Halevi, S.: The random oracle methodology, revisited. In: Advance in the 30th Anunal Symposium on the Theory of Computing, pp. 209–218. ACM, USA (1998)
11. Cao, F., Cao, Z.: An identity based proxy signature scheme secure in the standard model. In: GRC 10, pp. 67–72. IEEE, Los Alamitos (2010)
12. Gu, K., Jia, W., Jiang, C.: Efficient identity-based proxy signature in the standard model. Comput. J. **11**, 1–16 (2013)
13. Tian, H., Jiang, Z., Liu, Y., Wei, B.: A systematic method to design strong designated veri-fier signature without random oracles. Cluster Comput. **16**, 817–827 (2013)
14. Kang, L., Tang, X., Lu, X., Fan, J.: A short signature scheme in the standard model. Cryptology ePrint Archive: Report 2007/398 (2007)
15. Laskari, E., Meletiou, G., Stamatiou, Y., Vrahatis, M.: Cryptography and cryptanalysis through computational intelligence. In: Computational Intelligence in Information Assurance and Security Studies in Computational Intelligence, LNCS, vol. 57, pp. 1–49. Springer, Heidelberg (2007)
16. Sardha, W., Ma, M., Yue, K.: On-line signature verification using a computational intelligence approach. In: Computational Intelligence. Theory and Applications, LNCS, vol. 2206, pp. 699–711. Springer, Heidelberg (2001)

A New Metric for Proficient Performance Evaluation of Intrusion Detection System

Preeti Aggarwal and Sudhir Kumar Sharma

Abstract Intrusion Detection System (IDS) can be called efficient when maximum intrusion attacks are detected with minimum false alarm rate but due to imbalanced data, these two metrics are not comparable on the same scale. In this paper, a new NPR metric is suggested in view of the imbalanced data set to rank the classification algorithms for IDS which can help analyze and identify the best possible combination of high detection rate and low false alarm rate with maximum accuracy and F-score. The new NPR metric is used for comparison and ordering of ten classifiers simulated on KDD data set.

Keywords Intrusion detection system · Imbalanced data · KDD data set · False alarm rate · Detection rate · NPR

1 Introduction

The Intrusion attacks have become more frequent in recent times due to immense sharing of resources on Internet. The level of security needed at the host and network level is at the peak due to which ethics are required for sharing of information through the execution of intrusion detection system (IDS) [1]. The misuse and anomaly detection are delineated based on the characteristic of IDS whether it is capable of identifying the known attacks [2] only resulting in misuse detection or the identification of novel attacks giving rise to detection of anomaly in the system.

The data set frequently taken up for the empirical study of IDS is the DARPA data [3] whose updated adaptations are also available as KDDCUP'99 [4] and NSL-KDD [5] data set. KDDCUP'99 is the popular data set used for IDS but its

P. Aggarwal (✉) · S.K. Sharma
School of Engineering and Technology, Ansal University, Gurgaon, India
e-mail: preetaggarwal@gmail.com

S.K. Sharma
e-mail: sudhir.sharma@aitgurgaon.org

© Springer International Publishing Switzerland 2015
Á. Herrero et al. (eds.), *International Joint Conference*, Advances in Intelligent Systems and Computing 369, DOI 10.1007/978-3-319-19713-5_28

statistical analysis resulted in NSL-KDD data set that has many improvements over earlier versions like elimination of redundant records and reasonable number of records in train and test data set [6]. This dataset acts as a bench mark in testing IDS knowing the fact that it still suffers from some problems [7].

In a classic real world example, it is quite possible to face the problem of dealing with unbalanced data sets [8]. The data set is called imbalanced if the 'classes' are not equally distributed [9] but there are many techniques available to handle the problems of imbalanced datasets [10] like under sampling and oversampling. The imbalance in data can result in misinterpretation of results which can provide many different perspectives to the learning approach [11]. The method of classification is most suitable for the categorization of network observation as routine data or an instance of intended intrusion activity where the training data puts a benchmark on how the classifier is modeled for simulation [12]. The ensemble approach using multiple classifiers for testing the IDS [13] is quite acknowledged.

The performance evaluation is of utmost importance because IDS is chosen only on the basis of realization of designated performance characteristics anticipated from IDS [14]. Sokolova et al. [15] suggests that it is not sufficient to only identify the classes accurately but also give emphasis on failure avoidance and class discrimination. There also exists relationship between common performance measures for classifiers but it has been observed that the correlation between metrics within the family is not very high [16]. The established IDS evaluation metrics are accuracy, F-score which measure the overall potential of IDS and the Detection Rate (DR) with False Alarm Rate (FAR) which can help decide efficient IDS [17]. Receiver Operating Characteristic (ROC) curve analysis trades-off the true positive and false positive rate to compare the diverse models/classifiers giving the most favorable value of DR and processing cost of false positives hence Nagarajan et al. [18] suggest an integrated intrusion defense strategy for the same. The effort to reduce the FAR is continuously going on as proposed in [19] where two orthogonal and complementary approaches are presented for alert management system. A modular multiple classifier system is proposed in [20] which allows tuning of DR and FAR produced by each module to optimize the overall performance of IDS. ROC curve is a graphical evaluator of DR for IDS in conjunction with the FAR though it does not exactly tell the best possible combination of the two metrics [21]. Therefore, it is necessary to devise a metric that reflect on the selection of best classifier with regulated value for DR and FAR together.

1.1 Objective

The definitive goal of IDS is to achieve highest efficiency in terms of maximum DR and minimum FAR. The intrusion attacks actually happening on the system needs to be detected with immediate further action hence, requiring time, effort and cost. Therefore, it becomes the eventual requirement of IDS to not just attain maximum DR but realize acceptable FAR also.

This research paper presents a new metric that best assesses the performance of IDS thereby predicting its suitability in terms of DR and FAR. The imbalance in the data set used for simulation gives rise to biasing of the trained model towards a class with more number of instances resulting in misleading consequences like high accuracy value which might in fact be inferring high accuracy for negative class having maximum instances in the data set. Considering the case of binary classifier, the number of instances in each class should be equal but this is not true for imbalanced data set, hence the proposed metric directs the correct selection of a classification algorithm where the results show negligible biasing due to imbalanced data set. The proposed metric which is represented in terms of the correct predictions that is, true positive and true negative instances, though simple but can actually help decide whether to compromise on DR or FAR by providing a base of comparison for these two metrics and hence can best identify the correct operating point or the best classifier for IDS by ranking the classification algorithms based on the suggested metric.

1.2 Performance Metrics

Confusion matrix described in Table 1 suggests the basic attributes generated as part of the classification results. The result attributes in the confusion matrix are true negative (TN) defined as the number of instances correctly predicted as normal, true positive (TP) defined as the number of instances correctly predicted as anomalous, false negative (FN) defined as the number of instances incorrectly predicted as normal and false positive (FP) defined as the number of instance wrongly predicted as anomalous. Some of the well-known metrics used for the evaluation of IDS are: accuracy defined as the percentage of test data instances that are correctly classified; precision which assesses the accuracy of the prediction of positive instances; DR that measures the completeness of the same; F-score presenting the precision and DR together as one metric evaluating the harmonic mean of the two and FAR is the ratio of number of instances erroneously predicted as positive to the actual number of negative class instances [22].

Table 1 Confusion matrix for IDS		Predicted instances	
		Normal	Anomalous
Actual instances	Normal	TN	FP
	Anomalous	FN	TP

The rest of the paper is organized as follows: Sect. 2 illustrates the proposed metric NPR and Sect. 3 describes the experimental setup to execute the research plan. Section 4 presents the results and discussion with Sect. 5 communicating the conclusion.

2   Proposed NPR Metric

The imbalance in data deals with such a data set where distribution of data is not uniform amongst the classes. Considering the case of binary classes, imbalanced data means that the majority class is negative and the minority class is consisting of positive instances or vice versa [22]. The intrusion detection is one such system where the chances of having positive class instances is low resulting in unequal number of normal and anomalous instances therefore the classification results need to be studied and analyzed keeping in mind the small possibility of attacks transpiring on the system. Due to imbalance in the data set, the comparison of DR which owes to the positive instances with the FAR which is due to the negative instances becomes unrealistic. This section presents a proposed metric which keeps in mind the imbalance in data which definitely affects the learning of classifier and considers the behavior of testing data instances while interpreting the classification results.

While training the classification model, the behavior of the attributes of the data set is learned but the distribution of the classes is not considered as part of the learning but for this study, the distribution of classes is of key importance. The proposed metric focuses on this distribution of classes in the test data set which is given the name as NP Ratio (NPR). 'N' stands for True Negative (TN) and 'P' stands for True Positive (TP). For example, if assume that the test data set contains three anomalous and four normal data instances in a set of seven data instances, then the classifier's prediction results should best pursue the similar conduct.

$$NPR = \frac{Number\ of\ True\ Negative\ Instances\ (TN)}{Number\ of\ True\ Positive\ Instances\ (TP)} \tag{1}$$

The test data set is expected to check the correct prediction ability of the classification model and the NP ratio can help access the true interpretation of IDS evaluation metrics by ranking the algorithms based on the value of NPR. Let us consider a binary classifier with two classes: N meaning normal and A meaning anomalous. The actual number of instances in N and A class are represented by NI, AI respectively. Therefore, the NPR for the actual test data set is NPR_a.

$$NPR_a = \frac{NI_a}{AI_a} = \frac{TN_a}{TP_a} \tag{2}$$

Similarly, from the classification results, the correct predicted number of instances for class N and A can be identified as NI_p, AI_p respectively. Hence, the predicted NPR (NPR_p) is:

$$NPR_p = \frac{NI_p}{AI_p} = \frac{TNp}{TP_p} \tag{3}$$

Ideally speaking, the actual and predicted NP ratio should be same but practically this never happens because the classifier tends to classify some of the anomalous instances as normal and vice versa. Therefore, NPR gives the estimate of the deviation of predicted results from the actual expected results of the classifier. Considering the case when the test data is balanced, the actual and predicted NPR should be equal to one. To interpret the meaning of NPR, let us consider four cases where the predicted NPR is not equal to the actual NPR. To understand the significance of each case, DR and FAR is considered and analyzed whether the results are biased towards the negative class or the positive class.

1. $NPR_p > NPR_a$: A high predicted NPR as compared to actual NPR shows that the predicted negative (normal) instances are more than the actual negative instances which is the misclassification of positive (anomalous) instances. Hence this leads to low DR with comparatively better FAR.
2. $NPR_p \gg NPR_a$: Further high predicted NPR worsens the DR for positive (anomalous) class and hence it can be concluded that the classification results are more biased towards the negative (normal) class thereby reducing the FAR rates significantly.
3. $NPR_p < NPR_a$: A low value for predicted NPR as compared to expected actual NPR leads to a conclusion that either the predicted negative (normal) instances are low or the predicted positive (anomalous) instances are high. If number of predicted negative instances is low then it may lead to comparatively high FAR. Also, if number of predicted positive instances is high then it means that DR may be rather high.
4. $NPR_p \ll NPR_a$: A very low value for predicted NPR boosts the DR but worsens FAR at the same time.

As discussed earlier, the data set whether offline or online will have comparatively less number of attack (positive) instances, hence the aim of this metric is to look for a balanced point such that the classification results give equal importance to both the positive and negative class. This correspondence is required because not only maximum DR is required but also the minimum value for FAR. From the above analysis, the best case is when the actual and predicted NPR are equal signifying that there is no biasing due to imbalance in data and the results are accurate otherwise Case 1 provides the best point of consideration for the classifier results where DR and FAR are comparable and best possible at the same time. Hence, the NPR metric provides a base for comparison of DR with FAR for imbalanced data sets such that the classification algorithms can be ranked for the best combination of DR and FAR.

3 Experimental Setup

3.1 Research Methodology

This paper presents the calculation of the existing metrics and their assessment with the new performance evaluation metric known as NP Ratio (NPR). The KDD data is simulated for each of the ten selected algorithms and the results are generated in the form of confusion matrix. The suggested metric NPR and existing metrics are computed for each of the ten classification algorithm. Weka [23, 24] version 3.7.11 is used for simulation which is a freely available data mining tool.

3.2 KDD Data Set

The NSL-KDD [5] data set is used for this pragmatic study which has 42 attributes and classifies the data in either of the two classes: normal or anomalous. The data set comprises of two files used for training and testing purpose in the same order. Table 2 presents the distribution of data instances in training and test data file.

Table 2 Data instances

	Normal class instances	Anomalous class instances	Total
KDDTrain+_20Percent (training data)	13449	11743	25192
KDDTest-21 (test data)	2152	9698	11850

Here, the actual NPR (NPRa) for two test data sets can be calculated as:

$$NPR_a = \frac{2152}{9698} = 0.22 \tag{4}$$

3.3 Classifiers Used

The classification algorithms can be placed in the category of rule based, instance based, tree based, neural network based classifiers, and classifiers based on support vector machines [22]. This empirical study considers only those algorithms that require minimum or no parameter tuning for simulation and the default parameter setting is considered for each of the ten algorithms as specified in Weka. The ten algorithms selected for simulation under Weka are from the classifiers based on tree, rule, and support vector machine. Random Tree, Random Forest, J48,

Hoeffding are tree based classification algorithms, Naïve Bayes and Bayes Net are Bayesian networks, Decision Table and OneR are rule based algorithms with Classification via Regression as a meta class algorithm and SMO is an algorithm on support vector machines [23].

4 Results and Discussion

This section presents the observations for ten algorithms simulated on KDD data set with the various evaluation metrics calculated. The best entries are highlighted in bold print and worst entries are underlined. Column 1 of Table 3 shows the ten algorithms selected for the study, column 2, 3, 4, 5 lists TN, FN, FP, TP parameters respectively whereas column 6 presents the time to build the classification model. Values in columns 2 to 5 are used for calculation of all the other attributes listed in Table 3 which are Accuracy, FAR, DR, Precision and F-score. The NP ratio is calculated for each classification algorithm tabulated in Table 3 as a ratio of TN and TP whereas the actual value of NPR (0.22) is applicable for each algorithm since the same test data set is used for simulation on each of the ten selected classification algorithms. In the ideal case, the value of actual NPR is one wherein the data set under test is balanced meaning that the number of normal and anomalous instances is same. Here, in this case under test, the actual value of NPR which is not equal to one reflects that the test data is imbalanced hence there are chances of the results being biased towards the class with supplementary number of instances. Therefore, it becomes binding to maintain the same NPR for the predicted results also to be able to conclude negligible biasing in classification results.

Due to the imbalance in data set, DR (related to positive class) and FAR (related to negative class) are not comparable on the same scale, that is, let us say five percent drop-off in DR cannot be compensated by five percent improvement in FAR and vice versa. One way to classify good IDS is through testing it such that a moderate operating point is recognized where detection and false alarm rate are comparable and have best possible values. Referring Table 3, minimum FAR is observed for Random forest, Bayes Net followed by J48, Regression and Decision Table. The highest DR is monitored for Random Tree followed by Hoeffding Tree which has worst FAR. Rule based OneR is an algorithm that shows relatively decent values in terms of accuracy, DR, FAR and F-score. The other algorithms like Naïve Bayes and SMO show reduced values for accuracy, F-score, DR and FAR. It can be observed that the maximum time to build the classification model is taken by SMO.

From the discussion above, it is difficult to conclude the best possible classifier for IDS from the ten selected algorithms because some algorithms show very good DR but poor FAR and vice versa; hence it becomes very difficult to decide whether to compromise on DR or on FAR. Therefore, let us now reanalyze the classifiers in regard with the NP ratio (NPR) and understand the role of this metric in choosing best algorithm for IDS in view of imbalanced data set. Figure 1 shows the

Table 3 Summary of results

Algorithm	TN	FN	FP	TP	Time (sec)	Accuracy	FAR	DR	Precision	F-score	NPR
Hoeffding tree	1424	3330	728	6368	2.78	0.66	0.34	0.66	0.90	0.76	**0.22**
Random tree	1388	3008	764	6690	0.57	**0.68**	0.36	**0.69**	0.9	**0.78**	**0.21**
Naïve Bayes	1460	4549	692	5149	0.67	0.56	0.32	0.53	0.88	0.66	0.28
OneR	1836	3650	316	6048	0.56	0.67	0.15	0.62	0.95	0.75	0.30
SMO	1440	4893	712	4805	1082	0.53	0.33	0.50	0.87	0.63	0.30
J48	1879	3996	273	5702	5.71	0.64	0.13	0.59	0.95	0.73	0.33
Random forest	1887	4067	265	5631	2.1	0.63	**0.12**	0.58	**0.96**	0.72	0.34
Regression	1868	4602	284	5096	66.33	0.59	0.13	0.53	0.95	0.68	0.37
Decision table	1876	5429	276	4269	65.15	0.52	0.13	0.44	0.94	0.60	0.44
Bayes net	1890	5463	262	4235	2.13	0.52	**0.12**	0.44	0.94	0.60	0.45

distribution of accuracy, F-score and predicted NPR for the ten classifiers whereas
Fig. 2 presents the distribution of FAR, DR and predicted NPR for the same set of
classification algorithms. As shown in Fig. 1, it can be clearly inferred that as the
predicted NPR goes up compared to the actual NPR, F-score falls significantly.
Hoeffding Tree shows the value for predicted NPR exactly equal to the actual NPR
followed by Random Tree algorithm with little lower value.

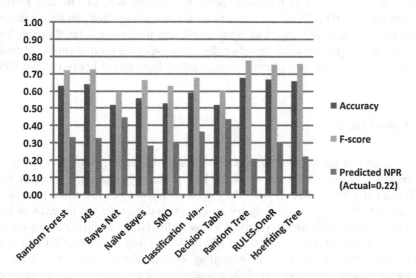

Fig. 1 Plot of accuracy, F-score and predicted NPR for ten classification algorithms

In Fig. 2, large variation in DR and FAR is observed for J48 and Random Forest
with comparatively low accuracy and F-score. For Random Tree algorithm, as the
predicted NPR is less than the actual NPR, DR increases but FAR worsens at the

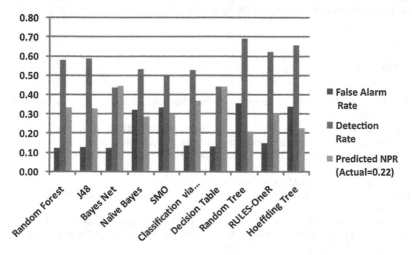

Fig. 2 Plot of detection rate, false alarm rate and predicted NPR for ten algorithms

same time. When the predicted NPR is more than the actual NPR, FAR improves but DR goes down.

Hence it can be stated that Hoeffding and Random Tree algorithms having predicted NPR closest to the actual NPR are the best algorithms out of the ten selected algorithms whose results show minimum biasing towards unbalanced data set. The NP ratio can help choose a moderate point of operation at which biasing in the classification results is minimum permitting comparison of DR and FAR to evaluate efficient IDS. Therefore, the results for Hoeffding Tree can be considered best with predicted NPR equal to actual NPR and then comes the Random Tree algorithm in comparison to all the other algorithms selected for this empirical study. In Table 3, the ten selected algorithms are ranked from top to bottom as per NPR.

5 Conclusion

The imbalance in the data set for Intrusion Detection System (IDS) gives rise to the incomparable status of the two popular IDS evaluation metrics, Detection Rate (DR) and False Alarm Rate (FAR). Any good IDS is not just identified in terms of accuracy which can be misleading for an imbalanced data set but in terms of high DR and low FAR. This paper presented a new metric NPR that formulated a leveled base for the comparison of DR and FAR by identifying a value of NPR for test results such that the learning is negligibly biased or not biased at all due to imbalance in data. The proposed NPR was applied for comparison of ten classifiers on KDD data set and hence ranked the algorithms by evaluating best computation point based on the number of true instances of each of the binary classes for the proficient IDS.

Acknowledgments The authors gratefully acknowledge the contribution of the anonymous reviewers' comments in improving the clarity of this work.

References

1. Tsai, C.-F. et al.: Intrusion detection by machine learning: a review. Exp. Syst. Appl. **36**(10), 11994–12000 (2009)
2. Gupta, S., Kumar, P., Abraham, A.: A profile based network intrusion detection and prevention system for securing cloud environment. Int. J. Distrib. Sens. Netw. (2013)
3. DARPA Intrusion Detection Evaluation, MIT Lincoln Labs. http://www.ll.mit.edu/mission/communications/ist/corpora/ideval/index.html
4. KDD Cup 1999. http://kdd.ics.uci.edu/databases/kddcup99/
5. NSL-KDD Data Set for Network-Based Intrusion Detection Systems. http://nsl.cs.unb.ca/NSL-KDD/
6. Tavallaee, M., Bagheri, E., Lu, W., Ghorbani, A.A.: A detailed analysis of the KDD CUP 99 data set. In: The Proceedings of IEEE Symposium on Computational Intelligence in Security and Defense Applications, pp. 1–6 (2009)

7. McHugh, J.: Testing intrusion detection systems: a critique of the 1998 and 1999 Darpa intrusion detection system evaluations as performed by Lincoln laboratory. ACM Trans. Inf. Syst. Secur. **3**(4), 262–294 (2000)
8. Qiang, Y., Wu, X.: 10 challenging problems in data mining research. Int. J. Inf. Technol. Decis. Making **5**(04), 597–604 (2006)
9. Chawla N.V.: Data mining for imbalanced datasets: an overview. Data Mining and Knowledge Discovery Handbook, pp. 875–886. Springer US (2010)
10. Kotsiantis, S., Kanellopoulos, D., Pintelas, P.: Handling imbalanced datasets: a review. GESTS Int. Trans. Comput. Sci. Eng. **30**(1), 25–36 (2006)
11. Hulse, V., Jason, Khoshgoftaar, T.M., Napolitano, A.: Experimental perspectives on learning from imbalanced data. In: Proceedings of the 24th International Conference on Machine learning. ACM, New York (2007)
12. Fernandez-Delgado, M., Cernadas, E.: Do we need hundreds of classifiers to solve real world classification problems? J. Mach. Learn. Res. **15**, 3133–3181 (2014)
13. González, S., et al.: Testing ensembles for intrusion detection: on the identification of mutated network scans. Computational intelligence in security for information systems, pp. 109–117. Springer, Berlin (2011)
14. Alhomouda, A., Munira, R., Dissoa, J.P., Awana, I., Al-Dhelaanb, A.: Performance evaluation study of intrusion detection systems. Proc. Comput. Sci. **5**, 173–180 (2011)
15. Sokolova, M., Lapalme, G.: A systematic analysis of performance measures for classification tasks. Inf. Process. Manag. **45**(4), 427–437 (2009)
16. Ferri, C., Hernández-Orallo, J., Modroiu, R.: An experimental comparison of performance measures for classification. Pattern Recogn. Lett. **30**(1), 27–38 (2009)
17. Cardenas, A.A., Baras, J.S., Seamon, K.: A framework for the evaluation of intrusion detection systems. IEEE Symp. Secur. Priv. 15–77 (2006)
18. Nagarajan, A., Quyen N., Banks, R., Sood, A.: Combining intrusion detection and recovery for enhancing system dependability. In: IFIP 41st International Conference on Dependable Systems and Networks Workshops (DSN-W) (2011)
19. Pietraszek, T., Tanner, A.: Information security technical report. **10**(3), 169–183 (2005) (Elsevier)
20. Giacinto, G., Perdisci, R., Del Rio, M., Roli, F.: Intrusion detection in computer networks by a modular ensemble of one-class classifiers. Special Issue on Applications of Ensemble Methods, Information Fusion, Vol. 9, no 1, pp. 69–82 (2008)
21. Chandola, V., Banerjee, A. and Kumar, V.: Anomaly detection: a survey. ACM Comput. Surv. (CSUR) **41**(3), 15 (2009)
22. Han, J., Kamber, M.: Data mining: concepts and techniques, 3rd edn. Morgan Kaufmann, San Francisco (2012)
23. Witten, I.H., Frank, E., Hall, M.A.: Data mining- practical machine learning tools and techniques. Morgan Kaufmann, San Francisco (2011)
24. Waikato Environment for Knowledge Analysis (weka) version 3.7.11. http://www.cs.waikato.ac.nz/ml/weka/

Clustering and Neural Visualization for Flow-Based Intrusion Detection

Raúl Sánchez, Álvaro Herrero and Emilio Corchado

Abstract To secure a system, potential threats must be identified and therefore, attack features are understood and predicted. Present work aims at being one step towards the proposal of an Intrusion Detection System (IDS) that faces zero-day attacks. To do that, MObile VIsualisation Connectionist Agent-Based IDS (MOVICAB-IDS), previously proposed as a hybrid-intelligent visualization-based IDS, is being upgraded by adding clustering methods. To check the validity of the proposed clustering extension, it faces a realistic flow-based dataset in present paper. The analyzed data come from a honeypot directly connected to the Internet (thus ensuring attack-exposure) and is analyzed by clustering and neural tools, individually and in conjunction. Through the experimental stage, it is shown that the combination of clustering and neural projection improves the detection capability on a continuous network flow.

Keywords Network intrusion detection · Network flow · Neural projection · Clustering

R. Sánchez · Á. Herrero (✉)
Department of Civil Engineering, University of Burgos,
Avenida de Cantabria s/n, 09006 Burgos, Spain
e-mail: ahcosio@ubu.es

R. Sánchez
e-mail: rsarevalo@ubu.es

E. Corchado
Departamento de Informática y Automática, Universidad de Salamanca,
Plaza de la Merced, s/n, 37008 Salamanca, Spain
e-mail: escorchado@usal.es

© Springer International Publishing Switzerland 2015   333
Á. Herrero et al. (eds.), *International Joint Conference*, Advances in Intelligent
Systems and Computing 369, DOI 10.1007/978-3-319-19713-5_29

1 Introduction

Intrusion Detection (ID) is a field that focuses on the identification of attacks, in both networks and computers. The huge amount of previous work focused on network-based ID can be categorized by different criteria. One of them is the nature of the analysed data and according to that, there are two categories based on the source of data to be analyzed: packets or flows. Some network Intrusion Detection Systems (IDSs) analyze packets travelling along the network and then extracts the information from the different fields in the packet (headers and payload, mainly). On the other hand, some IDSs deals with flows, being defined as "a set of IP packets passing on an observation point in the network during a certain time interval and having a set of common properties" [1].

In flow-based IDSs, rather than looking at all packets in a network, they look at aggregated information of related packets of network traffic in the form of a flow, so the amount of data to be analyzed is summarized and then reduced. With the rise of network speed and number and types of attacks, existing IDSs, face challenges of capturing every packet. Hence a flow-based IDS has an overall lower amount of data to be process, therefore it is the logical choice for high speed networks [2, 3].

MObile VIsualisation Connectionist Agent-Based IDS (MOVICAB-IDS) was proposed [4] as a novel IDS comprising a Hybrid Artificial Intelligent System (HAIS). Its main goal was to apply an unsupervised neural projection model to extract traffic dataset projections and to display them through a mobile visualisation interface. One of its main drawbacks was its dependence on human processing; MOVICAB-IDS could not automatically raise an alarm. Additionally, human users could fail to detect an intrusion even when visualised as an anomalous one, when visually processing big amounts of data [5]. This IDS is being extended by the application of clustering techniques in conjunction with neural visualization, to overcome its limitations.

Based on successful results obtained by upgrading MOVICAB-IDS with clustering techniques to detect different attacks on packet-based data [6, 7], present work focuses on flow-based data. Hence, present work proposes the combination of MOVICAB-IDS and different clustering techniques to analyze a database of flow-based attack situations, generated by the University of Twente [8]. The experimental study in present paper tries to know whether clustering could be more informative applied over the projected data rather than the original flow data captured from the network.

Clustering and neural visualization have been previously applied to the identification of different anomalous situations (network scans, MIB transfer and community string searches) related to the SNMP network protocol [6].

Clustering has been previously applied to intrusion detection: [9] proposes an alert aggregation method, clustering similar alerts into a hyper alert based on category and feature similarity. From a similar perspective, [10] proposes a two-stage clustering algorithm to analyze the spatial and temporal relation of the network intrusion behaviors' alert sequence. [11] describes a classification of network traces through an improved nearest neighbor method, while [12] applies data mining algorithms for the same purpose and the results of preformatted data are visually displayed. Finally [13] discusses on how the clustering algorithm is applied to intrusion detection and analyzes intrusion detection algorithm based on clustering problems. Differentiating from previous work, the approach proposed in present paper, applies clustering to previously projected data (processed by neural models).

The remaining sections of this study are structured as follows: Sect. 2 discusses the combination of visualization and clustering techniques and describes the applied ones. Experimental setting and results are presented in Sect. 3 while the conclusions of this study are discussed in Sect. 4.

2 Proposed Solution

To better detect intrusions, an upgrade of MOVICAB-IDS, combining projection and clustering results is being proposed. In keeping with this idea, present study focuses on flow-based data for the first time.

MOVICAB-IDS [4] is based on the application of different AI paradigms to process the continuous data flow of network traffic. In order to do so, MOVICAB-IDS split massive traffic data into limited datasets and visualises them, thereby providing security personnel with an intuitive snapshot to monitor the events taking place in the observed computer network. The following paradigms are combined within MOVICAB-IDS.

2.1 Cooperative Maximum Likelihood Hebbian Learning

One neural implementation of Exploratory Projection Pursuit (EPP) [14] is Maximum Likelihood Hebbian Learning (MLHL) [15]. It identifies interestingness by maximising the probability of the residuals under specific probability density functions which are non-Gaussian.

An extended version of this model is the Cooperative Maximum Likelihood Hebbian Learning (CMLHL) [16] model. CMLHL is based on MLHL [15] adding lateral connections [16], which have been derived from the Rectified Gaussian Distribution [17]. The resultant net can find the independent factors of a data set but does so in a way that captures some type of global ordering in the data set.

Considering an N-dimensional input vector (x), and an M-dimensional output vector (y), with W_{ij} being the weight (linking input j to output i), then CMLHL can be expressed [16] as:

1. Feed-forward step:

$$y_i = \sum_{j=1}^{N} W_{ij}x_j, \forall i. \tag{1}$$

2. Lateral activation passing:

$$y_i(t+1) = [y_i(t) + \tau(b - Ay)] + . \tag{2}$$

3. Feedback step:

$$e_j = x_j - \sum_{i=1}^{M} W_{ij}y_i, \forall j. \tag{3}$$

4. Weight change:

$$\Delta W_{ij} = \eta.y_i.sign(e_j)|e_j|^{p-1}. \tag{4}$$

where: η is the learning rate, τ is the "strength" of the lateral connections, b the bias parameter, p a parameter related to the energy function [16] and A a symmetric matrix used to modify the response to the data [16]. The effect of this matrix is based on the relation between the distances separating the output neurons.

2.2 Clustering

Cluster analysis [18, 19] consist in the organization of a collection of data items or patterns (usually represented as a vector of measurements, or a point in a multi-dimensional space) into clusters based on similarity. Hence, patterns within a valid cluster are more similar to each other than they are to a pattern belonging to a different cluster.

Pattern proximity is usually measured by a distance function defined on pairs of patterns. A variety of distance measures are in use in the various communities [20, 21]. There are different approaches to clustering data [18], but given the high number and the strong diversity of the existent clustering methods, a representative technique for partitional as well as hierarchical clustering are applied in present study.

In general terms, there are two main types of clustering techniques: hierarchical and partitional approaches. Hierarchical methods produce a nested series of partitions (illustrated on a dendrogram which is a tree diagram) based on a similarity for merging or splitting clusters, while partitional methods identify the partition that optimizes (usually locally) a clustering criterion. Hence, obtaining a hierarchy of clusters can provide more flexibility than other methods. A partition of the data can be obtained from a hierarchy by cutting the tree of clusters at certain level.

Partitional clustering aims to directly obtain a single partition of the data instead of a clustering structure, such as the dendrogram produced by a hierarchical technique. Many of these methods are based on the iterative optimization of a criterion function that reflects the similarity between a new data and the each of the initial patterns selected for a specific iteration.

3 Experimental Study

As previously stated, the proposed approach is applied to analyze flow-based data. To check the performance of clustering and projection techniques, two different alternatives are considered: clustering on projected (dimensionality-reduced) data, and clustering on original (twelve-dimensional or nine-dimensional) data.

This section describes the dataset used for evaluating the proposed clustering methods and how they were generated. The experimental settings and the obtained results are also detailed.

As similarity is fundamental to the definition of a cluster, a measure of the similarity is essential to most clustering methods and it must be carefully chosen. Present study applies well-known distance criteria used for examples whose features are all continuous. Four different distance measures are applied in present study for K-means algorithm, namely: sqEuclidean, Cityblock, Cosine, and Correlation. For agglomerative clustering and based on the way the proximity matrix is updated in the second phase, a variety of linking methods can be designed. Present study has applied the following linking methods: Single, Complete, Ward, Median, Average, Centroid, and Weighted.

3.1 Datasets

The analyzed dataset contains flow-based information from traffic collected by the University of Twente [8], in September 2008. The honeypot was directly connected to the Internet and ran several typical network services, such as ftp, ssh, etc.

This data set, consisting of 14.2 M flows, has been collected by using a honeypot ensuring traffic to be realistic. A honeypot can be defined as an "environment where vulnerabilities have been deliberately introduced to observe attacks and intrusions" [22]. Present work focuses on two segments obtained from the above mentioned database, which has been split in different overlapping segments, as MOVICAB-IDS usually do with network traffic. Every segment contains all the flows whose timestamp is between the segment initial and final time limit. Segment length is stated as 782 s to cover the whole database, whose length is 539,520 s (from the beginning of the first flow to the end of the last one), that is amounts to 6 days and 709 segments. As defined for MOVICAB-IDS, there is a slight time overlap of 10 s between each pair of consecutive segments.

Two out of the 709 generated segments have been chosen for present study:

- Segment 59: two types of data can be found: ssh_conn and irc_sideeffect.
- Segment 545: two types of data can be found: ssh_conn and http_conn.

The following fourteen features were extracted from the database to define every single flow:

- **id**: the ID of the flow.
- **src_ip**: anonymized source IP address (encoded as 32-bit number).
- **dst_ip**: anonymized destination IP address (encoded as 32-bit number).
- **packets**: number of packets in the flow.
- **octets**: number of bytes in the flow.
- **start_time**: UNIX start time (number of seconds).
- **start_msec**: start time (milliseconds part).
- **end_time**: UNIX end time (number of seconds).
- **end_msec**: end time (milliseconds part).
- **src_port**: source port number.
- **dst_port**: destination port number.
- **tcp_flags**: TCP flags obtained by ORing the TCP flags field of all packets of the flow.
- **prot**: IP protocol number.
- **type**: alert type.

Two of the above listed features are not provided to the models: the first one (added to identify each single packet) and the last one (alert type).

This set of features has been processed in order to summarize the four features related to time, being joined in only one feature, named as flow length. By doing so, new datasets have been generated, whose nine features are also analyzed. All the data sets (detailed time information vs. flow length) have been studied but, as the results are pretty similar, only the results for the second data set are shown in present paper. The four different datasets can be described as follows:

- **Segment 59**: contains 1,215 flows where 56 are from type irc_sideeffect and the rest 1159 are from type ssh_conn. Every segment comprises 12 dimensions.
- **Segment 545**: contains 731 flows where 12 are from type http_conn and the rest 719 are from type ssh_conn. Every segment comprises 12 dimensions.
- **Segment 59 with segment length**: contains 1,215 flows where 56 are from type irc_sideeffect and the rest 1159 are from type ssh_conn. The four dimensions related to time has been joined in one which is segment length calculated from start time to end time and given in milliseconds. Hence, every segment comprises 9 dimensions.
- **Segment 545 with segment length**: contains 731 flows where 12 are from type http_conn and the rest 719 are from type ssh_conn. The four dimensions related to time has been joined in one which is segment length calculated from start time to end time and given in milliseconds. Hence, every segment comprises 9 dimensions.

3.2 Results

The best results obtained by applying the previously introduced techniques to the described datasets are shown in this section. The results are projected through CMLHL and further information about the clustering results is added to the projections, mainly by the glyph metaphor (different colors and symbols). The projections comprise a legend that states the color and symbol used to depict each packet, according to the original category of the data. The following subsections comprise the results obtained by the projection and clustering technique for two of the datasets. Although the four above mentioned datasets (Segment 59, Segment 545, Segment 59 with segment length, and Segment 545 with segment length), only results for the two last ones are shown in this section because the other two are similar.

Segment 59 with segment length.

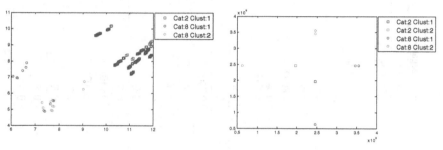

1.a K-means on projected data: $k=2$, sqEuclidean distance.　**1.b** K-means on original data: k=2, sqEuclidean distance.

Fig. 1 Some clustering results under the frame of MOVICAB-IDS through k-means for Segment 59 with segment length

For the irc_sideeffect flows on projected data (Cat. 8 in Fig. 1.a), the clustering splits all the data from this type on two different clusters (Clust. 1 and Clust. 2); the ssh_conn flows (Cat. 2 in Fig. 1.a) are grouped correctly in just one cluster (Clust. 1), however some of the flows from Cat. 8 are clustered together with them. Apart from these two projections, some more experiments have been conducted, whose details: performance, False Positive Rate (FPR) and False Negative Rate (FNR), values of k parameter, etc., can be seen in Table 1.

It can be seen from Table 1 that there is a non-zero False Negative Rate, but this value is worse on original data because the clusters are more mixed although the number of clusters (k parameter) is the same. The results on projected data in the majority of the cases probed, are better than on original (none projected) data.

Some of the run experiments for the agglomerative method with no clustering error are shown in Table 2. It can be seen that, in the case of projected data, the minimum number of clusters with no error is 3, while in the case of original data it

Table 1 K-means results for Segment 59 with segment length

Data	k	Distance criteria	False positive	False negative	Replicates/iterations	Sum of distances
Projected	2	sqEuclidean	0 %	1.4803 %	5/2	2498.16
Original	2	sqEuclidean	47.6974 %	2.3026 %	5/2	8.03752E+19
Projected	4	sqEuclidean	27.7961 %	0 %	5/7	786.499
Original	4	sqEuclidean	47.6974 %	0.7401 %	5/3	3.18651E+19
Projected	6	sqEuclidean	0 %	0 %	5/7	296.276
Original	6	sqEuclidean	0 %	0 %	5/3	7.1228E+16
Projected	2	Cityblock	59.2928 %	0 %	5/8	2065.86
Original	2	Cityblock	47.6974 %	1.8092 %	5/2	8.19316E+10
Projected	4	Cityblock	42.1053 %	0 %	5/10	955.305
Original	4	Cityblock	70.2303 %	0 %	5/2	6.98674E+10
Projected	6	Cityblock	13.8980 %	0 %	5/5	643.167
Original	6	Cityblock	29.6053 %	0 %	5/3	6.98583E+10
Projected	2	Cosine	70.2303 %	0.7401 %	5/2	1.16382
Original	2	Cosine	47.6974 %	2.3026 %	5/2	1.62372
Projected	4	Cosine	59.5395 %	0 %	5/5	0.119684
Original	4	Cosine	47.6974 %	0 %	5/4	0.0647986
Projected	6	Cosine	0 %	0 %	5/6	0.059721
Original	6	Cosine	0 %	0 %	5/5	0.000641182
Projected	2	Correlation	25.1645 %	0.7401 %	5/2	1.02283
Original	2	Correlation	47.6974 %	2.3026 %	5/2	1.95889
Projected	4	Correlation	47.8619 %	0 %	5/7	0.370916
Original	4	Correlation	47.6974 %	0 %	5/4	0.0823528
Projected	6	Correlation	47.8619 %	0 %	5/8	0.0810012
Original	6	Correlation	0 %	0 %	5/5	0.000810621

Table 2 Experimental setting of the agglomerative method for Segment 59 segment length

Data	Distance	Linkage	Cutoff	Cluster
Projected	Euclidean	Single	3	3
Projected	sEuclidean	Complete	6	3
Projected	Cityblock	Average	7	3
Projected	Minkowski p = 3	Weighted	4	3
Projected	Chebychev	Complete	5	3
Projected	Mahalanobis	Average	5	3
Projected	Cosine	Single	0.002	5
Projected	Correlation	Complete	0.005	6
Original	Euclidean	Complete	10×10^8	4
Original	Cityblock	Single	13×10^8	5
Original	Minkowski p = 3	Complete	15×10^8	4
Original	Chebychev	Complete	15×10^8	4
Original	Cosine	Average	0.007	6
Original	Correlation	Weighted	0.001	6

is 4, with appropriate distance method. In the case of original data, the sEuclidean distance and Mahalanobis distance can not be applied because the maximum recursion level has been reached in the first case, and the covariance matrix can not be computed in the second case.

Results for one of the experiments from Table 2 are depicted in Fig. 2, including flow visualization and the associated dendrogram on projected data. The chosen experiment parameters are: sEuclidean distance, complete linkage, cutoff: 6 and 3 groups with no clustering error.

Segment 545 with segment length.

Figure 3 shows the results obtained by k-means on this data. The data has been labeled as follows: ssh_conn flows (Cat. 2) and http_conn (Cat. 6).

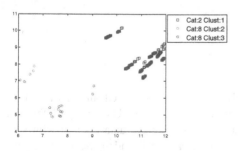
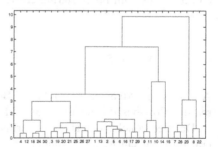

2.a Agglomerative clustering on projected data: sEuclidean, linkage: complete, cutoff: 6.   **2.b** Corresponding dendrogram.

Fig. 2 Best results of agglomerative clustering under the frame of MOVICAB-IDS for Segment 59 with segment length

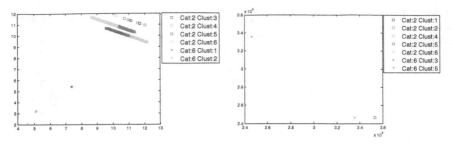

3.a K-means on projected data: k=6, Cityblock distance.

3.b K-means on original data: k=6, Cityblock distance.

Fig. 3 Some clustering result under the frame of MOVICAB-IDS through k-means for Segment 545 with segment length

Although very low, original data has a non-zero False Positive Rate value, while projected data has no error. For the ssh_conn flows on projected data (Cat. 2 in Fig. 3.a), the clustering technique groups data with no errors, even though the number of clusters (k parameter) is bigger than the categories, hence some clusters groups only data from the same category. Apart from these two projections, some more experiments have been run, whose details (performance, false positive and false negative rates, values of k parameter, etc.) can be seen in Table 3.

The run experiments for the agglomerative method with no error are shown in Table 4.

It can be seen that, in the case of projected data, the minimum number of clusters with no error is 3, while in the case of original data it is 4, with appropriate distance method. In the case of original data, the sEuclidean distance and Mahalanobis distance can not be applied because the maximum recursion level has been reached in the first case, and the covariance matrix can not be computed in the second case.

Results of one of the best experiments from Table 4 are depicted in Fig. 4, including traffic visualization and the associated dendrogram on projected data. The chosen experiment parameters are: sEuclidean distance, complete linkage, cutoff: 8 and 3 groups with no clustering error.

4 Conclusions

A clustering extension of MOVICAB-IDS has been proposed and applied to a real-life flow-based database obtained from a honeypot at the University of Twente.

Detailed conclusions about experiments on the different datasets and with several different clustering techniques and criteria can be found in Sect. 3. Experimental results show that some of the applied clustering methods obtain a good clustering performance on the analyzed data, according to false positive and false negative rates. The obtained results vary from the different analyzed datasets and

Table 3 K-means experiments with different conditions for Segment 545 with segment length

Data	k	Distance criteria	False positive	False negative	Replicates/iterations	Sum of distances
Projected	2	sqEuclidean	42.4863 %	0 %	5/12	1710.13
Original	2	sqEuclidean	49.1803 %	0.8197 %	5/2	3.65729E+17
Projected	4	sqEuclidean	15.9836 %	0 %	5/21	854.837
Original	4	sqEuclidean	0 %	0 %	5/3	3.88519E+10
Projected	6	sqEuclidean	0 %	0 %	5/20	121.144
Original	6	sqEuclidean	0 %	0 %	5/12	1.13974E+10
Projected	2	Cityblock	46.7213 %	0 %	5/8	814.403
Original	2	Cityblock	49.1803 %	0.8197 %	5/2	2.11713E+09
Projected	4	Cityblock	11.7486 %	0 %	5/21	593.5
Original	4	Cityblock	22.2678 %	0 %	5/14	1.05984E+09
Projected	6	Cityblock	0 %	0 %	5/16	402.065
Original	6	Cityblock	10.5191 %	0 %	5/12	1.0584E+09
Projected	2	Cosine	51.3661 %	0 %	5/9	1.37952
Original	2	Cosine	49.1803 %	0.8197 %	5/2	0.00345234
Projected	4	Cosine	25.1366 %	0 %	5/36	0.259027
Original	4	Cosine	0 %	0 %	5/3	1.04725E-09
Projected	6	Cosine	0 %	0 %	5/26	0.105203
Original	6	Cosine	0 %	0 %	5/12	3.07186E-10
Projected	2	Correlation	55.6011 %	0 %	5/8	53.3814
Original	2	Correlation	49.1803 %	0.8197 %	5/2	0.00437153
Projected	4	Correlation	24.8634 %	0 %	5/12	19.5017
Original	4	Correlation	0 %	0 %	5/3	1.14901E-09
Projected	6	Correlation	17.4863 %	0 %	5/27	13.6746
Original	6	Correlation	0 %	0 %	5/10	3.36364E-10

Table 4 Experimental setting of the agglomerative method for Segment 545 with segment length

Data	Distance	Linkage	Cutoff	Cluster
Projected	Euclidean	Single	8	3
Projected	sEuclidean	Complete	8	3
Projected	Cityblock	Average	12	3
Projected	Minkowski p = 3	Weighted	8	3
Projected	Chebychev	Single	5	3
Projected	Mahalanobis	Single	6	3
Projected	Cosine	Complete	0.08	3
Projected	Correlation	Average	0.05	8
Original	Euclidean	Single	1×10^8	4
Original	Cityblock	Complete	1×10^8	4
Original	Minkowski p = 3	Average	1×10^8	4
Original	Chebychev	Weighted	1×10^8	4
Original	Cosine	Single	0.0001	4
Original	Correlation	Complete	0.0001	4

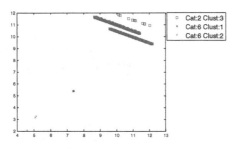

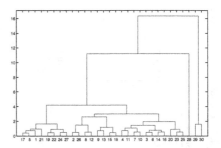

4.a Agglomerative clustering on projected data: sEuclidean, linkage: single, cutoff: 6.

4.b Corresponding dendrogram.

Fig. 4 Best results of agglomerative clustering under the frame of MOVICAB-IDS for Segment 545 with segment length

the behavior of the applied clustering techniques. These results are consistent with those previously obtained for other SNMP anomalous situations [6, 7].

There is no distance criterion which shows the best results, hence its selection will depend on the analyzed data. Comparing projected data results with the ones from original data, it can be said that projected data has better results (fewer number of groups with no errors).

Finally, it can be concluded that the applied methods are able to detect anomalous situations. It has been proven that clustering methods could help in intrusion detection over flow-based network data. On the other hand, using clustering, automatic response could be added to MOVICAB-IDS, to quickly abort intrusive actions while happening.

References

1. Quittek, J., Zseby, T., Claise, B., Zander, S.: Requirements for IP flow information export (IPFIX)
2. Sperotto, A., Schaffrath, G., Sadre, R., Morariu, C., Pras, A., Stiller, B.: An overview of IP flow-based intrusion detection. IEEE Commun. Surv. Tutor. **12**, 343–356 (2010)
3. Sperotto, A., Pras, A.: Flow-based intrusion detection. In: IFIP/IEEE International Symposium on Integrated Network Management (IM), 2011, pp. 958–963 (2011)
4. Corchado, E., Herrero, Á.: Neural visualization of network traffic data for intrusion detection. Appl. Soft Comput. **11**, 2042–2056 (2011)
5. Yom-Tov, E., Inbar, G.F.: Selection of relevant features for classification of movements from single movement-related potentials using a genetic algorithm. In: 23rd Annual International Conference of the IEEE Engineering in Medicine and Biology Society, 2001, vol. 2, pp. 1364–1366 (2001)
6. Sánchez, R., Herrero, Á., Corchado, E.: Clustering extension of MOVICAB-IDS to identify SNMP community searches. Logic J. IGPL **23**, 121–140 (2015)
7. Sánchez, R., Herrero, Á., Corchado, E.: Visualization and clustering for SNMP intrusion detection. Cybern. Syst. Int. J. **44**, 505–532 (2013)
8. Sperotto, A., Sadre, R., Vliet, F.v., Pras, A.: A Labeled Data Set For Flow-based Intrusion Detection, pp. 39–50. IP Operations and Management, Berlin (2009)
9. Zheng, Q.H., Xuan, Y.G., Hu, W.H.: An IDS alert aggregation method based on clustering. In: Zhang, H., Shen, G., Jin, D. (eds.): Advanced Research on Information Science, Automation and Material System, Pts 1-6, vol. 219–220, pp. 156–159. Trans Tech Publications Ltd, Stafa-Zurich (2011)
10. Qiao, L.B., Zhang, B.F., Lai, Z.Q., Su, J.S.: IEEE: Mining of Attack Models in IDS Alerts from Network Backbone by a Two-stage Clustering Method. In: 2012 IEEE 26th International Parallel and Distributed Processing Symposium Workshops & Phd Forum, pp. 1263–1269. IEEE, New York (2012)
11. Jiang, S., Song, X., Wang, H., Han, J.-J., Li, Q.-H.: A clustering-based method for unsupervised intrusion detections. Pattern Recogn. Lett. **27**, 802–810 (2006)
12. Cui, K.Y.: IEEE: Research on Clustering Technique in Network Intrusion Detection. IEEE Computer Society, Los Alamitos (2012)
13. Ge, L., Zhang, C.Q.: The application of clustering algorithm in intrusion detection system. In: Jin, D., Lin, S. (eds.) Advances in Future Computer and Control Systems, vol. 159, pp. 77–82. Springer, Berlin (2012)
14. Friedman, J.H., Tukey, J.W.: A projection pursuit algorithm for exploratory data-analysis. IEEE Trans. Comput. **23**, 881–890 (1974)
15. Corchado, E., MacDonald, D., Fyfe, C.: Maximum and minimum likelihood hebbian learning for exploratory projection pursuit. Data Min. Knowl. Disc. **8**, 203–225 (2004)
16. Corchado, E., Fyfe, C.: Connectionist techniques for the identification and suppression of interfering underlying factors. Int. J. Pattern Recognit. Artif.Intell. **17**, 1447–1466 (2003)
17. Seung, H.S., Socci, N.D., Lee, D.: The rectified Gaussian distribution. Adv. Neural Inf. Process. Syst. **10**, 350–356 (1998)
18. Jain, A.K., Murty, M.N, Flynn, P.J.: Data clustering: a review. ACM Comput. Surv. **31** (1999)
19. Xu, R., Wunsch, D.C.: Clustering. Wiley, New York (2009)
20. Andreopoulos, B., An, A., Wang, X., Schroeder, M.: A roadmap of clustering algorithms: finding a match for a biomedical application. Brief Bioinform **10**, 297–314 (2009)
21. Zhuang, W.W., Ye, Y.F., Chen, Y., Li, T.: Ensemble clustering for Internet security applications. IEEE Trans. Syst. Man Cybern. Part C-Appl. Rev. **42**, 1784–1796 (2012)
22. Pouget, F., Dacier, M.: Honeypot-based forensics. In: Proceedings of the AusCERT Asia Pacific Information Technology Security Conference 2004 (AusCERT2004), 23–27 May 2004, Brisbane, Australia (2004)

SW-SSS: Slepian-Wolf Coding-Based Secret Sharing Scheme

Kazumasa Omote and Tran Phuong Thao

Abstract A secret sharing scheme is a method for protecting distributed file systems against data leakage and for securing key management systems. The secret is distributed among a group of participants where each participant holds a share of the secret. The secret can be only reconstructed when a sufficient number of shares are reconstituted. Although many secret sharing schemes have been proposed, these schemes have not achieved an optimal share size and have not supported the share repair feature. This paper proposes a secret sharing scheme based on the Slepian-Wolf coding, named the SW-SSS, to obtain an optimal share size and to provide the share repair without recovering the secret. Furthermore, the share in the SW-SSS is constructed using the exclusive-OR (XOR) operation for fast computation. Unlimited parameters are also supported in the SW-SSS. To the best of our knowledge, we are the first applying the Slepian-Wolf coding to a secret sharing scheme.

Keywords Secret sharing scheme · Slepian-wolf coding · XOR network coding

1 Introduction

Distributed data mining has become very useful today with the increase in the amount of data. This in turn increases the need to preserve the privacy of the participants. Two approaches which can be used in privacy preserving data mining are encryption and secret sharing. The secret sharing is recently being considered as a more efficient approach because it does not require expensive operations (i.e., modular exponentiation of large numbers) and does not require key managements like the encryption approach [1–3]. Thus, improving secret sharing is our motivation in this paper.

K. Omote · T. P. Thao (✉)
Japan Advanced Institute of Science and Technology (JAIST),
1-1 Asahidai, Nomi, Ishikawa 923-1292, Japan
e-mail: tpthao@jaist.ac.jp

K. Omote
e-mail: omote@jaist.ac.jp

© Springer International Publishing Switzerland 2015
Á. Herrero et al. (eds.), *International Joint Conference*, Advances in Intelligent
Systems and Computing 369, DOI 10.1007/978-3-319-19713-5_30

Secret sharing schemes (SSS) are ideal for storing information that is sensitive. The basic ideas of SSS were introduced by Shamir [4] and Blakley [5]. A secret S is encoded into n shares. Each participant receives one share. This is known as the (m, n)-threshold SSS in which any m or more shares can be used to reconstruct the secret and in which the bit-size of a share is the same as the bit-size of the secret. The efficiency of the scheme is evaluated by the entropy of each share, and it must hold that $H(C_i) \geq H(S)$ where $H(S)$ and $H(C_i)$ are the entropies of a secret S and shares C_i ($i \in \{1, \cdots, n\}$) [6, 7]. To improve the efficiency of the Shamir-SSS, the Ramp-SSS was proposed [8–12] with a trade-off between security and coding efficiency. In a (m, L, n)-Ramp-SSS, the secret S can be reconstructed from any m or more shares; no information about S can be obtained from less than L shares; but a partial information of S can be leaked from any arbitrary set of $(m - t)$ shares with equivocation $\frac{t}{m}H(S)$ for $t \in \{1, \cdots, m - L\}$. It can attain that $H(C_i) = H(S)/(m - L)$, and thus the Ramp-SSS is more efficient than the Shamir-SSS [8, 9]. The drawback of these SSSs is the heavy computational cost because the shares are constructed using polynomials. An example of a polynomial is the Reed-Solomon code, which takes $O(n \log n)$ field operations for the secret distribution and $O(m^2)$ field operations for the secret reconstruction [13]. To address this drawback, the SSSs based on the XOR operation (XOR-SSS) were proposed to replace the polynomials. Examples include: the $(2, 3)$-SSS [14], the $(2, n)$-SSS [15], the $(3, n)$-SSS [16], and the (m, n)-SSS where $m = \{2, 3, 4, n - 3, n - 2, n - 1\}$ [13]. However, the threshold (m, n) in these XOR-SSSs is limited. Therefore, the (m, n)-SSSs where m and n are arbitrary were presented in [17–20]. Kurihara et al. then improved [18] to achieve the Ramp-SSS in [21].

Most of these schemes can support the secret reconstruction but cannot support the direct share repair feature. This means that when a share is corrupted, without the direct share repair, D must reconstruct the secret S before generating the new share to replace the corruption. If the direct share repair is supported, the corrupted share can be repaired directly from the remaining healthy shares without the need to reconstruct S. Breaking with the flow of previous schemes, the SSS based on the network coding (NC-SSS) was introduced [22, 23] to deal with the direct share repair. However, the schemes were constructed using a linear combination instead of the XOR. This paper shows that a special network coding based on the XOR [24–26], can be applied for SSS to address the drawbacks. The resulting scheme is called the XORNC-SSS. This paper then shows that another coding named the Slepian-Wolf coding (SWC) [27, 28], which is used to compress a data stream in a network, can be also applied for SSS to reduce the share size while still satisfying the benefits of the XORNC-SSS.

Contribution. This paper revisits the XOR network coding and applies it to the SSS (called XORNC-SSS). The XORNC-SSS has the following four advantages: (i) the shares are constructed using the XOR for fast computation; (ii) (m, n) can be chosen arbitrarily; (iii) the direct share repair is supported; (iv) the size of a share is smaller than the size of the secret.

This paper then mainly proposes the SSS based on the Slepian-Wolf coding (called SW-SSS). The share size in the SW-SSS is surprisingly less than the share size in the XORNC-SSS. In other words, the SW-SSS improves the fourth advantage of the XORNC-SSS while still satisfying the first three advantages of the XORNC-SSS.

Roadmap. The backgrounds of SSS, network coding and Slepian-Wolf coding are described in Sect. 2. The XORNC-SSS and SW-SSS schemes are presented in Sects. 3 and 4. The properties and the efficiency analysis are given in Sects. 5 and 6. The conclusion and future work are drawn in Sect. 7.

2 Background

2.1 Secret Sharing Scheme (SSS)

Shamir-SSS. The Shamir-SSS [4, 5] consists of n participants $P = \{P_1, \cdots, P_n\}$ and a dealer D. Two algorithms ShareGen and Reconst are run by D. The share generation algorithm ShareGen inputs a secret S and outputs a set $C = \{c_1, \cdots, c_n\}$. c_i ($i \in \{1, \cdots, n\}$) is called a *share* and is given to a participant P_i. The secret reconstruction algorithm Reconst inputs a set of m shares and outputs the secret S. A (m, n)-Shamir-SSS has the following properties:

- perfect SSS: any $m \leq n$ participants or more can reconstruct the secret S and no $(m - 1)$ participants or less can learn any information of the secret S. Formally, let $H(S)$ and $H(A)$ denote the entropy of the secret S and a set of shares $A \subseteq C$, respectively.

$$H(S|A) = \begin{cases} 0, & \text{if } |A| \geq m \\ H(S), & \text{if } |A| < m \end{cases}$$

- ideal SSS: the size of a share is the same as the size of the secret: $|c_i| = |S|$.

Ramp-SSS. The Ramp-SSS [8–12] was proposed to improve the coding efficiency of the Shamir-SSS. The Ramp-SSS has three parameters (m, L, n) and is constructed in a way that any m shares or more can reconstruct the secret S. Any set of $(m - t)$ shares where $t \in \{1, \cdots, m - L\}$ can learn a partial information of the secret S. Any L shares or less cannot obtain any information of the secret S. Formally,

$$H(S|A) = \begin{cases} H(S), & \text{if } |A| < L \\ \frac{m - |A|}{m}, & \text{if } L \leq |A| < m \\ 0, & \text{if } |A| \geq m \end{cases}$$

The Ramp-SSS is more efficient than the Shamir-SSS because in any (m, L, n)-Ramp-SSS, $H(C_i) = H(S)/(m - L)$ [8, 9].

2.2 Network Coding (NC)

The NC [29–32] was proposed to improve the data transmission efficiency. Suppose that a source node wants to send a message M to a receiver node. The source node firstly divides M into m blocks: $M = v_1 || \cdots || v_m$. $v_i \in \mathbb{F}_q^l$ ($i = \{1, \cdots, m\}$) where $\mathbb{F}_q^l$ denotes a l-dimensional finite field $\mathbb{F}$ over a large prime q. Before transmitting, the source node augments v_i with the vector of length m which contains a single '1' in the i-th position and '0' elsewhere. A resulting augmented block has the following form:

$$w_i = (\overbrace{0, \cdots, 0, \underbrace{1}_{i}, 0, \cdots, 0}^{m}, v_i) \in \mathbb{F}_q^{l+m}.$$

Then, the source node sends the m augmented blocks as packets to the network. Each node in the network linearly combines the packets it receives and transmits the resulting linear combination to its adjacent nodes. Suppose that an intermediate node receives t packets. The intermediate node randomly chooses t coefficients $\alpha_1, \cdots, \alpha_t \in \mathbb{F}_q$ and linearly combines the received packets by $c = \sum_{i=1}^{t} \alpha_i w_i$. The intermediate node then sends c to the network. Therefore, the receiver node can receive the linear combinations of the augmented blocks. Assume that the receiver node receives m packets $y_1, \cdots, y_m \in \mathbb{F}_p^{l+m}$ which are the linear combinations of $w_1, \cdots, w_m$. The receiver node can find $v_1, \cdots, v_m$ using the accumulated coefficients contained in the first m coordinates of y_i.

2.3 Slepian-Wolf Coding (SWC)

The SWC [27] was proposed to compress data in a network. The SWC has several approaches: syndrome-based, binning idea, LDPC-based and parity-based [28, 33, 34]. For the most efficient computation, this paper uses the *binning idea*. Suppose that the source has two data b_1 and b_2 which have the same size. To compress, b_1 is divided into a number of bins. During encoding, the index of the bin that the input belongs to is transmitted to the receiver instead of the input itself. For example, $|b_1|$ is divided into k bins. Each bin contains $\frac{|b_1|}{k}$ elements. If the SWC is not used, $\log_2 |b_1|$ bits are required to transmit the input to the receiver. If the SWC is used, only $\log_2 k$ bits are required. The receiver cannot decode b_1 if b_2 does not present because the receiver cannot know the corresponding element of the bin index. If b_2 is obtained, the receiver can decode b_1 by picking the element in the bin that is best matched with b_2.

2.4 Notations

Throughout this paper, the following notations are used:

- S denotes the secret.
- D denotes the dealer.
- m denotes the number of secret blocks (the first threshold).
- b_i denotes a secret block where $i \in \{0, \cdots, m-1\}$.
- n denotes the number of participants (the second threshold).
- L denotes the number of participants whose shares are collectively constructed from m secret blocks.
- P_i denotes a participant where $i \in \{0, \cdots, n-1\}$.
- c_i denotes a share stored in P_i.
- s_i denotes the XOR used to construct c_i ($s_i = b_j \oplus b_t \oplus b_z$).
- d_i denotes the metadata of c_i (the number of '1' bits in s_i).
- j denotes the index of the first operand of s_i.
- t denotes the index of the second operand of s_i.
- z denotes the index of the third operand of s_i.
- $|b|$ denotes the bit-size of a secret block ($|b| = \frac{|S|}{m}$).
- $\oplus$ denotes the XOR operator.
- $||$ denotes the concatenation operator.

3 The XORNC-SSS Scheme

The XOR-based NC is revisited in this section. Although the XOR-based NC exits in previous network coding schemes [24–26], the new thing here is that the construction is applied to a (m, n)-SSS instead of network coding. No previous scheme considers this applying.

D firstly divides S into m blocks: $S = b_0 || \cdots || b_{m-1}$ ($|b_i| = \frac{|S|}{m}$) and encodes S into n shares. Each participant P_i holds a share c_i ($i \in \{0, \cdots, n-1\}$). To compute c_i, D chooses secret blocks randomly and combines them using the XOR. D then pads that XOR with a vector of length m which contains a '1' bit in the index of each chosen secret block and $(m-1)$ '0' bits elsewhere. The padded vector is called the coefficient of c_i. Suppose that c_i is constructed from t secret blocks $b_{i_0}, \cdots, b_{i_{t-1}}$. Let $s_i = b_{i_0} \oplus \cdots \oplus b_{i_{t-1}}$. c_i has the following form:

$$c_i = (\underbrace{a_0, a_2, \cdots, a_{m-1}}_{m}, \underbrace{s_i}_{\frac{|S|}{m}})$$

where $a_i = 1$ if $i \in \{i_0, \cdots, i_{t-1}\}$ and $a_i = 0$ elsewhere. The share size is $|c_i| = m + \frac{|S|}{m}$. The ideal property of a SSS is $|c_i| = |S|$ (Sect. 2.1). The XORNC-SSS achieves a better share size if $|c_i| \le |S|$. From this inequality, $(m - \frac{|S|}{2})^2 \le \frac{|S|^2}{4} - |S|$.

Because $|S|$ is large in a real system, $\frac{|S|^2}{4} - |S| \approx \frac{|S|^2}{4}$. Therefore, $m \leq |S|$. In other words, if the parameters are chosen s.t. $m \leq |S|$, the scheme can reduce the share size. Furthermore, the coefficients are chosen s.t. the matrix consisting of the coefficients of any m shares has rank m. This condition is to ensure that m secret blocks can be reconstructed from any m shares. To reconstruct S, D chooses any m shares to find m secret blocks, then concatenates them together. To repair a corrupted share, D requires m healthy shares to reconstitute using the XOR.

Example. Suppose that $S = b_0||b_1||b_2$ and $n = 4$. D creates the shares $c_0 = (1, 1, 1, b_0 \oplus b_1 \oplus b_2)$, $c_1 = (1, 1, 0, b_0 \oplus b_1)$, $c_2 = (1, 0, 1, b_0 \oplus b_2)$, $c_3 = (1, 0, 0, b_0)$. D sends $\{c_0, \cdots, c_3\}$ to the participants $\{P_0, \cdots P_3\}$, respectively. To reconstruct S, D chooses $m = 3$ shares (suppose c_0, c_2, c_3) and constructs the following equation system:

$$\begin{cases} s_0 = b_0 \oplus b_1 \oplus b_2 \\ s_2 = b_0 \oplus b_2 \\ s_3 = b_0 \end{cases}$$

Then, $\{b_0, b_1, b_2\}$ are computed using the Gaussian elimination. $S = b_0||b_1||b_2$. Suppose that P_2 is corrupted, D requires P_0, P_1 and P_3 to provide s_0, s_1 and s_3. D repairs s_2 by $s_2 = s_0 \oplus s_1 \oplus s_3$.

4 The Proposed SW-SSS Scheme

In this scheme, a share c_i does not have the same form as in the XORNC-SSS. Instead, c_i is the index of the bin that the XOR belongs to. This scheme focuses on the share generation, secret reconstruction and share repair. Checking a corrupted participant is beyond the scope of this paper. Several existing schemes can be used: homomorphic MAC [32, 35] or homomorphic signature [36, 37].

4.1 Share Generation (ShareGen)

Each XOR is constructed from three different secret blocks. From m secret blocks, there are $\binom{m}{3}$ XORs. Because only n out of $\binom{m}{3}$ XORs are required for n participants, these n XORs are chosen s.t. the index sequence of the three secret blocks is a permutation of the proper set $\{0, \cdots, m - 1\}$ in an ascending order. Each XOR itself is also sorted in an ascending order. Namely, D chooses the shares for n participants from the following XORs respectively, until C has enough n XORs:

$$(b_0 \oplus b_1 \oplus b_2), (b_0 \oplus b_1 \oplus b_3), \cdots, (b_0 \oplus b_1 \oplus b_{m-1}),$$
$$(b_0 \oplus b_2 \oplus b_3), (b_0 \oplus b_2 \oplus b_4), \cdots, (b_0 \oplus b_2 \oplus b_{m-1}),$$
$$\cdots$$
$$(b_1 \oplus b_2 \oplus b_3), (b_1 \oplus b_2 \oplus b_4), \cdots, (b_1 \oplus b_2 \oplus b_{m-1}),$$
$$\cdots$$

Algorithm 1: ShareGen

Input : $m, n, |S|$
Output: $\{c_0, d_0\}, \cdots, \{c_{n-1}, d_{n-1}\}$

1   $S = b_0 || \cdots || b_{m-1}$;
2   $count \leftarrow 0$;
3   **for** $j \leftarrow 0$ **to** $m - 3$ **do**
4   **for** $t \leftarrow j + 1$ **to** $m - 2$ **do**
5   **for** $z \leftarrow t + 1$ **to** $m - 1$ **do**
6              $s_i \leftarrow b_j \oplus b_t \oplus b_z$;
7              $d_i \leftarrow s_i.count('1')$;
8              $M_i \leftarrow$ ListAnagram$(|b|, d_i)$;
9              $c_i \leftarrow index(M_i, s_i)$;
10              $count + +$;
11   **if** $(count == n - 1)$ **then**
12   **return**
                     $\{c_0, d_0\}, \cdots, \{c_{n-1}, d_{n-1}\}$
13   **end**
14   **end**
15   **end**
16 **end**

Given $m, n, |S|$, the ShareGen (Algorithm 1) outputs n pairs of share c_i and its metadata d_i. $\mathcal{D}$ firstly divides $S = b_0 || \cdots || b_{m-1}$ (line 1). The block size is $|b| = \frac{|S|}{m}$. $\mathcal{D}$ computes the XOR $s_i = b_j \oplus b_t \oplus b_z$ for each share (line 6). $\mathcal{D}$ finds the number of '1' bits in s_i, denoted by d_i (line 7). $\mathcal{D}$ constructs a set M_i consisting of all permutations of each XOR (line 8) using ListAnagram (the pseudo code of this function is omitted because it is supported in many programming languages). The elements in M_i are sorted in an ascending order. $\mathcal{D}$ finds the corresponding index of s_i in M_i, denoted by c_i (line 9). Observe that the share is not s_i but the index of s_i in the set M_i. The number of elements in M_i is $|M_i| = \binom{|b|}{d_i}$. The number of bits for representing a share is at most $\log_2 |M_i|$. The bandwidth and the storage cost can be reduced because the size of an index is less than the size of a XOR. ShareGen finally returns $\{c_0, d_0\}, \cdots, \{c_{n-1}, d_{n-1}\}$. $\mathcal{D}$ distributes $\{c_i, d_i\}$ to the participant $\mathcal{P}_i$.

Example 4-1. Suppose that $S = 10100111001110110001$. S is divided into $m = 5$ blocks: $b_0 = 1010, b_1 = 0111, b_2 = 0011, b_3 = 1011$ and $b_4 = 0001$ ($|S| = 20$, $|b_i| = 4$). Suppose that $n = 8$, the shares are $\{c_0, \cdots, c_7\}$. To construct c_0, $s_0 = b_0 \oplus b_1 \oplus b_2 = 1110$ is used. Because the number of '1' bits in s_0 is 3, $d_0 = 3$. Because

Table 1 Example 4-1

P_0	$s_0 = b_0 \oplus b_1 \oplus b_2 = 1110$	$d_0 = 3$	$M_0 = \{0111, 1011, 1101, 1110\}$	$c_0 = 3_{dec} = 11$
P_1	$s_1 = b_0 \oplus b_1 \oplus b_3 = 0110$	$d_1 = 2$	$M_1 = \{0011, 0101, 0110, 1001, 1010, 1100\}$	$c_1 = 2_{dec} = 10$
P_2	$s_2 = b_0 \oplus b_1 \oplus b_4 = 1100$	$d_2 = 2$	$M_2 = \{0011, 0101, 0110, 1001, 1010, 1100\}$	$c_2 = 5_{dec} = 101$
P_3	$s_3 = b_0 \oplus b_2 \oplus b_3 = 0010$	$d_3 = 1$	$M_3 = \{0001, 0010, 0100, 1000\}$	$c_3 = 1_{dec} = 1$
P_4	$s_4 = b_0 \oplus b_2 \oplus b_4 = 1000$	$d_4 = 1$	$M_4 = \{0001, 0010, 0100, 1000\}$	$c_4 = 3_{dec} = 11$
P_5	$s_5 = b_0 \oplus b_3 \oplus b_4 = 0000$	$d_5 = 0$	$M_5 = \{\}$	$c_5 = 0_{dec} = 0$
P_6	$s_6 = b_1 \oplus b_2 \oplus b_3 = 1111$	$d_6 = 4$	$M_6 = \{1111\}$	$c_6 = 0_{dec} = 0$
P_7	$s_7 = b_1 \oplus b_2 \oplus b_4 = 0101$	$d_7 = 2$	$M_7 = \{0011, 0101, 0110, 1001, 1010, 1100\}$	$c_7 = 1_{dec} = 1$

$d_0 = 3$ and $|b_i| = 4$, $M_0 = \{0111, 1011, 1101, 1110\}$. The elements in M_0 are sorted in an ascending order and are indexed as $\{0, \cdots, 3\}$. Because $|M_0| = \binom{4}{3} = 4$, at most $\log_2 4 = 2$ bits are required to represent c_0 instead of 4 bits of s_0. Because the index of s_0 in M_0 is 3, $c_0 = 3_{(decimal)} = 11$. $\{c_0, d_0\}$ are sent to the participant P_0. Similarly, $\{c_1, \cdots c_7\}$ are computed as in Table 1.

4.2 Secret Reconstruction (Reconst)

To reconstruct S, D requires m participants to provide their shares (suppose $c_{k_0}, \cdots, c_{k_{m-1}}$). These shares are chosen s.t. the binary matrix consisting of the coefficient vectors of the XORs has full rank. Given m pairs of $\{c_{k_i}, d_{k_i}\}$, the Reconst algorithm (Algorithm 2) outputs m secret blocks $\{b_0, \cdots, b_{m-1}\}$. For each c_{k_i}, D firstly lists all permutations given $|b|$ and d_{k_i} (line 2). D finds the XOR s_{k_i} by picking the c_{k_i}-th element in the set M_{k_i} (line 3). D executes LocateIndices to find the indices of three operands of s_{k_i} (line 4). D constructs a vector v_{k_i} consisting of $m+1$ elements: m first elements are the binary coefficients of m secret blocks and the finally element is s_{k_i}. Namely, $v_{k_i} = [e_0, e_1, \cdots, e_{m-1}, s_{k_i}]$ where $e_x \in \{0, 1\}$ for $x = 0, \cdots, m - 1$. $e_x = 1$ when x is the index of each operand in s_{k_i} (j_{k_i}, t_{k_i} and z_{k_i}). $e_x = 0$ elsewhere. All v_{k_i}'s are constructed (line 5-15) and are combined into a matrix Q (line 16). D executes the Gaussian elimination on Q to obtain a matrix Q' (line 17) in order to solve m unknowns $b_0, \cdots, b_{m-1}$. The explanation of GaussEliminate is omitted because it is a common function. D filters $b_0, \cdots, b_{m-1}$ from Q' (line 18). D reconstructs $S = b_0 || \cdots || b_{m-1}$ (line 19).

Algorithm 2: Reconst

Input : $\{c_{k_0}, d_{k_0}\}, \cdots, \{c_{k_{m-1}}, d_{k_{m-1}}\}$
Output: $\{b_0, \cdots, b_{m-1}\}$

1 **for** $i \leftarrow 0$ **to** $m - 1$ **do**
2     $M_{k_i} \leftarrow \text{ListAnagram}(|b|, d_{k_i})$;
3     $s_{k_i} \leftarrow M_{k_i}[c_{k_i}]$;
4     $j_{k_i}, t_{k_i}, z_{k_i} \leftarrow \text{LocateIndices}(m, k_i)$;
5     $v_{k_i} \leftarrow [\ ]$;
6   **for** $x \leftarrow 0$ **to** $m - 1$ **do**
7   **if** $(x == j_{k_i})$ **or** $(x == t_{k_i})$ **or** $(x == z_{k_i})$ **then**
8           $v_{k_i}[x] \leftarrow 1$;
9   **end**
10   **else**
11          $v_{k_i}[x] \leftarrow 0$
12   **end**
13   **end**
14     $v_{k_i}[m] \leftarrow s_{k_i}$;
15 **end**
16 $Q \leftarrow [v_{k_0}, v_{k_1}, \cdots, v_{k_{m-1}}]^T$;
17 $Q' \leftarrow \text{GaussEliminate}(Q)$;
18 $\{b_0, \cdots, b_{m-1}\} \leftarrow \text{filter}(Q')$;
19 $S \leftarrow b_0 || \cdots || b_{m-1}$;
20 **return** S

Algorithm 3: LocateIndices

Input : m, k_i
Output: $j_{k_i}, t_{k_i}, z_{k_i}$

1 $count \leftarrow -1$;
2 **for** $j \leftarrow 0$ **to** $m - 3$ **do**
3   **for** $t \leftarrow j + 1$ **to** $m - 2$ **do**
4   **for** $z \leftarrow t + 1$ **to** $m - 1$ **do**
5           $count + +$;
6   **if** $(count == k_i)$ **then**
7              $j_{k_i} \leftarrow j$;
8              $t_{k_i} \leftarrow t$;
9              $z_{k_i} \leftarrow z$;
10   **end**
11   **end**
12   **end**
13 **end**
14 **return** $j_{k_i}, t_{k_i}, z_{k_i}$

Example 4-2. Following the Example 4-1, suppose that $\{c_1, c_3, c_5, c_6, c_7\}$ are chosen to reconstruct S because the matrix consisting of the coefficient vectors of $\{s_1, s_3, s_5, s_6, s_7\}$ has full rank. Because $|b| = 4$ and $d_1 = 2, M_1 = \{0011, 0101, 0110,$ $1001, 1010, 1100\}$. Because the element whose index in M_1 is $c_1 = 10_{bin} = 2_{dec}$ is 0110, $s_1 = 0110$. Similarly, $s_3 = 0010$, $s_5 = 0000$, $s_6 = 1111$ and $s_7 = 0101$.

A vector v_{k_i} is then constructed for each c_{k_i}. Because $s_1 = b_0 \oplus b_1 \oplus b_3 = 0110$, $(j_1, t_1, z_1) = (0, 1, 3)$. Therefore, $v_1 = [1, 1, 0, 1, 0, 0110]$. Similarly, $v_3 = [1, 0, 1, 1, 0, 0010]$, $v_5 = [1, 0, 0, 1, 1, 0000]$, $v_6 = [0, 1, 1, 1, 0, 1111]$ and $v_7 = [0, 1, 1, 0, 1, 0101]$. The matrix Q and Q' are constructed as follows:

$$
Q = \begin{pmatrix} v_1 \\ v_3 \\ v_5 \\ v_6 \\ v_7 \end{pmatrix} = \left(\begin{array}{ccccc|c} 1,1,0,1,0, & 0110 \\ 1,0,1,1,0, & 0010 \\ 1,0,0,1,1, & 0000 \\ 0,1,1,1,0, & 1111 \\ 0,1,1,0,1, & 0101 \end{array} \right) \xrightarrow[\rightarrow]{Gauss-elimination} Q' = \left(\begin{array}{ccccc|c} 1,0,0,0,0, & 1010 \\ 0,1,0,0,0, & 0111 \\ 0,0,1,0,0, & 0011 \\ 0,0,0,1,0, & 1011 \\ 0,0,0,0,1, & 0001 \end{array} \right)
$$

From Q', $b_0 = 1010, b_1 = 0111, b_2 = 0011, b_3 = 1011$ and $b_4 = 0001$. Finally, $S = b_0 || \cdots || b_4$.

4.3 Share Repair (ShareRepair)

When a participant is corrupted, D performs the ShareRepair algorithm (Algorithm 4) to repair the corrupted share.

Algorithm 4: ShareRepair

Input : P_{corr}
Output: c_{corr}, d_{corr}

1 $j_{corr}, t_{corr}, z_{corr} \leftarrow$ LocateIndices$(m, corr)$;
2 Choose $\alpha, \beta \in \{0, \cdots, m-1\}$ s.t. $\alpha, \beta \neq j_{corr}, t_{corr}, z_{corr}$;
3 $\{j_{r_1}, t_{r_1}, z_{r_1}\} \leftarrow$ AscendingSort$(j_{corr}, t_{corr}, \alpha)$;
4 $\{j_{r_2}, t_{r_2}, z_{r_2}\} \leftarrow$ AscendingSort$(j_{corr}, z_{corr}, \beta)$;
5 $\{j_{r_3}, t_{r_3}, z_{r_3}\} \leftarrow$ AscendingSort$(j_{corr}, \alpha, \beta)$;
6 $P_{r_1} \leftarrow$ LocateParticipant$(j_{r_1}, t_{r_1}, z_{r_1})$;
7 $P_{r_2} \leftarrow$ LocateParticipant$(j_{r_2}, t_{r_2}, z_{r_2})$;
8 $P_{r_3} \leftarrow$ LocateParticipant$(j_{r_3}, t_{r_3}, z_{r_3})$;
9 $P_{r_1}, P_{r_2}, P_{r_3}$ are required to provide $\{c_{r_1}, d_{r_1}\}, \{c_{r_2}, d_{r_2}\}$, $\{c_{r_3}, d_{r_3}\}$ to D;
10 $M_{r_1} \leftarrow$ ListAnagram$(|b|, d_{r_1})$;
11 $M_{r_2} \leftarrow$ ListAnagram$(|b|, d_{r_2})$;
12 $M_{r_3} \leftarrow$ ListAnagram$(|b|, d_{r_3})$;
13 $s_{r_1} \leftarrow M_{r_1}[c_{r_1}]$;
14 $s_{r_2} \leftarrow M_{r_2}[c_{r_2}]$;
15 $s_{r_3} \leftarrow M_{r_3}[c_{r_3}]$;
16 $s_{corr} \leftarrow s_{r_1} \oplus s_{r_2} \oplus s_{r_3}$;
17 $d_{corr} \leftarrow s_{corr}.count('1')$;
18 $M_{corr} \leftarrow$ ListAnagram$(|b|, d_{corr})$;
19 $c_{corr} \leftarrow$ index(M_{corr}, s_{corr});
20 **return** $\{c_{corr}, d_{corr}\}$

Algorithm 5: LocateParticipant

Input : $j_{r_i}, t_{r_i}, z_{r_i}$
Output: $\mathcal{P}_{r_i}$

1 *count* $\leftarrow -1$;
2 **for** $j \leftarrow 0$ **to** $m-3$ **do**
3 | **for** $t \leftarrow j+1$ **to** $m-2$ **do**
4 | | **for** $z \leftarrow t+1$ **to** $m-1$ **do**
5 | | | *count* $++$;
6 | | | **if** $(j == j_{r_i})$ **and** $(t == t_{r_i})$
   | | | **and** $(z == z_{r_i})$ **then**
7 | | | | **return** *count*
8 | | | **end**
9 | | **end**
10 | **end**
11 **end**
12 **return** *count*

Suppose that the participant $\mathcal{P}_{corr}$ is corrupted, $\mathcal{D}$ uses three healthy participants to repair $\mathcal{P}_{corr}$. $\mathcal{D}$ firstly finds the indices of $b_{j_{corr}}$, $b_{t_{corr}}$ and $b_{z_{corr}}$ by LocateIndices (line 1). $\mathcal{D}$ chooses $\alpha, \beta \in \{0, \cdots, m-1\}$ s.t. $\alpha, \beta \neq j_{corr}, t_{corr}, z_{corr}$ (line 2). The idea to find three shares to repair $\mathcal{P}_{corr}$ is that:

$$b_{j_{corr}} \oplus b_{t_{corr}} \oplus b_{z_{corr}} = (b_{j_{corr}} \oplus b_{t_{corr}} \oplus b_\alpha) \oplus (b_{j_{corr}} \oplus b_{z_{corr}} \oplus b_\beta) \oplus (b_{j_{corr}} \oplus b_\alpha \oplus b_\beta)$$

$\{j_{corr}, t_{corr}, \alpha\}$, $\{j_{corr}, z_{corr}, \beta\}$ and $\{j_{corr}, \alpha, \beta\}$ are sorted in an ascending order using AscendingSort (line 3-5). The pseudo code of the AscendingSort is omitted because it is a simple function. Let $\{j_{r_1}, t_{r_1}, z_{r_1}\}$, $\{j_{r_2}, t_{r_2}, z_{r_2}\}$ and $\{t_{r_3}, t_{r_3}, z_{r_3}\}$ denote the results of these sorts, respectively. $\mathcal{D}$ finds the three participants who store the three XORs using LocateParticipant (line 6-8). $\mathcal{D}$ obtains the XORs $\{s_{r_1}, s_{r_2}, s_{r_3}\}$ by picking the c_{r_1}-th, c_{r_2}-th and c_{r_3}-th in the set M_{r_1}, M_{r_2} and M_{r_3}, respectively (line 10-15). s_{corr} is recovered by $s_{corr} = s_{r_1} \oplus s_{r_2} \oplus s_{r_3}$ (line 16). $\mathcal{D}$ finds the metadata d_{corr} by counting the number of '1' bits in s_{corr} (line 17). $\mathcal{D}$ computes the share c_{corr} (line 18-20) as the ShareGen algorithm.

Example 4-3. Following the Examples 4-1 and 4-2, suppose that $\mathcal{P}_4$ is corrupted. $\{j_{corr}, t_{corr}, z_{corr}\} = \{0, 2, 4\}$ because $\mathcal{P}_4$ uses $b_0 \oplus b_2 \oplus b_4$. Choose $\alpha = 1$ and $\beta = 3$ because $1, 3 \neq 0, 2, 4$.

$$b_0 \oplus b_2 \oplus b_4 = (b_0 \oplus b_2 \oplus b_1) \oplus (b_0 \oplus b_4 \oplus b_3) \oplus (b_0 \oplus b_1 \oplus b_3).$$

Let $\{j_{r_1}, t_{r_1}, z_{r_1}\} = \{0, 1, 2\}$, $\{j_{r_2}, t_{r_2}, z_{r_2}\} = \{0, 3, 4\}$, $\{j_{r_3}, t_{r_3}, z_{r_3}\} = \{0, 1, 3\}$. Given $\{j_{r_1}, t_{r_1}, z_{r_1}\} = \{0, 1, 2\}$, $\mathcal{P}_0$ is chosen because $\mathcal{P}_0$ uses $(b_0 \oplus b_1 \oplus b_2)$. Given $\{j_{r_2}, t_{r_2}, z_{r_2}\} = \{0, 3, 4\}$, $\mathcal{P}_5$ is chosen because $\mathcal{P}_5$ uses $(b_0 \oplus b_3 \oplus b_4)$. Given $\{j_{r_3}, t_{r_3}, z_{r_3}\} = \{0, 1, 3\}$, $\mathcal{P}_1$ is chosen because $\mathcal{P}_1$ uses $(b_0 \oplus b_1 \oplus b_3)$. $\mathcal{P}_0, \mathcal{P}_5$ and $\mathcal{P}_1$ are then required to provide $\{c_0, d_0\}, \{c_5, d_5\}$ and $\{c_1, d_1\}$ to $\mathcal{D}$. Given $\{c_0, d_0\} = \{11, 3\}$, $\mathcal{D}$ finds $s_0 = 1110$. Given $\{c_5, d_5\} = \{0, 0\}$, $\mathcal{D}$ finds $s_5 = 0000$. Given

$\{c_1, d_1\} = \{10, 2\}$, $\mathcal{D}$ finds $s_1 = 0110$. $\mathcal{D}$ computes $s_{corr} = s_0 \oplus s_5 \oplus s_1 = 1000$. Because the number of '1' bits in 1000 is 1, $d_{corr} = 1$. Because $d_{corr} = 1$ and $|b| = 4$, $M_{corr} = \{0001, 0010, 0100, 1000\}$. Because $|M_{corr}| = 4$, at most $\log_2 4 = 2$ bits are required to represent c_{corr}. The share c_{corr} is the index of s_{corr} in M_{corr}, which is $c_{corr} = 3_{dec} = 11_{bin}$.

5 Property Analysis

This section analyses two properties of the main scheme SW-SSS.

5.1 Secrecy

We firstly consider the secret reconstruction condition. Let *epoch* be a time step in which the participants are checked. If a corrupted participant is detected, it will be repaired in the epoch.

Theorem 1 *The secret can be reconstructed if in any epoch, at least m out of n participants are healthy, and the matrix consisting of the coefficient vectors of the XORs has full rank.*

Proof $S = b_0 || \cdots || b_{m-1}$. To reconstruct S, $b_0 || \cdots || b_{m-1}$ are viewed as the unknowns that need to be solved. To solve these unknowns, at least m shares $c_{i_1}, \cdots, c_{i_{m-1}}$ along with their metadata $d_{i_1}, \cdots, d_{i_{m-1}}$ are required to make the matrix have full rank.

$$\begin{cases} s_{i_0} = b_{j_0} \oplus b_{t_0} \oplus b_{z_0} \\ s_{i_1} = b_{j_1} \oplus b_{t_1} \oplus b_{z_1} \\ \cdots \\ s_{i_{m-1}} = b_{j_{m-1}} \oplus b_{t_{m-1}} \oplus b_{z_{m-1}} \end{cases}$$

In other words, the number of required participants is at least m, in order to ensure that the equation system is solvable. Theorem 1 leads to a constrain that $n > m$. Because each of n shares is constructed from any three out of $\binom{m}{3}$ secret blocks, another constraint is $n \leq \binom{m}{3}$. Therefore, m and n should be chosen s.t. $m < n \leq \binom{m}{3}$. □

The secrecy is now analysed as follows. Let $H(S)$ be the entropy of the random variable which is induced by S. Let L denote the number of participants whose shares are collectively constructed from m secret blocks. The SW-SSS satisfies the property of the Ramp-SSS as the following theorem:

Theorem 2 *The secrecy of t random variables $\{C_{i_1}, \cdots, C_{i_t}\}$ representing any t shares $\{c_{i_0}, \cdots, c_{i_{t-1}}\}$ and t random variables $\{D_{i_0}, \cdots, D_{i_{t-1}}\}$ representing any t metadata $\{d_{i_0}, \cdots d_{i_{t-1}}\}$ is:*

$$H(S|(C_{i_0}, D_{i_0}), \cdots, (C_{i_{t-1}}, D_{i_{t-1}})) = \begin{cases} H(S), & \text{if } t < L \\ \frac{m-t}{m}H(S), & \text{if } L \le t < m \\ 0, & \text{if } m \le t \end{cases}$$

Proof s_i is the original coded sequence that can be uniquely determined by the share c_i and its metadata d_i. From the property of the conditional entropy, $H(S|(c_{i_0}, d_{i_0}),$ $\cdots, (c_{i_{t-1}}, d_{i_{t-1}})) \le H(S|s_{i_0}, \cdots, s_{i_{t-1}})$. The equality holds if S is uniformly distributed. For each case of t, the secrecy is given as follows:

- Case 1 ($t < L$): $\{s_{i_0}, \cdots, s_{i_{t-1}}\}$ are constructed from inadequate m secret blocks $b_0, \cdots, b_{m-1}$. The matrix consisting of the coefficient vectors of s_{i_j} does not have full rank. Thus, $H(S|s_{i_0}, \cdots, s_{i_{t-1}}) = H(S)$. This yields $H(S|(C_{i_0}, D_{i_0}), \cdots, (C_{i_{t-1}}, D_{i_{t-1}})) = H(S)$.
- Case 2 ($L \le t$): the matrix consisting of the coefficient vectors of s_{i_j} has rank t. Thus, $H(S|s_{i_0}, \cdots, s_{i_{t-1}}) = \frac{m-t}{m}H(S)$. This yields: $H(S|(C_{i_0}, D_{i_0}), \cdots, (C_{i_{t-1}}, D_{i_{t-1}})) = \frac{m-t}{m}H(S) < H(S)$.
- Case 3 ($m \le t$): from (2), $\frac{m-t}{m}H(S) = 0$. This yields: $H(S|(C_{i_0}, D_{i_0}), \cdots, (C_{i_{t-1}}, D_{i_{t-1}})) = 0$. $\qquad\square$

5.2 Share Size

The comparison is given in Table 2. In the previous schemes, the share size is $|S| = m \cdot |b|$ (ideal SSS). In the XORNC-SSS, the share size is $m + |b|$ (Sect. 3). In the SW-SSS, the share size is at most $\log_2 \binom{|b|}{d_i}$ (Sect. 4.1). For $\forall |b|$ and $\forall d_i \in [0, |b|]$, $\log_2 \binom{|b|}{d_i} < |b|$. Thus, $\log_2 \binom{|b|}{d_i} = \frac{|b|}{x}$ for some $x > 1$. It is clear that $m + |b| > \frac{|b|}{x}$.

Table 2 The share size comparison

	Previous schemes (Shamir-SSS, XOR-SSS, NC-SSS)	XORNC-SSS	SW-SSS
Share size	$m \cdot |b|$	$m + |b|$	$\frac{|b|}{x}$

6 Efficiency Analysis

Let $(\times)$ and $(\oplus)$ denote the complexity of a multiplication operation in a finite field and the complexity of a XOR, respectively. The $\oplus$ operation is much faster than $\times$ operation, $(\times) \gg (\oplus)$. The efficiency comparison is given in Table 3.

6.1 Storage Cost

In the previous schemes and the XORNC-SSS scheme, the storage cost is the same as the share size because each P_i only stores a share. In the SW-SSS scheme, each P_i stores the share c_i ($|c_i| = \log_2\binom{|b|}{d_i}$) and the metadata d_i ($|d_i| = \log_2|b|$ because $d_i \in [1, |b|]$). Therefore, the storage cost is $O(\log_2\binom{|b|}{d_i} + \log_2|b|)$. For $\forall|b|$ and $\forall d_i \in [0, |b|]$, $\log_2\binom{|b|}{d_i} < |b|$. Thus, $\log_2\binom{|b|}{d_i} = \frac{|b|}{x}$ for some $x > 1$, and $(\frac{|b|}{x} + \log_2|b|) < (|b| + m)$ if $\log_2|b| < m$. This inequality holds if the parameters are chosen s.t. $|b| < 2^m$. If S is divided s.t. any three blocks are different in the same number of bits ($d_0, \cdots, d_{n-1}$ are the same), P_i does not need to store d_i because it becomes a shared information. The storage cost is thus the same as the share size.

6.2 Computation Cost

ShareGen. In the Shamir-SSS, the cost is $O(n \log n)$ because each share is computed from a polynomial. In the XOR-SSS scheme, the cost is $O(n_p n)$ where n_p is the smallest prime s.t. $n_p \geq n$. In the NC-SSS, the cost is $O(mn)$ because n shares are computed from a linear combination of m secret blocks. The XORNC-SSS also computes the shares as the NC-SSS. However, it uses the XOR instead of a linear combination over field multiplications. In the SW-SSS, the cost is $O(n)$ because n shares are computed from the XORs of a tuple of three secret blocks.

Reconst. The costs in the previous schemes, XORNC-SSS and SW-SSS schemes are $O(m^2)$ times $(\times)$ or $(\oplus)$ because the schemes use Gaussian elimination to solve m secret blocks (or to solve the secret and $(m-1)$ coefficients in the Shamir-SSS). Only in the XOR-SSS scheme, the dimension of the matrix used for the Gaussian elimination is $(n_p \times n_p)$, not $(m \times m)$ where n_p is the smallest prime s.t. $n_p \geq n$. Hence, the cost of the XOR-SSS scheme is $O(n_p^2)$ times $(\oplus)$.

ShareRepair. The costs in the NC-SSS and XORNC-SSS schemes are $O(m)$ because a corrupted share is repaired using m healthy shares. The difference between the two schemes is that the NC-SSS uses the field linear combinations while the XORNC-SSS uses the XORs. In the SW-SSS, the cost is $O(1)$ because a new share is computed from three healthy shares.

Table 3 The efficiency comparison

Feature		Previous schemes			Proposed schemes	
		Shamir-SSS [4]	XOR-SSS [21]	NC-SSS [22]	XORNC-SSS	SW-SSS
Feature	XOR-based	No	Yes	No	Yes	Yes
	Arbitrary threshold	Yes	Yes	Yes	Yes	Yes
	Direct share repair	No	No	Yes	Yes	Yes
Storage		$O(m\|b\|)$	$O(m\|b\|)$	$O(m\|b\|)$	$O(m+\|b\|)$	$O(\frac{\|b\|}{x}+\log_2\|b\|)$
Computation	ShareGen	$O(n\log n)(\times)$	$O(n_p n)(\oplus)$	$O(mn)(\times)$	$O(mn)(\oplus)$	$O(n)(\oplus)$
	Reconst	$O(m^2)(\times)$	$O(n_p^2)(\oplus)$	$O(m^2)(\times)$	$O(m^2)(\oplus)$	$O(m^2)(\oplus)$
	ShareRepair	N/A	N/A	$O(m)(\times)$	$O(m)(\oplus)$	$O(1)(\oplus)$
Communication	ShareGen	$O(nm\|b\|)$	$O(nm\|b\|)$	$O(nm\|b\|)$	$O(n(m+\|b\|))$	$O(n(\frac{\|b\|}{x}+\log_2\|b\|))$
	Reconst	$O(m^2\|b\|)$	$O(m^2\|b\|)$	$O(m^2\|b\|)$	$O(m(m+\|b\|))$	$O(m(\frac{\|b\|}{x}+\log_2\|b\|))$
	ShareRepair	N/A	N/A	$O(m^2\|b\|)$	$O(m(m+\|b\|))$	$O(\frac{\|b\|}{x}+\log_2\|b\|)$

6.3 Communication Cost

ShareGen. D distributes n shares to n participants. In the previous schemes, the cost is $O(nm|b|)$ because the share size is $m|b|$. In the XORNC-SSS, the cost is $O(n(m + |b|))$ because the share size is $(m + |b|)$. In the SW-SSS scheme, the cost is $O(n(\frac{|b|}{x} + \log_2|b|))$ because the share size is $(\frac{|b|}{x} + \log_2|b|)$.

Reconst. In the previous schemes and the proposed schemes, m participants are required to provide their shares to D, the costs are thus m times the storage cost of each scheme.

ShareRepair. The share repair is not supported in the Shamir-SSS and XOR-SSS schemes. In the NC-SSS and XORNC-SSS schemes, the costs are $O(m^2|b|)$ and $O(m(m + |b|))$ because m healthy participants are required to provide their shares to D for the share repair. In the SW-SSS scheme, the cost is $O(\frac{|b|}{x} + \log_2|b|)$ because only three healthy shares are required for the share repair.

6.4 Computation Evaluation

This section presents the simulation of the SW-SSS scheme to show that is is applicable to a real system. The program that is written by Python 2.7.3 is executed using a computer with Intel Core i5, 2.40 GHz, 4 GB of RAM, Win 7 64-bit OS. Each secret block, b_i, is set to be 2^{10} bits. Each result is the average of 100 runs. The gmpy2 library is used. The simulation results in Fig. 1 are observed with three sets of the computation performance: ShareGen, Reconst, and ShareRepair by varying the number of secret blocks, m. The number of participants, n, is set to be $n = m + 1$. The graphs reveal that the computation time of ShareRepair is almost constant and is independent on m. The computation time of ShareGen and Reconst linearly increases

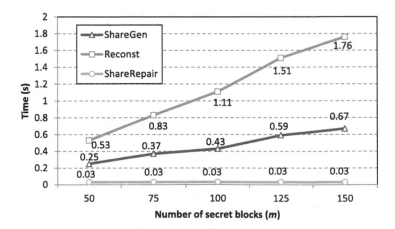

Fig. 1 The computation performance of SW-SSS

with m. The average slopes of increment in ShareGen and Reconst are 0.004 and 0.012, respectively. From these results, if the secret size, $|S|$, is 2^{26} bits ($m = 65,500$), which is almost an upper bound of the size of a secret (e.g., secret key, signature), the computation time of ShareGen, Reconst and ShareRepair is merely 374.72 s (6 min), 905.28 s (15.1 min) and 0.031 s, respectively.

7 Conclusion and Future Work

This paper firstly revisits the XOR-based NC to apply it for the SSS (called XORNC-SSS) in order to support the share repair, to obtain the arbitrary threshold, and to reduce the share size. This paper then proposes the main SW-SSS scheme to optimize the share size. The key idea is based on the binning idea of the Slepian-Wolf coding, which is commonly used in data compression in a network. The security analysis is provided based on the entropy theory. The efficiency analysis is discussed based on the complexity theory. The simulation results of the SW-SSS scheme reveal that it is applicable to a real SSS system. Future research is required to investigate the implementation of the previous schemes.

References

1. Pedersen, T.B, Saygin, Y., Savas, E.: Secret Sharing vs. Encryption-based techniques for privacy preserving data mining, pp. 17–19. Sciences, New York (2007)
2. Nirali, R.N., Devesh, C.J.: A game theory based repeated rational secret sharing scheme for privacy preserving distributed data mining. SECRYPT, pp. 512–517 (2013)
3. Kaya, S.V., Pedersen, T.B., Savaş, E., Saygıýn, Y.: Efficient privacy preserving distributed clustering based on secret sharing. In: Washio, T., Zhou, Z.-H., Huang, J.Z., Hu, X., Li, J., Xie, C., He, J., Zou, D., Li, K.-C., Freire, M. (eds.) PAKDD 2007. LNCS (LNAI), vol. 4819, pp. 280–291. Springer, Heidelberg (2007)
4. Shamir, A.: How to share a secret. Proc. Commun. ACM **22**(11), 612–613 (1979)
5. Blakley, G.R.: Safeguarding cryptographic keys. AFIPS National Comput. Conf. **48**, 313–317 (1979)
6. Karnin, E., Greene, J.W., Hellman, M.E.: On secret sharing systems. IEEE Trans. Inform. Theory **29**(1), 3541 (1983)
7. Capocelli, R.M., Santis, A.D., Gargano, L., Vaccaro, U.: On the size of shares for secret sharing schemes. J. Cryptology **6**, 157167 (1993)
8. Blakley, G.R., Meadows, C.: Security of ramp schemes. In: Blakely, G.R., Chaum, D. (eds.) CRYPTO 1984. LNCS, vol. 196, pp. 242–268. Springer, Heidelberg (1985)
9. Yamamoto, H.: On secret sharing systems using (k, L, n) threshold scheme. IEICE Trans. Fundam. **68**(9), 945–952 (1985) (Japanese Edition)
10. Kurosawa, K., Okada, K., Sakano, K., Ogata, W., Tsujii, T.: Non perfect secret sharing schemes and matroids. In: Workshop on the Theory and Application of Cryptographic Techniques (EUROCRYPT). LNCS 765, Springer, pp. 126–141 (1993)
11. Ogata, W., Kurosawa, K.: Some basic properties of general nonperfect secret sharing schemes. J. Univ. Comput. Sci. **4**(8), 690–704 (1998)

12. Okada, K., Kurosawa, K.: Lower bound on the size of shares of nonperfect secret sharing schemes. In: Conference on the Theory and Applications of Cryptology (ASIACRYPT). LNCS 917, p. 3441. Springer (1994)
13. Wang, Y.: Efficient LDPC Code Based Secret Sharing Schemes and Private Data Storage in Cloud without Encryption. Technical report, UNC Charlotte (2012)
14. Ishizu, H., Ogihara, T.: A study on long-term storage of electronic data. IEICE Gen. Conf. 125(1), 9–10 (2004)
15. Hosaka, N., Tochikubo, K., Fujii, Y., Tada, M., Kato, T.: $(2, n)$-threshold secret sharing systems based on binary matrices. In: Symposium on SCIS, pp. 2D1-4 (2007) (in Japanese)
16. Kurihara, J., Kiyomoto, S., Fukushima, K., Tanaka, T.: A fast (3, n)-threshold secret sharing scheme using exclusive-OR operations. IEICE Trans. E91-A(1), 127–138 (2008)
17. Shiina, N., Okamoto, T., Okamoto, E.: How to convert 1-out-of-n proof into k-out-of-n proof. In: Symposium on SCIS (in Japanese), pp. 1435–1440 (2004)
18. Kurihara, J., Kiyomoto, S., Fukushima, K., Tanaka, T.: A new (k,n)-threshold secret sharing scheme and its extension. In: Wu, T.-C., Lei, C.-L., Rijmen, V., Lee, D.-T. (eds.) ISC 2008. LNCS, vol. 5222, pp. 455–470. Springer, Heidelberg (2008)
19. Chunli, L., Jia, X., Tian, L., Jing, J., Sun, M.: Efficient ideal threshold secret sharing schemes based on EXCLUSIVE-OR operations. In: 4th Conference on Network and System Security (NSS), pp. 136–143 (2010)
20. Wang, Y., Desmedt, Y.: Efficient Secret Sharing Schemes Achieving Optimal Information Rate. In: Information Theory Workshop (ITW) (2014)
21. Kurihara, J., Kiyomoto, S., Fukushima, K., Tanaka, T.: A fast (k-L-N)-threshold Ramp secret sharing scheme. IEICE Trans. Fundam. (2009). doi:10.1587/transfun.E92.A.1808
22. Kurihara, M., Kuwakado, H.: Secret sharing schemes based on minimum bandwidth regenerating codes. In: Symposium on Information Theory and its Applications (ISITA), pp. 255–259 (2012)
23. Liu, J., Wang, H., Xian, M., Huang, K.: A secure and efficient scheme for cloud storage against eavesdropper. In: Qing, S., Zhou, J., Liu, D. (eds.) ICICS 2013. LNCS, vol. 8233, pp. 75–89. Springer, Heidelberg (2013)
24. Cai, N., Raymond, W.Y.: Secure network coding. In: IEEE International Symposium on Information Theory (2002)
25. Katti, S., Rahul, H., Hu, W., Katabi, D., Medard, M., Crowcroft, J.: XORs in the air: practical wireless network coding. Trans. Netw. 16(3), 497–510 (2008)
26. Yu, Z., Wei, Y., Ramkumar, B., Guan, Y.: An efficient scheme for securing XOR network coding against pollution attacks. In: 28th Conference on Computer Communication (INFOCOM), pp. 406–414 (2009)
27. Slepian, D., Wolf, J.K.: Noiseless coding of correlated information sources. IEE Trans. Inf. Theory 19(4), 471–480 (1973)
28. Cheng, S.: Slepian-Wolf Code Designs. http://tulsagrad.ou.edu/samuel_cheng/information_theory_2010/swcd.pdf (2010)
29. Ahlswede, R., Cai, N., Li, S.Y.R., Yeung, R.W.: Network information flow. IEEE Trans. Inf. Theory 46(4), 1204–1216 (2000)
30. Ho, T., Medard, M., Koetter, R., Karger, D.R., Effros, M., Shi, J., Leong, B.: A random linear network coding approach to multicast. IEEE Trans. Inf. Theory 52(10), 4413–4430 (2006)
31. Li, S.-Y.R., Raymond, W.Y., Cai, N.: Linear network coding. IEEE Trans. Inf. Theory 49(2), 371–381 (2003)
32. Agrawal, S., Boneh, D.: Homomorphic MACs: MAC-based integrity for network coding. In: Abdalla, M., Pointcheval, D., Fouque, P.-A., Vergnaud, D. (eds.) ACNS 2009. LNCS, vol. 5536, pp. 292–305. Springer, Heidelberg (2009)
33. Stankovic, V., Liveris, A.D., Xiong, Z., Georghiades, C.N.: On code design for the Slepian-Wolf problem and lossless multiterminal networks. IEEE Trans. Inf. Theory 52(4), 1495–1507 (2006)
34. Stankovi, V., Liveris, A.D., Xiong, Z., Georghiades, C.N: Design of Slepian-Wolf codes by channel code partitioning. In: Data Compression Conference (DCC), pp. 302–311 (2004)

35. Cheng, C., Jiang, T.: An efficient homomorphic MAC with small key size for authentication in network coding. IEEE Trans. Comput. **62**(10), 2096–2100 (2012)
36. Johnson, R., Molnar, D., Song, D., Wagner, D.: Homomorphic Signature Schemes. In: Preneel, B. (ed.). CT-RSA 2002. LNCS, vol. 2271, pp. 244–262. Springer, Heidelberg (2002)
37. Freeman, D.M.: Improved security for linearly homomorphic signatures: a generic framework. In: Fischlin, M., Buchmann, J., Manulis, M. (eds.) PKC 2012. LNCS, vol. 7293, pp. 697–714. Springer, Heidelberg (2012)

An Efficient RFID Distance Bounding Protocol

Li Zhai and ChuanKun Wu

Abstract RFID System is vulnerable for man-in-the-middle attack. Distance bounding protocols are designed for authentication while taking into consideration the distance measure. Distance bounding protocols solve the problem of man-in-the-middle attack in practical. In the known protocols from public literatures, the mafia fraud probability is usually very high, or the memory requirement is large. In this paper, we propose an efficient distance bounding protocol, which uses a single bit to challenge, and it does not require the final signature as many existing such protocols do. The protocol uses a linear memory and has a good probability in resisting against mafia fraud attack.

Keywords RFID · Man-in-the-middle attack · Distance bounding · Internet of things

1 Introduction

Radio Frequency Identification (RFID) is a technology for automated identification of objects. RFID systems have been widely deployed in our daily lives. Typically, RFID systems consist of RFID tags and RFID readers. Many authentication protocols have been proposed to secure RFID systems. However, most of them are vulnerable to location based attacks.

Desmedt et al. [1, 2] introduced an attack that could defeat any authentication protocol in 1987. This attack is called mafia fraud [3]. Mafia fraud attack is a kind of man-in-the-middle attack. In public literatures, this attack is also called relay attack. Brands and Chaum [4] proposed the first distance bounding protocol in 1993 to thwart this attack. Since then, many distance bounding protocols have been published.

L. Zhai (✉) · C. Wu
State Key Laboratory of Information Security, Institute of Information Engineering,
Chinese Academy of Sciences, 100093 Beijing, China
e-mail: zhaili@iie.ac.cn

L. Zhai
University of Chinese Academy of Sciences, 100190 Beijing, China

© Springer International Publishing Switzerland 2015
Á. Herrero et al. (eds.), *International Joint Conference*, Advances in Intelligent
Systems and Computing 369, DOI 10.1007/978-3-319-19713-5_31

Distance bounding protocols have many different methods for implementation. One method is to measure distance by round trip time of electronic signals. This method is based on the theory that no information can propagate faster than at the speed of light. In a limited period of time, the maximal distance can be measured by light speed. When a reader sends a bit to a tag and receives a bit from the tag, the communication time can be recorded and the distance can be measured. If the distance exceeds an expected limit, then the communication may be a fraud, and should be ignored. On the other hand, if the distance does not exceed the limit, the communication should be safe.

We will give an application to explain the mafia fraud attack. Suppose our tag is a Credit Card using RFID, then the reader is a POS terminal, the payment can be successful when the credit card and the reader are in a restrictive distance. The purpose of the mafia fraud adversary is to use a legal credit card to make a payment. First, the adversary uses a fake POS terminal and a fake credit card which are linked in communication, and then use the fake POS terminal to approach a legal credit card. Because the technology of RFID is non-contact, so it can be realized easily. Second, the adversary should use the fake credit card to approach a legal POS terminal. Finally, the adversary can make the payment successfully which is actually made by the legal credit card remote from the legal POS terminal.

Hancke and Kuhn [5] proposed the first RFID distance bounding protocol in 2005. The protocol is simple and effective. The HK protocol has two phases, slow phase and fast phase. In the slow phase, the protocol generates random data for the fast phase. The fast phase is the most important parts on the protocol. The fast phase is an n-round challenge response protocol. In this phase, there are $2n$ registers in this phase, every register stores a single bit. However, the mafia fraud attack can succeed with probability $(3/4)^n$, which is not optimal. Thus, many protocols [6–10] tried to improve the HK protocol.

MP protocol [6] requires three physical states: 0, 1, void. MUSE protocol [7] uses more states than MP protocol. This technique needs modulation to encode multiple states into a signal. Thus, these protocols are hard to be implemented on low cost tags. AT protocol [11] is more secure than HK protocol, but the memory requirement is exponential in the number of communication rounds. The authors in [11] proposed another protocol AT3. The memory requirement of AT3 is reduced, but it significant decreases the security level of the protocol.

In this paper, we propose a new secure distance bounding protocol. It is using single bit challenge, and do not require a final signature. Our protocol achieves a strong security against mafia fraud attacks. Also, the memory requirement of our protocol is linear. It is efficient for implementation in low cost devices.

2 Our Protocols

In this section, we present a new distance bounding protocol. Our distance bounding protocol is based on Hancke and Kuhn's scheme. Our protocol works as follows:

Protocol Parameters Assume that a reader and a tag share a secret key k and a pseudo-random function (PRF). The protocol has n rounds, in every rounds, the protocol uses 4 registers. Let Δt_{max} denote the max round-trip time (RTT). Let Δt_i denote the round-trip time at the ith round.

Protocol Execution Our protocol has two stages: slow phase and fast phase. In the slow phase, the protocol generates some random bits and uses them to initialize the register. In the fast phase, the reader checks the distance with the tag. The fast phase has n rounds. In every round, the reader sends a single bit to the tag. Upon receipt of the message, the tag sends a single bit response to the reader. The reader verifies the response and checks the round-trip time. Figure 1 illustrates our protocol.

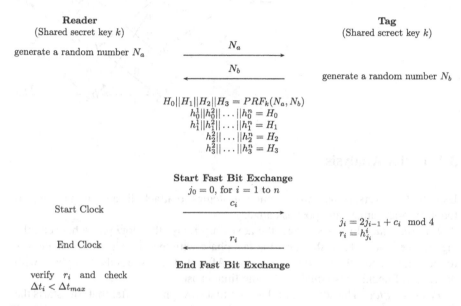

Fig. 1 Our protocol

Slow phase Reader generates a random string N_a, sends it to tag. Upon the receipt of the message N_a, the tag generates a random string N_b and sends it to the reader. The reader and the tag use the shared secret key k and N_a, N_b to generate some bits and use them to fill the registers. On the first round of fast phase, the protocol uses 2 bits. On the other rounds of fast phase, the protocol uses 4 bits. Thus, the total number of the register is $4n - 2$.

Fast phase This phase is also called the fast bit exchange phase. This phase is an n-round challenge-response protocol. In the every round, the reader starts a timer and sends a random bit c_i to the tag. At first, the position on the reader and the tag is 0. The position of this round j_i is determined by the last round's position j_{i-1} and the challenge c_i. The tag sends the response $r_i = h^i_{j_i}$. Upon receipt of the response r_i, the reader stops his timer, and compute the round-trip time Δt_i. The reader verifies the $\Delta t_i < \Delta t_{max}$ and checks if $r_i = h^i_{j_i}$.

We use Fig. 2 to illustrate the structure of the registers. Every node $h^i_{j_i}$ on the Fig. 2 is a register. The edge on the Fig. 2 is a challenge. If the challenge is 0, then the position j_{i-1} move to the next position j_i according to the edge labeled as 0. If the challenge is 1, then the position j_{i-1} move to the next position j_i according to the edge 1.

Fig. 2 Fast bit exchange

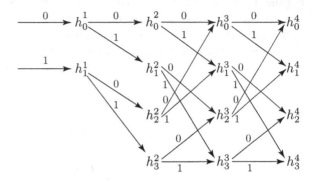

3 Security Analysis

Usually the adversary has two common strategies to attack distance bounding protocols: pre-ask attack and post-ask attack.

Pre-ask attack In this strategy, the adversary relays the slow phase between the tag and the reader. Then she starts the fast phase with the tag. She sends challenges to the tag, and gets the responses from tag. After that, she starts the fast phase with the reader. She may also finally relay the final phase.

Post-ask attack The adversary relays the first slow phase. After that, she starts the fast phase with the reader without asking the tag. Then, she sends the challenges to the tag. Finally, she replays the final phase. This strategy is useful when such a final slow phase exists [12]. Because our protocol only has one slow phase, the post-ask attack is not effective for our protocol.

We can proof the following to be true: whatever message the adversary sends in querying the tag, it cannot increase the success probability for our protocol.

Theorem 1 *The adversary relays the slow phase between the tag and the reader. Then she starts the fast phase. She sends a sequence of challenges $(\tilde{c}_1, \ldots, \tilde{c}_i)$ to the tag and gets the response $(r_1, \ldots, r_i)$. Then, she interacts with the reader, and answers the challenges from the reader. The reader sends a sequence of challenges $(c_1, \ldots, c_i)$ to the adversary, the adversary sends the response $(\tilde{r}_1, \ldots, \tilde{r}_i)$ back to the reader. Let W_i be the event that adversary passes the protocol until the ith round. Then, $\Pr[W_i]$ is independent of the sequence of challenges $(\tilde{c}_1, \ldots, \tilde{c}_i)$.*

Proof The adversary uses the pre-ask strategy. First, adversary needs to query the tag to gain information. Adversary chooses a sequence of challenges $(\tilde{c}_1, \dots, \tilde{c}_i)$ and sends it to the tag. Then the tag moves the position of the register according to the sequence of challenges. After that, adversary answers the challenges from the reader. Let the reader's challenge sequence be $(c_1, \dots, c_i)$. Let the random variable X_i be the position of the register on the reader after the challenges. Then, $\Pr[W_i]$ can be computed as follows:

$$\Pr[W_i] = \sum_{t=0}^{3} \Pr[W_i, X_i = t] \tag{1}$$

Let $f(\tilde{c}_1, \dots, \tilde{c}_i)$ be the register's position on the tag after the sequence of challenges $(\tilde{c}_1, \dots, \tilde{c}_i)$. When $X_i = f(\tilde{c}_1, \dots, \tilde{c}_i)$, then the tag and the reader are on the same position. Thus adversary can send the correct response at the ith round. When $X_i \neq f(\tilde{c}_1, \dots, \tilde{c}_i)$, adversary must guess the response. Then we have:

$$\Pr[W_i, X_i = t] = \begin{cases} \Pr[W_{i-1}, X_i = t] & \text{if } f(\tilde{c}_1, \dots, \tilde{c}_i) = t \\ \frac{1}{2}\Pr[W_{i-1}, X_i = t] & \text{if } f(\tilde{c}_1, \dots, \tilde{c}_i) \neq t \end{cases} \tag{2}$$

As $X_i = 2X_{i-1} + c_i \bmod 4$, thus we can compute $\Pr[X_i = t | W_{i-1}, X_{i-1} = s]$ as follows:

$$\begin{aligned} &\Pr[X_i = t | W_{i-1}, X_{i-1} = s] \\ &= \Pr[2X_{i-1} + c_i \bmod 4 = t | W_{i-1}, X_{i-1} = s] \\ &= \Pr[2s + c_i \bmod 4 = t | W_{i-1}, X_{i-1} = s] \\ &= \begin{cases} \frac{1}{2} & \text{if } t \in \{2s \bmod 4, 2s + 1 \bmod 4\} \\ 0 & \text{if } t \notin \{2s \bmod 4, 2s + 1 \bmod 4\} \end{cases} \end{aligned} \tag{3}$$

Using the Eqs. (2), (3), we can compute the probability $\Pr[W_i, X_i = t | W_{i-1}, X_{i-1} = s]$:

$$\begin{aligned} &\Pr[W_i, X_i = t | W_{i-1}, X_{i-1} = s] \\ &= \begin{cases} \Pr[X_i = t | W_{i-1}, X_{i-1} = s] & \text{if } f(\tilde{c}_1, \dots, \tilde{c}_i) = t \\ \frac{1}{2}\Pr[X_i = t | W_{i-1}, X_{i-1} = s] & \text{if } f(\tilde{c}_1, \dots, \tilde{c}_i) \neq t \end{cases} \\ &= \begin{cases} 0 & \text{if } t \notin \{2s \bmod 4, 2s + 1 \bmod 4\} \\ \frac{1}{2} & \text{if } t \in \{2s \bmod 4, 2s + 1 \bmod 4\} \wedge f(\tilde{c}_1, \dots, \tilde{c}_i) = t \\ \frac{1}{4} & \text{if } t \in \{2s \bmod 4, 2s + 1 \bmod 4\} \wedge f(\tilde{c}_1, \dots, \tilde{c}_i) \neq t \end{cases} \end{aligned} \tag{4}$$

As $0 \leq X_{i-1} \leq 3$ then:

$$\Pr[W_i, X_i = t] = \sum_{s=0}^{3} \Pr[W_i, X_i = t | W_{i-1}, X_{i-1} = s] \Pr[W_{i-1}, X_{i-1} = s] \tag{5}$$

If the adversary's challenges sequence $(\tilde{c}_1, \ldots, \tilde{c}_i)$ is determined, then the position $f(\tilde{c}_1, \ldots, \tilde{c}_i)$ is determined. From the Eqs. (1), (4), (5), the probability $\Pr[W_i]$ is depended on $(\tilde{c}_1, \ldots, \tilde{c}_i)$. Let us define a function $g(\tilde{c}_1, \ldots, \tilde{c}_i)$:

$$g(\tilde{c}_1, \ldots, \tilde{c}_i) = \begin{pmatrix} \Pr[W_i, X_i = 0] \\ \Pr[W_i, X_i = 1] \\ \Pr[W_i, X_i = 2] \\ \Pr[W_i, X_i = 3] \end{pmatrix} \tag{6}$$

We have the following statement: For two different challenge sequence $(\tilde{c}_1, \ldots, \tilde{c}_i)$ and $(\tilde{c}_1', \ldots, \tilde{c}_i')$, we have $g(\tilde{c}_1, \ldots, \tilde{c}_i) = P_\pi g(\tilde{c}_1', \ldots, \tilde{c}_i')$, where P_π is a permutation matrix. More specifically, they have the following relationship:

$$g(\tilde{c}_1, \ldots, \tilde{c}_i) = \sigma_v \sigma_u g(\tilde{c}_1', \ldots, \tilde{c}_i') \tag{7}$$

where $\sigma_0 = I$, $\sigma_1 = (12)$, $\sigma_2 = (13)$, $\sigma_3 = (14)$ and $u = f(\tilde{c}_1, \ldots, \tilde{c}_i)$, $v = f(\tilde{c}_1', \ldots, \tilde{c}_i')$.

We proof the result by mathematical induction:

Basis step: We first show that the above statement is true when $i = 2$. There are four different sequences of challenges: $(0, 0), (0, 1), (1, 0), (1, 1)$. It is easy to verify that all the cases hold.

Inductive step: We assume that the above statement is true when $i = k$. Without loss of generality, we assume that $f(\tilde{c}_1', \ldots, \tilde{c}_k') = 0$. Let $g(\tilde{c}_1', \ldots, \tilde{c}_k')$ be $(p_0 \ p_1 \ p_2 \ p_3)^T$. We analyze the problem by the following cases:

Case 1 $(f(\tilde{c}_1, \ldots, \tilde{c}_k) = 0, c_{\widetilde{k+1}} = 0$ and $c_{\widetilde{k+1}}' = 0)$. Because $f(\tilde{c}_1', \ldots, \tilde{c}_k') = f(\tilde{c}_1, \ldots, \tilde{c}_k) = 0$, according to the assumption, then $g(\tilde{c}_1, \ldots, \tilde{c}_k) = g(\tilde{c}_1', \ldots, \tilde{c}_k')$. Also, $c_{\widetilde{k+1}} = c_{\widetilde{k+1}}' = 0$, it is easy to see that:

$$g(\tilde{c}_1, \ldots, \tilde{c}_k, c_{\widetilde{k+1}}) = \sigma_0 \sigma_0 g(\tilde{c}_1', \ldots, \tilde{c}_k', c_{\widetilde{k+1}}')$$

Case 2 $(f(\tilde{c}_1, \ldots, \tilde{c}_k) = 0, c_{\widetilde{k+1}} = 0$ and $c_{\widetilde{k+1}}' = 1)$. Similar to Case 1 we have, $g(\tilde{c}_1, \ldots, \tilde{c}_k) = g(\tilde{c}_1', \ldots, \tilde{c}_k') = (p_0 \ p_1 \ p_2 \ p_3)^T$. Using the Eq. (5), we can calculate $g(\tilde{c}_1, \ldots, \tilde{c}_i, c_{\widetilde{k+1}})$ from $g(\tilde{c}_1, \ldots, \tilde{c}_k)$. In this case, $u = f(\tilde{c}_1, \ldots, \tilde{c}_k, c_{\widetilde{k+1}}) = 0$, $v = f(\tilde{c}_1', \ldots, \tilde{c}_k', c_{\widetilde{k+1}}') = 1$, we can verify that:

$$g(\tilde{c}_1, \ldots, \tilde{c}_k, c_{\widetilde{k+1}}) = \sigma_1 \sigma_0 g(\tilde{c}_1', \ldots, \tilde{c}_k', c_{\widetilde{k+1}}')$$

Case 3–16 $(f(\tilde{c}_1, \ldots, \tilde{c}_k) = 0\ldots3, c_{\widetilde{k+1}} = 0\ldots1$ and $c_{\widetilde{k+1}}' = 0\ldots1)$. Similar to Case 1, using the Eqs. (4), (5), we can verify that the these cases satisfy the relationship (7).

Therefore, the statement is hold when $i = k + 1$. Due to the properties of g, the sum of its elements is the same, so the success probability is the same.

According to the Theorem 1, the adversary can choose a whatever sequence of challenges, the success probability is the same. We choose the challenge sequence $(0, 0, \ldots, 0)$ to calculate the adversary's success probability. The following theorem gives the adversary's success probability of our protocol.

Theorem 2 *The probability* $\Pr[W_i]$ *can be computed as follows:*

$$\Pr[W_i] = \sum_{i=0}^{3} p_i$$

and $(p_0, p_1, p_2, p_3)^T = M^i \cdot l,\ l = (1,0,0,0)^T,\ M = \begin{pmatrix} \frac{1}{2} & 0 & \frac{1}{2} & 0 \\ \frac{1}{4} & 0 & \frac{1}{4} & 0 \\ 0 & \frac{1}{4} & 0 & \frac{1}{4} \\ 0 & \frac{1}{4} & 0 & \frac{1}{4} \end{pmatrix}$

Proof At first, the adversary chooses 0 to challenge tag in all rounds, so the register's position $f(\tilde{c}_1, \ldots, \tilde{c}_i)$ on the tag is always 0. As $f(\tilde{c}_1, \ldots, \tilde{c}_i) = 0$, using the Eq. (4), we have:

$$\Pr[W_i, X_i = t \mid W_{i-1}, X_{i-1} = s]$$

$$= \begin{cases} 0 & \text{if } t \notin \{2s \bmod 4, 2s+1 \bmod 4\} \\ \frac{1}{2} & \text{if } t \in \{2s \bmod 4, 2s+1 \bmod 4\} \wedge t = 0 \\ \frac{1}{4} & \text{if } t \in \{2s \bmod 4, 2s+1 \bmod 4\} \wedge t \neq 0 \end{cases} \tag{8}$$

Let be $M = (M_{ts})$ and $M_{ts} = \Pr[W_i, X_i = t \mid W_{i-1}, X_{i-1} = s]$, $0 \le t, s \le 3$, using the Eq. (5), we have:

$$\begin{pmatrix} \Pr[W_i, X_i = 0] \\ \Pr[W_i, X_i = 1] \\ \Pr[W_i, X_i = 2] \\ \Pr[W_i, X_i = 3] \end{pmatrix} = M \begin{pmatrix} \Pr[W_{i-1}, X_{i-1} = 0] \\ \Pr[W_{i-1}, X_{i-1} = 1] \\ \Pr[W_{i-1}, X_{i-1} = 2] \\ \Pr[W_{i-1}, X_{i-1} = 3] \end{pmatrix} \tag{9}$$

Using the Eq. (8), we can compute M as follows:

$$M = \begin{pmatrix} \frac{1}{2} & 0 & \frac{1}{2} & 0 \\ \frac{1}{4} & 0 & \frac{1}{4} & 0 \\ 0 & \frac{1}{4} & 0 & \frac{1}{4} \\ 0 & \frac{1}{4} & 0 & \frac{1}{4} \end{pmatrix} \tag{10}$$

The probability $\Pr[W_i \mid X_i = t]$ can be recursively computed as follows:

$$\begin{pmatrix} \Pr[W_i, X_i = 0] \\ \Pr[W_i, X_i = 1] \\ \Pr[W_i, X_i = 2] \\ \Pr[W_i, X_i = 3] \end{pmatrix} = M^i \begin{pmatrix} \Pr[W_0, X_0 = 0] \\ \Pr[W_0, X_0 = 1] \\ \Pr[W_0, X_0 = 2] \\ \Pr[W_0, X_0 = 3] \end{pmatrix} = M^i \cdot l \tag{11}$$

The position X_0 is 0 at the beginning, so the l is equal to $(1, 0, 0, 0)^T$. Use matrix M multiply vector l for i times, we get the probability distribution on ith round. According to Eq. (1), the sum of the element of $l \cdot M^i$ is the adversary success probability $Pr[W_i]$.

4 Distance Bounding Protocols Comparisons

In this section, our protocol is compared with the existing distance bounding protocols in terms of memory consumption and security level. Table 1 depicts the space requirement and success probability. The memory requirement of HK protocol is $2n$. The adversary's success probability is $(\frac{3}{4})^n$. The AT protocol is a tree based protocol, the memory requirement of the AT protocol is the number of nodes in the tree. The number of nodes in the tree is $2^{n+1} - 2$. The adversary's success probability is $(\frac{1}{2})^n(\frac{n}{2} + 1)$. The AT3 protocol is based on the AT protocol, it uses multiple trees. The memory consumption in AT3 is $\frac{14n}{3}$. The adversary's success probability of AT3 protocol is $(\frac{5}{16})^{\frac{n}{3}}$. The MUSE protocol use 4 physical states, the reader and the tag need to send $2n$ bits in n rounds. The communication cost of the MUSE protocol are $4n$ bits. The adversary's success probability is $(\frac{5}{8})^n$. The memory requirement of our protocol is $4n - 2$. Theorem 2 gives the adversary's success probability of our protocol (Table 2).

Table 1 Comparison of Distance Bounding Protocol

Protocol	Memory	Adversary's success probability	Commutation cost (bits)
HK	$2n$	$(\frac{3}{4})^n$	$2n$
AT	$2^{n+1} - 2$	$(\frac{1}{2})^n(\frac{n}{2} + 1)$	$2n$
AT3	$\frac{14n}{3}$	$(\frac{5}{16})^{\frac{n}{3}}$	$2n$
MUSE	$4n$	$(\frac{5}{8})^n$	$4n$
Ours	$4n - 2$	Theorem 2	$2n$

Table 2 Number of rounds required to obtain $Pr[W_n] < 10^{-10}$

Protocol	Number of rounds (n)	Commutation cost (bits)
HK	81	162
AT3	60	120
MUSE	49	196
Ours	55	110

In Fig. 3 we show the adversary's successful probability achieved by our protocol, HK protocol, MUSE protocol and AT3 protocol. The memory requirement of AT protocol is exponential in the number of rounds. It is not practical, even for the reader. We do not compare it with the other DB protocols. The HK protocol has the highest success probability. The MUSE protocol has the lowest success probability. But the communication cost is usually more important than the memory cost for low cost devices. But on the same security level, the MUSE communication cost has the highest communication cost (192 bits). Our protocol uses more memory than HK, but we reduce the communication cost. As it can be seen in Table 1, our protocol is the best protocol when considering memory consumption, communication cost and adversary's success probability at the same time.

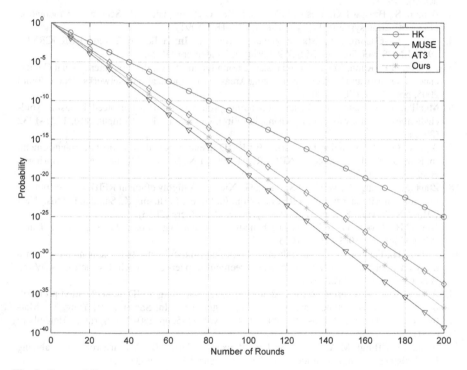

Fig. 3 Protocol Compare

5 Conclusion

In this paper, we proposed a secure, efficient distance bounding protocol. Due to its efficiency, it is suitable for low-cost RFID tags. Compared with some other known protocols, our protocol does not require the final signature and multiple physical states. Our protocol uses more memory than the HK protocol, but it greatly improves the security level, which can be a good trade-off between the memory and the security level, making it more practical.

References

1. Desmedt, Y.: Major security problems with the unforgeable(feige)-fiat-shamir proofs of identity and how to overcome them. In: SecuriCom, vol. 88, pp. 15–17 (1988)
2. Desmedt, Y.G., Goutier, C., Bengio, S.: Special uses and abuses of the fiat shamir passport protocol. In: Pomerance, C. (ed.) CRYPTO 1987. LNCS, vol. 293, pp. 21–39. Springer, Heidelberg (1988)
3. Bengio, S., Brassard, G., Desmedt, Y.G., Goutier, C., Quisquater, J.J.: Secure implementation of identification systems. J. Cryptol. 4(3), 175–183 (1991)
4. Brands, S., Chaum, D.: Distance bounding protocols. In: Helleseth, T. (ed.) EUROCRYPT 1993. LNCS, vol. 765, pp. 344–359. Springer, Heidelberg (1994)
5. Hancke, G.P., Kuhn, M.G.: An rfid distance bounding protocol. In: First International Conference on Security and Privacy for Emerging Areas in Communications Networks. SecureComm 2005, pp. 67–73. IEEE (2005)
6. Munilla, J., Peinado, A.: Distance bounding protocols for rfid enhanced by using void-challenges and analysis in noisy channels. Wirel. Commun. Mob. Comput. 8(9), 1227–1232 (2008)
7. Avoine, G., Floerkemeier, C., Martin, B.: RFID distance bounding multistate enhancement. In: Roy, B., Sendrier, N. (eds.) INDOCRYPT 2009. LNCS, vol. 5922, pp. 290–307. Springer, Heidelberg (2009)
8. Zhuang, Y., Yang, A., Wong, D.S., Yang, G., Xie, Q.: A highly efficient RFID distance bounding protocol without real-time PRF evaluation. In: Lopez, J., Huang, X., Sandhu, R. (eds.) NSS 2013. LNCS, vol. 7873, pp. 451–464. Springer, Heidelberg (2013)
9. Kim, C.H., Avoine, G.: RFID distance bounding protocols with mixed challenges. IEEE Trans. Wirel. Commun. 10(5), 1618–1626 (2011)
10. Jannati, H., Falahati, A.: Mutual distance bounding protocol with its implementability over a noisy channel and its utilization for key agreement in peer-to-peer wireless networks. Wirel. Pers. Commun. 1–23 (2013)
11. Avoine, G., Tchamkerten, A.: An efficient distance bounding RFID authentication protocol: balancing false-acceptance rate and memory requirement. In: Samarati, P., Yung, M., Martinelli, F., Ardagna, C.A. (eds.) ISC 2009. LNCS, vol. 5735, pp. 250–261. Springer, Heidelberg (2009)
12. Avoine, G., Bingl, M.A., Karda, S., Lauradoux, C., Martin, B.: A framework for analyzing RFID distance bounding protocols. J. Comput. Secur. 19(2), 289–317 (2011)

A New Approach to Compressed File Fragment Identification

Khoa Nguyen, Dat Tran, Wanli Ma and Dharmendra Sharma

Abstract Identifying the underlying type of a file given only a file fragment is a big challenge in digital forensics. Many methods have been applied to file type identification; however the identification accuracies of most of file types are still very low, especially for files having complex structures because their contents are compound data built from different data types. In this paper, we propose a new approach based on the deflate-encoded data detection, entropy-based clustering, and the use of machine learning techniques to identify deflate-encoded file fragments. Experiments on the popular compound file type showed high identification accuracy for the proposed method.

Keywords File fragment classification · Compressed file fragment classification · SVM · Shannon entropy

1 Introduction

File fragment identification is one of the most serious problems in the process of file carving applied for digital forensics [1–3]. Some data fragments are quite easy to identify their file types, such as fragments that belong to files having only unified data such as ASCII text. Conversely, there is no powerful method to identify fragments that are data portions of high entropy files [3, 4]. High entropy files normally contain compressed data, such as zip archives. Furthermore, other high entropy files that contain different data types – called compound files such as XML Microsoft Office files [5, 6], or Adobe portable document format (PDF) files [7].

Compound files have been the most popular formats for decades. Therefore, they have been included in the data set for many research works. Most of current

K. Nguyen (✉) · D. Tran · W. Ma · D. Sharma
Faculty of Education Science Technology and Mathematics, University of Canberra, Canberra, ACT 2601, Australia
e-mail: Khoa.Nguyen@canberra.edu.au

© Springer International Publishing Switzerland 2015
Á. Herrero et al. (eds.), *International Joint Conference*, Advances in Intelligent Systems and Computing 369, DOI 10.1007/978-3-319-19713-5_32

approaches provide low identification rates to file fragments of compound files which are less than 30 % as reported in [8, 9]. Other works bring better identification rates but with much smaller size or small number of file types as considered in [10, 11]. Roussev and Garfinkel [1] stated that the precedent approaches to file fragment identification became irrelevant because they treated compound files as unified data files. In compound files, a fragment may not be distinguished from a standalone image file fragment [4].

In this paper, we propose a new method based on the deflate-encoded data detection, entropy-based clustering, and a machine learning technique – Support Vector Machine (SVM) [12] to identify the deflate-encoded file fragments. Firstly, a deflate-encoded [13] data proportion is identified from a file fragment, and then this data proportion is decompressed to get the underlying data – inflate data. Special methods such as the approaches in [1, 14] enable deflate-encoded data fragment detection and decompression. Secondly, the inflate data is clustered into one of three group depending on its entropy value which is low, medium or high. Thirdly, SVM with byte frequency distribution as feature vector can be used to efficiently identify these inflate data. Moreover, it is demonstrated that SVM can provide high accuracy and performance to recognize the data fragments which have low or medium entropy values [9, 11]. In addition, it is showed that entropy-based clustering before feeding data into machine learning techniques might bring higher accuracy [15].

The rest of the paper is organized as follows. Section 2 presents some related studies in detecting file fragments, especially high entropy file fragments. Section 3 introduces our approach. Section 4 presents our experiments and results. Finally, Sect. 5 includes conclusion and future work.

2   Related Works

Identifying compound file fragments has been received interests from researchers for over a decade. However, most of solutions providing high identification rates for compound fragments were working on small and private data sets. All the works on large and public data sets provided very low identification rates.

In [1], Roussev and Quates pointed out two critical problems of file fragment identification: There was no solution that was in line with real-world file formats; and no appropriate evaluation framework and reproducible results were provided. Moreover, they also claimed that researchers had continually confused the notions of file types and data encoding. Dissimilar to simple primitive files such as TXT or HTML, compound file formats might have different contents, thus look different at the bit stream level. For example, a MS word document file might be primary text, or might contain photos, audio/video or embedded spreadsheets. Due to different components inside, the layout of the MS word file content is very different in each proportion. In order to make the arguments more convinced, the authors did a survey on a data set of 15293 MS Office 2007 files called MSX-13 consisting of

DOCX, XLSX, and PPTX. MS Office 2007 files bases on the Office Open XML (OOXML) format [16].

MS Office 2007 files are actually zip achievers in which different data contents are stored. Text data are compressed by the deflate algorithm with the aim to reduce the size of files. Other image formats such as JPG, PNG, GIF and TIFF are stored in the original forms. Figure 1 below depicts the content breakdown of MSX-13 files.

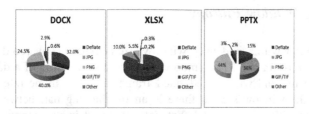

Fig. 1 Contribution of different data types in MSX-13 files

In the OOXML, text and some other parts are compressed by deflate algorithm, the images file are stored in original format. It is impossible if all data excerpts of those files are treated in the same way. There is no solution to distinguish an image excerpt of an OOXML file with an excerpt of an original image file. Therefore, they developed a tool called zsniff which tries to detect deflate header and decompress the compressed data in a file fragment. This method is quite similar to the bit-by-bit process in [14]. Moreover, in order to get a better opportunity in identifying at least one deflate-encoded block data, the data fragment size of 18 KiB was recommended. However, the PDF file type was not included in their work, even though deflate compressed data are the major part of PDF content [4].

In [4], the authors analyzed a data set of 85 GB of 131,000 PDF files downloaded from the Internet. From top level, a PDF file has a simple structure containing a number of autonomous objects encoding text and images. In addition, fonts, font program, layout, formatting and other information are used to display the content of PDF on screen. Parsing the data content in PDF files, they also drew a breakdown of data objects and their volume relative to overall file size as seen in Fig. 2.

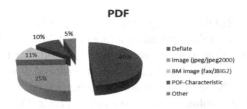

Fig. 2 Structural composition of PDF files [4]

It is very clear to see from Figs. 1 and 2 that deflate-encoded data contributes the largest part in compound files. Therefore, identification of the file type of which a deflate-encoded file fragment belongs to is critical.

3 The Proposed Approach

3.1 Sliding Window Entropy

A sliding window with a predefined window size is used to convert a data fragment into a sequence of frames. This window starts at the beginning of a data fragment and moves to the end with a predefined frame rate. The frame rate is normally smaller than the window size so there is an overlapping part between two consecutive frames. This is to model that data observed in the current observation (current frame) is dependent on data observed in the previous observation. Figure 3 shows how frames are created from a file fragment.

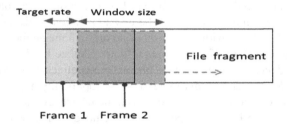

Fig. 3 A data fragment is depicted as a frame sequence by sliding window technique

Shannon entropy will then be applied on each frame of data fragment. Therefore, each file fragment has a sequence of Shannon entropy values. For a random variable X: $\{x_1, \ldots, x_n\}$, the Shannon entropy is used to measure the uncertainty of X and denoted by $H(X)$, is defined:

$$H(X) = - \sum_{i=1}^{n} p(x_i) \log_b p(x_i) \tag{1}$$

where $p(x_i)$ is the probability density function of x_i, and b is logarithm base.

Entropy is chosen due to the special structure of PNG format. According to [17], PNG data are filtered before they are compressed. Data filtering helps to provide the higher compressibility, so the size of PNG files are reduced. PNG defines 5 filtering algorithms, including "None" which denotes no filtering. Different filter type might be applied to different scanline in one PNG image (Table 1).

Table 1 Five filter types

Type	Name	Formula
0	None	
1	Sub	$Sub(x) = Raw(x) - Raw(x - bpp)$
2	Up	$Up(x) = Raw(x) - Prior(x)$
3	Average	$Average(x) = Raw(x) - floor(\frac{Raw(x-bpp)+Prior(x)}{2})$
4	Paeth	$Paeth(x) = Raw(x) - PaethPredictor(Raw(x - pbb), Prior(x), Prior(x - bpp))$

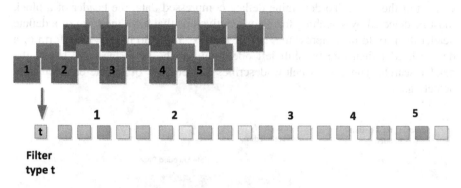

Fig. 4 Serializing and filtering a scanline

where x is from zero to the number of bytes representing the scanline minus one, $Raw(x)$ indicates the raw data byte at the byte position in the scanline, and bbp is defined as number of byte per complete pixel, rounding up to one (Fig. 4).

By running experiments on the data set of PNG files, it is shown that a large part of filtered data of PNG files with the entropy values from 3.5 to 6.5 occupy 94.91 %. The low entropy data contribute 5.08 % and the high entropy data only occupy 0.006 %.

3.2 Proposed Method

In order to identify the deflate-encoded file fragments, we use two phases: deflate-encoded detection phase and inflate (decompressed) data fragment classification phase, as seen in Fig. 5 below.

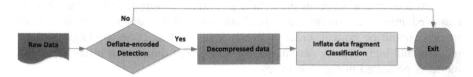

Fig. 5 The proposed model

- **First phase: Deflate-encoded data detection.**

In the first phase, deflate-encoded data in file fragments are detected using the approach in [1, 14]. A deflate-encoded data proportion is formed by a sequence of compressed blocks; each block includes header, Huffman table, and compressed data. The deflate header is established by 3 bits: the first bit indicates that whether this block is the last block in this portion, the next two consecutive bits show how data is coded: 00 – raw data (data is not compressed), 01 – data is compressed with static Huffman codes, and 10 – data is compressed with dynamic Huffman codes. Compressed data is a string of variable-length Huffman codes that describe the content of the block. To determine deflate compressed data, the header of a block must be detected by searching for 3 consecutive bits that have the form of a deflate header then try to decompress the data. If the first 3 bits do not have the form of a deflate header then one bit shift left operation will be performed and repeat the header search. The Fig. 6 below describes the process of deflate-encoded data detection.

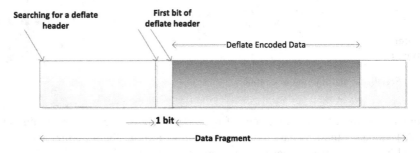

Fig. 6 Deflate-encoded data detection

- **Second phase: PNG data fragment classification.**

Due to compression algorithms, statistical properties of data cannot be used to classify the deflate-encoded data from different file formats. This fact is the reason why previous approaches that exploit statistical properties of compressed data as feature vectors brought low accurate rate [9, 18]. Even from the empirical approach of [1], the authors took the advantage of compression properties such as Huffman table size, the detection rate is still low. In this phase, the underlying data to be classified are uncompressed, which have different characteristics from deflate-encoded data. Machine learning techniques are now used to classify efficiently inflate data.

According the analysis in 3.1 (sliding window entropy), inflate data can be classified into three groups: low, medium, and high entropy. Therefore, three models should be built in order to detect PNG content correspondingly to low, medium and high entropy data in network flow. The diagram of inflate detecting model is presented in Fig. 7 below.

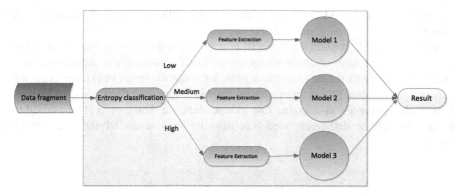

Fig. 7 Diagram of executable data detection

1. Training phase

Data fragments of the dataset are put into three groups depending on their sliding window entropies. Model 1 is created from SVM using low entropy data set. Suitable feature vectors are extracted from medium entropy fragments, and used for SVM to build Model 2. Model 3 is created as the same way with Model 1 but based on high entropy data fragments.

2. Detection phase

In the detection phase, a data fragment is specified as a low, medium or high entropy data fragment. Depending on its entropy value, the corresponding feature vector will be extracted from the data fragment then fed into one of three models (Model 1, 2, or 3) in order to identify whether or not this data fragment belongs to a PNG file.

4 Experiments

4.1 Data Set

With the goal to detect compressed file fragments which are zipped by the deflate algorithm, we only consider files which have deflate-encoded data consisting PDF, MS Office files, and PNG. We downloaded 14827 PDF files from the Govdocs1 (publicly available), with overall size of 16.3 GB. We inherited the data set in [1] for the MS Office files (DOCX, XLSX, and PPTX), this data set contained over 7000 files for each file type, with the total size of 24 GB. PNG files were downloaded from Govdocs1 and other sources from the Internet as well.

Embedded data objects in PDF [7] are stored between a pair of patterns starting with obj.<< and ending with .endobj. Objects that are compressed by deflate algorithm are pointed by the phrase/FlateDecode. Consequently, deflate-encoded data proportions are extracted by looking for those three key words. A program was created to search and extract deflate-compressed data pieces in PDF files. The total number of deflate-encoded objects extracted from PDF files is 458988, and each object was stored as one binary file.

MSX files might contain a number of images inside themselves, thus we develop a program to get rid of all those images from MSX-13 data set in order to get non image deflate-encoded data from office files. By using the result of [1], we extracted deflate-encoded data pieces from non-image MSX files and PNG files, each compressed data excerpt was stored as one independent file. The tool in [1] was used to decompress all the deflate-encoded files in order to seize underlying data (inflate data).

4.2 Feature Vectors

Feature vectors were extracted from the inflate data portions which were the output of the first phase. There are 256 features which are the byte distribution frequency values for the inflate data. These features and their variations were efficiently used by [9, 11, 18] to classify the file types of file fragments. On the other hand, we used bigram frequency as feature in our solution. The bigram is described by a two dimensional matrix A of [256, 256]. Assuming that b_i and b_{i+1} are two consecutive bytes in a data fragment with corresponding values v_i and v_{i+1}, the frequency of $b_i b_{i+1}$ appearances is represented by the number of the element of matrix [256, 256] at row v_i and column v_{i+1}, $A(v_i, v_{i+1})$. In our approach, each data fragment is represented by a matrix of bigram, and then the matrix is converted into a vector with 65536 elements ($256 \times 256 = 65536$) used as a feature vector for SVM. The reason for choosing those well-known features is the special structure of the underlying data of PNG inflate data which actually is filtered data by one of five filtering algorithm [17].

4.3 Experiments

We ran four experiments with two chosen features – BFD (Byte Frequency Distribution) and BFCC (Byte Frequency Cross Correlation) for inflate data fragments. For each experiment, we selected uniformly at random 12000 data fragments for each file type (MSX, PDF, and PNG). During experiments, the set of data fragments were split into a training set and a testing set. We assured that no data fragments appearing in both the training set and testing set.

The training set was used to train SVMs in linear kernel mode. The linear kernel was demonstrated as the most effective when classifying file fragments represented by feature vectors including BFD [2, 9]. We proceeded to train SVMs for both deflate and inflate data sets. Due to entropy has no effect on the deflate-encoded data (compressed data), therefore deflate-encoded data are not clustered into three groups by entropy value. Conversely, inflate data are clustered into three group of fragments which have low, medium and high entropy values. Three SVM models are created corresponding to three groups of data. In addition, following the recommendations from [11, 12], our feature vectors were scaled to decrease training time and improve the classification performance. Once models were created from the training data sets, we used SVM to classify the data fragments in the testing sets. In order to validate our results, we repeated each experiments ten times, each time picking data fragments uniformly at random for the data set.

4.4 Results

After running our experiments, we found that the accuracy for classifying deflate-encoded (compressed) data using SVM was not good enough. This result was another case to illustrate the statement that it is very hard to classify high entropy file fragment [1–4, 8, 9]. The previous approaches that used machine learning techniques and statistical feature vector also got poor identification rates for high entropy file fragments, especially for ones that were encoded by compression algorithms.

For the case of BFD as feature vector, we achieved the average identification accuracy of 85.23 % for deflate-encoded data. In the case of inflate data, the result is 89.89 % for the case of not clustering by entropy values and 94.56 % for the case of entropy clustering.

For the case of bigram as feature vector, the results were achieved 84.98 %, 88.92 %, and 93.25 % for deflate, inflate without entropy clustering, and inflate with entropy clustering data. The results show that the recognition rates are better, when entropy is used to cluster inflate data before machine learning techniques applied. In the case of inflate data, the results for BFD and BFCC are slightly different. However, the computational cost for BFCC is much larger than BFD; therefore, BFD is recommended as a good feature vector (Fig. 8).

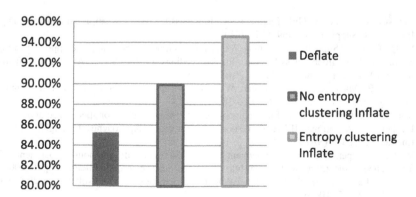

Fig. 8 BFD classification accuracies

5 Conclusion and Future Work

File fragment identification is a very important task in digital forensic. Although many research efforts have been done, some problems still remain. Compound file fragment identification is one of those problems. Normally, compound files contain different data encodings inside; therefore in order to correctly identify the compound file fragments, specialized approach needs to be proposed.

The largest portion of compound files is deflate-encoded data. Classifying compression data is always a very difficult task. It is stated that previous approaches cannot be used to identify high entropy file fragment consisting compression file fragments. This statement motivates us to propose a new approach where deflate encoded data detection technique is used in combination with machine learning technique to recognize the deflate-encoded data of compound file fragments.

It is stated that there is no discernable pattern in compression data that machine learning techniques can use to identify high entropy file fragments [1, 4]. Therefore, our approach that decompress the deflate data in order to get the underlying data fragments which have clear format such as text object format in PDF files, is the right choice before applying machine learning techniques.

The experiments show that our approach can overcome the shortcomings of previous solutions and provide very good identification accuracies. Particularly, we did not consider all types of data in a compound file the same way, different data encodings in a compound file were treated in different ways. Moreover, our solution can be applied efficiently in the real world. For the future work, other data encodings such as CCITFax-encoded and JBIG2-encoded bi-tonal images, ASCII85/ASCIIHex, and other formats in compound files will be investigated. We strongly believe that other different specialized approaches must be addressed for those data encodings.

References

1. Roussev, V., Quates, C.: File fragment encoding classification—an empirical approach. Digit. Investig. **10**(Supplement), S69–S77 (2013)
2. Li, Q., Ong, A., Suganthan, P., Thing, V.: A novel support vector machine approach to high entropy data fragment classification. In: Proceedings of the South African Information Security Multi-conference (SAISMC 2010), 2010
3. Penrose, P., Macfarlane, R., Buchanan, W.J.: Approaches to the classification of high entropy file fragments. Digit. Investig. **10**, 372–384 (2013)
4. Roussev, V., Garfinkel, S.L.: File fragment classification-the case for specialized approaches. In: Fourth International IEEE Workshop on Systematic Approaches to Digital Forensic Engineering, 2009 (SADFE '09), pp. 3–14
5. Rentz, D.: OpenOffice. org's documentation of the microsoft compound document. The Spreadsheet Project, OpenOffice.org. http://sc.openoffice.org/compdocfileformat.pdf (2007)
6. Park, B., Park, J., Lee, S.: Data concealment and detection in Microsoft Office 2007 files. Digit. Investig. **5**, 104–114 (2009)
7. Meehan, J., Rose, T.S.C.C.: PDF reference. Adobe Portable Doc. Format Vers. **1**, 1 (2001)

8. Axelsson, S.: The Normalised Compression Distance as a file fragment classifier. Digit. Investig. **7**(Supplement), S24–S31 (2010)
9. Fitzgerald, S., Mathews, G., Morris, C., Zhulyn, O.: Using NLP techniques for file fragment classification. Digit. Investig. **9**(Supplement), S44–S49 (2012)
10. Wei-Jen, L., Ke, W., Stolfo, S.J., Herzog, B.: Fileprints: identifying file types by n-gram analysis. In: Proceedings from the Sixth Annual IEEE SMC Information Assurance Workshop, 2005 (IAW '05), pp. 64–71
11. Sportiello, L., Zanero, S.: File block classification by support vector machine. In: 2011 Sixth International Conference on Availability, Reliability and Security (ARES), 2011, pp. 307–312
12. Chang, C.-C., Lin, C.-J.: LIBSVM: a library for support vector machines. ACM Trans. Intell. Syst. Technol. (TIST) **2**, 27 (2011)
13. Deutsch, L.P.: DEFLATE Compressed Data Format Specification Version 1.3 (1996)
14. Park, B., Savoldi, A., Gubian, P., Park, J., Lee, S.H., Lee, S.: Data extraction from damage compressed file for computer forensic purposes. Int. J. Hybrid Inf. Technol. **1**, 89–102 (2008)
15. Khoa, N., Dat, T., Wanli, M., Sharma, D.: An approach to detect network attacks applied for network forensics. In: 2014 11th International Conference on Fuzzy Systems and Knowledge Discovery (FSKD), 2014, pp. 655–660
16. Rice, F.: Introducing the office (2007) open XML file formats. Microsoft Developer Network (2006)
17. Boutell, T.: PNG (Portable Network Graphics) Specification Version 1.0 (1997)
18. Calhoun, W.C., Coles, D.: Predicting the types of file fragments. Digit. Investig. **5** (Supplement), S14–S20 (2008)

The Evolution of Permission as Feature for Android Malware Detection

José Gaviria de la Puerta, Borja Sanz, Igor Santos Grueiro
and Pablo García Bringas

Abstract Over the last few years, the presence of mobile devices in our lives has increased, offering us almost the same functionality as personal computers. Since the arrival of Android devices, the amount of applications available for this operating system has increased exponentially. Android has become one of the most popular operating systems in these devices. In fact, malware writers insert malicious applications into Android using the Play store and other alternative markets. Lately, many new approaches have been made. Sanz et al., for instance, presented PUMA, a method used to detect malicious apps just by taking a look at the permissions. In this paper, we present the differences between that interesting approach and a newer and bigger dataset. Besides, we also present an evolution in the permissions along the years.

Keywords Android · Malware · Permissions · Detection · Machine learning

1 Introduction

Nowadays, smartphones are becoming increasingly popular. These small computers come with us everywhere, allowing us to check our email, browse the Internet or play games with our friends. However, in order to be able to use all the possibilities that these devices offer, we need to install some applications in our smartphones.

When the first smartphones appeared, some users had lots of problems when installing those apps, as it was really difficult to find a centralized place where they

J.G. de la Puerta (✉) · B. Sanz · I.S. Grueiro · P.G. Bringas
DeustoTech Computing, University of Deusto, Bilbao, Spain
e-mail: jgaviria@deusto.es

B. Sanz
e-mail: borja.sanz@deusto.es

I.S. Grueiro
e-mail: isantos@deusto.es

P.G. Bringas
e-mail: pablo.garcia.bringas@deusto.es

© Springer International Publishing Switzerland 2015
Á. Herrero et al. (eds.), *International Joint Conference*, Advances in Intelligent Systems and Computing 369, DOI 10.1007/978-3-319-19713-5_33

389

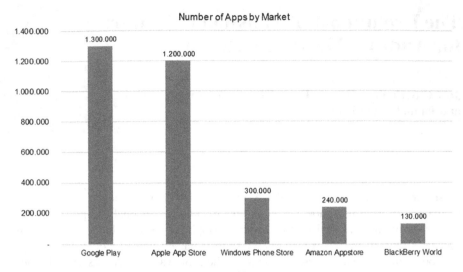

Fig. 1 Mobile apps by operating system in July 2014

could obtain them. That is why, if they wanted to do that, the only alternative they had was to search for them on the Internet.

When the users finally found the application they wanted to install, problems began. In order to protect the device and avoid piracy, several operating systems, such as Symbian, employed an authentication system based on certificates that caused several inconveniences for the users (e.g., they could not install applications even if they had bought them).

Nowadays, there are many new methods to distribute these applications. Thanks to the deployment of Internet connections in mobile devices, users can install any application without even connecting the mobile device to the computer. Apple's AppStore[1] was the first store to implement this new model and was so successful that other manufacturers, such as Google, RIM and Microsoft, have followed the same business model developing application stores accessible from the device. Now, users just need to create an account for an application store to buy and install new applications.

This new way of selling apps in a market created a new factor for the community of developers. It was a new world for them, so they started to create apps for every existing mobile platform. Furthermore, those factors have drawn developers' attention to these platforms. According to Statista,[2] the number of available applications on the App Store is over 1,200,000, whereas the Play Store of Android, which is the operating system of Google, has over 1,300,000. Figure 1 represents all the apps by operating system in July 2014.

---

[1]http://www.apple.com/iphone/features/app-store.html.

[2]http://www.statista.com.

In the same way, malicious software has arrived to both platforms. There are several applications whose behaviour is, at least, suspicious of trying to harm the users. There are other applications that are definitively malware. That is the reason why, those platforms have used different approaches to protect against this type of software.

According to their response to the US Federal Communication Commission's July 2009,[3] Apple applies a rigorous review process made by at least two reviewers. In contrast, Android relies on its security permission system and on the user's sound judgment. Unfortunately, users have usually no security consciousness and they do not read the required permissions before installing an application. This is one of the most important things that the security teams want the users to learn.

In spite of the fact that both AppStore and Play Store include clauses in the terms of services that urge developers not to submit malicious software, both have hosted malware in their stores. One of the solutions that these enterprises have developed, is to remove remotely the malicious applications that are installed in the devices. Sadly, the usage of these models is insufficient to ensure the user's security, and that is why the new models have been included.

It is demonstrated that the usage of machine learning techniques for generic malware detection and classification is widely applied in the scientific community [1]. Besides, several approaches [2, 3] have been proposed to classify applications specifying the malware class; e.g., trojan, worms, virus; and even the malware family.

Section 2 explains the permission security system that is used by Android. In Sect. 3 the both dataset, malware and goodware, are exposed. We can also see how we obtained the goodware dataset.

The next section is the evolution of the permissions. In this section we can see a graph with the difference between permissions in different times.

Section 4 shows us all the experimentation of the paper. Here, all the algorithms that we use in the approach are included. Moreover, the results and the discussion are included in it. Section 5 presents the conclusions of the paper.

2 Permission in Android

An additional security system that Android provides the operating system is the "permission" mechanism. With it, the S.O. enforces the restrictions of each app in the mobile device.

The normal functioning of these permissions is quite simple: for example, if one app wants to use the connectivity through Internet, it has to request the permission associated with the connection to it. These permissions are usually given at installation time, being the user the person that has to take the decision of installing the app after reading and understanding the permissions it requests.

---

[3]http://online.wsj.com/public/resources/documents/wsj-2009-0731-FCCApple.pdf.

```
<manifest android:versionCode="28" android:versionName="2.8.9" package="1xl.Live360.org"
  xmlns:android="http://schemas.android.com/apk/res/android">
    <uses-permission android:name="android.permission.INTERNET" />
    <supports-screens android:anyDensity="false" android:smallScreens="true" android:normalScreens="true" android:largeScreens="true" android:resizeable="false" />
    <uses-permission android:name="android.permission.WRITE_EXTERNAL_STORAGE" />
    <uses-permission android:name="android.permission.VIBRATE" />
</manifest>
```

Fig. 2 Permissions in a *Manifest.xml* file for one application

Figure 2 is an example of a Manifest.xml with different permissions for an application. All the permissions must appear in the clause entitled "uses-permission". If an app wants to access the Internet and does not have the "android.permission. INTERNET", it will not be able to do it.

3 Information Gathering

In this paper, we have chosen to compare two different datasets to see their evolution over the last few years in terms of permissions selected by the apps.

3.1 Older Dataset

The first dataset is the one that was selected by Sanz et al. in their paper called "PUMA: Permission Usage to Detect Malware in Android" [4]. This dataset is composed by a total of 357 applications of goodware, and 249 apps of malicious software. Malware samples were gathered by means of *VirusTotal*,[4] which is an analysis tool for suspicious files. We have used their service called Virus-Total Malware Intelligence Services, which is available for researchers to perform queries to their database.

With the goal of developing the goodware dataset, Sanz et al. gathered a collection of 1811 Android applications of different types. In order to classify them properly, they chose to follow the same naming as the official Android market. To this end, they used an unofficial Android Market API[5] to connect with the Android market and, therefore, obtain the classification of the applications.

They selected the number of applications within each category according to their proportions in the *Play Store*, before it was called *Android Market*. Despite of Android has got some types of applications, they did not make distinctions to choose one, or more, of these types. So, in that dataset all types are represented. Then, they selected randomly the apps from different categories.

---

[4]http://www.virustotal.com.
[5]http://code.google.com/p/android-market-api/.

3.2 New Dataset

The second dataset has been generated using two different techniques. Goodware samples have been gathered using the Selenium[6] application for automating web browser. With this automation we use the application web APIfy[7] to download Android applications from different categories.

Totally, we have downloaded 7.062 applications from Google Play. These apps have been gathered from different available categories in that market. That is exactly what we see in Table 1.

Table 1 Number of App by category

Category	Number	Category	Number
BUSINESS	179	GAME ACTION	156
LIFESTYLE	168	GAME PUZZLE	170
SHOPPING	163	GAME SIMULATION	165
BOOKS AND REFERENCE	173	ENTERTAINMENT	157
MEDICAL	176	GAME TRIVIA	161
GAME STRATEGY	163	GAME ROLE PLAYING	159
GAME ADVENTURE	161	NEWS AND MAGAZINES	170
GAME CASUAL	160	FINANCE	173
MEDIA AND VIDEO	156	GAME WORD	170
GAME CARD	176	APP WALLPAPER	163
LIBRARIES AND DEMO	185	PRODUCTIVITY	178
WEATHER	172	MUSIC AND AUDIO	156
GAME EDUCATIONAL	170	COMMUNICATION	177
GAME BOARD	175	HEALTH AND FITNESS	172
EDUCATION	179	COMICS	61
GAME FAMILY	157	GAME ARCADE	169
GAME SPORTS	168	GAME RACING	168
SOCIAL	163	TRAVEL AND LOCAL	174
GAME CASINO	175	PHOTOGRAPHY	159
TRANSPORTATION	174	TOOLS	177
SPORTS	161	PERSONALIZATION	111
GAME MUSIC	162		

Total: 7.062

The choice of these lists for obtaining samples was completely random, as what we wanted was to have a heterogeneous applications dataset from Google's store. To verify that the samples did not contain malicious code, we used the online platform VirusTotal. This platform uses 43 antivirus engines to scan the sample that

---

[6]http://docs.seleniumhq.org/.

[7]http://apify.ifc0nfig.com/.

has been sent previously. This analysis returns the total number of engines that have been detected as malicious and malware is for that engine. Since the experiment was desired to have all the possible clean malware samples, we made the decision that if a sample was detected as malicious by at least one of the antivirus engines, it would be immediately separated from the rest of the dataset. Furthermore, the ones that were detected as adware,[8] were also separated from that dataset right away.

After finishing that analysis, we chose a total of 5.511 goodware apps. This result implied that 21.96 % of the downloaded apps were considered malware by at least one antivirus engine.

The malware dataset was gathered from the Drebin malware dataset [5], which is composed by a total of 5,560 samples of malware and which is divided into different families.

3.3 Permissions Through the Years

Previously, it has been mentioned what a manifest file is. In it, we can see all the permissions that an application requests to the user. Android, for instance, has got a total of 150 permissions defined in its API.

Despite of this, some companies, like, for example, Samsung, have made their own group of permissions for using their terminal. In this experiment we do not study this type of permissions, we only look at the permissions of the Android Platform.

Figure 3 shows that there is one permission that every app in the market requests: "Internet". Besides, we can also see that, nowadays, the number of permissions required by many malware has increased greatly over the last few years, something that indicates that malware is more sophisticated now.

4 Experimental Validation

To compare the two datasets, we have employed supervised machine learning methods to classify Android applications into malware and benign software. To this extent, we have used Waikato Environment for Knowledge Analysis (WEKA).[9] In particular, we used the classifiers specified in Table 2. To evaluate the performance of machine-learning classifiers, k-fold cross validation is usually used [6]. Thereby, for each classifier we tested, we performed a k-fold cross validation [7] with k = 10. In this way, our dataset was split 10 times into 10 different sets for learning (90 % of the total dataset) and testing (10 % of the total data).

---

[8] Adware is a type of action hidden in applications, which send targeted advertisements to our device when you run an application.

[9] http://www.cs.waikato.ac.nz/ml/weka/.

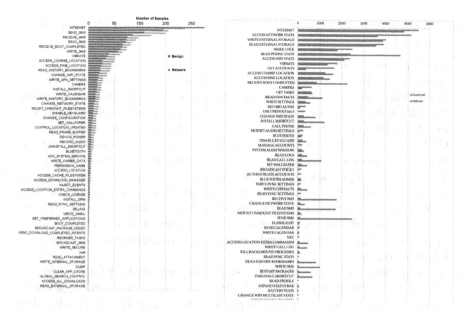

Fig. 3 Difference between permissions request in PUMA dataset and this experiment dataset

Table 2 Algorithms that are used for classification	Algorithms
	IBK 1
	IBK 3
	IBK 5
	SimpleLogistic
	NaiveBayes
	BayesNet K2
	BayesNet TAN
	SMO Poly
	SMO NPoly
	J48
	RandomTree
	RandomForest 10
	RandomForest 50
	RandomForest 100

4.1 Machine Learning Algorithms

In this research, we chose to compare the performance of different classification algorithms given the occasionally notable differences in effectiveness that can be observed in similar experiments conducted in other areas [8]. The algorithms used for the tests were the following: Random Forest, J48, Bayes Theorem-based algorithms, K-Nearest Neighbor (KNN), Sequential Minimal Optimization (SMO) and Simple Logistic.

- Random Forest. Random Forest is an aggregation classifier developed by Leo Breiman [9] which is formed by a bunch of decision trees considered in a way in which the introduction of a stochastic component improves the Accuracy of the classifier, either in the construction of the trees or either in the training dataset.
- J48. J48 is an open source implementation for Weka of the C4.5 algorithm [10]. C4.5 creates decision trees given an amount of training information making use of the concept *information entropy* [11]. The training data consist of a group $S = s_1, s_2, \ldots, s_n$ of already classified samples $s_1 = x_1, x_2, \ldots, x_m$ in which $x_1, x_2, \ldots, x_m$ represent the attributes or characteristics of each sample. In each node of the decision tree, the algorithm will choose the attribute in the data that most efficiently divides the dataset in enriched choruses of a given class using the entropy difference or the already mentioned *normalized information gain* as selective criteria.
- Bayes Theorem-based algorithms. The Bayes Theorem, the base for the Bayesian inference, is a statistical method which determines, based on a number of observations, the probability of a certain hypotheses being true. For the classification needs here exposed, this is the most important capability of Bayesian networks: in our case, the probability of an app being malicious or benin The theorem is capable of adjusting the probabilities as soon as new observations are performed. Thus, Bayesian networks conform a probabilistic model that represents a collection of randomized variables and their conditional dependencies by means of a directed graph. We have trained our models with three different search algorithms: K2, Hill Climbing and TAN.
  In this group we have also considered the inclusion of Naïve Bayes. The idea is that if the number of independent variables managed is too big, it does not make any sense to make probability tables [12]. Then, the reduced model with simplified datasets give to the algorithm the appellative of *Naïve*.
- K-Nearest Neighbor (KNN). The KNN algorithm is one of the most simple classification algorithms amongst all of those available for the machine learning techniques. It takes decisions based on the results of the k closest neighbours to the analysed sample in the experimental n-dimensional space ($\forall k \in \mathbb{N}$). In this case, and taken into account the simplicity of the algorithm, we have explored even more values ($k = 1, 3, 5$) so as to determine if this enlargement would throw any kind of additional advantage.
- Sequential Minimal Optimization (SMO). SMO, invented by John Platt [13], is an iterative algorithm used for the solution of the optimization problems that appear when training *Support Vector Machines* (SVM). Basically, SMO divides the problem into a series of smaller subproblems which are analytically solved lately.
  At this point, we have selected different kernels with these algorithms: a polynomial kernel, a normalized polynomial kernel.
- Simple Logistic. This algorithm is used to predict the result of a variable function of the independent variables, or predictor. The logistic regression formula is:

$$Y_i = \frac{1}{1 + e^{-(\beta_0 + \beta_1 X_{1,i} + \cdots + \beta_d X_{d,i})}} \tag{1}$$

Being Y_i the classification to be predicted by the model, in our case it would be *goodware* or *malware*. The variable X is the vector with the extracted permissions for a specific application, finding that $X_{d,i}$ is the value assigned to an n-gram of permissions in d enforcement position that is in row i. Parameters *beta* $_{ast}$ are determined by the algorithm in the training phase.

4.2 Validation Employed Parameters

This evaluation was performed according to the following parameters, usually employed to compare the performance of different algorithms in the field of machine learning:

- *True Positive Ratio (TPR)*, which is calculated by dividing the number of bening apps correctly classified *(TP)* between the total samples taken *(TP + FN)*. The formula is:

$$TPR = \frac{TP}{(TP + FN)} \tag{2}$$

- *False Positive Ratio (FPR)*, which is calculated by dividing the number of samples corresponding to malicious apps whose classification *(FP)* were missed by the total number of samples *(FP + TN)*. The formula is:

$$FPR = \frac{FP}{(FP + TN)} \tag{3}$$

- *Accuracy (P)*, which is calculated by dividing the total hits by the total number of instances in the dataset. The formula is:

$$P = \frac{TP + TN}{TP + FP + TN + FN} \tag{4}$$

- *Area Under ROC Curve (AUC)* [8], that establishes the relationship amongst the false negatives and the false positives. The ROC Curve it is usually used to generate statistics that represent the performance or the effectiveness in a wider sense of a classifier.

4.3 Results and Discussion

Using the algorithms and the parameters above mentioned to compare the performance, the Table 3 shows us the different results recovered with both datasets. The first question that we solved was if the usage of a bigger dataset improved the

Table 3 Result for the classification algorithms using the PUMA dataset and the Drebin dataset

PUMA Dataset

Algorithm	TPR	FPR	AUC	Accuracy
IBK 1	0.92	0.21	0.90	85.55%
IBK 3	0.90	0.22	0.89	83.96%
IBK 5	0.87	0.24	0.88	81.91%
SimpleLogistic	0.91	0.23	0.89	84.08%
NaiveBayes	0.50	0.15	0.78	67.64%
BayesNet K2	0.45	0.11	0.77	67.07%
BayesNet TAN	0.53	0.16	0.79	68.51%
SMO Poly	0.91	0.26	0.83	82.84%
SMO NPoly	0.91	0.19	0.86	85.77%
J48	0.87	0.25	0.86	81.32%
RandomTree	0.90	0.23	0.85	83.32%
RandomForest 10	0.92	0.21	0.92	85.82%
RandomForest 50	0.91	0.19	0.92	86.41%
RandomForest 100	0.91	0.19	0.92	86.37%

DREBIN Dataset

Algorithm	TPR	FPR	AUC	Accuracy
IBK 1	0.96	0.04	0.99	95.66%
IBK 3	0.96	0.05	0.99	94.93%
IBK 5	0.95	0.06	0.99	94.30%
SimpleLogistic	0.97	0.06	0.99	93.91%
NaiveBayes	0.93	0.14	0.96	86.81%
BayesNet K2	0.96	0.16	0.97	85.83%
BayesNet TAN	0.97	0.10	0.98	90.55%
SMO Poly	0.97	0.07	0.95	93.45%
SMO NPoly	0.97	0.05	0.96	94.70%
J48	0.96	0.05	0.98	95.06%
RandomTree	0.95	0.05	0.96	94.89%
RandomForest 10	0.97	0.04	0.99	95.63%
RandomForest 50	0.97	0.04	0.99	96.00%
RandomForest 100	0.97	0.04	0.99	96.05%

classification or not. Seeing the result in Table 3, all the algorithms improve their accuracy and all their parameters. Once again, it is also demonstrated that the best algorithm is *Random Forest* but using a different number of trees.

Something curious is the values of the FPR parameter in Drebin dataset. Moreover, the good values of AUC are also important, being of 0.99 in more than one case.

Using the new dataset, all the methods achieved accuracy rates higher than 85 %. The best classifier, in terms of accuracy, is *Random Forest*, with 100 trees.

Also in terms of TPR the best result that we have obtained is 0.97 in the classifiers *Random Forest* with 10, 50 and 100 trees, *SMO* with PolyKernel and Normalized PolyKernel, *Bayes Net* with TAN and *Simple Logistic*. The lowest FPR goes for *Random Forest* with 10, 50 and 100 trees and *IBK* with K = 1.

5 Conclusions

Permissions are the most recognisable security features in Android. The user must accept them in order to install any application. In this paper, we validate the viability of the usage of permissions as a mechanism to detect malware using machine-learning.

In order to validate the previous scope, we collected a total amount of 5,511 good-ware samples. Also, we used the Drebin dataset as malware dataset. Using different classifiers, we generated the models and evaluated their configuration with the Area Under ROC Curve (AUC). We obtained a 0.99 of AUC using the Random Forest classifier.

In light of these results and in spite of the fact that in the future this approach could change, the viability of Sanz's approach has been demonstrated. Nowadays, Android uses a group of permissions, so, with this approach we do not know if a malicious application is using all the permissions that it requests or not. However, this could be used as a first step before a more extensive analysis.

In future lines, as Drebin dataset is composed by families of malware, using their permissions and machine learning techniques, we could obtain their family. Besides, as the malware in Android is a problem that is growing exponentially every day,we believe that the creation of a dynamic tool by the community would be essential.

References

1. Schultz, M.G., Eskin, E., Zadok, E., Stolfo, S.J.: Data mining methods for detection of new malicious executables. In: Proceedings of 2001 IEEE Symposium on Security and Privacy, S&P 2001, pp. 38–49. IEEE (2001)
2. Rieck, K., Holz, T., Willems, C., Düssel, P., Laskov, P.: Learning and classification of malware behavior. In: Detection of Intrusions and Malware, and Vulnerability Assessment. Springer, pp. 108–125 (2008)
3. Tian, R., Batten, L., Islam, M., Versteeg, S.: An automated classification system based on the strings of trojan and virus families. In: 2009 4th International Conference on Malicious and Unwanted Software (MALWARE), pp. 23–30. IEEE (2009)
4. Sanz, B., Santos, I., Laorden, C., Ugarte-Pedrero, X., Bringas, P.G., Álvarez, G.: Puma: permission usage to detect malware in android. In: International Joint Conference CISIS'12-ICEUTE' 12-SOCO' 12 Special Sessions, pp. 289–298. Springer (2013)
5. Arp, D., Spreitzenbarth, M., Hübner, M., Gascon, H., Rieck, K., Siemens, C.: Drebin: effective and explainable detection of android malware in your pocket. In: Proceedings of NDSS (2014)
6. Bishop, C.M., et al.: Pattern Recognition and Machine Learning, vol. 4. Springer, New York (2006)

7. Kohavi, R., et al.: A study of cross-validation and bootstrap for accuracy estimation and model selection. IJCAI **14**, 1137–1145 (1995)
8. Singh, Y., Kaur, A., Malhotra, R.: Comparative analysis of regression and machine learning methods for predicting fault proneness models. Int. J. Comput. Appl. Technol. **35**(2009), 183–193 (2009)
9. Breiman, L.: Random forests. Mach. Learn. **45**, 5–32 (2001). doi:10.1023/A:1010933404324
10. Quinlan, J.: C4. 5: Programs for Machine Learning. Morgan kaufmann (1993)
11. Salzberg, S.L.: C4.5: programs for machine learning by J. Ross Quinlan. Morgan Kaufmann Publishers, inc., 1993. Mach. Learn. **16**, 235–240 (1994). doi:10.1007/BF00993309
12. Jiang, L., Wang, D., Cai, Z., Yan, X.: Survey of improving naive bayes for classification. In: Alhajj, R., Gao, H., Li, X., Li, J., Zaïane, O. (eds.) Advanced Data Mining and Applications. Lecture Notes in Computer Science, Vol. 4632, pp. 134–145. Springer, Berlin/Heidelberg (2007)doi:10.1007/978-3-540-73871-8_14
13. Platt, J.C.: Sequential minimal optimization: a fast algorithm for training support vector machines (1998)

Pollution Attacks Detection in the P2PSP Live Streaming System

Cristóbal Medina-López, L.G. Casado and Vicente González-Ruiz

Abstract This work studies the *pollution attack*: a challenging security-related problem in peer-to-peer streaming platforms. Different variations of this attack and its combinations are addressed. In order to mitigate such attacks, two different strategies have been proposed in the context of the P2PSP live streaming system, a peer-to-peer streaming platform with free access (i.e. it is not necessary to provide an identification and only endpoints are stored). The first prevention strategy is based on the existence of anonymous trusted peers that detect and report attackers, which are finally expelled. The second strategy increases the security level in the overlay at the cost of a higher computation and communication overhead. Both strategies are analyzed theoretically for several attack configurations.

Keywords P2PSP · Content integrity · Live streaming · Pollution attacks · Peer-to-peer networks

1 Introduction

The problem with scalability in client-server based live video streaming systems is well known: the server is forced to upload a copy of the content per client. In a peer-to-peer system the number of copies uploaded by the server is usually orders of magnitude lower than the number of peers. The collection of peers are in charge of sharing the content among them using their upload bandwidth excess. For that reason, the peer-to-peer (P2P) model is an interesting alternative to Client-Server model, specially since clients see how their available upload bandwidth grows year after year.

C. Medina-López (✉) · L.G. Casado · V. González-Ruiz
University of Almería (CeiA3), Almería, Spain
e-mail: cristobalmedina@ual.es

L.G. Casado
e-mail: leo@ual.es

V. González-Ruiz
e-mail: vruiz@ual.es

© Springer International Publishing Switzerland 2015
Á. Herrero et al. (eds.), *International Joint Conference*, Advances in Intelligent Systems and Computing 369, DOI 10.1007/978-3-319-19713-5_34

Because peers obtain most of the content from other peers rather than from the content source, it is easy for an attacker to alter and resend the data packets it receives. Therefore, it is necessary to design defense mechanisms. This paper focuses on fighting against pollution attacks [5], which are those that change the delivered content in some way. Here, two defense strategies are proposed in the context of the P2PSP streaming system [14, 17]. The first one keeps the communication overhead as low as possible but presents security risks that will be discussed. The second strategy mitigates those risks at the price of a higher communication overhead. Moreover, this work presents an analysis of the types of attacks that each strategy can bear or at least mitigate, as well as the best configurations to perform them.

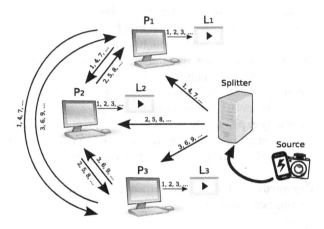

Fig. 1 A P2PSP team. Numbers in arrows refer the transmitted chunk numbers

The rest of this paper is organized as follows. Brief overviews of the P2PSP system and different pollution attacks are shown in Sects. 2 and 3, respectively. Related work is presented in Sect. 4. In Sect. 5 two strategies to prevent pollution and related attacks, introduced in Sect. 3, are presented. Analytical results for both strategies are shown in Sect. 6. The Sect. 7 shows the difficulties to determine who is the attacker, and the last section shows the conclusions and future work.

2 Description of P2PSP

The Peer-To-Peer Straightforward Protocol (P2PSP) is an application-layer protocol designed for real-time broadcasting of media over a peer-to-peer overlay. P2PSP establishes a push-based [10] fully connected mesh scheme [11] where every peer is connected with each other (see Fig. 1). A basic P2PSP network consists of a *splitter* (S) and a collection of *peers* (P_i) named *team*. The splitter receives a media stream from a *source* (O), divides the stream in data chunks of the same size and sends them to the team following a Round-Robin scheme. Next, each peer forwards those chunks received from the splitter to the rest of peers (chunks received from other peers are not forwarded). Finally, each peer sends the reconstructed stream to a *listener* (L), which is usually a media player.

As a consequence of this simple communication algorithm, the P2PSP is very suitable for execution in nodes with constrained computational resources [13]. Moreover, a peer failing or corrupting its data chunk before forwarding it, produces the loss of the same chunk for the other peers. This feature can be used to efficiently implement error concealment techniques [16], which are less effective when lost chunks are consecutive. For example, if the number of attackers or free riders is 2 on a team with 200 peers, only 1 % of chunks would not be played for some peers (those being attacked or unserved).

3 Pollution Attacks

Content integrity is one of the major issues in most P2P systems when trying to guarantee the quality of service in live-stream transmissions. Pollution attacks [5], also known as stream spoiling, basically consist of a peer or a set of peers modifying the content of the stream. Pollution attack can be done in different ways. Some of them, and other related attacks, are described next:

Persistent attack [5]: This attack consists of doing the highest possible amount of damage in the shortest period of time. Therefore, an attacker poisons every chunk received from the splitter and sends them to the entire team.

On-off attack [7]: In order to improve the *Persistent attack* and to avoid to be quickly expelled from the team, the attacker only poisons some chunks (for instance, 10 % of the total sent to the team). By acting this way, the attacker is resilient to detection systems using trust management methods, since the attacker would be assigned a high enough level of trust by the staying network.

Selective attack: This attack consists of poisoning chunks intended for only one peer or a small subset of peers. As in the previous attack, the reason behind this behavior is to be unnoticed by most of the peers and thus avoid to be expelled if a voting system is used by the overlay.

Collaborative attack [8]: Sometimes a single attacker is not able to produce a big damage; however several attackers may collaborate to produce *Selective* and *On-off* attacks to a large set of peers. By doing so, the amount of information obtained by a pollution detection system about an individual attacker is smaller than in single attacker variants. This is the most difficult attack to deal with.

Hand-wash attack [9]: Those attackers that have been discovered or think that they could have been discovered, leave the team and return to continue the attack with another alias.

Bad-mouth attack [8]: It can be used when peers can complain about others in an attacker detection system. Attackers do bad-mouthing by intentionally blame others regular peers for sending poisoned chunks or not sending chunks, with the intention of expelling them.

4 Related Work

Vulnerability and malicious peer behavior are some of the major issues in P2P overlays. For this reason, these topics have been studied extensively. In the specific scenario of live streaming, the dominant approaches to fight against attacks are: (1) trust management with reputation systems [7, 12, 18], (2) hashing/signature schemes [5] and (3) the use of trusted peers [2], to detect selfish peers, also known as free riders. A selfish peer relies smaller amount of data than it receives, or even no data. Following, a brief overview of previous defense mechanisms is shown:

Trust management: Each peer assigns a trust value for each other peer in the overlay. Moreover, each peer has its own perspective about its trust on the rest. Usually, a peer is expelled from the overlay when its trust value is smaller than a threshold. The main difficulty is how to compute a fair trust value for a peer. The common attack to trust management systems is the *On-off* one. *Trust management* and a method based on direct and indirect trust to fight against *On/off attack* is shown in [8].

Hashing/signature: They are used to detect poisoned chunks at the cost of an additional communication and computation overhead. The most common attack to this defense is the *Bad-mouth* attack. Non-repudiation methods can be a solution, but they still present some drawbacks for P2P networks [3].

Trusted peers: A set of trusted peers to monitor the bandwidth usage by other peers can be deployed [2]. Free rider peers detected by a trusted peer are expelled.

5 Proposals

This section presents two different strategies aiming to mitigate the impact of pollution and related attacks by combining trust management, hashing/signatures and trusted peers. Each strategy defines a set of rules, specially designed for the P2PSP system. The first strategy, denoted by STrPe (Strategy based on Trusted Peers), is a simple approach with a low data overhead in the overall operation of the team. The only difference with respect to an pollution-unaware P2PSP system is an extension of the splitter functionality and the inclusion of anonymous trusted peers (TP) who transparently monitor the behavior of regular peers in the team. Regular peers see a TP as another regular peer.

The second strategy, denoted by STrPe-DS (Strategy based on Trusted Peers and Digital Signatures) is an extension of the first one, where a digital signature of a chunk allows to detect attackers. STrPe-DS generates more data overhead than STrPe but the performance of the defense is greatly improved.

Those attacks detected by STrPe are also detected by STrPe-DS, but not the other way around. Therefore, the most adequate strategy should be used depending on the risk to be assumed. This decision is usually made before the deployment but it is also possible to change it once the P2PSP overlay is running.

5.1 Strategy Based on Trusted Peers (STrPe)

STrPe has been designed to maintain the simplicity characterizing the P2PSP system. By including trusted peers (TPs) into the team, an attacker sending a poisoned chunk to some TP is detected and expelled from the team. For this reason it is important to maintain the anonymity of TPs among regular peers but the splitter. The behavior rules are:

1. Only the splitter knows who are the TPs in the team. It is possible that all peers are TPs except the attacker.
2. Each TP creates a hash for each received chunk, including the chunk number and the endpoint of the source of the chunk. Depending on the computational power available in the TPs, all chunks or a random subset of them may be processed in each round.
3. TPs send the hashes to the splitter, who checks whether the chunks have been altered by comparing them with those hashes calculated when the chunks were delivered to the team. The splitter only listens to TPs for this task.
4. The splitter knows the peer in charge of relaying a given chunk and it knows who altered a chunk when an invalid hash has been received from a TP. Any exposed attacker is expelled from the team by removing it form the list of peers in the splitter (this implies that no more chunks will be send to it). In order to ensure that the attacker is removed from all lists of peers as soon as possible, the splitter sends a expulsion message containing the endpoint of the attacker to all peers in the team.

5.2 Strategy Based on Trusted Peers and Digital Signatures (STrPe-DS)

The STrPe-DS has been designed to mitigate the *Selective attack* and to identify poisoned chunks by using digital signatures. So, any peer is able to know if a chunk was poisoned and/or relayed by a different peer than the one in charge of it. Rules defining STrPe-DS are:

1. A peer receives the public key of the splitter, plus the other necessary information, when a peer joins the team.
2. For each chunk, the splitter sends a message with the chunk, its number ($nChunk$), the destination address (dst) and a digital signature $\{chunk, nChunk, dst, S_{priv}$ $(H(chunk + nChunk + dst))\}$, where H is a hash function and S_{priv} is the private key of the splitter.
3. When a peer receives a message, it performs the following steps:

   (a) Check whether dst matches the address of the sender. Notice that this action is vulnerable to the well known *Spoofing attack* [1]. The next step is performed only if this one is successful.

(b) Check the correctness of the hash value in the message.

If any of previous checks fails, the sender is removed from the list of peers of the current peer.

4. The splitter periodically requests (and the peers serve) the list of removed peers (since the previous request). In this way, a *Deny of Service (DoS) attack* by sending removed peers to the splitter at high rate is avoided. TPs are the only ones that can send the list of removed peers to the splitter as soon as they are detected.
5. Peers removed by any TP are directly expelled by the splitter (see point 4 in Sect. 5.1).
6. The splitter can decide to expel a peer based on the information received from well-intended or attackers peers. So, a typical approach here is to establish a trust value for each peer depending on several aspects, such as the number of complaints received about a peer during a given period of time, the number of affected peers by a possible attack, the age of the peers in the team, etc. (Fig. 2).

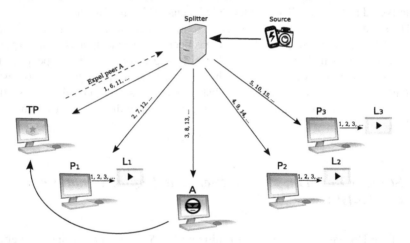

Fig. 2 A P2PSP using STrPe-DS, where a poisoned chunk is received by the TP, which immediately informs to the splitter

Any exchange of information between the splitter and peers should be authenticated by a digital signature in order to avoid *spoofing* attacks but we are interested in free access system where each peer is determined by its endpoint.

6 Analytical Results

Defense mechanisms in P2P overlays can be evaluated in several ways. The preferable is to test the system in a real scenario, but the cost of these experiments can be unaffordable. Another possibility is to analyze a set of simulations with different combinations of attack/defense schemes. We will address them in a future work. Here we perform a theoretical study of the advantages and drawbacks of STrPe and STrPe-DS strategies.

6.1 STrPe

Regarding the communication overhead, only the communications between the trusted peers and the splitter suffer it in this strategy. TPs need to send to the splitter a message digest for each received chunk from other peers, this implies a total overhead $O = sH \times Cps \times nTP$, where sH is the size of the hash, nTP is the number of trusted peers and Cps are the chunks per second. For instance, let suppose that a video, whose bit-rate is 1,024 Kbps, is being streamed and that the chunk size is 1,024 bytes. So, at least 125 chunks are being transmitted per second. If hash function generates a digest of 224 bit, the overhead per chunk is 224 bits. Thus, the communication overhead between a trusted peer and the splitter is $125 \times 224 = 28,000$ bps $\simeq 27.3$ Kbps. In this example, the total overhead in the team due to the defense strategy would be 27.3 Kbps $\times nTP$.

Concerning the probability of detecting an attacker, Fig. 3 shows it as $P = nTP/(N-A)$, where N is the team size and A is the number of attackers. The number of attackers and TPs represents a percentage of the size of the team. In general, the chance to detect an attacker increases with the number of trusted peers in the team.

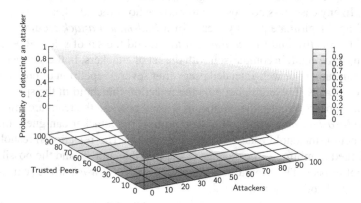

Fig. 3 Probability to detect an attacker depending on the percentages of trusted peers and attackers in the team

A *Persistent* and *On-off attacks* are quickly detected by the splitter. A *Bad-mouth attack* makes little sense in STrPe since only TPs are able to inform about hashes to the splitter. However, the main vulnerability of STrPe is that TPs must remain anonymous. In case of being discovered, the system will not be resilient to *Selective attacks* where attackers do not pollute chunks for TPs. Finally, regular peers do not know if a chunk has been polluted because the chunks are not digitally signed.

6.2 STrPe-DS

Whereas only TPs have communication overhead in previous strategy, the whole team suffer from it in this strategy, because all messages contains not only the chunk but also additional information (see Sect. 5.2). Using the same example of the previous section, if a RSA signature with 1,024 bit key is used (it seems to be fast enough for our purpose, according standards signature times [4]), the overhead per chunk is 1,072 bits (1,024 bits for signature + 48 bits for *dst*). Thus, the communication overhead is $125 \times 1,072 = 134,000$ bps $\simeq 130.8$ Kbps. However, in this strategy, the total overhead in the team is more difficult to calculate because it depends on the size of the removed peers request made by the splitter, the frequency of such requests, the number of attackers and the number of the trusted peers.

Regarding the detection of attacks, when a *Persistent attack* is carried out over the entire team, it is quickly detected by any TP (see Fig. 2).

A *On-off attack* is meaningless because peers remove the attackers from their lists as soon as they receive an incorrect message, according to the digital signature of the splitter. This attack becomes a *Persistent attack* in a relatively large period of time.

Performing a *Bad-mouth attack* by only one attacker is not usually enough to expel a regular peer in trust based systems. Additionally, the attacker will usually have more complains than any attacked peer and could finally be detected. According to the P2PSP system, a peer A removes a peer B from its list of peers because it does not receive chunks from B or the chunks received are incorrect. This results in B also removes A from its list of peers given that B is not receiving chunks from A anymore. In any case, it is not possible to know who is the attacker.

Expelling a legitimate peer by means of a *Bad-mouth attack* requires a combination with *Selective* and *Collaborative attacks*, and the set of selected peers to be attacked must be small in comparison with the set of attackers. In the absence of TPs, the team can run out of well-intended peers after several repetition of previous combination of attacks. Therefore, not only the rejection-threshold in the splitter or the trust value of peers but also the number of trusted peers in the team are important.

A *Hand-wash attack* is difficult to deal with, if the attacker can guess the value of the rejection-threshold established by the splitter. The rejection-threshold must be established to minimize the damage of the possible attacks and the possibility to expel a polite peer. TPs should perform a kind of *Hand-wash strategy* in order to reduce the probability of be detected.

To the best of our knowledge, there is no effective defense against a combined *Selective*, *Collaborative*, *Bad-mouth* and *Hand-wash attacks* in free access P2P live streaming protocols.

7 Expelling Peers from the Team

Digital signature of chunks allows to each peer receiving a poisoned chunk to detect the attacker, who is removed from its peer list. As mentioned before, a trusted peer receiving a poisoned chunk reports it to the splitter to expel the attacker from the team. In the StrPe-DS strategy peers can complain to the splitter. The splitter is in charge to decide who is the attacker based on the gathered information. This is a difficult task, because for instance, if five peers are complaining about one peer, it is difficult for the splitter to know whether there are five attackers trying to expel one well-intended peer (*Bad-mouth* attack) or if it is actually an attacker poisoning five well-intended peers. To address this problem there exists two alternatives:

Non-repudiation methods. Most of the current non-repudiation system are based on the existence of a Trusted Third Party (TTP). TTPs are not in consonance with the P2P philosophy because they reduce the distributed computing level in P2P systems. There are some proposals [3] where standard peers act as TTP, but they could be malicious, as well.

There are solutions without TTPs, but they are inefficient since the number of necessary messages is usually high [6]. Additionally, these solutions consider that both parties are interested in the content. This is not the case for P2P streaming systems where the attacker is interested in poisoning the content but not in the content itself. Thus, currently there is not a suitable non-repudiation system allowing the splitter to decide who is the attacker.

Trust-based methods. Due to the absence of a suitable non-repudiation system, this is usually the most used solution. In the StrPe-DS strategy, the splitter gathers all complaints from peers. Based on this and other possible information about peers, the splitter has to establish a decision method in order to determine who will be expelled from the team. As we shown before this is a difficult task.

As future work, we are interested in to study an efficient method for non-repudiation in P2P live streaming systems and a fair trust-based method [15].

8 Conclusions

In this paper two different strategies to mitigate pollution attacks in the P2PSP system are studied: (1) a strategy with low computation and communication overhead, which is designed for resource constrained devices and (2) an extension of the first strategy that increases the defense level against pollution attacks at the expense of introducing a higher overhead. It is necessary to develop a fair trust-based method in order to make the last strategy effective.

There are several problems that must be addressed in order to develop a system against pollution attacks in P2P live streaming networks with free access. A non-repudiation methods without Trusted Third Party can be a good solution but unfortunately the existing proposals are not suitable for P2P live streaming systems. We are also interested in study this topic in a future work.

Acknowledgments This work has been funded by grants from the Spanish Ministry (TIN2012-37483) and Junta de Andalucía (P11-TIC-7176), in part financed by the European Regional Development Fund (ERDF).

References

1. Bellovin, S.M.: Security problems in the TCP/IP protocol suite. SIGCOMM Comput. Commun. Rev. **19**(2), 32–48 (1989)
2. Conner, W., Nahrstedt, K.: Securing peer-to-peer media streaming systems from selfish and malicious behavior. In: Proceedings of the 4th on Middleware Doctoral Symposium, p. 13. ACM (2007)
3. Conrad, M.: Non-repudiation mechanisms for peer-to-peer networks: enabling technology for peer-to-peer economic markets. In: Proceedings of the 2006 ACM CoNEXT Conference, CoNEXT '06, pp. 30:1–30:2. ACM, New York, NY, USA (2006)
4. Dai, W.: Speed Comparison of Popular Crypto Algorithms. http://www.cryptopp.com/benchmarks.html
5. Dhungel, P., Hei, X., Ross, K.W., Saxena, N.: The pollution attack in P2P live streaming: measurement results and defenses. In: Proceedings of the 2007 Workshop on Peer-to-peer Streaming and IP-TV, pp. 323–328. ACM (2007)
6. Gouda, M.G.: Keynote talk: communication without repudiation: the unanswered question. In: Networked Systems, pp. 1–8. Springer (2014)
7. Hu, Bo., Zhao, H.V.: Pollution-resistant peer-to-peer live streaming using trust management. In: 2009 16th IEEE International Conference on Image Processing (ICIP), pp. 3057–3060, Nov 2009
8. Xin, K., Wu, Y.: A trust-based pollution attack prevention scheme in peer-to-peer streaming networks. Comput. Netw. **72**, 62–73 (2014)
9. Lin, W.S., Zhao, H.V., Liu, K.J.R.: Attack-resistant collaboration in wireless video streaming social networks. In: 2010 IEEE Global Telecommunications Conference (GLOBECOM 2010), pp. 1–4. IEEE (2010)
10. Cigno, R.L., Russo, A., Carra, D.: On some fundamental properties of P2P push/pull protocols. In: ICCE 2008 Second International Conference on Communications and Electronics, pp. 67–73. IEEE (2008)
11. Magharei, N., Rejaie, R.: Understanding mesh-based peer-to-peer streaming. In: Proceedings of the 2006 international Workshop on Network and Operating Systems Support for Digital Audio and Video, p. 10. ACM (2006)
12. Marti, S., Garcia-Molina, H.: Identity crisis: anonymity vs reputation in P2P systems. In: Proceedings of Third International Conference on Peer-to-Peer Computing, (P2P 2003). pp. 134–141. IEEE (2003)
13. Medina-López, C., García-Ortiz, J.P., Naranjo, J.A.M., Casado, L.G., González-Ruiz, V.: IPTV using P2PSP and HTML5+WebRTC. In: The Fourth W3C Web and TV Workshop. Munich, Germany, March 2014
14. Medina-López, C., Naranjo, J.A.M., García-Ortiz, J.P., Casado, L.G., González-Ruiz, V.: Execution of the P2PSP protocol in parallel environments. In: Guillermo Botella y Alberto A. Del Barrio Garcia, editor, Actas XXIV Jornadas de Paralelismo (http://www.congresocedi.es/images/site/actas/ActasParalelismo.pdf), pp. 216–221. Madrid, Septiembre 2013
15. Liu, Y., Yang, S., Guo, L., Chen, W., Guo, L.: A distributed trust-based reputation model in p2p system. In: 2007 Eighth ACIS International Conference on Software Engineering, Artificial Intelligence, Networking, and Parallel/Distributed Computing, SNPD 2007, vol. 1, pp. 294–299, July 2007
16. Salama, P., Shroff, N.B., Coyle, E.J., Delp, E.J.: Error concealment techniques for encoded video streams. In: International Conference on Image Processing, vol. 1, pp. 9–9. IEEE Computer Society (1995)
17. P2PSP Team. Peer to Peer Straightforward Protocol. http://p2psp.org/en/p2psp-protocol
18. Vieira, A.B., Campos, S., Almeida, J.: Fighting attacks in P2P live streaming. simpler is better. In: IEEE INFOCOM Workshops 2009, pp. 1–2. IEEE (2009)

CISIS 2015-SS01: Digital Identities Management by Means of Non-Standard Digital Signatures Schemes

Group Signatures in Practice

V. Gayoso Martínez, L. Hernández Encinas and Seok-Zun Song

Abstract Group signature schemes allow a user to sign a message in an anonymous way on behalf of a group. In general, these schemes need the collaboration of a Key Generation Center or a Trusted Third Party, which can disclose the identity of the actual signer if necessary (for example, in order to settle a dispute). This paper presents the results obtained after implementing a group signature scheme using the Integer Factorization Problem and the Subgroup Discrete Logarithm Problem, which has allowed us to check the feasibility of the scheme when using big numbers.

Keywords Cryptography · Digital signature · Group signature · Java

1 Introduction

The concept of group signatures was first proposed by Chaum and van Heyst in 1991 [1]. Group signatures allow a certain group to sign a message such that only one member of the group computes the signature on behalf of the whole group.

The main properties that must satisfy a group signature scheme are the following:

(i) Only one member signs the message on behalf of the group.
(ii) The receiver of the message can verify that its associated signature was generated by a member of the signer group, but he or she cannot determine which member of the group was the actual signer.
(iii) In case it is necessary, it must be possible to determine which group member was the actual signer of the message.

V.G. Martínez · L.H. Encinas (✉)
Institute of Physical and Information Technologies (ITEFI),
Spanish National Research Council (CSIC), Madrid, Spain
e-mail: luis@iec.csic.es

V.G. Martínez
e-mail: victor.gayoso@iec.csic.es

S. Song
Department of Mathematics, Jeju National University, Jeju, Korea
e-mail: szsong@jejunu.ac.kr

© Springer International Publishing Switzerland 2015
Á. Herrero et al. (eds.), *International Joint Conference*, Advances in Intelligent
Systems and Computing 369, DOI 10.1007/978-3-319-19713-5_35

There exist several proposals involving group signatures and their applicability to different scenarios (e.g., [2–7]). Some of these proposals need a Key Generation Center (KGC), denoted by C, or a Trusted Third Party (TTP), denoted by $\mathcal{T}$, at least for the initialization process. Other schemes, however, allow any user to create the group they choose to belong to. The actions performed by C and $\mathcal{T}$ are similar and, for this reason, the roles of both entities are usually considered equivalent.

As a general rule, the cryptographic primitives used by those proposals base their security on computationally-intractable mathematical problems such as the Integer Factorization Problem (IFP) and the Subgroup Discrete Logarithm Problem (SDLP). As it is well known, the IFP can be described as follows [8]: Given a positive integer n, find its prime factorization; that is, write $n = p_1^{e_1} p_2^{e_2} \dots p_k^{e_k}$ where the p_i are pairwise distinct primes and each $e_i \geq 1$. Besides, the SDLP is defined as follows [8]: Let p be a prime and q a prime divisor of $p - 1$. Let us consider g a generator of the unique subgroup G of $\mathbb{Z}_p^*$ of order q, and y an element in G. The problem is that of computing the integer x, $0 \leq x \leq q - 1$, such that $y = g^x \pmod{p}$.

For example, the schemes described in [9–14] base their security in the random oracle model, the strong RSA problem and the Decisional (bilinear) Diffie-Hellman problem. Other schemes, such as [15–17], are identity-based group signature protocols. They mainly use bilinear maps, such as bilinear pairings, which tend to be heavy in terms of computational load. In our case, we have employed a construction based on the SDLP which is different to those proposed in the aforementioned references.

This work presents the results obtained when implementing a modified version of a group signature scheme previously designed by one of the authors [18]. The goal of the modification, as it will become clear when describing the scheme, is to facilitate the implementation in devices with limited resources [19, 20]. In our scheme, $G = \{U_1, U_2, \dots, U_t\}$ is the group of users allowed to perform signatures. Those users share a public key, and at the same time they have different private keys. When a signature is needed, the element playing the role of the KGC, C, randomly chooses a member of the group so that member can perform the signature on behalf of G. After that, the verifier of the signature can check if the signature was performed by one of the members of G by using the public key that all members share. Without further information, the verifier will not be able to identify who was the original signer, unless the verifier is the PKC.

The rest of this paper is organized as follows: In Sect. 2, a detailed description of the group signature scheme is included. Section 3 describes the Java application developed in order to test the feasibility of the scheme. In Sect. 4, we offer to the readers the experimental results obtained with that application. Finally, our conclusions are presented in Sect. 5.

2 Description of the Scheme

Let $G = \{U_1, U_2, \ldots, U_t\}$ be the group of users allowed to perform signatures, and C the element acting as the Key Generation Center. The following subsections describe all the details of the group signature scheme.

2.1 Setup Phase

In this phase, C generates the system parameters, its own private key, the public key shared by the group, and the private keys of all members of G [21]. The steps that C must complete are the following ones:

1. C chooses two large primes p and q, such that $p = u_1 r p_1 + 1$ and $q = u_2 r q_1 + 1$, where r, p_1, q_1 are prime numbers and $u_1, u_2 \in \mathbb{Z}$ with $\gcd(u_1, u_2) = 2$; that is, $u_1 = 2v_1, u_2 = 2v_2$, where v_1 and v_2 are prime numbers. In the original version [18], v_1 and v_2 could be composite numbers; we have introduced this modification so that the number of factors of $\lambda(n)$ (see next step) does not depend on v_1 and v_2, which improves the iteration through the divisors of $\lambda(n)$ in the third step.
   In order to guarantee the security of the scheme, the bit length of r is selected so that the SDLP of order r in $\mathbb{Z}_n^*$ is computationally infeasible.
2. C computes the values $n = pq$, $\phi(n) = (p-1)(q-1) = u_1 u_2 r^2 p_1 q_1$, and $\lambda(n) = \mathrm{lcm}(p-1, q-1) = 2v_1 v_2 r p_1 q_1$, where $\phi(n)$ is the Euler function and $\lambda(n)$ is the Carmichael function.
3. C selects an element $\alpha \in \mathbb{Z}_n^*$ with multiplicative order r modulo n, such that $\gcd(\alpha, \phi(n)) = 1$. The element α can be efficiently computed as C knows the factorization of n and consequently it knows $\phi(n)$ and $\lambda(n)$.
   In practice, it is enough to find a random value $g \in \mathbb{Z}_n^*$ such that $g^{\lambda(n)} \equiv 1 \pmod{n}$ and check that none of the 62 non-trivial divisors of $\lambda(n)$ are the actual order of g [21]. By non-trivial divisor we mean a divisor of $\lambda(n)$ different from 1 or $\lambda(n)$. The number of non-trivial divisors of $\lambda(n)$ is derived from the fact that $\lambda(n) = 2v_1 v_2 r p_1 q_1$ and all the factors are prime numbers. Once the value g is found, the generator is obtained through the following computation [21]:

$$\alpha = g^{\lambda(n)/r} \pmod{n}.$$

4. C generates a secret random number $s \in \mathbb{Z}_r^*$ and determines

$$\beta = \alpha^s \pmod{n}.$$

5. C publishes the values n, r, α, and β, while the elements p, q, and s are kept secret.
6. C sets its private key by generating four random numbers $a_0, b_0, c_0, d_0 \in \mathbb{Z}_r^*$.

7. C determines the shared public key for G by computing

$$P = \alpha^{a_0}\beta^{b_0} \pmod{n} = \alpha^{a_0+sb_0} \pmod{n},$$
$$Q = \alpha^{c_0}\beta^{d_0} \pmod{n} = \alpha^{c_0+sd_0} \pmod{n}.$$

8. C computes the integers $h, k \in \mathbb{Z}_r$ such that $h = a_0 + sb_0 \pmod{r}$ and $k = c_0 + sd_0 \pmod{r}$.

9. C determines the private key for each signer $U_i \in G$, $1 \le i \le t$, where each private key is the tuple (a_i, b_i, c_i, d_i) and $a_i, b_i, c_i, d_i \in \mathbb{Z}_r$.

In order to do that, C first generates t pairs of random numbers, $b_i, d_i \in \mathbb{Z}_r$. Then, it obtains the remaining elements by using the following equations:

$$a_i = h - sb_i \pmod{r}, \quad c_i = k - sd_i \pmod{r}.$$

Once C has obtained the private keys of all the users, it distributes them to the signers via some secure channel.

2.2 Parameter and Key Verification

Each member of the signer group, U_i, $1 \le i \le t$, may check the parameters of the system by verifying that $\alpha \ne 1 \pmod{n}$ and $\alpha^r = 1 \pmod{n}$.

Moreover, each signer, U_i, $1 \le i \le t$, may verify that their private key is related to the shared public key, by checking:

$$P = \alpha^{a_i}\beta^{b_i} \pmod{n}, \quad Q = \alpha^{c_i}\beta^{d_i} \pmod{n}. \tag{1}$$

2.3 Group Signature Generation

Let M be the message to be signed by a member of G. By using, for example, a public hash function of the SHA-2 family [22], either the signing user or C compute $\mathfrak{h}(M) = m$, where m represents the hash output.

In order to calculate the signature, U_i must obtain the values f_i and g_i that compose the signature in this way:

$$f_i = a_i + c_i m \pmod{r}, \quad g_i = b_i + d_i m \pmod{r}. \tag{2}$$

After that, the group signature for the message M, which is $(f, g) = (f_i, g_i)$, can be published.

2.4 Group Signature Verification

Any verifier knowing the message, M, the hash function, $\mathfrak{h}$, the public key of the group G, (P, Q), and the group signature, (f, g), can check if the signature is valid through the following computation:

$$PQ^m = \alpha^f \beta^g \pmod{n}. \tag{3}$$

Equation (3) can be justified from expressions (1)–(2):

$$\alpha^f \beta^g \pmod{n} = \alpha^{a_i + mc_i} \beta^{b_i + mc_i} = \alpha^{a_i} \beta^{b_i} \left(\alpha^{c_i} \beta^{d_i} \right)^m = PQ^m.$$

2.5 Disclosure of the Signing User

In case it is necessary, C is able to verify the signature and determine the actual signer, as C knows all the private keys of the users belonging to the group G.

If the original message is M, its associated hash code is $m = \mathfrak{h}(M)$, and the corresponding group signature is the pair (f, g), C needs to iterate the following loop:

$$\left. \begin{array}{l} \bar{f}_i = a_i + c_i m \pmod{r} \\ \bar{g}_i = b_i + d_i m \pmod{r} \end{array} \right\} \quad 1 \leq i \leq t, \tag{4}$$

stopping whenever it finds an index j, such that $(\bar{f}_j, \bar{g}_j) = (f, g)$. Using this procedure, C determines that the actual signer was U_j.

3 Java Implementation of the Scheme

The group signature scheme presented in this contribution has been implemented as a Java application using Java SE 8. The application is composed of three panels which are described in detail in the next subsections. In each panel, the user has the option of converting the data from decimal (or text, in the case of the message to be signed) to hexadecimal and vice versa.

In all the cases where a random number is needed, the application uses the standard Java classes BigInteger [23] and Random [24], so the requested number is obtained through the following code:

```
Random random = new Random();
BigInteger number = new BigInteger(numBits,random);
```

In the previous code, the element numBits indicates that the desired number must be uniformly distributed over the range 0 to $2^{numBits} - 1$. Regarding the Random class, it uses a 48-bit seed which is modified using a linear congruential formula according to the method described in Sect. 3.2.1 of [25].

Whenever a random prime number is needed, the following code is used after obtaining a random number:

```
BigInteger prime = number.nextProbablePrime();
```

By calling the method nextProbablePrime() over the element number, the application obtains the first integer greater than number that is probably prime, where the probability that the number returned is composite does not exceed 2^{-100} [23].

3.1 Parameters Panel

This panel includes the general parameters, the KGC's private key and the group's public key, as it can be seen in Fig. 1. More specifically, it includes text boxes for the private elements p, q, s, a_0, b_0, c_0, and d_0 and the public elements n, r, α, β, P, and Q.

There are four buttons available in this panel:

- *Generate*: It computes all the parameters according to the steps 1–8 of the procedure described in Sect. 2.1.
- *Save*: It allows the user to save either the public data or all the data included in this panel. The information is stored in a file using an XML structure.
- *Load*: It allows the user to overwrite the data existing in the text boxes with the information stored in the XML file selected by the user.
- *Clear*: It deletes the content of all the text boxes pertaining to this panel.

3.2 Users Panel

This panel includes the private keys of the four users managed by this application. It is important to point out that the number of users implemented in this version of the application is not a limitation of the scheme, but a figure selected in order to simplify the usage of the application. For each user from $i = 1$ to 4, a set consisting of the associated values a_i, b_i, c_i, and d_i is displayed, as it can be seen in Fig. 1.

The four buttons available in this panel implement the following functionality:

- *Generate*: It generates all the private elements associated to the private keys of the users according to the step 9 of the procedure described in Sect. 2.1.
- *Save*: It allows the user to save the private elements of the four users in a file using an XML structure.

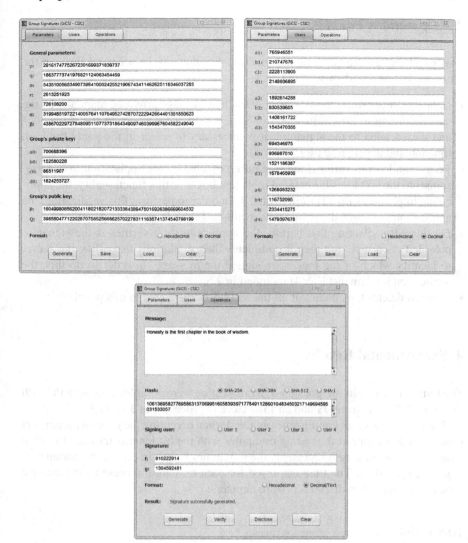

Fig. 1 View of the three panels of the application

- *Load*: It allows to overwrite the data existing in the text boxes with the information stored in the XML file selected by the user.
- *Clear*: It deletes the content of all the text boxes displayed in this panel.

3.3 Operations Panel

This panel includes the operational functionality that can be accessed through the following buttons, as displayed in Fig. 1:

- *Generate*: It generates the signature of the text message provided manually by the user according to Eq. (2) included in Sect. 2.3. In order to obtain the elements f and g associated to the signature, it is necessary to select in the panel the hash function and the signing user.
- *Verify*: It allows to verify if the signature provided by the application user corresponds to the text message entered in its text box. For this functionality, the application implements Eq. (3) from Sect. 2.4. In order to perform this verification, the application only uses the public data available in the *Parameters* panel.
- *Disclose*: By selecting this button, the application first verifies if the signature provided is correct, and then it completes the calculations that allow the server to identify which of the four users signed the message. In order to do this, the application implements Eq. (4) included in 2.5.
- *Clear*: It deletes the content of all the text boxes displayed in this panel.

4 Experimental Results

The tests whose results are presented in this section were completed using a PC with Windows 7 Professional OS and an Intel Core i7 processor at 3.40 GHz.

Table 1 includes the running time obtained when executing the general parameters generation procedure in the testing computer with the bit lengths indicated in each case, where the bit length represents the maximum length in bits of the parameters r, p_1, q_1, v_1, and v_2. The time displayed for each bit length represents the average time of the generation of 100 sets of parameters.

Table 1 General parameters generation running time

Length (bits)	32	64	96	128	160	192
Time (seconds)	0.30	4.24	25.21	57.19	128.83	300.57

As expected, the running time has an exponential shape, as it can be seen in Fig. 2.

Even though the running time obtained for large bit lengths is considerable, it is important to point out that the process of generating the general parameters is executed only once during the life cycle of that set of parameters. In the rest of operations (signature generation and verification, and signature disclosure) the running time obtained is less than 5 milliseconds even for parameters of 192 bits.

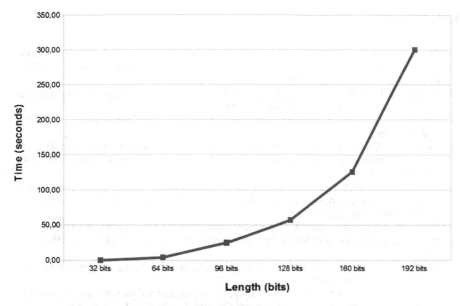

Fig. 2 General parameters' generation running time

5 Conclusions

In this contribution we present a modification of the first group signature scheme described in [18]. In order to implement the scheme as a Java application, we have modified the scheme by adding a new requirement which mandates v_1 and v_2 to be both prime numbers, as explained in Sect. 2.1. With this modification, we force the number of non-trivial divisors of $\lambda(n)$ to be exactly 62, which facilitates the implementation in devices with limited resources as the application does not need to factor v_1 and v_2 in order to determine the actual number of non-trivial divisors of $\lambda(n)$.

The tests performed with the application allow us to confirm the expected difficulty in generating the system parameters for bit lengths greater than 64 bits. Nevertheless, as the system parameters generation procedure is only executed once by the PKC, it is not a limitation for deploying the group signature service in devices with limited resources, as they must only perform the signature generation and verification procedures.

Acknowledgments This work has been partially supported under the framework of the international cooperation program managed by National Research Foundation of Korea (NRF-2013K2A1 A2053670) and by Comunidad de Madrid (Spain) under the project S2013/ICE-3095-CM (CIBER-DINE).

References

1. Chaum, D., van Heyst, E.: Group signatures. Lect. Notes Comput. Sci. **547**, 257–265 (1991)
2. Camenisch, J., Stadler, M.: Efficient group signature schemes for large groups. Lect. Notes Comput. Sci. **1296**, 410–424 (1997)
3. Camenisch, J., Michels, M.: Separability and efficiency for generic group signature schemes. Lect. Notes Comput. Sci. **1666**, 413–430 (1999)
4. Bresson, E., Stern, J.: Efficient revocation in group signature. Lect. Notes Comput. Sci. **2001**, 190–206 (1992)
5. Chung, Y.F., Chen, T.L., Chen, T.S., Chen, C.S.: A study on efficient group-oriented signature schemes for realistic application environment. Int. J. Innovative Comput. Inform. Control **8**(4), 2713–2727 (2012)
6. Ogawa, K., Ohtake, G., Fujii, A., Hanaoka, G.: Weakened anonymity of group signature and its application to subscription services. IEICE Trans. Fundam. Electron. Commun. Comput. Sci. **E97-A**(6), 1240–1258 (2014)
7. Emura, K., Miyaji, A., Omote, K.: An r-hiding revocable group signature scheme: group signatures with the property of hiding the number of revoked users. J. Appl. Math. **2014**, 1–14 (2014)
8. Menezes, A.J., van Oorschot, P.C., Vanstone, S.A.: Handbook of Applied Cryptography. CRC Press Inc, Boca Raton (1996)
9. Ateniese, G., Camenisch, J., Joye, M., Tsudik, G.: A practical and provably secure coalition-resistant group signature scheme. Lect. Notes Comput. Sci. **2000**, 255–270 (1880)
10. Ateniese, G., de Medeiros, B.: Efficient group signatures without trapdoors. Lect. Notes Comput. Sci. **2894**, 246–268 (2003)
11. Nguyen, L., Safavi-Naini, R.: Efficient and provably secure trapdoor-free group signature schemes from bilinear pairings. Lect. Notes Comput. Sci. **3329**, 89–102 (2004)
12. Shoup, V., Gennaro, R.: Securing threshold cryptosystems against chosen ciphertext attack. J. Cryptol. **15**(2), 75–96 (2002)
13. Camenisch, J., Lysyanskaya, A.: Signature schemes and anonymous credentials from bilinear maps. Lect. Notes Comput. Sci. **3152**, 56–72 (2004)
14. Boneh, D., Boiyen, X., Shacham, H.: Short group signatures. Lect. Notes Comput. Sci. **3152**, 41–55 (2004)
15. Han, S., Wang, J., Liu, W.: An efficient identity-based group signature scheme over elliptic curves. Lect. Notes Comput. Sci. **3262**, 417–429 (2004)
16. Tan, Z.: An improved identity-based group signature scheme. Lect. Notes Comput. Sci. **3262**, 417–429 (2004)
17. Li, L., De-gong, D., Ying-liang, D.: An improved identity-based group signature scheme. In: International Conference on Information Technology, Computer Engineering and Management Sciences 2011 (ICM 2011). Vol. 2, pp. 269–271 (2011)
18. Díaz, D.R., Hernández, E.L., Muñoz, M.J.: Two proposals for group signature schemes based on number theory problems. Logic J. IGPL **21**(4), 630–647 (2013)
19. Potzmader, K., Winter, J., Hein, D., Hanser, C., Teufl, P., Chen, L.: Group signatures on mobile devices: practical experiences. Lect. Notes Comput. Sci. **7904**, 47–64 (2013)
20. Spreitzer, R., Schmidt, J.M.: Group-signature schemes on constrained devices: the gap between theory and practice. In: First Workshop on Cryptography and Security in Computing Systems (CS2'14), 31–36 (2014)
21. Susilo, W.: Short fail-stop signature scheme based on factorization and discrete logarithm assumptions. Theoret. Comput. Sci. **410**(8), 736–744 (2009)
22. NIST: Secure Hash Standard. National Institute of Standard and Technology, Federal Information Processing Standard Publication, FIPS, pp. 180–4 (2012)

23. Oracle Corporation: BigInteger (Java Platform SE 8). http://docs.oracle.com/javase/8/docs/api/java/math/BigInteger.html. (2014)
24. Oracle Corporation: Random (Java Platform SE 8). http://docs.oracle.com/javase/8/docs/api/java/util/Random.html. (2014)
25. Knuth, D.E.: The Art of Computer Programming, Vol. 2 (3rd Ed.): Seminumerical Algorithms. Addison-Wesley Longman Publishing Co., Inc, Boston (1997)

Non-conventional Digital Signatures and Their Implementations—A Review

David Arroyo, Jesus Diaz and Francisco B. Rodriguez

Abstract The current technological scenario determines a profileration of trust domains, which are usually defined by validating the digital identity linked to each user. This validation entails critical assumptions about the way users' privacy is handled, and this calls for new methods to construct and treat digital identities. Considering cryptography, identity management has been constructed and managed through conventional digital signatures. Nowadays, new types of digital signatures are required, and this transition should be guided by rigorous evaluation of the theoretical basis, but also by the selection of properly verified software means. This latter point is the core of this paper. We analyse the main non-conventional digital signatures that could endorse an adequate tradeoff between security and privacy. This discussion is focused on practical software solutions that are already implemented and available online. The goal is to help security system designers to discern identity management functionalities through standard cryptographic software libraries.

1 Introduction

Identity management is one of the most challenging matters in communication networks. Although it is possible to reach a conclusion about physical identity, it is not so easy to establish a relationship between a physical identity and a digital identity. Cryptography provides a means to associate a digital identity to an user on the grounds of asymmetric cryptography. In this guise, a pair of keys (the public and the private keys) are generated such that each component of this pair is connected to the other, but it is not computationally possible to use one of them to obtain the other.

D. Arroyo (✉) · J. Diaz · F.B. Rodriguez
Grupo de Neurocomputacion Biologica, Departamento de Ingenieria Informatica,
Escuela Politecnica Superior, Universidad Autonoma de Madrid, Madrid, Spain
e-mail: david.arroyo@uam.es

J. Diaz
e-mail: j.diaz@uam.es

F.B. Rodriguez
e-mail: f.rodriguez@uam.es

© Springer International Publishing Switzerland 2015
Á. Herrero et al. (eds.), *International Joint Conference*, Advances in Intelligent
Systems and Computing 369, DOI 10.1007/978-3-319-19713-5_36

If one uses her private key, then the encrypted information can be only decrypted by means of the related public key. Correspondingly, if an user sends a (*hashed*) message and the message encrypted with her private key, the previous procedure leads to a verification of both the integrity of the message and the correctness of the private key. Loosely speaking, this procedure depicts the way the basic digital signatures are generated. Digital signatures are *publicly verifiable* and *transferable* cryptographic primitives, and they also have the property of non-repudiation [33, Chap. 1]. These are the main properties of what we can call *conventional digital signatures*.

The combination of non-repudiation and transferability are not always required and they are even replaced by a *deniability* commitment. For example, this is the case of scenarios as e-voting and e-coin where a user is interested on proving the authenticity of a piece of information to a certain receiver, but she wants to prevent the sender from proving this fact to other parties. Besides, in other situations it is demanded to design a signature scheme where a message or a document can be signed by multiple users (for example, if we are dealing with a committee that has to endorse as a whole a document). As it is underlined in [12], there are more than 60 digital signatures models. The classification of those models is not an easy task, since the discern between the underlying properties is far from a straightforward operation. Regarding the implementation of the different digital signatures models, this fact incorporates an additional risk. Certainly, one of the major problems in the design of security software is drawn by non-complete description of basic assumptions and their implications in concrete application contexts. This task should be based on the identification of standard products offering the functionalities that our design demands [19]. Indeed, software standards are products that have been carefully evaluated, which implies that we can assume a reasonable certainty about their reliability.

In this paper we review the most relevant families of *non-conventional digital signatures*, being the core of our effort to identify software libraries sustaining each one of the considered schemes. We discuss the main properties and applications of the considered digital signatures. In some cases we show that there does not exist well-founded and properly evaluated software libraries. However, it is possible to establish a set of basic cryptographic primitives and software libraries as a means to finally implement the referred digital signatures.

2 Group Signatures

Group signatures allow members of a group of signers to issue signatures on behalf of the group that can be verified without telling which specific member issued it [15]. These schemes typically include a group manager (*GM*) responsible for setting up the group and, sometimes, managing it. The main functionality is issuing signatures (*sign*) and *verify*ing them. Additionally, *GM* can *open* a group signature to retrieve the identity of its issuer. These schemes endow users with anonymous authentication and unlinkability. However, there are also schemes to enable signatures *linkability*

[47] or *traceability* [34]. It is also possible to *revoke* a member's private key, preventing her to issue new signatures. This may be done by publishing the trapdoor used for *open*, publishing the trapdoor used as *trace*, or just enabling an authority for answering this kind of status requests. Some schemes modify this basic scenario, by dividing *GM* in multiple authorities for the tasks related to managing the group [6]; or add new authorities for performing other delicate tasks, like tracing revoked group members [34].

Group signatures provide an anonymity degree proportional to the group size. Thus, they are typically used for privacy-respectful authentication in anonymous certificates [5], e-voting [47], e-cash [38], and anonymous attestation [10].

Standards and Implementations. Group signatures have been standardised by the ISO/IEC [29] (general setting and main operations) and [30] (which defines a total of 7 schemes with opening and linking capabilities). In [5, 20] extended X.509 certificates are used to manage digital identities based on group signatures. Several implementations of group signature schemes are currently available online. [8] is implemented in C within the PBC_sig library[1] and using Python within the Charm framework[2] [3]; [6] is implemented in C in the FTMGS library[3]; [4, 11] are implemented in the libgs library using Java, as part of the PP2db project[4]; and the extensible libgroupsig C library[5] implements [8, 16, 34], and allows the addition of new group signature schemes.

3 Ring Signatures

As an alternative to group signatures, ring signatures can be considered to have a more flexible solution where anonymity revocation is not possible. Ring signatures were first introduced in [42], and further contributions incorporate additional controls on the original proposal. Thus, in some specific contexts it is necessary to determine whether two signatures have been created by the same group member, which is addressed by the so-called linkable ring signatures [37]. In other situations it is convenient to adopt traceable ring signatures to trace the origin of two signatures with respect to the same metainformation or tag [22]. The main difference between group and ring signatures is given by the initial setup and the possibility of conforming ad hoc groups. Ring signatures are not ruled by a central authority and there is no need

---

[1] http://crypto.stanford.edu/pbc/sig/.
[2] https://code.google.com/p/charm-crypto/.
[3] http://www.lcc.uma.es/~vicente/swprj/index.html#libftmgs.
[4] http://www.ing.unibs.it/ntw/tools/pp2db/.
[5] https://bitbucket.org/jdiazvico/libgroupsig.

for an initial setup. Moreover, if one adheres to ring signatures, then groups can be generated without an extra cost derived from re-organization (i.e., a new setup to include the new members of the group). This characteristic is of major importance to tackle non-closed groups in e-cash, e-voting, and e-token systems [47]. In fact, here we have to acknowledge the efforts from the bitcoin-related community. Ring signatures are key components of P2P electronic cash infrastructures where the existence of a trusted central authority cannot be taken for granted. Privacy preserving social networks also demand procedures to validate a piece of information as originated by a certain group and to avoid identity forgery attacks (as sockpuppetry or sybil attacks), and correspondingly there are some recent proposals applying linkable ring signatures for such a goal [39].

Standards and Implementations. Along with group signatures, ring signatures are one of the cryptographic means to manage users' anonymity. This being the case, the most relevant standard for ring signatures is given by [29, 30]. Regarding software implementation of ring signatures, we have to underline the variant of [23] provided in cryptonote's library,[6] the inclusion of a Python version of [17] in Charm, the C implementation given in the PBC library for [49], and the software counterpart of [36] in the Crypto-book prototype.[7]

4 Blind Signatures

A blind signature scheme allows a user U to obtain a signature from a signer S over any arbitrary message m, but without S learning anything about m [14]. There are variants of this basic behavior, like partially blind signatures [1], allowing to include a message known by both signer and user; restrictive blind signatures [9], which only allow the issuance of a blind signature for messages that comply certain rules; finally, in fair blind signatures [45] an authority has privileged information allowing the signer to link message and signature pairs. Usually, a blind signing protocol is a three step process. First, during the *blinding* step, the user blinds the message to be signed with the help of a random value. This blinded value is then sent to the signer, who performs some verifications upon it, *signs* the blinded message and sends the result to the user. Finally, the user generates the final signature applying the random value used to blind the message to *unblinds* the received token. Blind signatures offer a distinction between authentication and token assignment, which is of major importance for creating privacy respectful protocols, like e-cash [32] and e-voting [35].

---

[6]https://github.com/AlbertWerner/cryptonotecoin.
[7]https://github.com/jyale/crypto-book/.

Standards and Implementations. For blind signatures, there is currently an undergoing effort by the ISO/IEC to standardise the general setting, entities and processes [27], along with the discrete logarithm based mechanisms [28]. As for available implementations, the Bouncy Castle Java library[8] includes the class `RSABlinding` `Engine` for [14] blind signatures. Many current systems use the basic variant of blind signatures (like OpenCoin[9]) and, since it is quite simple to program given a working RSA implementation, there seems to be a lack of independent open source libraries. There also does not seem to exist open source implementations of any of the derivatives of blind signatures.

5 Multi and Aggregate Signatures

In a multisignature n signers create a signature over a message m, such that it is possible to verify that all of them have participated in the signing process [31]. While one naive way for achieving this will be to have each signer attach her conventional signature (e.g., using RSA), this has the drawback of producing multisignatures that are of linear size in the number of participants and with linear verification time (also depending on the participants). In a multisignature scheme there are n signing steps (one for each signer), and a verification process, which is run independently of the number of signers.

Standards and Implementations. Harn et al. [24] highlights that the ISO/IEC 14888-2 standard [26] can be used to build identity-based multisignatures. Besides, naive multisignatures can be implemented using conventional digital signatures.

6 Threshold Signatures

In these schemes, at least t signers out of n need to collaborate in order to produce a valid signature, composing what is called a (t, n)-threshold signature scheme [18]. Group signatures and multisignatures can be seen as $(1, n)$ and (n, n) threshold signature schemes, respectively. Besides signers and verifiers, it is also frequent to find a special entity, the *combiner*, who gathers all the shares and joins them to produce the final signature. Therefore, the processes in a threshold signature scheme are: a *signing* algorithm through which each of the signers produces a signature share; a *combination* process (which may be performed by the signers, verifier, or by the combiner), which merges all the available shares (that must be at least t in a (t, n)

---

[8]https://www.bouncycastle.org/.
[9]http://opencoin.org/.

scheme); and the *verify* algorithm, determining whether the signature produced after combining the shares is valid.

Standards and Implementations. *Threshsig*[10] implements [44] in Java, and *Threshold_ECDSA*[11] implemented in Java an ECDSA based threshold scheme, although it does not seem to be available at present. Finally, Bitcoin uses a simple approach for threshold signatures for contract signing.[12]

7 Proxy Signatures

Proxy signatures allow a user U_1 (the delegate) to sign a message on behalf of another user U_2 (the delegator), if a trusted proxy cooperates [40]. Proxy re-signatures [7] allow a proxy to convert a valid signature by U_1 over a given message into a valid signature by U_2 over the same message. In proxy signature and proxy re-signature schemes, there are delegators, delegates and a proxy required to convert signatures. The proxy and the delegatee must run two separate processes *psign* and *dsign*, respectively, in order to produce the final signature on behalf of the delegator. In proxy re-signatures, the equivalent to these two processes is named *resign*, and it is executed by the proxy. Also, the function *rekey* creates the necessary keys for the proxy to be able to perform the transformation. The main application of proxy signatures is on delegating signing capabilities. In [25] proxy re-signatures are also proposed for authenticated routing.

Standards and Implementations. While we have found publications reporting analysis on implementation of specific schemes (like [46], but which does not make the code available), we have not been able to locate either standards or implementations worth mentioning for this primitive.

8 Signatures of Knowledge

A conventional digital signature proves a statement of the form *"The issuer of this signature, with public key PK, knows the corresponding private key SK"*. Signatures of knowledge extend this, allowing to issue digital signatures proving knowledge

---

[10]https://code.google.com/p/threshsig. As a work of an undergraduate student, however, it is no longer maintained. See http://www.metzdowd.com/pipermail/cryptography/2013-November/018674.html.

[11]http://nssl.eew.technion.ac.il/files/Projects/Threshold_ECDSA/html/doc/ECDssSignature.html.

[12]https://en.bitcoin.it/wiki/Contracts. In Bitcoin *multisig* transactions are specified through the CHECKMULTISIGVERIFY opcode and demand n valid signatures out of m for a transaction to be approved. This is actually an approach for threshold signatures, although it requires n separate signatures (instead of just one).

of witnesses for any NP statement [13]. Specifically, for any NP language L, given a statement $x \in L$ and a witness w proving it, a signature of knowledge provides a signing algorithm $\sigma = sign(m, w, x, L)$ which creates a signature σ of m over $x \in L$, that can be read as "*Someone knowing a witness to* $x \in L$ *is sending the message* m". This can be verified with the verification counterpart algorithm $verify(\sigma, m, x, L)$. Signatures of knowledge allow creating privacy respectful signatures, since there is no need to *leak* any additional information besides the knowledge of some fact. In turn, this enables important functionalities demanded, for example, to implement delegate credentials.

8.1 Standards and Implementations

There are no standards for this primitive. Additionally, we have not been able to find direct implementations either. However, it is worth noting that signatures of knowledge can be easily constructed from Zero-Knowledge proof systems using the technique explained in [21] (this is usually called Non-Interactive Zero-Knowledge proofs, or NIZK proofs).

9 Identity-Based Signatures

Identity-based signatures eliminate the need of distributing public keys, allowing the verification of digital signatures just from the identity that the signer claims to own [43]. For this, initial schemes relied on a trusted Private Key Generators (PKG), which are basically trusted entities generating the private keys used for signing that produce public key independence. However, recent proposals have reduced the trust placed in this entity, by allowing the detection of dishonest actions on its behalf (thus, addressing the key escrow problem) [48]. Identity-based signature schemes include *signing* and *verifying* processes, the latter requiring the identity of the signer instead of her public key. Additionally, the mentioned schemes reducing the trust in the PKG add another process for checking if the signature has been generated by a dishonest PKG.

Standards an Implementations. In [24] it is highlighted that the standard ISO/IEC 14888-2 [26] can be applied to derive identity-based multisignatures. Concerning implementations, there is an implementation of the [41] scheme in the JPBC library[13] and implementations of several schemes in Cayrel's website.[14]

---

[13]http://gas.dia.unisa.it/projects/jpbc/schemes/ibs_ps06.html.
[14]http://cayrel.net/?Implementation-of-code-based-zero.

10 Conclusion

The laws of identity have a plethora of implications in its general scope, but even more in the specific context of the digital realm. Along this paper we have distinguished some of the most relevant properties of digital identities in today communication networks. We have conducted a survey to expose some important contributions both from the theoretical and practical point of view. In Table 1 we show a summary of the standards and implementations cited along this paper. This list in some cases leads to highlight the lack of software proposals and/or formal standards. We hypothesise that the lack of implementations may be due to the fact that system designers (usually computer science engineers, not cryptographers) are not typically aware of these new digital signature schemes. Consequently, system designers resort to conventional methods to implement the required functionality, creating a circular dependency that could be problematic unless those conventional primitives are efficiently implemented and rigorously tested.

However, since the transition from formal definition of cryptographic primitives to cryptography engineering should be done through a rigorous evaluation process, standardisation is not an option but a commitment. Consequently, we should claim the absence of software libraries for certain digital signatures schemes as a call and an urging. This need should be guided by a rigorous evaluation process on the grounds of formal and computational models, but also taking as bottom line basic cryptographic libraries validated by the cryptographic community and broadly accepted. This is the case of GMP,[15] MIRACL,[16] Cryptopp,[17] and Ben Lynn's library for Pairing Based Cryptography.[18] These libraries contain the most fundamentals symmetric and asymmetric cryptosystems, but there also exist libraries providing implementations of Zero-Knowledge proofs,[19] crytographic commitments[20] and Oblivious Transfers.[21] Finally, it is necessary to comment on current efforts to adequate digital signatures to low computational-power environments. Certainly, before incorporating any cryptographic library we should realised a performance study using benchmarks in the vein of [2]. Here, it is relevant the NaCL library,[22] since it contains some important lightweight digital signatures implementations for ARM architectures.

---

[15] https://gmplib.org/.

[16] http://docs.certivox.com/docs/miracl.

[17] http://www.cryptopp.com/.

[18] http://crypto.stanford.edu/pbc/.

[19] https://www.peloba.de/index.php/zk-library/?lang=en.

[20] git://git-crysp.uwaterloo.ca/polycommit, http://scapi.readthedocs.org/.

[21] https://github.com/JHUISI/charm/releases.

[22] http://nacl.cr.yp.to/.

Table 1 Reviewed signature types and related standards and implementations. If there are standards or implementations that we have not located/referenced, please contact us.

Signature type	Standardization efforts	Implementations
Group signatures	[5, 20, 29, 30]	Extensible libraries
Ring signatures	[5, 29, 30]	Specific schemes
Blind signatures	[27, 28]	Specific schemes
Multi signatures	[26] (related)	Unknown
Threshold signatures	Unknown	Specific schemes
Proxy signatures	Unknown	Unknown
Signatures of knowledge	Unknown	Unknown
Identity-based signatures	[26] (related)	Specific schemes

Acknowledgments This work was supported by Comunidad de Madrid (Spain) under the project S2013/ICE-3095-CM (CIBERDINE) and the Spanish Government project TIN2010-19607.

References

1. Abe, M., Fujisaki, E.: How to date blind signatures. In: ASIACRYPT (1996)
2. Abusharekh, A.: Comparative analysis of software libraries for public key cryptography
3. Akinyele, J.A., Garman, C., Miers, I., Pagano, M.W., Rushanan, M., Green, M., Rubin, A.D.: Charm: a framework for rapidly prototyping cryptosystems. J. Cryptographic Eng. 3(2), 111–128 (2013)
4. Ateniese, G., Camenisch, J.L., Joye, M., Tsudik, G.: A practical and provably secure coalition-resistant group signature scheme. In: Bellare, M. (ed.) CRYPTO 2000. LNCS, vol. 1880, pp. 255–270. Springer, Heidelberg (2000)
5. Benjumea, V., Choi, S.G., Lopez, J., Yung, M.: Anonymity 2.0 – X.509 extensions supporting privacy-friendly authentication. In: Bao, F., Ling, S., Okamoto, T., Wang, H., Xing, C. (eds.) CANS 2007. LNCS, vol. 4856, pp. 265–281. Springer, Heidelberg (2007)
6. Benjumea, V., Choi, S.G., Lopez, J., Yung, M.: Fair traceable multi-group signatures. In: Financial Cryptography, pp. 231–246 (2008)
7. Blaze, M., Bleumer, G., Strauss, M.J.: Divertible protocols and atomic proxy cryptography. In: Nyberg, K. (ed.) EUROCRYPT 1998. LNCS, vol. 1403, pp. 127–144. Springer, Heidelberg (1998)
8. Boneh, D., Boyen, X., Shacham, H,.: Short group signatures. In: Franklin, M, (ed.) CRYPTO 2004. LNCS, vol. 3152, pp. 41–55. Springer, Heidelberg (2004)
9. Brands, S.: Untraceable off-line cash in wallets with observers. In: Stinson, D.R. (ed.) CRYPTO 1993. LNCS, vol. 773, pp. 302–318. Springer, Heidelberg (1994)
10. Brickell, E.F., Camenisch, J., Chen, L.: Direct anonymous attestation. In: Proceedings of the 11th ACM Conference on Computer and Communications Security, CCS 2004, pp. 132–145. Washington, DC, USA, 25–29 Oct 2004
11. Camenisch, J., Groth, J.: Group signatures: Better efficiency and new theoretical aspects. In: 4th International Conference Security in Communication Networks 2004, Italy, Sept 8–10, 2004, Revised Selected Papers. pp. 120–133 (2004)
12. Cao, Z., Liu, M.: Classification of signature-only signature models. Sci. China Ser. F: Inform. Sci. 51(8), 1083–1095 (2008)
13. Chase, M., Lysyanskaya, A.: On signatures of knowledge. In: Dwork, C. (ed.) CRYPTO 2006. LNCS, vol. 4117, pp. 78–96. Springer, Heidelberg (2006)

14. Chaum, D.: Blind signatures for untraceable payments. In: CRYPTO (1982)
15. Chaum, D., van Heyst, E.: Group signatures. In: Davies, D.W. (ed.) EUROCRYPT 1991. LNCS, vol. 547, pp. 257–265. Springer, Heidelberg (1991)
16. Choi, S.G., Park, K., Yung, M.: Short traceable signatures based on bilinear pairings. In: Yoshiura, H., Sakurai, K., Rannenberg, K., Murayama, Y., Kawamura, S. (eds.) IWSEC 2006. LNCS, vol. 4266, pp. 88–103. Springer, Heidelberg (2006)
17. Chow, S.S.M., Yiu, S.-M., Hui, L.C.K.: Efficient identity based ring signature. In: Ioannidis, J., Keromytis, A.D., Yung, M. (eds.) ACNS 2005. LNCS, vol. 3531, pp. 499–512. Springer, Heidelberg (2005)
18. Desmedt, Y.G., Frankel, Y.: Shared generation of authenticators and signatures. In: Feigenbaum, J. (ed.) CRYPTO 1991. LNCS, vol. 576, pp. 457–469. Springer, Heidelberg (1992)
19. Diaz, J., Arroyo, D., Rodriguez, F.B.: A formal methodology for integral security design and verification of network protocols. J. Syst. Softw. **89**, 87–98 (2014)
20. Diaz, J., Arroyo, D., Rodriguez, F.B.: New x.509-based mechanisms for fair anonymity management. Comput. Secur. **46**, 111–125 (2014)
21. Fiat, A., Shamir, A.: How to prove yourself: practical solutions to identification and signature problems. In: Odlyzko, A.M. (ed.) CRYPTO 1986. LNCS, vol. 263, pp. 186–194. Springer, Heidelberg (1987)
22. Fujisaki, E.: Sub-linear size traceable ring signatures without random oracles. IEICE Trans. Fundam. Electron. Commun. Comput. Sci. **95**(1), 151–166 (2012)
23. Fujisaki, E., Suzuki, K.: Traceable ring signature. In: Public Key Cryptography—PKC 2007, Proceedings of the 10th International Conference on Practice and Theory in Public-Key Cryptography, pp. 181–200. China, 16–20 April 2007
24. Harn, L., Ren, J., Lin, C.: Efficient identity-based GQ multisignatures. Int. J. Inf. Sec. **8**(3), 205–210 (2009)
25. Hohenberger, S.: Advances in Signatures, Encryption, and E-Cash from Bilinear Groups. Massachusetts Institute of Technology, Department of Electrical Engineering and Computer Science (2006)
26. ISO 148888-2: Information technology—Security techniques—Digital signatures with appendix—Part 2: Integer factorization based mechanisms (2014)
27. ISO/IEC 18370-1: Information technology—Security techniques—Blind digital signatures—Part 1: General (2015)
28. ISO/IEC 18370-2: Information technology—Security techniques—Blind digital signatures—Part 2: Discrete logarithm based mechanisms (2014)
29. ISO/IEC 20008-1: Information technology—Security techniques—Anonymous digital signatures—Part 1: General (2013)
30. ISO/IEC 20008-2: Information technology—Security techniques—Anonymous digital signatures—Part 2: Mechanisms using a group public key (2013)
31. Itakura, K., Nakamura, K.: A public-key cryptosystem suitable for digital multisignatures. NEC J. Res. Dev. (1983)
32. Juels, A., Luby, M., Ostrovsky, R.: Security of blind digital signatures. In: Kaliski, B.S., Jr (ed.) CRYPTO 1997. LNCS, vol. 1294, pp. 150–164. Springer, Heidelberg (1997)
33. Katz, J.: Digital Signatures. Advances in Information Security. Springer, US (2010)
34. Kiayias, A., Tsiounis, Y., Yung, M.: Traceable signatures. In: Cachin, C., Camenisch, J.L. (eds.) EUROCRYPT 2004. LNCS, vol. 3027, pp. 571–589. Springer, Heidelberg (2004)
35. Kucharczyk, M.: Blind signatures in electronic voting systems. In: Kwiecień, A., Gaj, P., Stera, P. (eds.) CN 2010. CCIS, vol. 79, pp. 349–358. Springer, Heidelberg (2010)
36. Liu, J.K., Wei, V.K., Wong, D.S.: Linkable spontaneous anonymous group signature for ad hoc groups. In: Information Security and Privacy. Springer (2004)
37. Liu, J.K., Wong, D.S.: Linkable ring signatures: security models and new schemes. In: Gervasi, O., Gavrilova, M.L., Kumar, V., Laganá, A., Lee, H.P., Mun, Y., Taniar, D., Tan, C.J.K. (eds.) ICCSA 2005. LNCS, vol. 3481, pp. 614–623. Springer, Heidelberg (2005)
38. Lysyanskaya, A., Ramzan, Z.: Group blind digital signatures: a scalable solution to electronic cash. In: Hirschfeld, R. (ed.) FC 1998. LNCS, vol. 1465, pp. 184–197. Springer, Heidelberg (1998)

39. Maheswaran, J., Wolinsky, D.I., Ford, B.: Crypto-book: an architecture for privacy preserving online identities. In: Twelfth ACM Workshop on Hot Topics in Networks, HotNets-XII, p. 14. College Park, MD, USA, 21–22 Nov 2013

40. Mambo, M., Usuda, K., Okamoto, E.: Proxy signatures for delegating signing operation. In: CCS '96, Proceedings of the 3rd ACM Conference on Computer and Communications Security, pp. 48–57. India, 14–16 March 1996

41. Paterson, K.G., Schuldt, J.C.N.: Efficient identity-based signatures secure in the standard model. In: Batten, L.M., Safavi-Naini, R. (eds.) ACISP 2006. LNCS, vol. 4058, pp. 207–222. Springer, Heidelberg (2006)

42. Rivest, R.L., Shamir, A., Tauman, Y.: How to leak a secret. In: Boyd, C. (ed.) ASIACRYPT 2001. LNCS, vol. 2248, pp. 552–565. Springer, Heidelberg (2001)

43. Shamir, A.: Identity-based cryptosystems and signature schemes. In: Blakely, G.R., Chaum, D. (eds.) CRYPTO 1984. LNCS, vol. 196, pp. 47–53. Springer, Heidelberg (1985)

44. Shoup, V.: Practical threshold signatures. In: Preneel, B. (ed.) EUROCRYPT 2000. LNCS, vol. 1807, pp. 207–220. Springer, Heidelberg (2000)

45. Stadler, M.A., Piveteau, J.-M., Camenisch, J.L.: Fair blind signatures. In: Guillou, L.C., Quisquater, J.-J. (eds.) EUROCRYPT 1995. LNCS, vol. 921, pp. 209–219. Springer, Heidelberg (1995)

46. Tang, S., Xu, L.: Proxy signature scheme based on isomorphisms of polynomials. In: Xu, L., Bertino, E., Mu, Y. (eds.) NSS 2012. LNCS, vol. 7645, pp. 113–125. Springer, Heidelberg (2012)

47. Tsang, P.P., Wei, V.K.: Short linkable ring signatures for e-voting, e-cash and attestation. In: Deng, R.H., Bao, F., Pang, H., Zhou, J. (eds.) ISPEC 2005. LNCS, vol. 3439, pp. 48–60. Springer, Heidelberg (2005)

48. Yuen, T.H., Susilo, W., Mu, Y.: How to construct identity-based signatures without the key escrow problem. Int. J. Inf. Sec. **9**(4), 297–311 (2010)

49. Zhang, F., Kim, K.: ID-based blind signature and ring signature from pairings. In: Zheng, Y. (ed.) ASIACRYPT 2002. LNCS, vol. 2501, pp. 533–547. Springer, Heidelberg (2002)

Using Smart Cards for Authenticating in Public Services: A Comparative Study

D. Arroyo Guardeño, V. Gayoso Martínez, L. Hernández Encinas and A. Martín Muñoz

Abstract Smart cards are well-known tamper-resistant devices, and as such they represent an excellent platform for implementing strong authentication. Many services requesting high levels of security rely on smart cards, which provide a convenient security token due to their portability. This contribution analyses two Spanish smart card deployments intended to be used for accessing eGoverment services, comparing their respective contents and capabilities.

Keywords Cryptography · Digital signature · Electronic prescription · Smart cards

1 Introduction

Secure electronic identification is an important enabler of data protection and the prevention of online fraud. These aspects have a great importance in areas such as eGovernment, which consists of the digital interactions between government, citizens, public agencies, and employees. In this scenario, the European Commission's eGovernment Action Plan 2011–2015 supports the provision of a new generation of eGovernment services, as well as the strengthening of services already deployed [1].

As a measure of its importance, only in Spain more than 480 million administrative procedures were conducted by citizens and companies with the central

D. Arroyo Guardeño
Grupo de Neurocomputación Biológica, Departamento de Ingeniería Informática,
Escuela Politécnica Superior, Universidad Autónoma de Madrid, Madrid, Spain
e-mail: david.arroyo@uam.es

V. Gayoso Martínez (✉) · L. Hernández Encinas · A. Martín Muñoz
Institute of Physical and Information Technologies (ITEFI),
Spanish National Research Council (CSIC), Madrid, Spain
e-mail: victor.gayoso@iec.csic.es

L. Hernández Encinas
e-mail: luis@iec.csic.es

A. Martín Muñoz
e-mail: agustin@iec.csic.es

© Springer International Publishing Switzerland 2015   437
Á. Herrero et al. (eds.), *International Joint Conference*, Advances in Intelligent
Systems and Computing 369, DOI 10.1007/978-3-319-19713-5_37

government in 2013, of which over 367 million (76.5 %) were conducted electronically and over 112 million (23.5 %) by other means. For enterprises, 94 % of administrative procedures were done electronically and for citizens 65 % [2]. Among the services for citizens most widely used in Spain we can find those related to income taxes (declarations, notifications of assessment, etc.), social security benefits (unemployment, pensions, etc.), the request of personal documents (passports, driving licences, etc.), and health related services (appointments for hospitals, etc.) [2].

This contribution analyses two smart card deployments for the authentication of users in public services, comparing their respective characteristics and capabilities: the Spanish electronic identity card (known as DNIe, *Documento Nacional de Identidad electrónico*), and the smart card delivered to part of the medical doctors working at the Community of Madrid for the electronic prescription service, which in this contribution will be referred to as the EPSC (Electronic Prescription Smart Card).

The rest of this paper is organized as follows: Sect. 2 provides a brief introduction to smart cards. Section 3 shows the details of the DNIe. In Sect. 4, the main features of the EPSC are included. Section 5 offers a comparison of both smart cards. Finally, our conclusions are presented in Sect. 6.

2 Smart Cards

A smart card is a plastic card with an embedded chip that controls the access to the stored data. The two most widespread communication models for smart cards are the byte-oriented, half duplex transmission protocol T=0 and the block-oriented, half duplex protocol T=1, both defined in ISO/IEC 7816-3 [3]. The T=1 protocol is newer, and implements error detection capabilities.

The elements known as APDU (Application Protocol Data Unit), built according to the ISO/IEC 7816-3 [3] and 7816-4 [4] specifications, are the data packets exchanged between the external application and the card by means of a smart card reader. The card operating system is responsible for analysing any incoming APDU and redirecting it to the application it is intended for. The operating system is also responsible for retrieving the response data from the card application and submitting it to the external application using the card reader.

There are two types of APDUs: command and response. Figure 1 shows the format of command APDUs, which consist of a header and optionally a body with the following elements:

- CLA (1 byte): Command class.
- INS (1 byte): Specific instruction within the class.
- P1 (1 byte): First parameter associated to the instruction. It can be used to give more information about the instruction, or as input data.

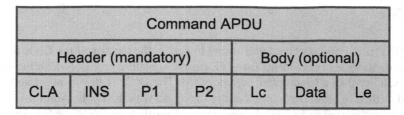

Fig. 1 Command APDU

- P2 (1 byte): Second parameter associated to the instruction. As in the previous case, it can be used to give more information about the instruction, or as input data.
- Lc (1 byte, optional): Number of bytes in the data field of the command. Since its highest value is 0xFF, the maximum data length is 255 bytes, although some cards allow to send 256 bytes using the value 0x00.
- Data (variable size, optional): Information to be processed by the applet.
- Le (1 byte, optional): Maximum number of bytes to be included in the data field of the response APDU.

In comparison, the format of any response APDU is simpler (see Fig. 2), as it only includes the following items:

Response APDU		
Body (optional)	Result (mandatory)	
Data	SW1	SW2

Fig. 2 Response APDU

- Data (variable length, optional): Information returned by the card application.
- SW1 (1 byte): First status byte, which provides general information about the result of the command execution.
- SW2 (1 byte): Second status byte.

Following the ISO/IEC 7816 notation, the smart card file structure is represented by means of two types of elements: DF (Dedicated File) and EF (Elementary File). While DFs can be interpreted as directories or folders of a standard file system, EFs can be considered to be data files, belonging either to the operating system or to other smart card applications.

3 DNIe

The DNIe is a T=0 smart card that allows to certify the identity of the DNIe holder and to digitally sign documents using electronic signature protocols with the same legal validity than a handwritten signature [5]. The DNIe is the soundest and preferred method to prove one's own identity in any act with the Public Administration. Since it was started to be issued, more than 43 millions of DNIe cards have been delivered to citizens [6].

In January 2015, it was announced a new version of the DNIe, called DNIe 3.0, witch incorporates an NFC (Near Field Communications) chip with the goal to facilitate its usage with smartphones and tablets, avoiding the limitation of delivering smart card readers to potential users.

Figure 3 (left) shows the file tree of the DNIe, where the Master File (typically represented as the DF 3F00) is the root directory of the file system.

The information stored in the chip is divided into three areas with different access levels and security conditions [7–9]:

- Public area: Reading access without restrictions. It includes, among others, the following files:

  - EF 601F: X.509 component certificate (each DNIe has a different component certificate associated to the actual smart card), with an RSA public key of 1024 bits.
  - EF 6020: X.509 certificate of the component intermediate CA (Certification Authority), with an RSA public key of 1024 bits.
  - EF 7006: X.509 certificate of the DGP (*Dirección General de la Policía*) intermediate CA, with an RSA public key of 2048 bits.

- Private area: Reading access allowed after validation of the citizen's PIN (Personal Identification Number) code. Some of the files included are the following:

  - EF 7004: X.509 user signing certificate with an RSA key of 2048 bits.
  - EF 7005: X.509 user authentication certificate with an RSA public key of 2048 bits.

- Security area: Reading access allowed only after biometric verification. In order to make this verification, the citizen must use the biometric devices located at the DNIe issuing offices. The files protected by this procedure are the following:

  - EF 7001: Citizen's filiation data (name, surname, date of birth, etc.).
  - EF 7002: Digitized image of the citizen's photograph.
  - EF 7003: Digitized image of the citizen's handwritten signature.

Given that the DNIe is a device linked to the identity of the citizens, its security, both physical and electronic, is of paramount importance. In that sense, the DNIe is a SSCD (Secure Signature Creation Device) compliant with the European standard EN 14890-1 [10], which defines how to establish a communication between the SSCD and an external application. Because of that, the operating system of the

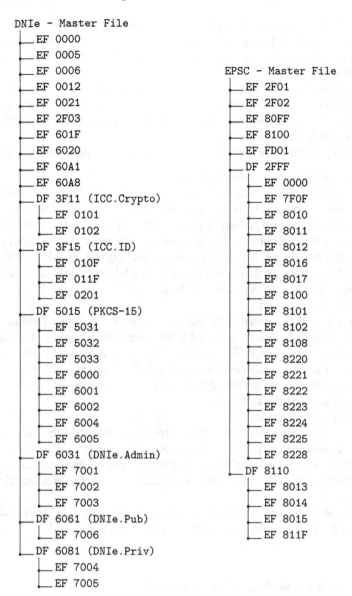

Fig. 3 File structure of the DNIe (left) and EPSC (right)

DNIe subordinates the sending of certain APDUs (like the Verify PIN command) to the establishment of a secure channel [11].

In order to establish the secure channel, it is necessary to exchange the public keys of the card and the external application that wants to communicate with the DNIe. After those certificates are verified by both parties, they must perform a mutual authentication protocol, including a seed exchange for the derivation of the encryption and MAC (Message Authentication Code) session keys. Once the secure channel is established, any command must be protected before its transmission using the session keys.

In the descriptions that follow, the word *terminal* represents the pair formed by the software application that intends to communicate with the DNIe and the physical device where the application is executed, while the word *card* is used as an alternative to the terms DNIe and smart card.

The establishment of the secure channel consists of the following four phases:

1. Certificate exchange and verification:
   Before starting the mutual authentication process, the terminal must send its authentication certificate to the card, so the DNIe can verify that the certificate is correct and has been properly signed by a trusted certification authority. If that is the case, the application will request the card component certificate in order to verify it. Once this exchange is completed, the application will have the public key and the certificate associated to the card, whilst the card will have obtained the public key and the certificate of the terminal.

2. Internal authentication:
   In this phase, the terminal must request the card to perform a valid authentication. In order to do this, the terminal must generate a random number that is sent to the card as a challenge. The card uses this value to generate in turn its contribution to the session key creation process. If the terminal is able to recover that value, this means that the data provided by the card was valid and that the card has been successfully authenticated.

3. External authentication:
   After the two previous phases, the terminal has identified the card as a valid DNIe. In order to complete the mutual authentication procedure, it is necessary to perform now the external authentication process, so the terminal is authenticated by the card, following a process similar to that of the previous phase.

4. Session key generation:
   The last step consists in calculating the encryption and MAC keys that will be used in the communication through the secure channel.

4 Electronic Prescription Smart Card

EPS (Electronic Prescribing and Dispensing) is the term that identifies the system and processes that allow all the stages of the prescribing, supply of medicine, and claiming process to be completed electronically, providing an alternative to the typical paper based prescription system in public health environments.

EPS enables prescribers (mainly medical doctors) to create, sign, and send prescriptions electronically to a dispenser (such as a pharmacy) of the patient's choice or to a central server, from where they can be electronically retrieved by any dispenser.

This makes the prescribing and dispensing process more efficient and convenient both for patients and the medical staff. EPS is a key initiative currently being implemented or already deployed in many countries (e.g. United Kingdom [12], Australia [13], etc.), and the European Union is focusing now in developing the interoperability of electronic prescriptions [14].

In Spain, the electronic prescription system being rolled out is not yet completed. As of December 2014, 89.58 % of general health centers, 52.56 % of local clinics, 66.21 % of specialized centers, and 89.35 % of pharmacies were already working with the new system nationwide, though the figures vary a lot between Autonomous Communities [15] (for example, at the beginning of 2014 it was already implemented in Communities such as Andalucía or Galicia, while in other Communities such as Murcia it is expected to be rolled out during 2015 [16]). In the Community of Madrid, where we have made our study, the new system started to work across all the region at the end of 2014 [17].

In the current phase of the EPS deployment in the Community of de Madrid, medical doctors sign the prescriptions electronically using their login credentials. However, in next phases this system is expected to be replaced by a strong authentication scheme based on smart cards, and with that goal medical doctors have received their own, individual smart cards. As mentioned in the Introduction, in this contribution we will refer to this smart card as EPSC, which is a T=1 smart card.

Figure 3 (right) shows the complete file structure of the EPSC. All the files included in the figure can be read without verifying the user's PIN. The most interesting files are the following ones:

- EF 2F02: It includes the serial number of the smart card.
- EF 8223: This file contains details about the user, mainly the name, surname, and NIF (*Número de Identificación Fiscal*, the identification number of each Spanish citizen consisting of a sequence of 8 digits and a letter associated with that sequence. The NIF is the identification number displayed at the DNIe).
- EF 8224: It includes details about the intermediate CA.
- EF 8228: This file stores all the elements in the certificate chain up to the user's certificate, as it is displayed in Fig. 4. Camerfirma is a company participated by more than 85 spanish Chambers of Commerce [18], and that is part of Chambersign, a European organization that provides support to national Chambers of Commerce from a supranational standpoint [19]. All three certificates shown in Fig. 4 are related to RSA keys of 2048 bits.

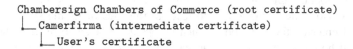

Fig. 4 EPSC certificate chain

Unlike the DNIe, it is not necessary to establish a secure channel before sending some APDUs like the Verify PIN command. Apart from that, it is interesting to point out that many files have an empty content (i.e., sequences of the byte 0x00). Presumably, the content of those files will be updated once the smart cards begin to be used.

Another difference between DNIe and EPSC is that, while the former follows the indications given by the PKCS #15 (Public-Key Cryptography Standards) standard [20], the file structure of the latter is not compliant with that specification.

5 Comparison

The chip mounted by the DNIe received the EAL5+ (Evaluation Assurance Level 5 augmented) accreditation in 2005 [21]. Besides, version 1.13 of the DNIe was evaluated as a smart card by the National Cryptologic Centre (CCN, *Centro Criptológico Nacional*), a government organization that belongs to the National Centre of Intelligence (CNI, *Centro Nacional de Inteligencia*), and obtained the EAL4+ (Evaluation Assurance Level 4 augmented) certification in 2007 [22].

The DNIe is equipped with several physical elements that allow to consider it a highly secure physical token. Among those items, we can highlight the following [7]: holograms, kinegrams, optically variable inks, changeable laser images, surface relief structures, inks visible only under infrared or ultraviolet light, citizen's photography and handwritten signature engraved with laser, etc. In comparison, the EPSC is not equipped with the physical countermeasures just mentioned, and the details about its accreditation are not public.

Regarding their cryptographic capabilities, both the DNIe and EPSC contain the user's X.509 certificate as well as an associated 2048-bit RSA key pair intended for performing digital signatures.

However, the security of the DNIe is much more robust, as it only allows to send certain commands after establishing a secure channel which authenticates both the smart card and the software communicating with it. In the DNIe, the retrieval of certain elements, such as the user's certificates, can only be accomplished after correctly entering the PIN code. As the APDU containing the Verify PIN command can only be sent through the secure channel, an attacker cannot access the content of the user's certificates unless he knows the PIN code and establishes the secure channel. Another consequence of this scheme is that no attacker can block the PIN code of the DNIe without completing the process that sets up the secure channel.

In comparison, the EPSC allows to read all the files of its file system without the need of entering the PIN code, which allows an attacker to retrieve the user's certificate if he has access to the smart card. Besides, the attacker could block the PIN code, which would render the legitimate user unable to make signatures unless he was in possession of the PUK (Personal Unlocking Key) code.

Regarding the Spanish legislation associated to digital signatures, we remind the readers that the Law 59/2003 establishes the following concepts [5]:

- Electronic signature: It is the set of electronic data that can be utilized as a means of identifying the signing user.
- Advanced electronic signature: It is the electronic signature that allows the signing user to be identified. The signature must be created by methods that the signing user can keep under his exclusive control.
- Qualified electronic signature: It is the advanced electronic signature based on a qualified certificate and generated by a secure signature creation device.

Based on those definitions, the DNIe can be considered as a device that allows to generate qualified electronic signatures. The EPSC, unless fully accredited, would have to be considered as a device allowing to generate only advanced electronic signatures which could not be used in another environments.

6 Conclusions

As described in the previous sections, the DNIe is a highly secure, certificated smart card that allows to generate qualified electronic signatures. Even though some technical information about the EPSC is not available to researchers, it can be considered a less robust authentication device.

If we take into account that the latest version of the DNIe includes support for NFC devices, it seems reasonable to suggest the use of the DNIe instead of the EPSC for the task of signing the electronic prescriptions. This decision would be doubly beneficial: on the one hand, it would allow to avoid the cost of purchasing the smart cards and delivering them to the medical doctors; on the other hand, it would allow doctors working with NFC-capable devices such as modern tablets and smartphones to avoid installing smart card readers, which additionally represent an avoidable cost for health centres that had not purchased them before.

Finally, from a standards perspective, the DNIe fully adapts to the PKCS #15 structure, which facilitates interoperability with future applications.

Acknowledgments This work has been partially supported by Comunidad de Madrid (Spain) under the project S2013/ICE-3095-CM (CIBERDINE).

References

1. European Commission: European eGoverment action plan 2011–2015. https://ec.europa.eu/digital-agenda/en/european-egovernment-action-plan-2011-2015 (2011)
2. European Commission: eGoverment in Spain. https://joinup.ec.europa.eu/sites/default/files/41/be/69/eGov%20in%20ES%20-%20February%202014%20v.16.0.pdf (2014)

3. ISO/IEC: Identification cards—Integrated circuit cards—Part 3: Cards with contacts—Electrical interface and transmission protocols. International Organization for Standardization/International Electrotechnical Commission. ISO/IEC 7816–3, 3rd edn. (2006)

4. ISO/IEC: Identification cards—Integrated circuit cards—Part 4: Organization, security and commands for interchange. International Organization for Standardization/International Electrotechnical Commission. ISO/IEC 7816-4, 3rd edn. (2013)

5. Fábrica Nacional de Moneda y Timbre: Electronic signature. https://www.sede.fnmt.gob.es/en/normativa/firma-electronica

6. Direccinó General de la Policía: Página oficial de la DGP—Notas de prensa. http://www.policia.es/prensa/20150112_1.html (2015)

7. Espinosa García, J., Hernández Encinas, L., Queiruga Dios, A.: The new Spanish electronic identity card: DNI-e. In: Proceedings of the International Conference on Information Technologies (InfoTech'2007), volume I: Technological Aspects of the e-Governance and Data Protection, pp. 77–82 (2007)

8. Ministerio del Interior de España: Portal oficial sobre el DNI electrónico. http://www.dnielectronico.es/ (2013)

9. Gayoso Martínez, V., Hernández Encinas, L., Martín Muñoz, A.: La tarjeta de identidad española como método de autenticación en redes sociales. In: VII Congreso Iberoamericano de Seguridad Informática 2013 (CIBSI 2013), pp. 32–44 (2013)

10. CEN: Application interface for smart cards used as secure signature creation devices—Part 1: basic services. European Committee for Standardization. UNE-EN 14890–1. http://www.cen.eu/cen/Sectors/TechnicalCommitteesWorkshops/CENTechnicalCommittees/Pages/Standards.aspx?param=6205Żtitle=CEN%2FTC+224. (2009)

11. Ministerio del Interior de España: Documento Nacional de Identidad electrónico: Guía de referencia técnica (2012)

12. Health and Social Care Information Centre: Electronic prescription service. http://systems.hscic.gov.uk/eps (2015)

13. Department of Human Services: Electronic prescribing and dispensing of medicines. http://www.medicareaustralia.gov.au/provider/pbs/pharmacists/dispense.jsp (2014)

14. eHealth Governance Initiative: Guidelines on ePrescription. http://www.ehgi.eu/Lists/Posts/Attachments/15/eHealth%20Network%205%20%5BAthens%202014%5D%20Topic%206%20-%20Discussion%20Paper%20-%20Guidelines%20on%20ePrescription.pdf (2014)

15. Oficina para la Ejecución de la Reforma de la Administración: Informe de Progreso de la Comisión para la Reforma de las Administraciones Públicas. http://www.seap.minhap.gob.es/dms/es/areas/reforma_aapp/proceso/CORA-Informe-anual-de-progreso--Diciembre-2014/CORA-Informe%20anual%20de%20progreso.%20Diciembre%202014.pdf (2015)

16. Sociedad Española de Farmacia Comunitaria: Estudio de implantacin de la receta electrónica en España. http://www.farmaceuticoscomunitarios.org/sites/default/files/wysiwyg/181_brizuelarodicioluis_recetaelectronica_recetaelectronica.pdf (2014)

17. Gaceta Médica: Madrid culmina la implantación de la e-receta. http://www.gacetamedica.com/noticias-medicina/articulo.aspx?idart=884387Żidcat=797Żtipo=2 (2015)

18. Camerfirma: About us. http://www.camerfirma.com/camerfirma/quienes-somos/ (2015)

19. Chambersign: About us. http://www.chambersign.com/about-us (2015)

20. ISO/IEC: Identification cards—Integrated circuit cards—Part 15: cryptographic information application. International Organization for Standardization/International Electrotechnical Commission. ISO/IEC 7816-15 (2004)

21. ANSSI: Produit certifié Critères Communs (réf: 2005/40). Agence Nationale de la Sécurité des Systèmes d'Information. http://www.ssi.gouv.fr/fr/produits-et-prestataires/produits-certifies-cc/certificat_2005_40.html (2005)

22. CCN: Informe de certificación INF-148. Centro Criptológico Nacional. http://www.dnielectronico.es/seccion_integradores/cc1_p.jpg (2007)

Reducing the Key Space of an Image Encryption Scheme Based on Two-Round Diffusion Process

Alberto Peinado and Andrés Ortiz García

Abstract In this work, we analyze the security of the image encryption scheme proposed by Norouzi et al. (Multimedia Syst 20:45–64, 2014) [1]. The encryption scheme is based on a hash function and a two-round diffusion process applied to each 8×8 patches that compose the entire image. The analysis reveals the existence of a huge number of weak keys, allowing us to compute the effective key length, thus reducing the key space from 2^{512} to 2^{195} keys.

Keywords Image encryption · Equivalent key · Effective key length

1 Introduction

In 2014, Norouzi et al. [1] propose an image encryption scheme that combines a hash function with two rounds of a diffusion process. Although it has been proposed for encryption, the operations defined over the 8×8 patches composing the entire image could also be applied for identification, since the encryption process is strongly dependent of the image content. However, although the scheme presents a good behavior, as the authors claim and justify in [1], we have analyzed the security of this scheme focusing our attention in the key space, since 512 bits secret keys are used, and no particular study appears regarding this aspect. The analysis we have performed is mainly applied to find weak and/or invalid keys.

Section 2 describes the encryption in [1]. Section 3 deals with the characterization of weak and invalid keys, allowing the computation, in Sect. 4, of the

A. Peinado (✉) · A.O. García
Universidad de Málaga, Andalucía Tech, E.T.S.Ingeniería de Telecomunicación,
Departamento Ingeniería de Comunicaciones, Campus de Teatinos, 29071 Málaga, Spain
e-mail: apeinado@ic.uma.es

A.O. García
e-mail: aortiz@ic.uma.es

Á. Herrero et al. (eds.), *International Joint Conference*, Advances in Intelligent
Systems and Computing 369, DOI 10.1007/978-3-319-19713-5_38

447

effective key length that reduces considerably the key space. Finally, conclusions are provided.

2 Norouzi's Image Encryption Scheme

The image encryption scheme proposed by Norouzi et al. [1] is a symmetric scheme based on the application of two rounds of a diffusion process that is applied to each 8×8 patch composing the entire image. It uses a 512 bits secret key and it is designed for 256 grayscale images (grayscale images with a bit depth of 8 bits/pixel), of any size and aspect ratio. However, the most-right column and bottom row must be padded to get a total number of rows and columns multiple of 8. Since the authors in [1] use 256×256 images for experimental purposes, we also consider, for simplicity reasons, 256×256 images to perform the security evaluation without loss of generality. We use the notations in [1].

2.1 Encryption Algorithm

Let $N = M = 256$ be the number of rows and columns, respectively. The encryption is applied in two rounds. The description follows the original definition in [1].

First round

Step 1: KEY GENERATION. The user's secret key is not directly applied for encryption. Instead, the 512 bit user's secret key is transformed on new 512 bits by means of salsa20 hash function [2]. The new 512 bits are rearranged in a 8×8 matrix, denoted by *hash.key*. In this way, the user's secret key is apparently isolated from the encryption process, in such a way that it cannot be deduced from the *hash.key*.

Step 2: INITIALIZATION. Since the main operation of the encryption scheme is applied on 8×8 patches, the plain-image and cipher-image are defined as arrays of 8×8 elements. Hence, the original (input) image I is transformed into a $8 \times NM/8$ array, denoted by P, whose content is partitioned into $NM/64$ blocks such as $P = \{Se_0, Se_1, ..., Se_{NM/64-1}\}$. In a similar way, the ciphered image will be composed by 8×8 blocks, all of them conforming a unique $8 \times NM/8$ array $C = \{c_0, c_1, ..., c_{NM/64-1}\}$, that finally will be reorganized to show a $N \times M$ image.

The ciphered image C is initialized with the content of P; that is $C = P$. Furthermore, the index i, defined for patch indexation, is initialized to 0.

Step 3: SECONDARY KEY GENERATION. Besides the global key *hash.key*, the encryption scheme applies different keys to different patches. These secondary keys depend on the global key and the current content of the other patches. The secondary keys are computed as follows

$$meanSe_i = \left(\left(\sum_{j=1}^{\frac{NM}{64}-1} \text{mean2}(c_j) \right) - \text{mean2}(c_i) \right) \frac{10^{14}}{4 \cdot 256^4} \tag{1}$$

$$keySe_i = \text{floor}(hash.key \cdot meanSe_i \bmod 256) \tag{2}$$

where $\text{mean2}(c_i)$ means the mean value of the block c_i. If $i = 0$, we go to step 4; else we go to step 5.

Step 4: FIRST BLOCK ENCRYPTION. The first block c_0 of the cipher-image is obtained using the following equation. Next go to step 6.

$$c_0 = Se_0 \oplus (hash.key + keySe_0 \bmod 256) \oplus keySe_0 \tag{3}$$

Step 5: BLOCKS ENCRYPTION. The blocks c_i are computed as

$$c_i = Se_i \oplus (hash.key + c_{i-1} \bmod 256) \oplus keySe_i \tag{4}$$

Step 6: $i \leftarrow i + 1$ and return to step 3 until i reaches $NM/64\text{-}1$. We set $d_1 = c_{NM/64-1}$.

Second round

Step 1: INITIALIZATION. In order to produce a better diffusion effect, the ciphered image C obtained at first round is transposed. The index i is reinitialized to 0. In this way, the 8×8 patches are processed in a different order.

Step 2: SECONDARY KEY GENERATION. This step is the same as that of first round. The keys $meanSe_i$ and $keySe_i$ are computed using Eqs. 1 and 2.

Step 3: FIRST BLOCK ENCRYPTION. If $i = 0$, the first block c_0 of the cipher-image is obtained using the following equation. Next go to step 5.

$$c_0 = c_0 \oplus (hash.key + d_1 \bmod 256) \oplus keySe_0 \tag{5}$$

Step 4: BLOCKS ENCRYPTION. If $i > 0$ then the blocks c_i are computed as

$$c_i = c_i \oplus (hash.key + c_{i-1} \bmod 256) \oplus keySe_i \tag{6}$$

Step 5: Let $i \leftarrow i + 1$ and return to step 3 until i reaches $NM/64\text{-}1$.

2.2 Decryption Algorithm

The decryption algorithm applied the same equations that encryption. The only difference is the order in which they are applied. Second round is applied first, beginning from $i = NM/64\text{-}1$ until $i = 0$. Next, the first round is applied beginning from $i = NM/64\text{-}1$.

3 Characterization of Weak Keys

The encryption algorithm described in [1] (Sect. 2) employs a secret key of 512 bits that is converted into a new 512 bits stream by means of salsa20 hash function. Since the output of salsa20 is uniformly distributed, we perform the security evaluation over the output of salsa20 hash function. Hence, keys identified in this section do not correspond to the user secret key. Nevertheless, the results are directly applicable to real cryptanalysis as one can see later. We consider key *hash. key* in the 8 × 8 matrix form with 8-bit coefficients.

3.1 Encryption Using All-Zero Key

Cryptanalysts use to guess the cryptographic keys using particular values, such as all-zero key, in order to detect flaws in the mathematical design. This is one of those cases. It is not probable that a user selects the all-zero value as his secret key, but it is very interesting to study the behavior of the encryption system for this particular key.

Apparently, nothing happens when encryption is applied (see Fig. 1). However, since the all-zero *hash.key* determines that $keySe_i = 0$, for all i (see Eqs. 1–2), the ciphered image loses much of its nonlinearity. The proof of this fact is that the cryptanalyst can decrypt the ciphered image using a different key.

Fig. 1 (a) 256 × 256 Lena image. (b) Encrypted image using all-zero key

a   b

We have performed experiments encrypting the "Lena" image with the all-zero key and trying to decrypt it using keys with nonzero coefficients (see Fig. 2). As one can observe, the recovered image presents a perceptible noise but the image is clearly recognizable by the observer. The greater number of nonzero coefficients the higher noise in the recovered image. The experiments reveal that more than 32 nonzero coefficients do not produce a usable image.

Fig. 2 Recovered images encrypted with all-zero key and decrypted using nonzero keys

3.2 Decryption Using All-Zero Key

The particular case described in Sect. 3.1 is not very interesting by itself, because it is not probable that a user selects the all-zero key. However, in this section, we show that the behavior of this image encryption scheme is symmetric in that case. That is, we find the reciprocal situation when the user selects a key with a considerable number of zero coefficients. As an example, if *hash.key* takes the value

$$\begin{pmatrix} 0 & 0 & 0 & 0 & 0 & 0 & 0 & 0 \\ 0 & 0 & 12 & 0 & 0 & 0 & 0 & 0 \\ 0 & 0 & 0 & 20 & 0 & 0 & 76 & 0 \\ 0 & 2 & 0 & 0 & 0 & 0 & 0 & 0 \\ 0 & 0 & 0 & 255 & 0 & 0 & 0 & 0 \\ 0 & 0 & 0 & 0 & 0 & 0 & 0 & 0 \\ 0 & 34 & 0 & 0 & 45 & 0 & 0 & 0 \\ 0 & 0 & 0 & 0 & 0 & 0 & 0 & 64 \end{pmatrix}, \tag{7}$$

the image can be decrypted using the all-zero value. Furthermore, this image can also be decrypted using different keys with certain level of sparsity. Hence, the encryption scheme shows the reciprocal behavior detected in Sect. 3.1.

Figure 3 illustrates the recovered images using keys with different number of nonzero coefficients to decrypt the image encrypted with the key of Eq. 7.

a

b

Fig. 3 (a) Encrypted Lena image using key in Eq. 7. (b) Decrypted image using all-zero key

It is important to note that this behavior is a strong weakness of the system because any image encrypted with sparse keys can be decrypted totally or partially using the all-zero key. As a consequence, all of those keys must be removed from the key space.

4 Computation of the Effective Key Length

The authors in [1] claim that the key space is 2^{512}, since 512 is the secret key bit length. Actually, the user chooses a 512 bit secret key that is internally converted on a different 512 bits stream by means of the hash function salsa20 [2]. Therefore, the user has 2^{512} possible values to select his secret key. On the other hand, the output of salsa20 hash function determines the internal key *hash.key* belonging to a theoretically key space of 2^{512} keys.

However, as it is described in the previous section, there is a huge number of weak keys that must be removed from the key space because the security is clearly compromised when they are applied. In other words, the weak keys characterized in Sect. 3 have to be considered invalid keys. In this way, the image encryption scheme could be still used.

The computation of the effective key length requires the identification (Sect. 3) and counting of all the invalid keys, so as to determine the number of valid ones. In general terms, the set of invalid keys is composed by all of them with a considerable level of sparsity at byte level, that is, with a considerable number of zero coefficients when the key is rearranged as 8×8 matrices of 8-bit coefficients. The experiments show that the keys having at least 32 zero coefficients can be considered invalid ones.

We can compute the invalid keys in the following way. We first consider only the keys with one nonzero coefficient. The cardinality of this set is 255×64 keys. The number of keys with exactly two nonzero coefficients is

$$255^2 \cdot \binom{64}{2} \text{ keys.} \tag{8}$$

For the general case, the number of keys with j nonzero coefficients is

$$255^j \cdot \binom{64}{j} \text{ keys.} \tag{9}$$

Therefore, the number of invalid keys, in the sense explained in Sect. 3, can be computed by the following equation.

$$\sum_{j=1}^{32} 255^j \binom{64}{j} \text{ keys,} \tag{10}$$

producing the value $1.8794e + 095$ that allows us to obtain the effective key length as

$$512 - \log_2(1.8794e + 095) = 195\text{bits.} \tag{11}$$

The effective length determines a new key space of 2^{195} keys.

5   Conclusions

The security analysis performed on the key space of the image encryption scheme proposed in [1] reveals the existence of a huge number of invalid keys. Furthermore, all of them can be considered equivalent keys, since any image encrypted with them can be decrypted by another key belonging to this group. In some cases, as one can observe in the figures, the image is not completely decrypted. The scheme must be modified in order to avoid the utilization of the invalid keys, because it is not possible to apply the restriction in the user secret keys. Note that the secret key is transformed into another key of the same length by means of salsa20 hash function. Further work is necessary to complete the characterization of invalid keys, and to provide an improvement of this scheme.

Acknowledgments This work has been supported by the MICINN under project "TUERI: Technologies for secure and efficient wireless networks within the Internet of Things with applications to transport and logistics", TIN2011-25452.

References

1. Norouzi, B., Seyedzadeh, S.M., Mirzakuchaki, S., Mosavi, M.R.: A novel image encryption based on hash function with only two-round diffusion process. Multimed. Syst. **20**, 45–64 (2014)
2. Bernstein, D.J.: "The Salsa20 Stream Cipher." SKEW2005, Symmetric Key Encryption Workshop, 2005, Workshop Record. http://www.ecrypt.eu.org/stream/salsa20p2.html

On the Difficult Tradeoff Between Security and Privacy: Challenges for the Management of Digital Identities

David Arroyo, Jesus Diaz and Víctor Gayoso

Abstract The deployment of security measures can lead in many occasions to an infringement of users' privacy. Indeed, nowadays we have many examples about surveillance programs or personal data breaches in online service providers. In order to avoid the latter problem, we need to establish security measures that do not involve a violation of privacy rights. In this communication we discuss the main challenges when conciliating information security and users' privacy.

1 Introduction

Along the second half of the twentieth century there was a technological revolution that has led to the so-called *information era*. Information technologies have evolved from having a marginal position to being a cornerstone of today's economical activities. Certainly, in these days it is almost impossible to conceive the daily work without the usage of technological means related to information management.

This capital prominence of Information and Communication Technologies (ICT) is consequently endorsed by the digital agenda of the national and transnational organisations. For example, in the digital agenda of the European Union it is underlined that economical progress is currently built upon the proper concretion and management of ICT [1]. In fact, the *health* of modern democracy is heavily dependent on the application of ICT to enlarge and ensure governability. The result of such an effort is oriented to what it is coined as *e-democracy* [26], which cannot be attained unless users trust and use ICT appropriately and respectfully.

D. Arroyo (✉) · J. Diaz
Departamento de Ingenieria Informatica, Escuela Politecnica Superior, Grupo de
Neurocomputacion Biologica, Universidad Autonoma de Madrid, Madrid, Spain
e-mail: david.arroyo@uam.es

J. Diaz
e-mail: jesus.diaz@uam.es

V. Gayoso
Institute of Physical and Information Technologies (ITEFI),
Spanish National Research Council (CSIC), Madrid, Spain
e-mail: victor.gayoso@iec.csic.es

© Springer International Publishing Switzerland 2015
Á. Herrero et al. (eds.), *International Joint Conference*, Advances in Intelligent
Systems and Computing 369, DOI 10.1007/978-3-319-19713-5_39

455

According to the cybersecurity strategy of the European Commission [1], it is demanded a supra-national effort to reinterpret the laws in *physical world* in the new coordinates of the *digital world*. The main goal is on shaping Internet as a democratic *agora* for sharing and discussing information on the grounds of free speech, data confidentiality and users' privacy protection. This complex objective can be only achieved by creating and using adequate access and authorisation rules and, of course, by applying robust and efficient security measures.

On this point, we have to take into account that the performance of security procedures and methods should not imply an erosion of citizens' rights. As it is pointed out by Hanni M. Fakhour (staff attorney of the Electronic Frontier Foundation), any control measure and/or security procedure should not pose a threat to fundamental citizen rights such as confidentiality and privacy [17]. Furthermore, the implementation of technological and normative regulations is not an exclusive responsibility of government institutions and private enterprises: citizens are required to use ICT in a responsible way, and thus we have to be aware that most *cyber attacks* are enabled by the information trace that we leave in the Internet. Indeed, our visits to online social networks [34] and our continuous search for information drop an almost indelible fingerprint of ourselves [25].

In this paper we present a study of the most significant aspects of privacy protection when using cryptographic tools. The concept of privacy is far from being easily captured by a closed definition. In this communication, we have interpreted privacy with respect to the expected requirements in security systems. Thus, we have focused our study in mechanisms to preserve privacy through confidentiality protection (privacy as confidentiality), by means of convenient access control services (privacy as control), and on the grounds of easy-to-use solutions (privacy as practice). Accordingly, we have distinguished six challenges that should be faced to accomplish a fully integration of the previous highlighted domains.

2 Challenge 1: Privacy Enhanced Secure Data Provenance

Data provenance is required to further protect information systems, since information assets cannot be protected just by assuming that confidentiality, integrity and availability are guaranteed. Certainly, in modern Information Security Management Systems (ISMS) it is not a good option to take for granted any expected behaviour from our systems. It is necessary to design and establish proper defences against information assets such as confidentiality, integrity and availability. After that, a continuous evaluation process is needed in order to confirm the likelihood of our protection measures and the suitability of the authentication and authorisation protocols. For such an examination underlying logging systems must be developed for capturing the different events that appear when using our ICT network to access and manage our information assets. In other words, auditability is another major concern in security engineering.

A critical step when creating audit trails consists in deciding the events to capture and the information needed to characterise each event. This definition is fundamental for intrusion detection, and thus the audit logs should be properly protected [32]. Here we have to consider both efficiency and privacy compliance as the current challenges with respect to data provenance in the scope of auditing. Regarding efficiency, we should bear in mind that the main operation to perform on audit trails is derived from information search processes. If we are dealing with encrypted files, it implies that we need to decrypt all logs in order to look for events. While this could be a bottleneck, in practice we can solve it by using searchable encrypted logs [4]. Furthermore, since the main concern when encrypting logs is to avoid the threat of non-trusted storage servers and the application of anti-forensics techniques, we could use peer-to-peer (P2P) schemes to create split trust domains in order to reduce the effect of *malicious* individual storage servers [33].

As for privacy-respectful audits, we should consider that anonymisation and proper event coding are not an option but a requirement [7]. In specific, secure data provenance relies on four elements: access control, privacy, integrity, and accountability. As it occurs with the protection of audit logs, conventional encryption and authentication procedures are not the best solution. Aligned with the claim presented in [7], we should adopt solutions coming from the field of the Privacy Enhancing Technologies (PET). This being the case, the challenge of secure data provenance should be tackled by considering homomorphic encryption [18], secure multi party computation [12], multi signatures [8], group signatures [10], ring signatures [11], etc.

3 Challenge 2: Client-Side Encryption

The emergence of *cloud computing* is one of the most relevant novelties in the technological scenario in the last five years. It entails a low cost opportunity to share and store data, which is specially important for Small and Medium Enterprises (SMEs). Actually, platforms such as Dropbox or Google Drive make possible to backup our information assets for free. Nevertheless, we have to take into account that the final storage servers cannot be assumed to be fully trusted.

Certainly, in most situations the adoption of a cloud computing solution implies some form of data outsourcing, so there is not a clear guarantee about the way our data are going to be treated and protected. This is the main reason behind the design of new procedures enabling an active role for the client of cloud computing services. These initiatives are intended to create software products to provide users with the proper encryption tools, so each item of data is encrypted before it is sent to the cloud [20]. As an important effort in client-side encryption, we have to pinpoint all the contributions to solve the main security problems (due to some form of code injection) in web applications [31]. In addition, we can find several meaningful proposals handling the shortcomings of homomorphic encryption to create privacy-respectful web platforms [27].

4 Challenge 3: Client-Side Integrity Verification

Client-side encryption is not the only means to manage zero trust models in cloud computing [37]. Indeed, data fragmentation is another important issue that should be considered when pondering the pros and cons of outsourcing data storage. The main problem with data fragmentation is that we need to have several copies of each data fragment (otherwise we cannot recover the original information), and an exhaustive protocol for data integrity verification is also demanded in order to guarantee data coherence and consistence. Integrity verification is a commitment in all the different schemes of cloud computing where we do not have a complete trust in the service provider [23]. Moreover, if digital signatures are used to carry out integrity validation, then schemes as group signatures could be integrated into our system to protect users' anonymity [9].

5 Challenge 4: Anonymity Management

Being privacy one of the basic rights in current societies, different approaches have been suggested in order to guarantee it. Anonymity is certainly one of the fundamental alternatives: if the identities of the parties inside a system are not known, it is harder to violate their privacy. However, as it happens with all privacy-respectful techniques, it may be misused for illegitimate purposes, giving institutions and governments a reason to ban it.[1] Accordingly, it would probably also make it harder for companies (who, as stated in [13] also possess private information of value to governments) to decline government requests and, of course, to trust anonymising systems. As matter of fact, this risk has not gone unnoticed to systems providing anonymity.[2]

One logical way to address this issue is to incorporate mechanisms that allow the detection and sanction of illegitimate anonymous actions. However, for this to be possible, anonymity management must be somehow included. Efforts in this direction have already been made, which allow to create X.509 anonymous identities based on group signatures [6, 10] and subsequently manage and revoke them if necessary [16]. Moreover, specific applications have been proposed for systems such as Tor [15], but also as an additional layer for these anonymising networks [21, 22, 35, 36]. However, this is a concern that needs to be carefully dealt with, since a significant reduction in anonymity would certainly be rejected by the users of anonymous communication systems.

---

[1]A recent example of privacy enhancing technologies being questioned by a government is that of Cameron in the UK who, after the attack on Charlie Hebdo in Paris, stated: *"are we going to allow a means of communications where it simply is not possible to do that [listen in on communications]?"* http://www.theguardian.com/technology/2015/jan/15/david-cameron-encryption-anti-terror-laws.

[2]See, for instance, the call made by Tor: https://blog.torproject.org/blog/call-arms-helping-internet-services-accept-anonymous-users.

6 Challenge 5: Digital Content Life-Cycle Management

The definition and implementation of Digital Rights Management (DRM) systems is one of the most difficult tasks in the current technological scenario. Although DRM is usually assumed to be just a mechanism to protect intellectual property, it also provides privacy protection as a result of avoiding information leakage in assets with Personal Identifiable Information (PII) [2]. DRM systems can (and, in some cases, should) be extended considering P2P privacy-respectful platforms [29] and Private Information Retrieval protocols (PIR) [3].

7 Challenge 6: Usable Identity Management Systems

The claim for assuming the security-by-design and privacy-by-design paradigms is a recurrent chant in recent contributions in cryptography engineering [19]. In this regard, standards and well-known technologies are a basic set to design new security systems, since they are continuous and carefully evaluated by the information security community. This initial selection of technologies should be further complemented with a explicit definition of security assumptions and a correct analysis and validation of the underlying security protocols [14].

However, the success of a security solution is not possible without the acceptance of end users. On this point, we have to bear in mind that the low users' acceptance of encryption [38] is even worse when considering the broader spectrum of PET [30]. Therefore, it is highly advisable to acknowledge and learn from recent contributions on creating usable authentication procedures [24] and enforcing privacy settings online [5]. In fact, this last concern is of paramount importance in online services, since the related providers collect and manage PII whereas the user generally has only partial knowledge about the proper and legitimate use of his or her data [28].

8 Conclusion

In this contribution we have summarised the main open problems in the crossed domain of security and privacy, providing an sketch of the most critical challenges in that field. We have provided a list of the limitations in current systems implementations, and we have also discussed some possible ways to solve those shortcomings.

As the main lesson of our study, we can conclude that an adequate collaboration between the cryptology community and the security engineering collective is necessary. There are plenty of theoretical solutions to handle the problems that we have underlined in this paper. However, in many occasions those theoretical solutions are difficult to implement and tend to incur in excesive computational costs.

Regarding this issue, software engineers could help to adapt theoretical proposals to already deployed infrastructures, taking into account the usability of the final products as a mandatory requisite. Moreover, this help should be built upon a thorough examination of each step of the design, implementation and maintenance phases of the corresponding software products. In this regard, software engineers could call for assistance from formal system analysts and information theory experts in order to better study and characterise data breaches.

Acknowledgments This work was supported by Comunidad de Madrid (Spain) under the project S2013/ICE-3095-CM (CIBERDINE).

References

1. EU Cybersecurity plan to protect open internet and online freedom and opportunity—cyber security strategy and proposal for a directive. http://ec.europa.eu/digital-agenda/en/news/eu-cybersecurity-plan-protect-open-internet-and-online-freedom-and-opportunity-cyber-security (February 2013). http://ec.europa.eu/digital-agenda/en/news/eu-cybersecurity-plan-protect-open-internet-and-online-freedom-and-opportunity-cyber-security
2. Aaber, Z.S., Crowder, R.M., Fadhel, N.F., Wills, G.B.: Preventing document leakage through active document. In: 2014 World Congress on Internet Security (WorldCIS), pp. 53–58 (Dec 2014)
3. Backes, M., Gerling, S., Lorenz, S., Lukas, S.: X-pire 2.0: A user-controlled expiration date and copy protection mechanism. In: Proceedings of the 29th Annual ACM Symposium on Applied Computing, pp. 1633–1640. SAC '14, ACM, New York, NY, USA (2014). doi:http://doi.acm.org/10.1145/2554850.2554856
4. Backes, M., Maffei, M., Pecina, K.: Automated synthesis of privacy-preserving distributed applications. In: Proceedings of ISOC NDSS (2012). http://www.lbs.cs.uni-saarland.de/publications/asosda-long.pdf
5. Balsa, E., Brandimarte, L., Acquisti, A., Diaz, C., Gurses, S.: Spiny CACTOS: OSN users attitudes and perceptions towards cryptographic access control tools. In: Proceedings 2014 Workshop on Usable Security (2014). https://www.internetsociety.org/doc/spiny-cactos-osn-users-attitudes-and-perceptions-towards-cryptographic-access-control-tools
6. Benjumea, V., Choi, S.G., Lopez, J., Yung, M.: Anonymity 2.0 - X.509 extensions supporting privacy-friendly authentication. In: Proceedings of Cryptology and Network Security, 6th International Conference, CANS 2007, pp. 265–281. Singapore, 8–10 Dec 2007. doi:10.1007/978-3-540-76969-9_17
7. Bertino, E., Ghinita, G., Kantarcioglu, M., Nguyen, D., Park, J., Sandhu, R., Sultana, S., Thuraisingham, B., Xu, S.: A roadmap for privacy-enhanced secure data provenance. J. Intell. Inf. Syst. **43**(3), 481–501 (2014)
8. Boyd, C.: Digital multisignatures. In: Cryptography Coding, pp. 241–246 (1989)
9. Camenisch, J.: Efficient anonymous fingerprinting with group signatures. In: Advances in Cryptology-ASIACRYPT 2000, pp. 415–428. Springer (2000)
10. Chaum, D., van Heyst, E.: Group signatures. In: Proceedings of Advances in Cryptology—EUROCRYPT'91, Workshop on the Theory and Application of of Cryptographic Techniques, pp. 257–265. Brighton, UK, 8–11 April 1991. doi:10.1007/3-540-46416-6_22
11. Chow, S.S., Yiu, S.M., Hui, L.C.: Efficient identity based ring signature. In: Applied Cryptography and Network Security. pp. 499–512. Springer (2005)
12. Damgård, I., Pastro, V., Smart, N., Zakarias, S.: Multiparty computation from somewhat homomorphic encryption. In: Advances in Cryptology-CRYPTO 2012, pp. 643–662. Springer (2012)

13. Díaz, C., Tene, O., Gürses, S.: Hero or villain: the data controller in privacy law and technologies. Ohio State Law J. **74** (2013)
14. Diaz, J., Arroyo, D., Rodriguez, F.B.: A formal methodology for integral security design and verification of network protocols. J. Syst. Softw. Accepted (In Press). doi:10.1016/j.jss.2013.09.020
15. Diaz, J., Arroyo, D., Rodriguez, F.B.: Fair anonymity for the Tor network. CoRR abs/1412.4707 (2014), http://arxiv.org/abs/1412.4707
16. Diaz, J., Arroyo, D., Rodriguez, F.B.: New x.509-based mechanisms for fair anonymity management. Comput. Secur. **46**, 111–125 (2014). doi:10.1016/j.cose.2014.06.009
17. Fakhoury, H.M.: Technology and privacy can co-exist. The New York Times (12 Dec 2012). http://www.nytimes.com/roomfordebate/2012/12/11/privacy-and-the-apps-you-download/privacy-and-technology-can-and-should-co-exist
18. Gentry, C.: A fully homomorphic encryption scheme. Ph.D. thesis, Stanford University (2009)
19. Gurses, S., Troncoso, C., Diaz, C.: Engineering privacy by design. Comput. Priv. Data Prot. **317**, 1178–1179. http://www.ncbi.nlm.nih.gov/pubmed/17761870
20. He, W., Akhawe, D., Jain, S., Shi, E., Song, D.: Shadowcrypt: Encrypted web applications for everyone. In: Proceedings of the 2014 ACM SIGSAC Conference on Computer and Communications Security, pp. 1028–1039. ACM (2014)
21. Henry, R., Henry, K., Goldberg, I.: Making a nymbler nymble using verbs. In: Privacy Enhancing Technologies, pp. 111–129 (2010)
22. Johnson, P.C., Kapadia, A., Tsang, P.P., Smith, S.W.: Nymble: anonymous ip-address blocking. In: Privacy Enhancing Technologies, pp. 113–133 (2007)
23. Juels, A., Kaliski Jr, B.S.: Pors: Proofs of retrievability for large files. In: Proceedings of the 14th ACM conference on Computer and communications security, pp. 584–597. ACM (2007)
24. Li, S., Sadeghi, A.R., Heisrath, S., Schmitz, R., Ahmad, J.: hpin/htan: a lightweight and low-cost e-banking solution against untrusted computers. In: Danezis, G. (ed.) Financial Cryptography and Data Security, Lecture Notes in Computer Science, vol. 7035, pp. 235–249. Springer, Berlin Heidelberg (2012). doi:10.1007/978-3-642-27576-0_19
25. Long, J., Skoudis, E., Eijkelenborg, A.V.: Google Hacking for Penetration Testers. Syngress Publishing, San Francisco (2004)
26. OECD: The E-Government imperative (Complete Edition—ISBN 9264101179), E-Government Studies, vol. 2003 (2003)
27. Popa, R.A., Stark, E., Valdez, S., Helfer, J., Zeldovich, N., Balakrishnan, H.: Building web applications on top of encrypted data using mylar. In: Proceedings of the 11th USENIX Symposium on Networked Systems Design and Implementation, NSDI 2014, pp. 157–172, 2014, Seattle, WA, USA, April 2–4(2014). https://www.usenix.org/conference/nsdi14/technical-sessions/presentation/popa
28. Preibusch, S., Peetz, T., Acar, G., Berendt, B.: Purchase details leaked to PayPal. In: Financial Cryptography (2015). https://lirias.kuleuven.be/handle/123456789/476251
29. Qureshi, A., MegÃas, D., RifÃ-Pous, H.: Framework for preserving security and privacy in peer-to-peer content distribution systems. Expert Syst. Appl. **42**(3), 1391–1408 (2015). http://www.sciencedirect.com/science/article/pii/S0957417414005351
30. Renaud, K., Volkamer, M., Renkema-Padmos, A.: Why doesn't jane protect her privacy? In: Privacy Enhancing Technologies, pp. 244–262. Springer (2014)
31. Ryck, P.D.: Client-side web security: mitigating threats against web sessions. Ph.D. thesis, University of Leuven (2014). https://lirias.kuleuven.be/bitstream/123456789/471059/1/thesis.pdf
32. Schneier, B., Kelsey, J.: Secure audit logs to support computer forensics. ACM Trans. Inf. Syst. Secur. **2**(2), 159–176 (1999)
33. Seneviratne, O., Kagal, L.: Enabling privacy through transparency. In: 2014 Twelfth Annual International Conference on Privacy, Security and Trust (PST), pp. 121–128. IEEE (2014)
34. Thomas, K., McCoy, D., Grier, C., Kolcz, A., Paxson, V.: Trafficking fraudulent accounts: The role of the underground market in twitter spam and abuse. In: Proceedings of the 22nd Usenix Security Symposium (2013)

35. Tsang, P.P., Au, M.H., Kapadia, A., Smith, S.W.: Blacklistable anonymous credentials: blocking misbehaving users without TTPs. In: ACM Conference on Computer and Communications Security, pp. 72–81 (2007)
36. Tsang, P.P., Kapadia, A., Cornelius, C., Smith, S.W.: Nymble: blocking misbehaving users in anonymizing networks. IEEE Trans. Dependable Sec. Comput. **8**(2), 256–269 (2011)
37. De Capitani di Vimercati, S., Erbacher, R., Foresti, S., Jajodia, S., Livraga, G., Samarati, P.: Encryption and fragmentation for data confidentiality in the cloud. In: Aldini, A., Lopez, J., Martinelli, F. (eds.) Foundations of Security Analysis and Design VII, Lecture Notes in Computer Science, vol. 8604, pp. 212–243. Springer International Publishing (2014). doi:10.1007/978-3-319-10082-1_8
38. Whitten, A., Tygar, J.D.: Why johnny can't encrypt: a usability evaluation of pgp 5.0. In: Proceedings of the 8th Conference on USENIX Security Symposium—Volume 8, pp. 14–14. SSYM'99, USENIX Association, Berkeley, CA, USA (1999). http://dl.acm.org/citation.cfm?id=1251421.1251435

CISIS 2015-SS02: User-Centric Security & Privacy

GOTCHA Challenge (Un)Solved

Ruxandra F. Olimid

Abstract Password-based authentication is common due to its high usability and simplicity to implement; however, it raises many security problems. This implies a continuous effort in designing new password-based authentication techniques. J. Blocki, M. Blum and A. Datta introduced GOTCHA (Generating panOptic Turing Tests to Tell Computers and Humans Apart), an innovative method to perform password-based authentication: a challenge-response mechanism that gives humans a great advantage over machines. The authors of GOTCHA proposed a public challenge to test its strength. We disclosed all 5 passwords of the first round, because of a leakage in the released code. In this paper, we present our attack: an improved brute-force that revealed each of the 7-digit password in less than 0.5 h and the 8-digit password in approximately 1.5 h on a personal laptop.

Keywords GOTCHA challenge · Password-based authentication · Hash functions · Offline attacks · Dictionary attacks

1 Introduction

Password-based authentication is a wide use method to allow access to resources and systems. It achieves high usability and it is easy to implement. At login, the user fills in his credentials: username and password. The system searches the username in a password file and checks if the given password equals the one that is stored. The authentication succeeds if the two values match; otherwise, the authentication fails.

The reverse is that password-based authentication raises many security problems. Famous companies are frequently hacked and hundred thousand passwords stolen.

A basic security rule states that passwords must never be stored in clear. This would immediately allow an attacker with physical access to the password file to learn all the passwords.

R.F. Olimid (✉)
Department of Computer Science, University of Bucharest, Bucharest, Romania
e-mail: ruxandra.olimid@fmi.unibuc.ro

© Springer International Publishing Switzerland 2015   465
Á. Herrero et al. (eds.), *International Joint Conference*, Advances in Intelligent
Systems and Computing 369, DOI 10.1007/978-3-319-19713-5_40

Hashing comes as a straightforward protection for the previous scenario. A hash is a one-way function: having the password it is easy to determine the hash value, but knowing the hash value it is computationally infeasible to determine the password.

Nevertheless, hashing alone cannot provide security. The attacker can mount a simple brute force attack: he lists all possible passwords, computes their hashes and tries to match them with the stored values. The attack succeeds when a match is found.

A solution to prevent a brute force attack is immediate: the space of all possible passwords must be large enough such that the attack becomes impractical (i.e. it requires an unacceptable amount of time to succeed). An adversary with unbounded computational power will eventually learn the password, but even then he gains nothing if, meanwhile, the password became invalid (for example, due to password expire policies).

However, security problems persist. Users tend to choose passwords that they can easily remember (e.g. "123456" or "password") [3]. Hence, the human behavior leads to another flaw - the dictionary attack: the adversary ignores the low-probable passwords and only checks the high-probable passwords (usually strings from a predefined dictionary). Salting mechanisms were introduced to defend against table rainbow attacks [4]: the authentication system concatenates a random string (called salt) to the password before hashing. For each user, the password file stores the username, the salt value and the hash. Salting brings an additional advantage: an attacker with access to the password file only cannot directly notice if multiple users select the same password (the hashes differ for each user, due to the distinct random selection of the salt).

Despite the security enhancements, password-based authentication raises still many security problems. This implies a continuous effort in designing new or improved techniques.

Our work analysis an alternate solution: GOTCHA Authentication. Section 2 briefly describes the technique. Section 3 presents the challenge released by the authors to test security. Section 4 gives very few details of the implementation. Section 5 details the attack that allowed the disclosure of the first five passwords of the challenge. Last section concludes.

2 GOTCHA Authentication

J. Blocki, M. Blum and A. Datta, from Carnegie Mellon University, introduce GOTCHA (Generating panOptic Turing Tests to Tell Computers and Humans Apart), a new password-based authentication technique whose main goal is to stand against offline dictionary attacks [1]. The authors propose an innovate method based on the interaction between the user and the authentication system that prevents automatic attacks (applications that run automatically and perform repetitive tasks). The idea is somehow similar to CAPTCHA (Completely Automated Public Turing test to tell Computers and Humans Apart), the known challenge-response test used to make the difference between a human and a computer [5]. Both constructions rely on puzzles that are easy to solve by humans, but difficult to solve by machines.

GOTCHA authentication technique implements 2 processes: (1) Create Account and (2) Authenticate. Figure 1 illustrates their interface to the user, as released in the GOTCHA Challenge [2].

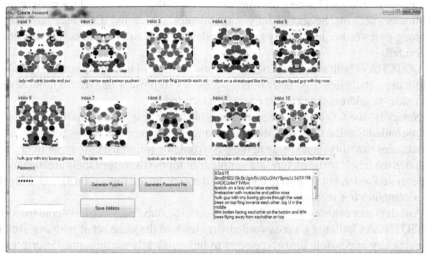

(a) Create Account

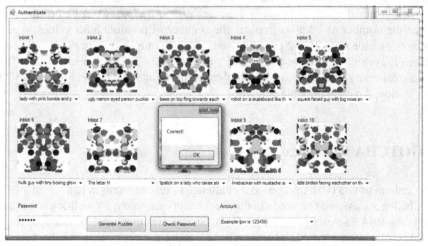

(b) Authenticate

Fig. 1 GOTCHA's phases [2]

To register, the user selects a password. The system responds with some inkblot images, generated deterministically based on the selected password. To complete the

Create Account step, the user labels each image with a representative description. Then, the system randomly selects a permutation, concatenates it to the password and stores the corresponding hash, together with the labels in permuted order. Note that the permutation is not kept in clear in the password file.

Each time the user wants to authenticate, he first fills in his password. The authentication system generates the corresponding inkblots and lists the labels as stored in the password file, challenging the user to recognize the correct label for each image. If the user succeeds, he successfully authenticates; contrarily, if the user introduces a wrong password or incorrectly labels the descriptions to the images, the authentication fails.

GOTCHA is built on a theoretical basis and benefits of a security analysis [1]. We skip more details here (as they are not necessary to present our results), but invite the reader to address the original paper if interested.

Naturally, the GOTCHA authentication mechanism is less user friendly than a simple authentication technique that only requires username and password. However, it maintains usability: in addition to the password, the user must only remember how to match the labels to the pictures - a simple task for the user (the labels are displayed in permuted order, but they were chosen by the user during registration), but difficult for a computer to execute.

An adversary cannot tell if distinct users own the same password. By construction (in GOTHCA Challenge), a password always leads to the same set of inkblots. But, it is practically impossible for two persons to independently set the same descriptions and for the authentication system to randomly choose the same permutation. Hence, the password file stores distinct hashes for each user, even if the passwords are equal.

The permuted order of labels is concatenated to the password before hashing, hence the number of inkblots impacts the number of possible hash values; e.g.: 2 inkblots provide only $2! = 2$ possible hash values (computed from password using either $(1, 2)$ or $(2, 1)$), while 10 inkblots provide $10! = 3628800$ possibilities. This means that for a system that uses 10 inkblots, a simple attack must performs 3628800 verifications for each possible password.

3 GOTCHA Challenge

The authors of GOTCHA proposed a challenge to test its strength [2].

Challenges are quite common in the world of cryptography. A well-known example is the RSA Factoring Challenge: RSA laboratories made public multiple RSA-numbers they considered hard to factor and asked the community to disclose the correct factoring [6].

Cryptographic challenges stand as a first measure to check resistance against practical cryptanalysis: if a system fails a correct setup challenge, then it is proved vulnerable. The opposite does not hold: if a system passes the challenge test, this does not prove its security.

GOTCHAs authors released first 5 challenges. They made public the source code for the Create Account and Authenticate processes (in C#) and 5 password files that are supposed to be physically accessible to an attacker. Each file contains the labels introduced by the user in the Authentication phase in permuted order and the hash. The first 4 passwords were uniformly random generated in the range $[0, 10^7]$, while the fifth password was uniformly random chosen in $[0, 10^8]$. The challenge asked for passwords disclosure.

We revealed all 5 passwords. Our work was possible due to a bug in the original challenge: the source folder included the 10 inkblots corresponding to each account. More, the inkblots were stored in the exact order as they were generated starting from the password (and not in a permuted order). This (apparently small) mistake made an improved brute force attack practical.

Independently from our work, P. Kosinar gave the solution for the first 4 challenges, using a similar attack.

We highlight that breaking the first 5 challenges does not directly imply the insecurity of the GOTCHA authentication technique; however, it does raise some security issues under specific circumstances, which we will refer to later in this paper.

GOTCHA challenge remains active. New challenges have been revealed and they do not disclose the corresponding inkblots.

4 The Implementation

We omit the theoretical aspects of GOTCHA and only resume to some aspects of the practical implementation that are necessary to describe our attack.

The authors use BCRYPT (level 15), a slow hash function that diminishes the chances of offline dictionary or brute force attacks [7].

By contrary, the inkblots generation from password is fast. The password seeds a pseudorandom generator and the output sequence is used to generate 10 inkblots: they give the coordinates and the colors of the plotted figures (40 big circles, 20 ellipses and 20 small circles, together with their symmetric, for each inkblot). Each figure is plot on top of the already existing ones, while the previous remain in the background. Figure 2 exemplifies the first inkblot image generated by the first challenge password $pwd1 = 1258136$. From construction, the first figure to be plot is a big circle that has no other figure underneath. We will use this observation to explain our attack.

5 The Attack

We succeeded to break the first 5 GOTCHA challenges by an improved brute force attack. We present next our method, exemplifying the particular case when the password lies in $[0, 10^7]$. The same approach applies to the fifth challenge also; the only difference is that the attack requires 10 times more attempts to find the password (in the worst-case scenario). A traditional brute force attack must enumerate more than 10^{13} possible combinations (10^7 possible passwords $\times 10! = 3628800$ possible permutations). Taking into consideration that the hash function is slow, the range can be considered large enough to make the attack impractical.

Fig. 2 The first inkblot image (pwd1 = 1258136) [2]

Fig. 3 The first inkblot (pwd1 = 1258136) processed to highlight the possible first plotted big circle

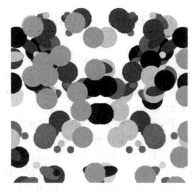

The inclusion of the inkblots into the source code gives the attacker a great advantage. Instead of computing the hash function for each possible password-permutation combination and compare it to the stored hash, the adversary can generate the inkblots from each possible password and compare it to the stored inkblots.

First, the space on which the attacker performs brute force significantly decreases from more than 10^{13} to 10^7, since the attack must only enumerate the passwords - the permutation has no role in the generation of the inkblots.

Second, computation is significantly faster, since the inkblots generation is fast. More, there is no need to generate the whole set of 10 inkblots, but only restrict to the first one (inkblots generation is deterministic for each password). We improved the attack even more, based on some simple observations we describe next.

Table 1   The list of possible 2 first pseudo-random numbers (pwd1 = 1258136)

X	Y	-X	-Y
(1st random no.)	(2nd random no.)	(1st random no.)	(2nd random no.)
31	31	409	31
201	61	239	61
149	20	291	20
108	113	332	113
59	132	381	132
30	171	410	171
73	170	367	170
70	234	370	234
155	216	285	216
166	267	274	267
206	152	234	152
466	226	-26*	226*
471	326	-31*	326*
170	342	270	342
450	398	-10*	398*
436	481	4	481
350	451	90	451
205	443	235	443
467	121	-27*	121*

From the inkblot generation algorithm, the first two numbers output by the pseudo-random generator give the coordinates of the first plotted big circle on the first inkblot. Possible candidates include the coordinates of the upper-left corner of the square circumscribing a big circle that has no figures underneath. We performed a manual inspection using GIMP graphical tool [8] for all 5 challenges and obtained approximately 20 possible circles to meet the criteria, ignoring their symmetric. This leads to almost a double number of possible coordinates (some coordinates are outside the image range, hence they are impossible candidates).

Figure 3 illustrates the manual processed image for the first challenge (the possible first plotted big circle are marked in light green). Table 1 lists their coordinates, which represents the set of the possible two first generated numbers in the pseudo-random sequence (coordinates outside of the image range are marked with *).

Hence, we obtain a simple condition to eliminate potential passwords: we seed the password into the pseudo-number generator, output the first two numbers in the sequence and check if they match a value in the set of possible coordinates. If the condition fails, we discard the password; otherwise we continue to generate the whole inkblot and compare it to the first stored inkblot.

We implemented our attack on a personal laptop (Intel Core 2 Duo CPU at 2GHz, 3GB RAM) running 64-bits Windows 7. Each of the first 4 challenges was broken in less than 30 minutes and the fifth challenge in approximately 1.5 h. It is straightforward that the attack is practical even on low cost machines.

6 Conclusions

We showed that GOTCHA authentication (under the particular construction released in the GOTCHA Challenge) is totally insecure when the attacker knows the inkblots. Our work explained the practical attack we have mounted against the first five puzzles of the challenge.

We highlight that the given implementation of the GOTCHA authentication mechanism becomes totally insecure if the adversary knows the set of inkblots. The exposed attack always succeeds, even in the absence of the password file. Despite the disclosure of the first 5 passwords, the GOTCHA challenge cannot be yet claimed broken - the exposed attack becomes useless when the inkblots are not available to the adversary. Hence, the authors have generated new challenges that keep the inkblots private. Under the new conditions, an AI (Artificial Intelligence) approach seems to suite best for a brute force attack, if an intelligent machine becomes able to distinguish between sets of inkblots that can or cannot correspond to the public known labels. The existence of such a machine remains an open problem.

Acknowledgments The author would like to thank Alex Gatej for informing about the GOTCHA challenge.

References

1. Blocki, J. Blum, M., Datta A.: GOTCHA password hackers!. In: AISec'13 Proceedings of the 2013 ACM workshop on Artificial Intelligence and Security, pp. 25–35 (2013)
2. GOTCHA Challenge. http://www.cs.cmu.edu/jblocki/GOTCHA-Challenge.html. Accessed Jan 2015
3. New York Times—If Your Password Is 123456, Just Make It HackMe. http://www.nytimes.com/2010/01/21/technology/21password.html?_r=0. Accessed Jan 2015
4. Oechslin, P.: Making a faster cryptanalytic time-memory trade-off. Adv. Crypt.—CRYPTO **2003**, 617–630 (2003)
5. CAPTCHA: Telling Humans and Computers Apart Automatically. http://www.captcha.net/. Accessed Jan 2015
6. RSA Laboratories—The RSA Factoring Challenge. http://www.emc.com/emc-plus/rsa-labs/historical/the-rsa-factoring-challenge.htm. Accessed Jan 2015
7. Provos, N., Mazieres, D.: A future-adaptable password scheme. In: USENIX Annual Technical Conference, FREENIX Track, pp. 81–91 (1999)
8. GIMP—The GNU Image Manipulation Program. http://www.gimp.org/. Accessed Jan 2015

Tracking Users Like There is No Tomorrow: Privacy on the Current Internet

Iskander Sánchez-Rola, Xabier Ugarte-Pedrero, Igor Santos
and Pablo G. Bringas

Abstract Since the beginning of the web, users have been worried about usability but not always about security or privacy. Nowadays people are starting to realize that sometimes it is important to protect their privacy not only in real life, but also in the virtual world. This paper analyzes the current privacy debate surrounding online web tracking and explains the most relevant techniques and defenses. It also presents the different companies involved and related standards and regulations.

Keywords Web privacy · Online tracking · Information security

1 Introduction

Privacy is a right often violated in the current Internet, sometimes due to the ignorance of the users, and other times because of the abuse of service providers. Therefore, it has become an issue of great concern for users. According to the Oxford English Dictionary [1], privacy is the state of being free from public attention. Based on that premise and considering the technological environment in which we find ourselves, it is harder than ever to preserve that right [2, 3]. For this reason, privacy is more important than ever.

Online privacy goes far beyond accepting some terms and conditions in social networks [4, 5] but continually collecting a large amount of data with or without our permission [6]. The data collected is as diverse as browser identificators and browsing history [7]. This can happen for various reasons, such as not reading the privacy

I. Sánchez-Rola (✉) · X. Ugarte-Pedrero · I. Santos · P.G. Bringas
S3lab, DeustoTech—Computing, University of Deusto, Bilbao, Spain
e-mail: iskander.sanchez@deusto.es

X. Ugarte-Pedrero
e-mail: xabier.ugarte@deusto.es

I. Santos
e-mail: isantos@deusto.es

P.G. Bringas
e-mail: pablo.garcia.bringas@deusto.es

© Springer International Publishing Switzerland 2015
Á. Herrero et al. (eds.), *International Joint Conference*, Advances in Intelligent
Systems and Computing 369, DOI 10.1007/978-3-319-19713-5_41

policies correctly, or simply because online advertisers collect more data than the strictly necessary. Although part of the information collected is not dangerous independently, if crossed, it can become a serious privacy invasion.

The intention of this paper is to familiarize computer security and privacy researchers with web tracking, examining and discussing all the different factors and privacy implications related to usual web browsing.

2 Privacy Attacks

Despite privacy-violating techniques are possible and even likely, sometimes we do not know how many different ways actually exist and to what extent are a threat. This section reviews the most common privacy attack vectors and explains the different techniques used.

2.1 Fingerprinting

Identifying someone unequivocally on the Internet is a common practice nowadays. Furthermore, fingerprinting allows to gather huge amounts of data related to user browsing, independently of where he is [8]. This technique raises serious privacy concerns for users.

All the data obtained could be used to protect users and web applications against malicious actors, for instance, by detecting the use of stolen credentials. However, it is also possible to use the information to conduct specific attacks against users. There are three main types of fingerprinting:

- **Browser recognition**: A fingerprint is a list of attributes that have disparate values between different web browsers but always have the same on each one. If those values are distinctive enough, their combination could be unique and work as an identifier [9]. These attributes consist of version number of the browser, screen resolution, and the list of used fonts, among others [10]. Canvas fingerprinting is another type of browser or device fingerprinting technique that leverages the Canvas API of the browser [11, 12], exploiting the differences in the rendering of the same piece of text in order to get an identifier.
- **Unique IDs**: Maybe the most effective and simple method of identification is creating a single javascript for each user, including a unique identifier in a variable. This javascript is cached and will always be used. Another interesting technique is to return images with unique and exclusive Etags [13]. During the next connection, the server will realize that there has not been any change in the image, which implies that it is the same user. Identifiers can also be sent in HTTP requests using redirections or javascript for the assignation.

- **Cognitive identification**: JavaScript allows a website to easily create a full itinerary of all the interactions of the user with the different parts of the web-page just making use of event handlers of mouse and keyboard [14, 15]. Mouse moves, scroll-behavior and highlighted texts are some of the obtainable data that can be used to identify certain browsing patterns.

2.2 Information Storage

External code included in a website, has access to many different parts of the host website. This information, susceptible of being leaked, include cookies and many other sensitive data. Even if external code could enhance the user experience, it could also be used for malicious purposes.

- **Cookies**: These are the most common option (both Flash and HTTP). It is very easy to get information about the user's browsing habits with this method and to combine with user-identifying data [16]. For instance, a server can relate different identifiers from the users with the information in the referer header of the request, sharing cookie values between websites (i.e., syncing). Although cookies can be deleted, accepted or blocked, they are the cause of many online privacy attacks to the user. An example is to store an HTTP and a Flash cookie, and if the user removes the HTTP cookie, copy the value from the Flash cookie (i.e., respawning).
- **HTML5**: Using local storage, websites have the possibility of storing informa-tion in the browser of the user [17]. Before, the only way of storing data was with cookies. This method is more secure, and allows websites to store many infor-mation locally (more than 5 MB), without slowing down browsing. Information is never transferred to the server and is domain dependent. All the sides from the same domain, can access the information or store new. This local storage technique presents the same problem as cookies.
- **Javascript**: Window.name, is a non-persistent property of Javascript (could be stored in cache), which is used to pass information between different website pages. Even if it is often used for setting targets for hyperlinks and forms, it has security drawbacks that can be used to store a session.

2.3 Data Sniffing

Most web browsers share access to a single browsing history and cache (file and DNS). This leads to history sniffing attacks, where a tracker can determine if the user has lately visited some other unconnected webpage.

- **Cache Timing**: The tracker can obtain this information calculating the time differ-ence between the execution of certain operations related to data caching [18, 19]. This is possible because all web browsers implement many types of caching and the time needed to obtain the data are related to the browsing history of the user.

– **Information Leakage**: Sometimes an attacker does not need to perform any type of privacy violations by himself, because some trackers send their information via HTTP (e.g., using 1 pixel × 1 pixel transparent graphic images). Sniffing the traffic on the network would allow to obtain all the data that is being sent. Another leakage attack, exploits the fact that browsers display links in a different way if the webpage was previously accessed [20]. Using JavaScript, the tracker just needs to create a hidden link to the target webpage and then, making use of the browser's DOM, check how the link was presented. Depending on the result, it is possible to determine if the website is in the user's history.

2.4 Discussion

All the techniques described are somehow used in the wild, but their biggest limitations resides on the fact that each of them gives the tracker only some specific information about the user. Independently, these techniques are not as dangerous as they can be if combined, because trackers only get a partial overview. Data exchange between trackers that use different techniques is indeed, one of the main problems.

In the fast changing environment of web development, these attacks may not work if some specifications or properties of HTML5 and Javascript vary. Moreover, modifications related to regulations in data collections and privacy could invalidate their use.

3 Implementations in the Wild

There are two different groups that implement privacy attacks in order to achieve specific objectives. Some of them use the extracted information to improve the quality of their service, others use the data with possible malicious intentions. Regardless of the objective, this information is being obtained without explicit acceptance of the end-user.

3.1 Advertising and Analytics Services

These services provide tools for websites to figure out the preferences of visitors, indicating demographics, browser, operating system, views and interactions [21]. They create usage profiles of the websites a user interacts with over time.

Although these implementations can differ from service to service, nearly all have adopted one of the two typical models. Some offer analytics as a paid service; they cannot use any client's analytics information and they protect the obtained information. Others offer a free analytics service, but they use the obtained data for ad targeting, market understanding or any other purpose. Advertising companies do not always depend on the data sold by the analytics services. They use their own techniques to understand the user. Information transference between banners is one of the most used techniques [22].

3.2 Self Implemented

We tend to think that websites only use pre-packaged solutions like the ones previously commented to obtain information of their users. However, some of them construct their own implementations [23]. Sometimes they are even obfuscated to evade detection systems. As these methods do not follow any specific information flow, they are much more difficult to detect and stop.

4 Standarizations

One possible solution to the problem of online privacy attacks is making an standard to control the information that is being transmitted. Two main projects have been advanced for giving users control over their personal data: Do Not Track (DNT) [24] and Platform for Privacy Preferences (P3P) [25].

4.1 Do Not Track

It is a proposal that combines technology and policies in order to send user's preferences on web tracking. This information is sent in a HTTP header, DNT. All modern browsers (Chrome, Firefox, Opera, Safari and Internet Explorer) support a Do Not Track opt-out preference (i.e., DNT: 1 header). The policy also indicates that websites must stop tracking the user for whatever reason when they receive a Do Not Track header.

4.2 Platform for Privacy Preferences

This project facilitates websites the task of communicating their privacy habits in a standard format that can be automatically obtained and understood by user agents. Users have the possibility of coming to a decision based on the privacy practices indicated by the website [26]. Thanks to that, users do not need to read the privacy policies of all the webpages they access, they just need to read it's practices. Sites implementing these policies have to make their habits public. Browsers can help the user to interpret those privacy habits with user-friendly interfaces.

4.3 Discussion

Although many stakeholders (policy makers, consumer advocates and researchers) think that Do Not Track could decidedly reduce tracking and data collection on the web, as the final decision of taking it into account only resides in websites, it is not followed as expected [27].

The case of P3P is similar, due to the lack of support from current browsers for the implementation, the P3P Specification Working Group suspended the project.

5 Defense Methods

Accessing certain websites can lead to information leakage that could harm the user on many levels [28], including their own privacy. Although some people think that they may not be the final objective of any privacy attack, information of millions of users is being collected everyday. In order to prevent some of those leaks, we describe the two main approaches proposed in the literature.

5.1 Identification and Control

To protect the end-user from the different privacy attacks, there are fully functional anti-tracking web browsers that implement a precise and general information analysis and control [29, 30]. In order to make browsing as normal as possible, they try to have a low performance overhead. There is some research that focuses all the analysis in user's browsing [31]. In that way, all the accessed pages could be analyzed without exception, taking into account that the user is the weakest link in the security chain. Applying taint analysis or dynamic controls and making use of determine policies, it is possible to detect privacy violations [32, 33].

5.2 Spoofing and Configurations

Spoofing the browser profile can guard against many attempts of user tracking. Although the best option is that all users have the same browser profile, it is impossible. Spoofing the data and having random browser profiles could help, because it eradicates the possibility of identifying a user for the uniqueness of his browser. Some of these properties are browser, platform, time zone or screen resolution.

Regarding to possible configurations that could avoid some privacy attacks, the most effective but less appropriate one is disabling Javascript. Most of the attacks described use or depend somehow in Javascript. Another option to protect the web privacy, is to browse in temporary modes such as private or guest mode, so the browser does not save or cache what you visit and download [34]. Some other key points are disabling certain font sets, cookies and plugins by default or blocking the requests to websites listed as tracking servers.

5.3 Discussion

The main problem of identification and control methods is that they only take into account certain fields and privacy attacks, forgetting about the rest. Understanding and controlling every type of privacy attack would enormously improve webbrowsers. Nevertheless, the biggest disadvantage of a general control, as taint analysis, is that they are not computationally efficient.

Regarding configurations, disabling Javascript would prevent many tracking approaches, but it would stop many websites from rendering correctly. Disabling other secondary aspects used in privacy attacks can be a better choice because the number of websites that rely on them is much smaller. Finally, spoofing could be counterproductive because these attempts to hide the identity may be fingerprintable [35].

6 Regulations

After understanding the magnitude of the problem, we should know the existing regulations of the area in the United States and European Union [36]. It was not until recently that these regulations appeared in order to restrict the ability of large-scale collection of personal data [37].

6.1 United States

One of the missions of the Federal Trade Commission (FTC) is the promotion of consumer protection. They can only prevent practices of businesses that are either unfair or deceptive under 15 U.S.C. § 45. First violations will incur on a small payment, but subsequent violations get big monetary penalties.

On 2012 the FTC issued its final report [38] establishing four best practices for companies to protect the privacy of all American consumers and give them the possibility to have more control of tracking options and personal information collection. The report expands on a preliminary report released in 2010 [39], which proposed a framework for consumer privacy control because of the new technologies that allow for information collection that is often not perceivable by consumers. The objective is to balance the personal data of consumers with innovation.

6.2 Europe

The Directive 2002/58/EC on Privacy and Electronic Communications, also known as E-Privacy Directive, indicates that the use of electronic communications networks to store information or to gain access to information stored in the terminal equipment of a user is only allowed on condition that the user concerned is provided with clear and comprehensive information about the purposes of the processing, and is offered the right to refuse such processing [40]. If the above indicatations are not met, penalties could be up to 2 % of the revenue.

The Article 29 Working Party (WP29) addresses the topic of device fingerprinting in the Opinion 9/2014 , which extends over the previous Opinion on Cookie Consent Exemption [41], and indicates that websites cannot process device fingerprints which are generated through the gaining of access to or the storing of information on the users terminal device if there is not a explicit consent of the user (unless some specific exemptions) [42].

6.3 Self-regulation

In 2009, many of the largest advertising and marketing companies and associations, supported by the Council of Better Business Bureaus, created a self-regulatory program with the principal objective of giving total control over the collection and use of private data to the users [43]. Websites should have clear options regarding to the data collection and use, letting the user decide if they want that collection or not. There should also be a limit on the specific data type obtained if it is sensitive information. Until that moment, all the different actors worked interdependently in this area. Nevertheless, it is only indicated for the data collection used to predict user interests to deliver online advertising. These principles do not apply to websites that collect that information for its own uses.

2 years later, the Digital Advertising Alliance (DAA) announced and expansion of the program in order to include the non-advertising businesses to the self-regulation [44]. These new principles prohibit third parties to collect, use or transfer any multi-site information. However, these data was mostly covered in the areas of insurance, credit, employment or health.

6.4 Discussion

Although many regulations exist, there is not a continuous control of the websites to check if they are actually following them. Creating a organization responsible for this would secure the compliance of regulations and therefore improve the privacy control of the users.

7 Conclusion

This paper analyzed and discussed the different factors related to online web tracking as of early 2015. Privacy is an area in continuous evolution that directly interferes in the end-users and need to be addressed in order to protect them.

We hope that the survey presented here provides security and privacy researchers with a good background in order to contribute to the field. Future work is oriented to developing new detection methods for privacy violations in order to enhance the results and the system performance of existing ones.

Acknowledgments This research was partially supported by the Basque Government under the pre-doctoral grants given to Iskander Sánchez-Rola and Xabier Ugarte-Pedrero.

References

1. Stevenson, A.: Oxford Dictionary of English. OUP, Oxford (2010)
2. Milanovic, M.: Human rights treaties and foreign surveillance: privacy in the digital age. Harvard Int. L. J. (Forthcoming) (2014)
3. Bernal, P.: Internet Privacy Rights: Rights to Protect Autonomy, vol. 24. Cambridge University Press, Cambridge (2014)
4. Squicciarini, A.C., Paci, F., Sundareswaran, S.: Prima: a comprehensive approach to privacy protection in social network sites. Annals of telecommunications-annales des télécommunications **69**(1–2), 21–36 (2014)
5. Wang, Y., Nepali, R.K., Nikolai, J.: Social network privacy measurement and simulation. In: International Conference on Computing, Networking and Communications (ICNC), pp. 802–806. IEEE (2014)
6. Cecere, G., Rochelandet, F.: Privacy intrusiveness and web audiences: empirical evidence. Telecommun. Policy **37**(10), 1004–1014 (2013)
7. Hayes, C.M., Kesan, J.P., Bashir, M., Hoff, K., Jeon, G.: Informed Consent and Privacy Online: A Survey. Available at SSRN 2418830 (2014)
8. Acar, G., Juarez, M., Nikiforakis, N., Diaz, C., Gürses, S., Piessens, F., Preneel, B.: FPDetective: dusting the web for fingerprinters. In: Proceedings of the 2013 ACM SIGSAC Conference on Computer & Communications Security, pp. 1129–1140. ACM (2013)
9. Eckersley, P.: How unique is your web browser? In: Atallah, M.J., Hopper, N.J. (eds.) PETS 2010. LNCS, vol. 6205, pp. 1–18. Springer, Heidelberg (2010)
10. Fifield, D., Egelman, S.: Fingerprinting web users through font metrics. In: Proceedings of the 19th International Conference on Financial Cryptography and Data Security (2015)
11. Acar, G., Eubank, C., Englehardt, S., Juarez, M., Narayanan, A., Diaz, C.: The web never forgets: persistent tracking mechanisms in the wild. In: Proceedings of the 21st ACM Conference on Computer and Communications Security (CCS 2014) (2014)
12. Mowery, K., Shacham, H.: Pixel perfect: fingerprinting canvas in html5. In: Proceedings of W2SP (2012)
13. Ayenson, M., Wambach, D.J., Soltani, A., Good, N., Hoofnagle, C.J.: Flash cookies and privacy ii: now with html5 and etag respawning. In: Social Science Research Network (2011)
14. Atterer, R., Wnuk, M., Schmidt, A.: Knowing the user's every move: user activity tracking for website usability evaluation and implicit interaction. In: Proceedings of the 15th International Conference on World Wide Web, pp. 203–212. ACM (2006)
15. Keromytis, A.: Darpa, active authentication program. http://www.darpa.mil/our_work/i2o/programs/active_authentication.aspx (2015)

16. Soltani, A., Canty, S., Mayo, Q., Thomas, L., Hoofnagle, C.J.: Flash cookies and privacy. In: AAAI Spring Symposium: Intelligent Information Privacy Management (2010)
17. West, W., Pulimood, S.M.: Analysis of privacy and security in html5 web storage. J. Comput. Sci. Coll. **27**(3), 80–87 (2012)
18. Felten, E.W., Schneider, M.A.: Timing attacks on web privacy. In: Proceedings of the 7th ACM Conference on Computer and Communications Security, pp. 25–32. ACM (2000)
19. Focardi, R., Gorrieri, R., Lanotte, R., Maggiolo-Schettini, A., Martinelli, F., Tini, S., Tronci, E.: Formal models of timing attacks on web privacy. Electron. Notes Theor. Comput. Sci. **62**, 229–243 (2002)
20. Weinberg, Z., Chen, E.Y., Jayaraman, P.R., Jackson, C.: I still know what you visited last summer: leaking browsing history via user interaction and side channel attacks. In: 2011 IEEE Symposium on Security and Privacy (SP), pp. 147–161. IEEE (2011)
21. Altaweel, I., Cabrera, J., Choi, H.S., Ho, K., Good, N., Hoofnagle, C.: Web Privacy Census: Html5 Storage Takes the Spotlight as Flash Returns (2012)
22. Roesner, F., Kohno, T., Wetherall, D.: Detecting and defending against third-party tracking on the web. In: Proceedings of the 9th USENIX Conference on Networked Systems Design and Implementation, pp. 12–12. NSDI'12, Berkeley, CA, USA, USENIX Association (2012)
23. Jang, D., Jhala, R., Lerner, S., Shacham, H.: An empirical study of privacy-violating information flows in javascript web applications. In: Proceedings of the 17th ACM Conference on Computer and Communications Security, pp. 270–283. ACM (2010)
24. Narayanan, A., Mayer, J.: Do not track, universal web tracking opt out. http://donottrack.us (2011)
25. World Wide Web Consortium: Platform for privacy preferences (p3p) project. http://www.w3.org/P3P (2002)
26. Byers, S., Cranor, L.F., Kormann, D., McDaniel, P.: Searching for privacy: design and implementation of a P3P-enabled search engine. In: Martin, D., Serjantov, A. (eds.) PET 2004. LNCS, vol. 3424, pp. 314–328. Springer, Heidelberg (2005)
27. Mayer, J.: Tracking the trackers: early results. http://cyberlaw.stanford.edu/blog/2011/07/tracking-trackers-early-results (2011)
28. Teltzrow, M., Kobsa, A.: Impacts of user privacy preferences on personalized systems. In: Designing Personalized User Experiences in eCommerce, pp. 315–332. Springer, Berlin (2004)
29. De Groef, W., Devriese, D., Nikiforakis, N., Piessens, F.: Flowfox: a web browser with flexible and precise information flow control. In: Proceedings of the 2012 ACM Conference on Computer and Communications Security, pp. 748–759. ACM (2012)
30. Pan, X., Cao, Y., Chen, Y.: I do not know what you visited last summer: protecting users from third-party web tracking with trackingfree browser. In: NDSS: Proceedings of the Network and Distributed System Security Symposium (2015)
31. Hedin, D., Birgisson, A., Bello, L., Sabelfeld, A.: Jsflow: Tracking information flow in javascript and its APIs. In: Proceedings of 29th ACM Symposium on Applied Computing (2014)
32. Sen, K., Kalasapur, S., Brutch, T., Gibbs, S.: Jalangi: A selective record-replay and dynamic analysis framework for javascript. In: Proceedings of the 2013 9th Joint Meeting on Foundations of Software Engineering, pp. 488–498. ACM (2013)
33. Chugh, R., Meister, J.A., Jhala, R., Lerner, S.: Staged information flow for javascript. In: ACM Sigplan Notices, vol. 44, pp. 50–62. ACM (2009)
34. Aggarwal, G., Bursztein, E., Jackson, C., Boneh, D.: An analysis of private browsing modes in modern browsers. In: USENIX Security Symposium, pp. 79–94 (2010)
35. Nikiforakis, N., Kapravelos, A., Joosen, W., Kruegel, C., Piessens, F., Vigna, G.: Cookieless monster: exploring the ecosystem of web-based device fingerprinting. In: 2013 IEEE Symposium on Security and privacy (SP), pp. 541–555. IEEE (2013)
36. Mayer, J.R., Mitchell, J.C.: Third-party web tracking: policy and technology. In: 2012 IEEE Symposium on Security and Privacy (SP), pp. 413–427. IEEE (2012)

37. Goldfarb, A., Tucker, C.E.: Privacy regulation and online advertising. Manag. Sci. **57**(1), 57–71 (2011)
38. Federal Trade Commission: Protecting consumer privacy in an era of rapid change: recommendations for businesses and policymakers. https://www.ftc.gov/reports/protecting-consumer-privacy-era-rapid-change-recommendations-businesses-policymakers (2012)
39. Federal Trade Commission: Protecting consumer privacy in an era of rapid change, a proposed framework for businesses and policymakers. https://www.ftc.gov/reports/preliminary-ftc-staff-report-protecting-consumer-privacy-era-rapid-change-proposed-framework (2010)
40. European Parliament: Directive 2002/58/ec. http://eur-lex.europa.eu/LexUriServ/LexUriServ.do?uri=CELEX:32002L0058:en:HTML (2002)
41. Article 29 Data Protection Working Party: Opinion 04/2012 on cookie consent exemption. http://ec.europa.eu/justice/data-protection/article-29/documentation/opinion-recommendation/files/2012/wp194_en.pdf (2012)
42. Article 29 Data Protection Working Party: Opinion 9/2014 on the application of directive 2002/58/ec to device fingerprinting. http://ec.europa.eu/justice/data-protection/article-29/documentation/opinion-recommendation/files/2014/wp224_en.pdf (2014)
43. Digital Advertising Alliance: Self-regulatory principles for online behavioral advertising, behavioral advertising. http://www.aboutads.info/resource/download/seven-principles-07-01-09.pdf (2009)
44. Digital Advertising Alliance: Self-regulatory principles for multi-site data. http://www.aboutads.info/resource/download/Multi-Site-Data-Principles.pdf (2011)

A Practical Framework and Guidelines to Enhance Cyber Security and Privacy

Michał Choraś, Rafał Kozik, Rafał Renk and Witold Hołubowicz

Abstract In this paper the practical framework to enhance cyber security and privacy is described. The major contribution of the paper is the framework and its description, comparison to other standards as well as practical aspects of its implementation. The framework is developed for ICT systems and for Privacy Held Information Systems in particular- this term will be used in the paper to describe ICT systems containing personal information and data.

Keywords Cyber security · Privacy · Framework · CIPHER

1 Introduction

In this paper, we present the methodological framework developed to help owners of the PHIS (Privacy Held Information Systems) to protect their systems, stored data and (more importantly) their customers privacy. The term PHIS is used here to describe the ICT (Information and Communication Technology) systems containing personal and private (notes, ideas, etc.) data [1].

The proposed framework consists of the set of practical guidelines (controls) to follow, in order to enhance cyber security of the system and the privacy of the users and their data.

M. Choraś (✉) · R. Kozik · W. Hołubowicz
University of Science and Technology, UTP, Bydgoszcz, Poland
e-mail: mchoras@itti.com.pl

M. Choraś · R. Renk · W. Hołubowicz
ITTI Sp. z o.o., Poznań, Poland

R. Renk · W. Hołubowicz
Adam Mickiewicz University, Poznań, Poland

© Springer International Publishing Switzerland 2015
Á. Herrero et al. (eds.), *International Joint Conference*, Advances in Intelligent
Systems and Computing 369, DOI 10.1007/978-3-319-19713-5_42

485

Currently the users of ICT networked systems should be careful while submitting or providing personal data to such systems. There are two major concerns (that the presented framework addresses):

- Cyber security – the systems can be compromised if protection means are not well implemented
- Privacy – very often there is no need to submit the personal data (users should be aware of concepts of proportionality, purposiveness and data minimization)

Some recent problems with cyber attacks and personal data were: publishing stolen LinkedIn credentials, attacks on financial sector and banks e.g. in Spain, attacks on medical sector in the Netherlands where medical records of the users were published on Cyberlab website, attack on the file sharing service dropbox etc., [2–4]. Furthermore, 77 million accounts (names, e-mails, bank details, etc.) were stolen from social network of Sony PlayStation users in 2011 [5].

One can notice, that even big companies or services are attacked, which means that it might be even more easy for hackers or cyber criminals to attack public administration or SMEs.

Therefore, we propose a practical framework consisting of guidelines to improve both cyber security and privacy of the personal data in ICT systems.

The paper is structured as follows: in Sect. 2 the framework is presented. In Sect. 3 we discuss it position with regards to other known standards or frameworks. Practical aspects of its usage are discussed in Sect. 4 while conclusions are given thereafter.

2 The Methodological Framework

The proposed framework is dedicated for ICT systems developers, owners and users. The methodological framework specifies clearly three types of direct beneficiaries, namely: PHIS developers, PHIS owners (administrators), Users (citizens or organisations).

A PHIS developer is an organisation or a person who designs, integrates and/or implements the PHIS system. A PHIS owner is an organisation or a person who owns the PHIS and provides services that exploit the functionality on top of it.

Additionally, we also provide recommendations for PHIS systems users (citizens). A user of the PHIS is a person or organisation who uses the services provided by the PHIS owner. A user provides personal data via dedicated components of the PHIS system. The data is the subject of further storage and/or processing performed by the system. Moreover, citizens also store data on their own computers and other devices (e.g. handhelds) and should be aware of the cyber threats.

The proposed methodological framework represents the unique approach by providing convergent cyber security, personal data protection and privacy aspects.

2.1 Main Characteristics

The main objective of the proposed methodological framework is to propose a set of methods and best practices for cyber security of Privately Held Information Systems (PHIS). PHIS (Privately Held Information Systems) are defined as computer systems that are owned by organisations, both public and private, and that acquire, store and process personal data collected from their customers.

The key characteristics of our methodological framework are:

- Technology independent (versatility) – this means that the framework can be applied for every organisation operating in every domain. Moreover, it also facilitates the non-aging feature, meaning that the framework can still be applied even if technologies are getting older or are replaced by new ones. In other words, the proposed framework is not sector-specific or size-specific. Furthermore, we do not name any technologies or particular solutions (e.g., we suggest that encryption should be used without pointing to any particular method/standard).
- User-centric – it means that the proposed framework explicitly focuses on the key users, namely: PHIS owners, PHIS developers and citizens. While developing the framework a user and experts perspective and requirements were taken into account.
- Practicality – it means that the framework lists practical guidelines and controls to follow in order to enhance or check if the organisation is protecting the data from cyber threats.
- Easy to use and user-friendly – not focusing on technologies and not requiring a special expertise from organisations and individuals.

The proposed framework consists of over 80 concrete and practical guidelines organised in 3 dimensions.

Each guideline is the practical advice on what to do and what to follow in order to improve cyber security of PHIS, data protection and privacy of the users/customers. The results of some guidelines control/check can be written down in the proposed registries.

We organised the guidelines according to the dimensions and the suggested flow of guidelines check/appliance.

However, the application of the guidelines is very versatile: they can be checked in any order, the organisations can prioritise them in accordance to their needs, as well as they can choose the frequency of the framework application. Our approach has its place alongside (not instead of) other information security standards and preferably should be used in conjunction with them.

2.2 Comprehensive Approach: Framework Dimensions

In this section, the rationale behind the proposed methodological framework is covered by the description of the identified three dimensions of the framework: Infrastructure dimension, Organisational dimension, Operational dimension.

Infrastructure dimension.
The infrastructure dimension of the proposed framework (shown in Fig. 1) aims at presenting associations between different entities in the cyber ecosystem. Typically, when a PHIS system is established, it already has different cyber security solutions deployed (e.g. firewalls, antivirus software, etc.). These solutions aim at protecting PHIS owners' services against the cyber-related attacks. Therefore, the whole infrastructure can be divided into two parts, namely:

- Protected infrastructure – services and solutions delivered by PHIS owners,
- Protecting infrastructure – all sorts of assets that serve the purpose of protecting the infrastructure against cyber attacks and cybercrime (depicted as cyber threats and vulnerabilities).

Moreover, the protected infrastructure usually engages solutions which can be classified as offline or online. In order to effectively protect computer systems and networks both, online and offline aspects must be considered.

The offline solutions are those which adapt different sorts of software for cyber security posture evaluation. Commonly such software analyses different aspects that are not particularly related to the network traffic, e.g. software versions and its vulnerabilities, configuration files, security policies, etc. Offline aspects include, for instance, simulations and prediction, visualisation, human factors (operator in the loop), risk management, vulnerability assessment, development of security policies. Online aspects include, for example, monitoring, intrusion detection and prevention, intrusion tolerance, resilience, reaction, and remediation as well as testing capabilities, audits, etc. It is obvious that, the common challenge here (either for offline or online solutions) is to update knowledge related to cyber security aspects.

For example, it may happen that due to the so-called zero-day exploits[1] the existing cyber security solutions will be inefficient. Therefore, it is always the responsibility of PHIS owners to identify such cyber security threats before they materialise. It can be achieved by:

- Delegating the responsibilities related to cyber security posture evaluation, risk management and administration,
- Monitoring the current market of cyber security solutions,
- Monitoring the information about cyber-related threats.

---

[1]Zero-days exploits are attacks that are previously unknown to the signature-based protection systems.

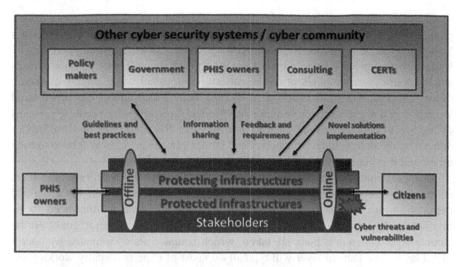

Fig. 1 Graphical representation of the infrastructure dimension

Typically, it may happen that zero-day exploits found in third-party software can be very critical and affect key functionalities provided by the PHIS system. In such cases, the PHIS owner needs to wait until a patch for the discovered zero-day exploit is provided. Therefore, it is often important to identify and to understand the connection and interdependencies with third-party providers.

Using the infrastructure, the PHIS owners deliver services and solutions that are used by citizens. The enhancement of the computer systems security can be achieved by collaboration with the cyber community, e.g. by the following means:

- Guidelines and best practices development/exchange,
- Information sharing (about cyber security related information, e.g. vulnerabilities),
- Novel solution implementation,
- Results feedback & new security protection requirements.

Organizational dimension.
For each framework actor, it is essential to recognize the organizational level at which such actor operates. This will allow for identifying the gaps and the overlaps in the spectrum of the stakeholders. Typically, the cyber security aspects can be addressed at different organizational layers, namely:

- Policy,
- Strategy,
- Operational, and
- Tactical.

The policy layer typically includes high level decisions that concern general action plans with the scope usually defined for several years. The policy usually defines "what" should be achieved, while the strategy usually defines "when" and "how". Usually, national and European bodies are involved at the policy level. Such entities usually define common goals, priorities, milestones or policies (e.g. directives in case of the European level) and coordinate their fulfilment. The strategy layer involves all these bodies that participate in defining the strategic direction, which typically consist of high-level actions that are necessary to achieve the objectives and milestones defined by the policy. In fact, the policy and strategy layers are closely related to each other, and it is hard to make clean cut between these two layers.

The operational and tactical layers involve those bodies that transform high-level actions into particular decisions and processes, which are planned for a defined period of time. At the operational layer, it is known what entities need to be involved and what resources need to be engaged to fulfil the strategic goals. At the tactical layer concrete actions are taken, which actually are coordinated by operational layer. The tactical layer will typically consist of experts (usually working for different organisations, like CERTs, security solution providers or telecommunication operators/providers) whose involvement will focus on low level technical aspects and actual problem solving.

Operational dimension.
Additionally, each role of the framework actors (stakeholders) will vary depending on time/phase, which usually is divided in the following phases:

- Preparedness – where no cyber crisis is observed,
- Crisis and Reaction – where the cyber system is partially (or not) operational or the system is fully operational but the crisis state is declared (e.g. an attacker breached security and the system may be threatened),
- Recovery – where root causes analysis takes place. Usually the analysis results in the best practices and the procedures that prevent similar crisis to occur in the future.

The proposed framework should be typically used in the preparedness and recovery phases, and of course it should be frequently re-run.

For different phases of crisis management cycle, the involvement of different actors, which were identified in the previous dimension is needed.

The most common phase will be the preparedness. It typically includes a wide variety of actions that involve:

- Risk management,
- Training activities,
- Security posture evaluation procedures,
- Cooperation initiatives (between different entities),
- Prevention activities,
- System stress testing.

During the preparedness phase, the activities aiming at identifying the supply chain and the interdependencies of the PHIS system with the third-party organisations should also be conducted. In some cases, this will require the involvement of consulting companies that will help the PHIS owners to analyse those relations, and to identify novel, more robust technologies for cyber security.

The preparedness phase can also be a good opportunity for establishing the communication channels with other key players within the cyber ecosystem (e.g. CERTs institutions). The proposed framework includes practical guidelines to assure that such communication channels are established.

For the crisis phase, the agility of response is fundamental. The procedures, action plans and communications channels established during the preparedness phase, play important role during the crisis. These allow the PHIS owners to resolve the problem quickly and to save time and resources. Moreover, the CERTs institutions (e.g. those established for a PHIS system), during the crisis phase, can verify the correctness of their action plans and procedures established during the training. During the crisis, it is fundamental for a PHIS owner:

- To know how to act,
- To know what are the responsibilities (not only within the organisation but in a whole cyber ecosystem),
- To have an action plan with possible scenarios,
- To have the contingency plans.

During the recovery phase, it is important for the PHIS owners to identify the lessons learnt. Typically, this will include:

- Procedures adjustment,
- Root cause analysis,
- Action plans improvement,
- Responsibilities re-assignment,
- Processes optimisation.

Obviously, it will depend on the scale of the PHIS system, but it may happen that the conducted analysis will reveal that closer cooperation with other actors within the cyber ecosystem is required. As the conducted research shows, more and more institutions are willing to share their knowledge and best practices related to cyber crisis management aspects. Such initiatives allow different institutions to improve their own procedures and eventually establish common standards that may be applied by any partner of the community within the ecosystem.

3 Comparison of the Proposed Framework to Other Methodologies

There are several standards that address information security and data management. Some of them vary in scope and in the level of details of the analysed aspects.

It must be noticed, that the studies presented in this section do not intend to show that our framework addresses the information management and the information security aspects better than other well-known and field-proven standards. Our intention is to show that the proposed framework focus and to explain how our framework extends some aspects that (in our opinion) are not sufficiently tackled by other standards.

In fact, our intention is to prove that our approach has its place alongside (not instead of) other information security standards and preferably should be used in conjunction with them. The standards that are analysed in this section are:

- ISO 27001:27002,
- ITIL,
- COBIT,
- PCI-DSS.

COBIT is an IT governance standard that is based on industry best practices. It addresses wide range of different aspects, like IT-related risk management, compliance, business continuity, as well as security and secure data management. ITIL is more focused on IT services. Therefore, it addresses a wide range of aspects related to IT services management and improvement. ITIL and COBIT are similar to each other. However, COBIT is focused on risk management and process-based approach, while ITIL focus is on IT service. Moreover ITIL provides comprehensive checklists, tasks and procedures and in result it gives more detailed guidelines telling "how" (and using "what" means) to achieve particular objectives. However, ITIL covers fewer domains than COBIT. ISO 27001 is a standard focused on the information security framework while ISO 27002 is a set of best practices telling how to implement ISO 27001. ISO adapts process-based approach to risk management and provides a set of controls that aims at mitigating the identified risk. The domains covered by ISO are strongly related to information security. Therefore, the spectrum of aspects covered by this standard is not as wide as in ITIL or in COBIT. The PCI-DSS standard is focused on a very specific domain which is related to payment card industry. It provides specific and technical controls that aim at reducing credit card frauds. This standard focuses on protecting card holder data, access control, and cyber security aspects of the systems that manage the data and payment transactions.

The proposed framework is complementary, not alternative to ISO, COBIT and ITIL, with slightly different objectives. The framework clearly focuses on cyber security, personal data protection and privacy of the users.

Summarizing our studies, it must be noticed that a variety of small and large organizations implement information security and management standards. Some of them are even certified with worldwide popular standards (e.g. ISO 27001, COBIT or ITIL). However, huge incidents in information security in such organizations still occur. It happens due to the fact that standards like ISO 27001, COBIT or ITIL will not solve cyber security problems, because these are management standards, not cyber security standards. Only dedicated cyber security standards can ensure that the security level of an organization is sufficient. Surprisingly, in some cases there is a possibility that ISO 27001-compliant information security management systems do not address the information security properly. It could be caused by inadequately chosen (or improperly implemented) security measures or lack of expertise in risk assessment and/or cyber security domains.

In comparison to ISO 27001, COBIT, PCI-DSS, and ITIL standards, the proposed framework:

- Comes from technological and cyber security foundations, but does not require vast expert knowledge to be used,
- Strongly addresses data protection and cyber security aspects,
- Provides cyber security and data privacy best practices,
- Navigates the organisations in what to implement and how to track cyber security posture improvements,
- Addresses both beneficiaries of information systems, namely the owner and the user (citizen).

Also some similarities of the proposed framework to ISO and other standards can be pointed out:

- The framework addresses different operational levels (in operational view) which resemble the multi layered structure of an organisation and process-based approach of ISO standard,
- Key characteristic of both our framework and ISO standards is they are generic in order to be applicable for wide variety of organisations, regardless of type, size and nature,
- Like the ISO standard, the proposed framework intends to identify and manage responsibilities in an organisation (in organisational view),
- All the mentioned standards (as well as our framework) define measures that indicate what must be achieved (so called control objectives).

4 Practical Evaluation and Validation

The Framework guidelines are presented in textual form of so called security controls. Over 80 practical guidelines are divided onto dimensions and subgroups so it is easier to follow them. These guidelines are presented in [1]. It is important to note, that they can be read/used in any order depending on the organization needs.

Apart from the guidelines, also the registries (to store the information suggested by the guidelines) were proposed and are prepared as xlsx files. Of course, those registries are only a suggestion – each organization can use their own databases, standard procedures etc. The important aspect is to consider and store the information indicated by the guidelines – how and where is the secondary aspect.

We have used several means of validation such as questionnaires, requirements check, face to face interviews, workshop with the experts (in Madrid) and assistance during the framework implementation attempts.

The general feedback was very positive. Firstly in a sense on the urgency and the need in the covered subject. Secondly with the framework content, form and practicality. The long overview and details of the framework validation approach and results is presented in [11].

Currently, we work on practical adoption and implementations e.g. in the medical sector.

5   Conclusions

If the organisation decides to use and follow the proposed framework, it can achieve many benefits and goals such as: to increase the cyber security and personal data related awareness of decision makers within the organisation, to stay compliant with well-known information security standards (like ISO 27001), to constantly improve (cyber) security posture, with the possibility to track the progress, to assess the (cyber) security of personal data in a methodological manner, to ensure the privacy and data protection of the PHIS users, to increase security and privacy awareness among citizens, to better protect their personal data stored and/or processed in PHIS, to empower individuals who make use of PHIS on how to deal with granting access to their personal information.

The proposed framework is general in a sense that it can be applied to almost each organization type (various domains, sectors, size, etc.).

Furthermore, the framework is practical, concrete advices and guidelines are provided and can be implemented by PHIS owners in their ICT systems owners and administrators. Additionally, the web-based, interactive version of the proposed framework is available.[2]

Moreover, by using and following the proposed framework, the organisation shows its commitment to protect data collected from users, and importance of cyber security and privacy for this organisation.

In other words, the users of such PHIS system (using the framework) will know that the organisation cares about cyber security and privacy.

Acknowledgment Authors warmly acknowledge the work, in terms of ideas and practical implementations, of a number of colleagues of the different institutions involved: for Everis Spain S.L.U., David de Castro Pérez and Fernando Sanchez Palencia. For Delft University of Technology, Mariëlle den Hengst-Bruggeling and Martijn Warnier. For Tecnalia, Erkuden Rios Velasco and Eider Iturbe Zamalloa.

---

[2]http://cipherproject.eu/cipher_webapp.

References

1. Choraś, M. (ed.): Methodological Framework for Enhancing Cyber Security, Data Protection, and the Privacy of the Users, CIPHER Project Report published as a book with ISBN. ISBN 978-83-64539-00-8, 2014
2. Paul, I., PC World: Update: LinkedIn Confirms Account Passwords Hacked. http://www.pcworld.com/article/257045/6_5m_linkedin_passwords_posted_online_after_apparent_hack.html
3. Kitten, T.: Bank Info Security, Eurograbber A Smart Trojan Attack. Hackers' Methods Reveal Banking Know-How. http://www.bankinfosecurity.com/eurograbber-smart-trojan-attack-a-5359/op-1
4. Brodkin, J.: Ars Technica, Dropbox Confirms It Got Hacked, Will Offer Two-Factor Authentication. http://arstechnica.com/security/2012/07/dropbox-confirms-it-got-hacked-will-offer-two-factor-authentication/
5. Quinn, B., Arthur, C.: PlayStation Network Hackers Access Data of 77 Million Users. The Guardian (2011)
6. COBIT: http://www.isaca.org/COBIT/
7. ITIL: http://www.itil-officialsite.com
8. ISO: http://www.iso.org/iso/home/store/catalogue_tc/catalogue_detail.htm?csnumber=54534
9. PCI-DSS: www.pcisecuritystandards.org/security_standards/
10. Directive 95/46/EC of the European Parliament and of the Council of 24 October 1995 on the protection of individuals with regard to the processing of personal data and on the free movement of such data, Official Journal L 281, 23.11.1995, P.0031 + 0050, http://ec.europa.eu/justice/newsroom/data-protection/news/120125_en.htm
11. CIPHER D3.1 Test Report (official project deliverable): http://cipherproject.eu/wp-content/uploads/2014/12/CIPHER_D3-1_FINAL.pdf

An Analysis of the Effectiveness of Personalized Spam Using Online Social Network Public Information

Enaitz Ezpeleta, Urko Zurutuza and José María Gómez Hidalgo

Abstract Unsolicited email campaigns remain as one of the biggest threats affecting millions of users per day. Spam filters are capable of detecting and avoiding an increasing number of messages, but researchers have quantified a response rate of a 0.006 % [1], still significant to turn a considerable profit. While research directions are addressing topics such as better spam filters, or spam detection inside online social networks, in this paper we demonstrate that a classic spam model using online social network information can harvest a 7.62 % of click-through rate. We collect email addresses from the Internet, complete email owner information using their public social network profile data, and analyzed response of personalized spam sent to users according to their profile. Finally we demonstrate the effectiveness of these profile-based templates to circumvent spam detection.

Keywords Spam · Security · Facebook · Personalized spam · Online social networks

1 Introduction

The mass mailing of unsolicited e-mails have been one of the biggest threats for years. Spam campaigns have been used both for the sale of products such as online fraud. Researchers are investigating many approaches that try to minimize this type of malicious activity that report billionary benefits.

Within the spam problem, most research and products focus on improving spam classification and filtering. According to Kaspersky Lab data, the average of spam in email traffic for the year 2014 stood at 66.9 % [2].

E. Ezpeleta (✉) · U. Zurutuza
Electronics and Computing Department, Mondragon University,
20500 Goiru Kalea, 2, Arrasate-Mondragón, Spain
e-mail: eezpeleta@mondragon.edu

U. Zurutuza
e-mail: uzurutuza@mondragon.edu

J.M.G. Hidalgo
Pragsis Technologies Manuel Tovar, 28034 43-53, Fuencarral, Madrid, Spain
e-mail: jmgomez@pragsis.com

© Springer International Publishing Switzerland 2015
Á. Herrero et al. (eds.), *International Joint Conference*, Advances in Intelligent
Systems and Computing 369, DOI 10.1007/978-3-319-19713-5_43

With the rise of online social networks (OSNs), specifically Facebook, which has more than 1.39 billion monthly active users as of December 2014 [3], the extraction of personal information that users leave public on their profiles multiplies spam success possibilities. Facebook provides a great opportunity for attackers to personalize the spam, so a much lower volume of messages would get a higher return on investment.

The main objective of this paper is to measure the consequences of displaying information publicly in OSNs. It also aims to demonstrate that advanced techniques for generating personalized email that evade current spam detection systems while increasing the click-through rate can be developed. These techniques can enable new forms of attacks. First we extracted email addresses while crawling the Internet. These addresses were then checked on Facebook to look for related profiles. Once stored a considerable quantity of user addresses, we extracted all the related public profile information and temporally stored it in a database. Then this information was analyzed in order to determine user profiles based on their main activities in Facebook. Email templates were generated using common information patterns. Finally, to demonstrate the effectiveness of these templates when systems circumvent spam detection, different experiments have been performed. We collected sufficient evidence to confirm that the goal has been achieved.

The remainder is organized as follows. Section 2 describes the previous work conducted in the areas of personalized spam, and social network spam. Section 3 describes the process of the aforementioned experiments, regarding data collection, data processing, and personalized spam testing. In Sect. 4, the obtained results are described, comparing typical spam results with the personalized ones. Section 5 gives a discussion of the countermeasures that can be applied to prevent personalized spam. Section 6 describes the ethical considerations about this research. Finally, we summarize our findings and give conclusions in Sect. 7.

2 Related Work

2.1 Personalized Spam

During the last years several works about the possibilities to create personalized spam or collect personal information from different OSNs have been proposed.

In 2009, researchers at University of Cambridge and Microsoft analyzed the difficulty of extracting user information from Facebook to create user profiles [4]. They described different ways of collecting user related data, and they demonstrated the efficiency of the proposed methods. Authors conclude that the protection of Facebook against information crawlers was low. They also they proved that big scale collection of data is possible. While it is true that Facebook has improved its systems' security since then, like limiting its own query language, the research proved that the option was effective.

In [5] researchers found a Facebook vulnerability giving attackers the possibility of searching people through email addresses in OSNs. Starting from a list of different emails, they managed to connect those email addresses with the account of their owners. After that, they collected all the information they could, and created different user profiles. This work left open doors for allowing attackers to launch sophisticated and specific attacks, but still did not realize about the potential of creating personalized spam campaigns. In the same direction, Polakis et al. demonstrated in [6] the risk that different OSNs suffer to create personalized spam campaigns.

2.2 Online Social Network Spam

Over the last few years, social networking sites have become one of the main ways to keep track and communicate with people. Sites such as Facebook and Twitter are continuously among the top 10 most-viewed web sites on Internet [7]. The tremendous increase in popularity of OSNs allows them to collect a huge amount of personal information about users. Unfortunately, this wealth of information, as well as the ease with which one can reach many users, also attracted the interest of malicious parties.

Researchers from the University of California proved that spam is a very big issue for OSNs [8]. In their research they created a large and diverse set of false profiles on three large social networking sites (Facebook, Twitter and MySpace), and logged the kind of contacts and messages they received. They then analyzed the collected data and identified anomalous behaviors of users who contacted their profiles. Based on the analysis of this behavior, they developed techniques to detect spammers inside OSNs, and they aggregated their messages in larger spam campaigns. Results show that it is possible to automatically identify accounts used by spammers, and block these spam profiles.

In Gao et al. [9] authors carried out a study to quantify and characterize spam campaigns launched from accounts on OSNs. They studied a large anonymized dataset of asynchronous "wall" messages between Facebook users. They analyzed all wall messages received by roughly 3.5 million Facebook users, and used a set of automated techniques to detect and characterize coordinated spam campaigns. This study was the first to quantify the extent of malicious content and compromised accounts in a large OSN. While they cannot determine how effective these posts are at soliciting user visits and spreading malware, their result clearly showed that OSNs are now a major delivery platform targeted for spam and malware. In addition, their work demonstrates that automated detection techniques and heuristics can be successfully used to detect social spam.

While most of the research focus on spam campaigns that might appear inside OSNs, we still think that a combination of typical spam and OSN spam exposes serious threats that needs to be addressed.

3 Creating a Personalized Spam Campaign

Our study was carried out in four different phases. In the first phase, we collected a large amount of public information from Facebook. To do this we used email addresses that were publicly available when crawling the Internet. In a second step of our research, we computed a number of interesting statistics from the collected information that will be shown later. As a result of the data analysis, different user profiles were identified, and used then as customizable email template. Once we had defined these templates, we developed an automatic email sending system and conducted two different experiments. Finally we analysed the results obtained in the experiments.

3.1 Collection of Data

At this stage, the main objective has been the collection of information. This process have been performed in three steps:

Email address collection. In this task we considered two options: the first one, obtain the email addresses using the techniques as explained in [6], where they get e-mail addresses using various combinations of public information from users OSNs' users. And the second, using applications to find emails on the Internet. In the end, we have chosen the second one.

Email address validation. First of all we have worked to obtain the greatest numbers of emails. After that, we have checked each email in Facebook, and we have obtained the amount of emails related with a Facebook account.

In order to carry out this work, we have seen that Facebook offers the option to find a personal account through the email. An application to automate the validation has been developed. It first authenticates a user to the OSN, and then searchers for a user corresponding to each email harvested before.

Once we found the account we have extracted and saved the user's ID and the full name.

Next URL is used to check if an specific email (shown in the URL as EMAIL) corresponds to a Facebook user.

http://www.facebook.com/search.php?init=s%3Aemail\Źq=EMAIL\Ztype=users

Collection of the information. On Facebook there is the possibility to extract information from the source code of all the pages. But to do this, it is obligatory to access directly to the page from which we want to extract the information. That is, when you enter Facebook, if you visit the Facebook page of a friend directly, you could get all the user's public information from the source code. On the contrary, if from that page of a friend you visit the profile of a third person, from the source code you cannot get any information. Therefore, in this program we have used user identifier from Facebook to connect directly to the user information page. Thus, we have visited all users pages and we have been able to extract all the public information that users have in the Facebook database.

Below is the address where can be found all public information of each user. 'USERUID' correspond to the ID that Facebook gives each user, which is stored in the database, because we have collected in the previous phase.

http://www.facebook.com/profile.php?id=USERUID\Źv=info

Results. We found that a 19 % of the email addresses in Internet have a corresponding Facebook account associated to it. We found 22,654 Facebook accounts using 119,012 email addresses (19.04 %).

3.2 Data Processing

At this stage the aim has been to treat the data stored in the database to extract user profiles. To facilitate this task we have decided to create a new table in the database. In this table we saved temporally the summarized information for each user. Using the statistical table, we have queried it to extract as much knowledge as possible. After that, we have analysed all the data to present the most important or relevant statistics.

Results. With the previous statistics, we can draw the following conclusions:

- Most Facebook users choose their favourite band and leave it public.
- 30 % of users who have some data entered and posted on Facebook, have at least one company with which they have been connected.
- The number of men is 12 percentage points larger than women on our study.

3.3 Personalized Spam

The objective of this phase is to create different templates that will be later used. With this templates it is possible to send personalized mail to all Facebook users stored in the database. Once the templates were designed and implemented, the next step was to create a way to count the number of users that "bite the hook" of the spam. For this we have implemented a website.

Mail templates. Before any other action, the first step was to define a template through which we were able to send personalized emails to the people.

To do so, we have taken the decision to create the templates after analyzing the data and the information stored in the database, and using statistics.

For the analysis we used the table in which it was possible to see the number of users that had inserted data in each variable. The Table 1 shows the five more introduced variables.

As it can be seen, the most abundant variables are those related to the gender. Although these data cannot be used for creating templates, it can be used for implementing a formal greeting according to their gender.

In contrast, we have used other three variables to create spam templates. That is, if a user has entered his favourite music group in the profile, it will receive a personalized **music template**. However, if the person has no any singer or group in Facebook and has added the university in which he or she has studied, it will received a personalized message with the **studies template**. And if none of those two had been added but the information refers to their current job or a company in which the user works, it will received a personalized the email using the **company template**.

For better customization, we have also used some profile fields such as the language, the name of each user, the gender, and the city in some of the templates.

It should also be added that all emails will include a URL through which users access to our website.

Table 1 Number of users who have entered each variable (total users: 22,654)

Variable	Amount	Percentage
Man	8,786	39 %
Woman	6,189	27 %
Music	5,788	26 %
Titles	5,612	25 %
Company	5,149	23 %

Website. Access to this site should only come from the users personal email. It has been necessary to define a system for it. We must also take into account that it should store information about which user, and by what type of mail has come to the site. Considering all these details, we decided that the most appropriate way was to introduce parameters in the URL which will be included in emails. When the user clicks on the URL and gets to the site, these parameters will be stored in the database. It also gives the user the possibility to write a comment or to unsubscribe from the system so that no longer will receive emails.

4 Experimental Results

We have performed two separate experiments. In the beginning, we sent typical spam from a classical spam text in order to measure the success rate. There could be different possibilities here. The spam could have been detected and filtered by the email service, Internet Service Provider, email client in the users computer, or it could have been ignored or deleted by the user. In the second experiment, we focused on personalized spam, in order to prove the click through rate obtained, sending a bigger amount of personalized spam. The results of each experiment, and explanation thereof will be explained in two different subsections. And the comparison of the results in the last subsection.

4.1 First Experiment: Typical Spam

Using multiple email accounts and sending a total of less than one hundred emails per day, we sent a typical spam email. The account change is due to a strategy to make things more difficult to spam detection systems. We sent one of those emails where spammers try to draw the receiver's attention to enter a web address. In total, we sent 972 typical spam emails. Only four users followed our website address. This means that the click-through of the typical spam in our experiment is 0.41 %.

4.2 Second Experiment: Personalized Spam

Once extracted the data shown in the previous experiment, it has conducted the second experiment. In this case, instead of sending typical spam, it was sent a personalized email to 2,889 Facebook users. We used the same strategy as in the first experiment sending the messages, and we sent each template from different email accounts. Always taking care to sent less than one hundred emails per day and account.

As we mentioned previously, we use three different templates in our study. Those templates had a personalized URL that could keep track and detail of each sent mail. Note that the website explained the experiment, apologizing for the damage caused, and left space for user comments.

The next table shows the amount of emails we sent, and the number of emails for each profile templates. There we can see that the most common template we used was the once we call 'Music'. More than the 60 % of the personalized mails were about music preferences of users. This is mainly because it is the first templates that our program try to use. And if it is not possible to use this template, the application try with the next one (Studies). And the last option is the 'Company' template (Table 2).

Table 2 Number of sent emails

Type	Amount	Percentage
Music	1,787	61.85 %
Studies	843	29.18 %
Company	259	8.97 %
Total	2,889	100 %

To analyze the responses, we must analyze the results automatically stored in the database. As we can see in the Table 3, 220 users have acceded the website. This is 7.62 % of the people that received a personalized email. Also note that 1.38 percent of people have discharged from the study.

Table 3 Website data

	Amount	Percentage of total shipments
Users who have accessed the website	220	7.62 %
Users who have been discharged	40	1.38 %
Users who have left comments	11	0.38 %

Moreover, we can also break down the answers taking into account the different templates. Because each user acceded to the website from the personalized template as it is shown in the next table together with the click-through of each template (Table 4).

Table 4 Information according to each template

	Access to the website	Percentage of total accesses	Click-through
Music	111	50.45 %	6.21 %
Studies	81	36.82 %	9.61 %
Company	28	12.73 %	10.81 %

As we can see in the table, most of the users, who entered in our website, had received musical spam. This can be considered as normal, because we sent more musical spam than the other two types. But it is important to see that the template with the highest click-through rate is the 'Company' template. Otherwise, the musical template had the worst results.

4.3 Comparison Between Experiments

Finally, if we want to analyze the results of the second experiment in the best way, we have to compare with the results of the first experiment. In this way we can see the difference between typical spam and personalized spam results.

Table 5 Comparison between results

	Sent	Answered	Percentage
Typical spam	972	4	0.41 %
Personalized spam	2.889	220	7.62 %

Table 5 summarizes the response rates obtained using the different spam types. If we analyze these data further back, the first interesting information that emerges is that only 4 people have gone through the typical spam. In contrast, 220 other people have come through personalized email. I.e. 0.41 percent compared to 7.62 percent.

5 Countermeasures

Completed the experiments, here are three ways to avoid spam customization. Two from the OSNs point of view, and the other from the users perspective.

- *Limiting users public information:* OSNs may limit public information from users. Thus, will be more difficult to extract information from users. And the attackers can not use this information in their attacks.
- *Changing the code of the website:* Today it is possible to collect information from the source code of the Facebook web page. If they change the website and do not leave the user information in a readable format, it will be more difficult to extract information for attackers.

- *Raising Awareness:* We must teach people how dangerous it can be to leave personal information publicly. If people minimize their profiles public information will be much more difficult to customize the spam.

6 Ethical Considerations

Some actions taken in this paper are ethically sensitive. For some people, collecting information from internet is not ethically correct. But as was said in [10, 11] and more recently in [5], the best way to do an experiment, it is to do as realistically as possible. We defend this mode of action for the following reasons.

First, we must be clear that we work to improve the safety of users, we use users information to protect them in the future. Second, we only use information that users displayed publicly in OSNs. This means that we never attacked any account, password or private area. Third, attackers use this information, if we use the same information and act in the same way, we will defend users better.

Finally, we have consulted to the general direction of our university and they have given us the approval. For this, before the experiment we proposed our intentions to the general direction of the university, where we showed them the ethical considerations for conducting the study. We also explained them the procedure we had designed to collect personal data and the way we had thought to send emails. Once the R&D Manager had gave us the approval, we started with the experiment.

7 Conclusions

This work makes clear the issue that could exist if spam campaign creators turn their spam templates into a personalized text based on user characteristics, interests, and motivating subjects. Attackers have millions of email addresses stored. We have demonstrated that a 19 % of the email addresses in Internet have a corresponding Facebook account associated to it. Moreover, even if not too much, basic public information can be extracted from those users, which is sufficient to create personalized email subject and bodies. This emails can have a click-through rate bigger than an 7.62 %, being it more than 1,000 times bigger than typical spam campaign rate. It is obvious that in parallel to the research of new techniques for spam detection inside OSNs, it is necessary to research beyond the state of art of classic spam filtering, taking into account the possibility of personalized spam campaign success.

Regarding the behavior of OSN users analyzed, we found that most of Facebook users choose their favorite music band and leave it public. We could also see that 30 % of users who have some data posted on Facebook, have at least one company with which they have been connected.

Also another interesting fact is the difference in the number of men and women, while the number of men is 12 percentage points higher than the number of women. This means that from all mails on the Internet, there are more men that are associated with their email to Facebook. The most probable reason is that men are less conscious of leaving their mail addresses public on Internet.

But the main conclusion to be drawn has been that it has achieved the main objective of the project: to show that we can develop advanced techniques for generating personalized mail that circumvent current spam detection systems. Clear examples of this are the results shown in the results section. In the first experiment, we can see that only the 0.41 % of users have bitten the bait. Whereas in the second 7.62 % of the users have entered to the project website. The second result rate is more than 18 times higher than the first one.

We can see that it is not a large number of people, but as a steady stream of visitors, which means that personalized emails reach their destination. Then, once the message is on the user's email inbox, it depends on each person's behavior to click on the link that is sent in the mail. This shows that spam is not blocked as it's customization have not been detected.

Acknowledgments This work has been partially funded by the Basque Department of Education, Language policy and Culture under the project SocialSPAM (PI_2014_1_102).

References

1. Kanich, C., Kreibich, C., Levchenko, K., Enright, B., Voelker, G.M., Paxson, V., Savage, S.: Spamalytics: an empirical analysis of spam marketing conversion. In: Proceedings of the 15th ACM Conference on Computer and Communications Security, pp. 3–14. CCS '08, New York, NY, USA. ACM (2008)
2. KasperskyLab: Spam and phishing in q3. http://www.kaspersky.com/about/news/spam/2014/iPhones-and-Ice-Buckets-Used-to-Promote-Junk-Mailings
3. Facebook: Facebook: Newsroom. http://newsroom.fb.com/company-info/
4. Bonneau, J., Anderson, J., Danezis, G.: Prying data out of a social network. In: International Conference on Advances in Social Network Analysis and Mining, pp. 249–254 (2009)
5. Balduzzi, Marco, Platzer, Christian, Holz, Thorsten, Kirda, Engin, Balzarotti, Davide, Kruegel, Christopher: Abusing social networks for automated user profiling. In: Jha, Somesh, Sommer, Robin, Kreibich, Christian (eds.) RAID 2010. LNCS, vol. 6307, pp. 422–441. Springer, Heidelberg (2010)
6. Polakis, I., Kontaxis, G., Antonatos, S., Gessiou, E., Petsas, T., Markatos, E.P.: Using social networks to harvest email addresses. In: Proceedings of the 9th Annual ACM Workshop on Privacy in the Electronic Society, pp. 11–20. WPES '10, New York, NY, USA, ACM (2010)
7. Alexa Internet, I.: Alexa top 500 global sites. http://www.alexa.com/topsites
8. Stringhini, G., Kruegel, C., Vigna, G.: Detecting spammers on social networks. In: Proceedings of the 26th Annual Computer Security Applications Conference, pp. 1–9. ACSAC '10, New York, NY, USA. ACM (2010)
9. Gao, H., Hu, J., Wilson, C., Li, Z., Chen, Y., Zhao, B.Y.: Detecting and characterizing social spam campaigns. In: Proceedings of the 17th ACM conference on Computer and Communications Security, pp. 681–683. CCS '10, New York, NY, USA. ACM (2010)
10. Jakobsson, M., Johnson, N., Finn, P.: Why and how to perform fraud experiments. IEEE Secur. Priv. **6**(2), 66–68 (2008)
11. Jakobsson, M., Ratkiewicz, J.: Designing ethical phishing experiments: a study of (ROT13) rOnl query features. In: WWW '06: Proceedings of the 15th International Conference on World Wide Web, pp. 513–522, New York, NY, USA. ACM (2006)

ICEUTE 2015: Domain Applications and Case Studies

European Computer Science (ECS)

Wolfgang Gerken

Abstract European Computer Science (ECS) is a 3 years double degree Bachelor course offered by 8 universities in 7 countries. Its main characteristic is that the students spend their third year not at their home university where they started their studies but go abroad to one of the partner universities. This paper describes the structure of the course and the experiences with it.

Keywords Bachelor · Double degree · Computer science · Student exchange

1 Introduction

This first section gives a short introduction to ECS and describes its benefits for the participating universities, the students and the companies who want to hire graduates of the ECS course.

1.1 What Is ECS?

European Computer Science (ECS) is a double degree Bachelor course. It has been developed in 2005 by a group of European universities. Though it exists meanwhile for 10 years. The first international publication about ECS was in 2010 [1]. Here we want to present the enhancements and the actual situation.

The ECS course is a cooperation between 8 universities in 7 countries (see sect. 2.1). The duration is 3 years with 180 ECTS. During the first two years (120 ECTS) the students study at the ECS partner university of their home country. In the last

W. Gerken (✉)
Department Computer Science, Hamburg University of Applied Sciences,
Hamburg, Germany
e-mail: wolfgang.gerken@haw-hamburg.de

© Springer International Publishing Switzerland 2015   509
Á. Herrero et al. (eds.), *International Joint Conference*, Advances in Intelligent
Systems and Computing 369, DOI 10.1007/978-3-319-19713-5_44

year they go abroad to another partner university of the network. Here they finish their studies (60 ECTS) and finally, if they succeed, they receive a double degree of both countries (with some exceptions because of national regulations in Spain and Finland).

During the design, implementation and first running time of ECS, the European Union supported this project 3 times. Without this financial support the actual status, the experiences and results of ECS would not have been achieved.

1.2 The Benefit for the Students

The aim of ECS is to prepare students for leading roles in computer science projects with international partners or with customers in foreign countries. They study abroad and therefore they are familiar with a foreign language and the cultural specialities of their host country. This is combined with fundamental knowledge in the area of computer science. In addition, they receive a double degree. Alberto Oleaga, Erasmus Mundus local coordinator for engineering programmes, Faculty of Engineering Bilbao, Universidad del Pais Vascos, states that his experience is "… that engineering students who have taken part in double degree programmes have a very good opportunity to be employed in internationalised companies at home or at companies in the other country in which they have studied; and they can easily develop an international career in a third country. The quality of their jobs is also higher than those of their fellows who have not taken part in these programmes. [2]".

1.3 The Benefit for the Companies

Especially in computer science, an increasing cooperation between companies in different countries can be noticed, either as project partners or as supplier resp. customers. Therefore, companies need graduates of higher education computer science courses with international experiences and background. ECS graduates have shown their ability to stay and "survive" in a foreign country and speak a foreign language, combined with a good education in theory and practice of computer science.

1.4 The Benefit for the Universities

Universities try to increase the number of students who stay during their studies for one or two semester in a foreign country. This is supported by the European Union, since 2014 by ERASMUS + [3]. In Germany e.g. the definition of joint and double degree courses is supported by the German Academic Exchange Service DAAD [4]. A special programme of the European Union for supporting Master courses is ERAMUS MUNDUS [5].

The advantage of ECS in comparison to many other double degree courses is that beside the language modules no extra lectures are necessary. ECS students attend the lectures of other existing courses. Secondly they can start their studies in any institution of the network and in the third year they can go to any partner university in another country for finishing their studies.

2 The Course Structure

Different to many other double degree courses ECS has no predefined pathway. After listing the members of the ECS network this section describes its key features: the core modules of the first two years and the selectable specialisations in the third year.

2.1 The Partners of the Network

Actually, the following partners belong to the ECS network (in alphabetic order):

- Turku University of Applied Sciences, Turku (Finland)
- Université François Rabelais, Tours/Blois (France)
- Université de Lorraine, Metz (France)
- Hamburg University of Applied Sciences, Hamburg (Germany)
- Universita Ca' Foscari Venezia, Venice (Italy)
- Instituto Politécnico de Coimbra, Coimbra (Portugal)
- Universitatea de Vest, Timisoara (Romania)
- Universidad de Burgos, Burgos (Spain)

All of them have signed a cooperation agreement that contains and describes the regulations of the cooperation and student exchange. Usually computer science degree courses do not support going abroad because the study plans mismatch. Within ECS this problem is solved by the concept of core modules and specialisations.

2.2 The Core Modules

During the first 2 years, the students study at their home university. These years contain the basics of a higher education computer science course and prepare for the study of a specialization in the 3rd year at one of the partner universities. The course content in the years 1 and 2 is insofar harmonized that all the agreed topics must be covered by the local ECS implementations in the partner universities with at least a minimum amount of credits. These mandatory topics are bundled to modules. That means we have focused on harmonized lecture content/topics which can be spread upon different local modules and not on common lecture titles. There is no one-to-one mapping of the ECS core modules to existing modules of the partner universities. This structure allows the students to study the 3rd year in another university without any lack of topics. The following table contains the different modules and minimum numbers of credits.

Table 1 ECS core modules with requested minimum number of credits

Module	Min. number of credits
Programming	20
Software engineering and project management	10
Information systems	5
Computer networks	5
Computer architecture and operating systems	10
Mathematics and computer science theory	13
Computer science application	5
European languages, culture and communication	20
Special computer science topics	≤32

Special feature is the module "European languages, culture and communication". To be able to follow the lectures in the 3rd year in another country and language, the language education is an important part of the program. Students should reach the B2 level of the European language framework to be able to follow the lectures in a foreign language.

The last module titled 'Special Computer Science Topics' contains the local specialities of the first 2 years. It is up to the partner to decide what belongs to it and may be smaller if some of the other module have more than the minimum number of ECTS.

2.3 The Specialisations in the Third Year

In the 3rd year, the students study abroad at one of our partner universities. Every partner offers at least one specialisation for this year. Therefore, by choosing a host university, the students choose the specialisation. To be well prepared for this mandatory stay at the partner university, the students have to pass the already mentioned language modules in the first 2 years, because the lectures – with exceptions in Finland, Romania and Portugal - are given in the native language of the guest university. That means that they have to make their choice very early. The actual specialisations are as follows:

Turku University of Applied Sciences: Embedded Software
Embedded systems refer to devices that are controlled with in-built microprocessors. The devices include e.g. mobile phones, bio-technical instruments, digital TVs and control devices for various engines. The offered specialisation emphasises software engineering of these devices. The studies deal with design and implementation of real-time systems using C as the programming language. The lectures are given in English.

Université François Rabelais: Information Systems and Software Development
This year consists of subjects among advanced IT and CS skills, including also additional topics on system programming, formal languages, complexity, local area networks and TCP/IP protocols. This year also includes a mandatory 12-weeks industrial placement.

Université de Lorraine: Computer Science
ECS guest students can choose some of the lectures of the regular French 3rd year students or other lectures like Business Intelligence, .NET Development etc. They have IT Security or Cryptology, Operations Research or Statistics, an Industrial Placement and others.

Hamburg University of Applied Sciences: Technical Computer Science
The course covers technology fundamentals of computer systems and distributed systems, design aspects of software and the implementation of applications. Students learn to work in project groups.

Universita Ca' Foscari: Web Interfaces and Web Software Technologies
This 3rd year specialisation includes several courses that cover different aspects of Web programming and software development. In particular, Web technologies, Web interfaces, and software engineering principles. Students learn to work in project groups. Each course is taught in one of the two semesters. Every ECTS credit point corresponds to 8 h of lessons.

Instituto Politécnico de Coimbra: Artificial Intelligence and Databases
The aim of this specialisation is to prepare the students to apply the most recent techniques of Artificial Intelligence in the management and extraction of knowledge on Databases, with a high degree of technical and scientific competence. They have lectures like Intelligent Systems, Operations Research and an industrial placement. The lectures are in Portuguese with English support.

Universitatea de Vest: Artificial intelligence and Distributed Computing
This specialisation focuses on theoretical concepts and practical tools used in solving computational problems and in designing software products, with a particular emphasis on intelligent systems and distributed computing. The lectures are given in English.

Universidad de Burgos: Artificial Intelligence and Knowledge Management
The courses included in this specialisation provide the student with a deep and wide knowledge about modern techniques and paradigms lying in the field of Artificial Intelligence and its application in modern paradigms of workflows in companies (i.e. Big Data, Decision Support, Knowledge Sharing...).

The course consists of interrelated and up-to-date topics in the field of Artificial Intelligence and Data Analysis like Algorithms, Artificial Intelligence, Neural Networks, Evolutionary Computation, Expert Systems and Data Mining. It is complemented by other topics related with the application of these techniques, such as Knowledge Management or Strategic Management.

2.4 The Organisational Framework

For running a course like ECS some rules are necessary. As mentioned above there exists a cooperation agreement as a basis for the partner universities. Furthermore we have defined a set of documents which support the student's exchange. This is especially necessary for the different national regulations concerning the range of marks, the transformation of the ECS core modules to the real taught modules of the different universities etc.

Last not least we have regular meetings – at least one each year – to discuss the actual situation, possible problems and future activities. For instance we have to check regularly if the content of the core modules is still up-to-date. Sometimes there are some changes of the national laws which may have effects on our ECS regulations and the cooperation agreement.

3 Actual Situation and Next Activities

This section gives an overview about the actual number of ECS students and highlights the next activities.

3.1 ECS in 2014/15

Actually, the ECS course has the following number of students and graduates.

Obviously the number of applicants for the ECS course could be higher. The foreign language seems to be an obstacle. Furthermore the student exchange is not balanced. Students prefer universities with lectures in English to go to in the 3rd year. Experiences of other European double/joint degrees in higher education show, that students are more interested in international course content and not in the language of the partner countries [6].

Table 2 Number of ECS students and graduates in 2014/15

Partner university in	Beginners in 2014 for ECS	3rd year guest students in 2014/15	Students studying their 3rd year abroad 2014/15	Own ECS graduates in 2014
Blois	7	1	0	4
Burgos	no own intake	0	1	1
Coimbra	20	0	4	3
Hamburg	2	2	2	2
Metz	no own intake	1	1	1
Timisoara	no own intake	1	0	0
Turku	no own intake	3	0	0
Venice	7	2	0	1

We are just defining a questionnaire to find out if the ECS graduates are satisfied with the structure, content and stay at the chosen guest university or if they see any problems. But we need some time for this evaluation to have reliable results.

Another problem is the different number of credits to receive a Bachelor degree in Europe. In Spain and Finland students need 240 credit points for a national degree. In France, Germany, Italy, Portugal and Romania they only need 180 credits.

3.2 Marketing Activities

Some marketing activities are planned to increase the number of applicants. Graduates shall be requested to produce a short video clip where they talk about their experiences abroad. Example are the interviews with graduates of the Franco-German institute ISFATES [7]. The purpose is to reduce the barrier for pupils applying for ECS.

Furthermore we want to make a relaunch of our ECS website. The actual website (see [8]) is not modern and doesn't invite to surf. It should be suitable for smart-phones, too. And social networks like Facebook could be a channel for getting in contact with applicants and students.

4 Conclusions

The core module concept insures that all students have a minimum and common state of knowledge if they study their selected specialisation in the third year. This allows that the ECS students can finish their studies successfully.

Nevertheless the number of students is very poor. The language problem still exists. Increased marketing activities are necessary.

Beside the personal engagement of the ECS coordinators in the partner universities a financial support from the European Union, especially for the exchange students in the third year, is necessary to keep a network such as ECS alive.

References

1. Gerken W.: The european bachelor degree course ECS. In: Proceedings of the 1st International Conference on European Transnational Education, Burgos 2010
2. Oleaga A.: International double degrees: are they worth the trouble?. http://www.eaie.org/blog/international-double-degrees-trouble/. Accessed 03 Mar 2015
3. European Commission, Erasmus plus. http://ec.europa.eu/programmes/erasmus-plus/index_en.htm. Accessed 18 Dec 2014
4. DAAD, Integrierte internationale Studiengänge mit Doppelabschluss, German. https://www.daad.de/hochschulen/programme-weltweit/studiengaenge/de/23193-integrierte-internationale-studiengaenge-mit-doppelabschluss/. Accessed 03 Mar 2015
5. EACEA, Erasmus Mundus Programme. http://eacea.ec.europa.eu/erasmus_mundus/index_en.php
6. Nickel S., Zdebel T., Westerheijden D.: Joint degrees in european higher education. http://www.jointdegree.eu/uploads/media/Che_Joint_Degrees_in_European_Higher_Education.pdf. Accessed 03 Mar 2015
7. Université de Lorraine, Témoignages d'anciens étudiantes de l' ISFATES. http://videos.univ-lorraine.fr/index.php?act=view&id_col=81. Accessed 18 Dec 2014
8. ECS partners, European Computer Science. http://ecs.ecs-emacs.net/ecs-project. Accessed 12 Mar 2015

Better Careers for Transnational European Students?

A Case Study for French-German Diploma Alumni

Gabriel Michel, Véronique Jeanclaude and Sorin Stratulat

Abstract We present recent surveys about alumni of transnational degrees. The surveys have been conducted within two French–German university structures, ISFATES-DFHI and UFA-DFH, that organise regularly surveys about the future of their alumni. We compare the results of Computer Science alumni with that for alumni of other fields, in particular Mechanical Engineering, and conclude that transnational diplomas are a clear added-value on the job market, especially for Computer Science alumni. Finally, we discuss about how to develop similar European programs at Bachelor and Master levels.

Keywords Student mobility · Double degree · Master · Computer science · Europe · Intercultural · Learning · Professional skills · Surveys · ERASMUS

1 Introduction

Student mobility is a critical issue for the future of Europe.[1] Acquiring mobility skills is often considered as an important factor for any student development. In fact, the knowledge of several foreign languages and cultures as well as the

---

[1] http://ec.europa.eu/education/programmes/socrates/erasmus/evalcareer.pdf and http://www.touteleurope.eu/actualite/la-mobilite-des-etudiants-erasmus.html.

G. Michel (✉) · V. Jeanclaude
ISFATES, University of Lorraine-Metz, Metz, France
e-mail: gabriel.michel@univ-lorraine.fr

V. Jeanclaude
e-mail: veronique.jeanclaude@univ-lorraine.fr

V. Jeanclaude
LEM3, University of Lorraine-Metz, Metz, France

S. Stratulat
LITA, University of Lorraine-Metz, Metz, France
e-mail: sorin.stratulat@univ-lorraine.fr

G. Michel
PERSEUS, University of Lorraine-Metz, Metz, France

© Springer International Publishing Switzerland 2015
Á. Herrero et al. (eds.), *International Joint Conference*, Advances in Intelligent
Systems and Computing 369, DOI 10.1007/978-3-319-19713-5_45

517

living experience in different international environments are very useful for the professional life [5]. Moreover, it has been proved that international mobility implies higher salaries after several years of working experience [1]. It is therefore necessary to aim at a wide student mobility. In this respect, student mobility via ERASMUS exchanges is unanimously recognised as a big success: more than 3 million students have benefited since its creation in 1987. Only during the academic year 2012/2013, the ERASMUS program was followed by 322,000 European students. The official statistics have shown that the number of ERASMUS students increased steadily during the last 7 years.[2] They have also shown that Spain, France, Germany, United Kingdom and Italy are among the European countries that sent and attracted the most part of the ERASMUS students. For example, during the same year 2012/2013, 35,000 out of 2 million French students have studied abroad in the frame of ERASMUS exchanges, which represents 6 % more than the previous year. In average, the exchange period was 6.8 months. However, the student mobility via ERASMUS exchanges is not enough[3] and it can also be improved by creating transnational diplomas. The number of French-German diplomas has increased, too, as confirmed by the evolution of the French-German University UFA-DFH (Université franco-allemande/Deutsch-Französische Hochschule). Consequently, its number of students augmented from 4800 in 2010 to about 6000, as it can be noticed today.

In this paper, we present surveys performed within two French-German university structures, ISFATES-DFHI and UFA-DFH, both proposing French-German diplomas. The surveys concern alumni from different disciplines of these structures, involving much more disciplines for UFA-DFH than ISFATES-DFHI but having access to more detailed answers from ISFATES-DFHI. Then, we analyse the results of these surveys. Finally, we discuss about how to develop similar European programs at Bachelor and Master levels.

2 Survey of the UFA-DFH Alumni

2.1 Presentation of UFA-DFH

UFA-DFH (Université franco-allemande/Deutsch-Französische Hochschule)[4] consists of a network of French-German higher education establishments. Its administrative center is located in Saarbrücken (Germany) and its campus is spread over several French and German university establishments that offer joint or double French-German diplomas. It proposes a great number of curricula covering a wide range of topics: engineering sciences, natural sciences, mathematics, computer science, medicine, economics/management, law, social and human sciences as well as training for teachers.

---

[2]http://fr.statisticsforall.eu/ and http://europa.eu/rapid/press-release_IP-14-821_fr.htm.

[3]http://www.contrepoints.org/2014/11/17/188498-erasmus-la-mobilite-etudiante-nouvelle-fracture.

[4]http://www.dfh-ufa.org/.

During the academic year of 2009/2010, UFA-DFH counted 4800 students. Today, 6000 students are enrolled and follow one of the 166 curricula delivering over 1100 diplomas per year. Its main objective is to overpass 10,000 students in 2020.

2.2 The Survey

The survey took place during February 7–April 6, 2014 and concerned all the UFA-DFH alumni starting from year 2000. Globally, 1806 alumni have been involved in a survey[5] available in French and German. Its conclusions are based on the pertinent answers given by 1582 among them which can be divided as:

- 41.5 % men and 58.5 % women, as well as
- 40 % of French nationality, 52 % of German nationality, 5 % of both French and German nationality, 3 % other European nationalities.

The major of their curricula are: 29 % social and human sciences, 26 % engineering sciences/architecture, 23 % economics/management, 8 % natural sciences/mathematics/computer science, the remaining 14 % other domains. We will present here only a part of this survey.

One of the interesting questions of the survey was the search period for a job.

How long you needed to find a job?
Almost 71 % of the participants found in less than 3 months a job appropriate to their qualifications (35 % received a job offer before graduation), 17 % took three to 6 months to find a job and 12 % more than 6 months. Most of the participants graduating engineering sciences (79 %), law (76 %) and economics (75 %) succeeded to find a job in less than 3 months.

In which country did you find the first job?
The answer was: Germany (57 %), France (26 %), others (17 %).

Would you recommend a French-German curricula to future students?
Almost 90 % of the participants would strongly recommend a French-German curricula, while only 3.4 % would not recommend any of them.

3 The Survey of the ISFATES-DFHI Alumni

3.1 Presentation of ISFATES-DFHI

History of a French-German institute. ISFATES-DFHI[6] was created on September 15, 1978 during a meeting at the French-German summit of Aachen (Germany) in

---

[5]http://www.dfh-ufa.org/uploads/media/DFH_Alumni_Studie_D_web_01.pdf.
[6]http://www.isfates-dfhi.eu.

order to pose the first stepping stones of a European education. The institute was created between the towns of Metz (France) and Saarbrüscken (Germany). Since its creation, it has not ceased to develop and adapt. Today, nearly 2500 French and German students followed the bi-national training of ISFATES-DFHI and obtained a double (French and German) university diploma. Since 1997, the institute belongs to UFA-DFH.

Currently, ISFATES-DFHI (or ISFATES, for short) has 6 branches: Computer Science, Logistics, Management Sciences, Civil Engineering and Infrastructures, Mechanical Engineering and Industrialized Manufacturing, and Engineering Systems. For each branch, a group of 15–30 French and German students "will travel" together each year from one university to the other. The students of a given year also have some "inter-branch" courses in order to maintain a certain cohesion. The structure of the programs is as follows:

- *for the Bachelor degree*: three semesters take place in Metz and three semesters in Saarbrücken; at the end, the students obtain a French-German Bachelor Degree. It is also possible to spend one semester using ERASMUS exchanges in another country. In the case of Civil Engineering students, they spend the first year in France, the second in Luxemburg and the third year in Germany; at the end of the 3 years, they get a triple diploma recognized by the three countries.
- *for the Master degree*: three semesters of study followed by a 6-month placement in a third country. For the three semesters of study, the first one takes place in Saarbrücken, the second in a third country or Saarbrücken, and the third one in Metz. In the end, a French-German Master degree is delivered.

3.2 The ISFATES-DFHI Alumni Survey

This survey has been conducted in January 2013 and concerned ISFATES-DFHI alumni. Since 2006, ISFATES-DFHI proposed a Bachelor-Master curricula for all of its 6 specialities, based on the conclusions of a previous survey [4] that suggested more management and soft skills-related courses. The first Bachelor alumni date from 2009 and the first Master alumni from 2011. This implies that the number of participants to this survey was significantly smaller than for the previous survey.

In all, 91 alumni have participated to the survey. Issued from different curricula, they can be divided as:

- 69 % men and 31 % women (in the first survey, we can notice a higher percentage of women which can be explained by the participation of social and human science disciplines), as well as
- 67 % French, 20 % German, 2 % other Europeans and 11 % Africans.

Among them, 40 % had benefited of financial aids as member of underprivileged social classes.

We will present in the following only a part of this survey, concerning only Computer Science and Mechanical Engineering alumni.

Presentation of the participants

- In Computer Science: 19 participants (18 men, 1 woman). Nationality: 15 French, 2 German, 2 other nationalities. Graduate level: 16 Bachelor, 7 Master (among which 3 alumni did not graduate a Bachelor curriculum at ISFATES).
- In Mechanical Enginnering: 16 participants (12 men, 4 women). Nationality: 8 French, 3 German, 4 both French and German, 1 from Cameroon.

It is worth to notice that, due to the big number of questions of this survey, there were participants that did not answer to all of them.

Here we detail the answers to some of the main interesting questions:

How did you find your first job and how long did it take?

For this question, we distinguished the participants according to their degree level (Bachelor or Master). In fact, in France, the first jobs in Computer Science and Mechanical Engineering are difficult to reach after a Bachelor degree. For this reason, many of the French Bachelor students continue their studies in Master programs. The survey shows that this is not the case for the bi-national Bachelor diplomas.

- In Computer Science, only 13 Bachelor and Master alumni answered to this question. 6 received a job offer before graduation, and 4 within a month after graduation. Other participants have answered that they have the feeling that having a higher or lower degree level does not change the final result.

  More exactly, among the 6 Master alumni that answered to this question, 3 received a job offer before graduation, and 3 have found a job within a month after graduation.

  Also, among the 7 Bachelor alumni, 4 received a job offer before graduation, 1 within a month after graduation, 1 got the job after three to 6 months and 1 in less than 1 year (the participant from Cameroon).

- In Mechanical Engineering, there were 11 Bachelor and Master participants that answered to this question. 2 received a job offer before graduation, 2 in less than 1 month after graduation, 4 in a period between one and 3 months, 2 in a period between 3 and 6 months, and 1 in less than 9 months.

  When detailed according the study level, only 4 Master alumni out of 6 answered, as follows. 2 received a job offer before graduation, 2 within 3 months from graduation, 1 in a period between three to 6 months, and 1 in less than 9 months (this alumni has benefited of a collaboration with Polytechnique Montréal to continue his studies by following a Master curricula of Technologic Project Management, then found his first job in Canada shortly after graduation).

  7 out of the 10 Bachelor alumni have answered to this question, as follows. 2 received a job offer before graduation, 2 found a job within a month after graduation, 2 during a period between one to 3 months.

In which country did you find the first job?

- In Computer Science: Germany (8), France (2), Luxemburg (1), and Canada (1).
- In Mechanical Engineering: Germany (6), France (4), and Canada (1).

How do you perceive the future (of the job offers)?

- In Computer Science: ordinary (2), good (3), very good (3).
- In Mechanical Engineering: good (2), very good (5).

And if it were to do it again, would you do it at ISFATES-DFHI?

- In Computer Science: 100 % of positive answers.
- In Mechanical Engineering: 4 out of 6 alumni answered and the answers were 100 % positive.

4 Discussion

For sake of simplicity, we denote by:

- E1, the UFA-DFH alumni survey, and
- E2, the ISFATES-DFHI alumni survey.

These two surveys about transnational French-German diplomas are different with respect to:

- the number of alumni (91 for E2 and 1802 for E1),
- the kind of diploma (engineering and management for E2 and any category for E2),
- the kind of questions (some questions have been formulated differently).

However, the surveys' results seem to converge on the following particular points:

- the French-German diplomas allow to find rather quickly a job adequate to the qualification level (less than 3 months for 71 % of participants in E1, and 92 % of the Computer Science alumni and 72 % Mechanical Engineering alumni in E2, respectively).
- the country of the first job is mostly Germany for both surveys (57 % for E1, and 67 % for computer scientists and 54 % for mechanical engineers in E2, respectively), then France (26 % for E1, and 16,5 % for computer scientists, 36 % for mechanical engineers, respectively), and finally other countries (26 % for E2, and 16,5 % for computer scientists and 9 % for the mechanical engineers, respectively).
- great satisfaction has been noticed for making the choice of a French-German curriculum since the question *"Would you recommend a French-German curricula to future students?"* received 90 % of positive answers in E1, which is comparable to the answers received for the question in E2 *"And if it were to do it again, would you do it at ISFATES-DFHI?"* (100 % of positive answers for the involved computer scientists and mechanical engineers). We didn't notice big differencies between nationalities.

To our knowledge, surveys conducted for alumni of national degrees did not include similar questions. For this reason, we have found difficulties to compare our results with them. Our results show the added-value of the French-German transnational diplomas, particularly, for majors from the domain of engineering sciences, as we have seen for mechanical engineers and, even more predominantly, for computer scientists. In fact, there exist to our knowledge only four French-German curricula in Computer Science: (1) at ISFATES-DFHI, (2) at UFA-DFH, involving INSA from Lyon (France) and University of Passau (Germany), (3) in the frame of the ECS (European Computer Science) network [2], between University of Lorraine (France) and HAW of Hamburg (Germany), and (4) in the frame of an ERASMUS Mundus Master in Information Technology for Business Intelligence.[7]

Going back to the Computer Science alumni, the conclusions of our surveys can be explained by the small number of alumni with respect to the needs of the French-German job market (recall that Germany is the first economy partner for France). There are few of such diplomas and many potentially interested enterprises may ignore their existence. On the other hand, it seems that only some students are interested by such curricula: the demand of computer scientists is still strong on the market and the added-value of transnational diplomas is not necessary for most of the cases. There are other reasons that may explain the lack of popularity of French-German diplomas:

- the lack of interest to learn a foreign language for those interested in following a scientific career, in particular the learning of the German language in France or of the French language in Germany, which diminishes year after year;
- the little attractiveness for France or Germany: other countries are currently more demanded, as the United States of America or Canada;
- the lack of information to be disseminated in high schools about the job opportunities offered by the French-German curricula in Computer Science and, more generally, in other engineering domains.

The main question that we can ask is how to augment the number of candidates for French-German curricula? Its answer depends on how effective is the learning of the partner language (only 7 % of French pupils learn German and 8 % of German pupils learn French). But this is all about political choices to be taken in both countries.

5 Conclusions

Even if the two surveys concerned only French-German joint diplomas, we have confirmed once more that transnational curricula imply a great added-value to students. The conducted surveys have shown that bi-lingual Computer Science alumni are among the most demanded on the job market.

---

[7]http://www.fib.upc.edu/en/masters/it4bi.html.

In particular, the transnational diplomas are a chance for all of us: for students (acquisition of intercultural competences, achievement of maturity and autonomy in a way more effective than for no-mobile students [4]), for the entreprises operating in France as well as in Germany (speaking the local language may limit the cultural problems), and for universities (having bi-cultural groups may benefit to no-mobile students and the teaching experience is more lively).

The main question is how to augment the number of bi-national curricula students, in France, Germany and, more generally, in Europe. Clearly, a more effective political and financial help from UE is needed to advertise the transnational diplomas at European level. We think that there are other solutions that could be effective, too. Here are some of them:

- to advertise them in high schools by the means of, for example, short video movies (to witness alumni experience or to share the student experience in a foreign country, ...) that can be remotely accessed by internet using smartphones or tablets.
- to make aware the German language teachers in France and French language teachers in Germany about the opportunities offered by the French-German curricula.
- to send in high-schools alumni and current students already convinced by the benefits of transnational diplomas and talk about them in front of pupils, to become in some way 'ambassadors' of these curricula.

References

1. Bracht, O., Engel, C., Janson, K., Over, A., Schomburg, H., Teichler, U.: The professional value of ERASMUS mobility. Technical Report. European Comission (2006)
2. Gerken, W.: The European bachelor degree course ECS. In: International Conference on European Transnational Education (ICEUTE 2010), pp. 70–75 (2010)
3. Harfi, M., Mathieu, C.: Mobilité internationale et attractivité des étudiants et des chercheurs. Horizons stratégiques 1, 28–42 (2006)
4. Michel, G.: Why mobility is important for european students in computer science: review of 18 years of a franco-german university training with a dual degree. In: ACM-IFIP Informatics Education Europe III (EEIII), pp. 5–11 (2008)
5. Michel, G., Stratulat, S.: Good reasons to implement transnational European diploma programs in Computer Science. In: ICEUTE'2010 (First International Conference on EUropean Transnational Education), pp. 135–143 (2010)
6. Terrier, E., Séchet, R.: Les étudiants étrangers : entre difficultés de la mesure et mesures discriminantes. Une application à la Bretagne. Norois-Environnement, aménagement, société 203, 67–85 (2007)

Impact Level Within the English Friendly Program in Virtual Degrees at the University of Burgos. A Real Case Scenario of Transnational Education

Ana María Lara Palma, José María Cámara Nebreda
and Elena María Vicente Domingo

Abstract Virtual teaching has become an innovative challenge at the University of Burgos (UBU). It has been founded under a continuous quality process perspective. Globalization and technology have been a solid support for the Bologna Programs and the online degrees. University of Burgos has set up for the first time during the academic year 14-15 four virtual courses, such as, Virtual Degree of Tourism, Virtual Degree of Spanish: Language and Literature, Virtual Degree in Politic Sciences and Public Management and Virtual Degree in Computer Science Engineering. All of them are worth 240 ECTS (European Credit Transfer and Accumulation System) and have been accredited by the University Council (Ministry of Education) after receiving a positive evaluation from the Agency of Quality for the University System in Castilla and León (ASUCYL). Although very recent too, the English Friendly Program was put in place in our university to assist foreign students lacking a sufficient command of the Spanish language. This is done by a series of mandatory actions for all the courses offered under the English Friendly tag. The purpose of this paper is to prove that these two unrelated programs can complement one another and become a powerful tool to promote our courses internationally.

Keywords Transnational education · Virtual learning · English friendly program

A.M.L. Palma (✉)
Department of Civil Engineering, University of Burgos, Burgos, Spain
e-mail: amlara@ubu.es

J.M.C. Nebreda
Department of Electromechanical Engineering, University of Burgos, Burgos, Spain
e-mail: checam@ubu.es

E.M.V. Domingo
Vicerrector of Internationalization and Cooperation, University of Burgos, Burgos, Spain
e-mail: evicente@ubu.es

© Springer International Publishing Switzerland 2015
Á. Herrero et al. (eds.), *International Joint Conference*, Advances in Intelligent Systems and Computing 369, DOI 10.1007/978-3-319-19713-5_46

1 From the Bologna Process to the Transnational Education

The Bologna Process was the starting point of a comprehensive process meant to achieve the convergence of the European Higher Education Area. Long before, the Erasmus Program, now Erasmus+, had become a successful framework for student mobility within the European boundaries. In spite of the differences in educational models, the diversity of diplomas and the large number of languages spoken in the continent, the convergence, assisted by mobility, was before an imperative need and it is now a reality [1].

Since its creation, back in 2010, our innovation group has focused its efforts on the promotion of the usage of English language in regular teaching. Two main reasons led us to this endeavor: the improvement of the language skills of our students and the increase of incoming students from other countries [2].

The results of our previous research provided interesting findings about Spanish student's attitudes and behavior towards English language. Furthermore, in [3] we proposed a series of activities to improve students' mobility. These actions, although effective to improve outgoing mobility, had no significant impact on incoming mobility figures. This is not only the case of the Higher Polytechnic School, but a reality in the whole Institution. In academic year 2013/14, the number of outgoing mobility students in our university was 246, whereas we were only visited by 184 foreign students.

The English Friendly Program, proposed by our innovation group and running since academic year 2012/13, is aimed to encourage foreign students to come to the University of Burgos. Although lectures are provided in Spanish, documentation and exams are available in English. This recent program has already yielded some significant results.

In academic year 2014/15 our university has started a whole new program: online education. It is being progressively introduced in several bachelor and master degrees. The goal is to make it easier for students to reconcile their studies and personal life. The program involves both teaching and evaluation so the students don't need to attend any event at the University premises.

In this virtual scenario, the University of Burgos points out a series of parameters that could make more attractive this wide range of learning to the students. They are summarized next below:

The time parameter: all the resources are online twenty four hours per day, seven days per week. Therefore, students have total access and disposal of their practical cases, theory, seminars, lectures, marks and tutorials.

The distance parameter: students do not need to attend lectures in the institution because the explanations come through the intranet and the virtual net-platform of the University of Burgos (UBUNET). The benefits, among others, are that the students save time and money as they do not need to live in the city.

The personal-circumstance parameter: each student can decide how distribute time. The advantages are many in this sense, because if the students have a job or any other commitment, they can arrange their timetable accordingly.

The socialization parameter: virtual teaching is not an impairment to socialization between teachers and classmates because teachers encourage students to work in groups, attend debates and share their experiences and doubts through the chat.

The innovation parameter: the web site of the University of Burgos, in its "Campus of International Excellence" link, shows crystal clear the idea that "the educational model of the UBU places the students in the heart of the educational process guarantying the acquisition of all competences that are demanded by the European Higher Education Area (EHEA). This model tries to apply the Information Technologies (IT) to the learning world, strengthening the abilities and virtual skills of the students in order to help them adopt an active role in their whole lives". To achieve this target, the UBU has made a big effort investing in new technologies, creating a new intranet as common place to work with the students and other colleagues.

The initiative has been successful enough in the first year so it will presumably be maintained and probably extended over the years to come.

Apparently, these two programs, English Friendly and on-line teaching are not related to one another but, what we want to prove in this work is that they are highly compatible and is on the interest of the university and the students to combine them.

2   A Real Case Scenario. Virtual Degrees and the English Friendly Program. An Empirical Analysis

This paper reflects the research carried out to analyze how virtual degrees and the English Friendly Program can match and to find out the impact this could have on our on-line students, both Spanish and foreigners. More specifically, this research is deeply focused on the group of students registered in the subject "Deontology and Legal Fundamentals of IT" in the Virtual Degree of Compute Science Engineering. In order to achieve the purpose of the study, the authors conducted a pilot survey during the winter semester 2014/15. Field research and sample present following characteristics:

- 18 survey responses[1] (18 out of 31 students registered in the first academic year of Virtual Degree in Computer Science Engineering), all of them valid.
- The data collected for the study is representative of the total population, as standard error estimation[2] is around 84.75 %. For this mathematical formula (1) it has been chosen the standard error [4]:

---

[1]Number of survey responses is limited to the students registered in the Virtual Degree in Computer Science Engineering. .

[2]Standard Error Estimation has been selected in order to understand how significant are the results considering the sample size and sampling fraction. Among others, [4] define this correlation.

$$\text{Standard Error Estimation} = [(N - n)/N]^{0.5} * (1/n)^{0.5} = 0.1524 \qquad (1)$$

- The survey displays 11 questions meant to collect students' interest on attending courses within the English Friendly Program (EFP). The design of the survey obeys to the criteria of feasibility of the EFP and the relationship with the academic and professional success of the students. The outcomes of the research are presented in the next section.

3 EFP. Perceived Interest and Benefits

The results of this survey provided interesting findings about student's interest towards EFP in Virtual Degree on Computer Science Engineering. Figure 1 gives an overview of them. A majority of respondents acknowledged the importance of bilingual teaching (67 %). However, 33 % of respondents were reluctant to participate. Although the positions are not strongly polarized for and against EFP in virtual degrees, it is required to analyze strategies to enhance student motivation, such as, break monotony, active learning, encourage them to participate in the feedback learning cycle or help them to get their competences efficiently. These recommendations are worth highlighting because provide opportunities not only to the institution to maximize the effectiveness of learning with higher implications, but also the students to achieve their course learning goals.

As far as the survey results show, few recognize that participate in the bilingual teaching would improve their academic performance (11 %), whereas a majority of respondents felt that EFP was a challenge in their academic performance (89 %).

These results lead us to conclude that many students associate the effort of enrolling in a bilingual teaching with positive and beneficial consequences.

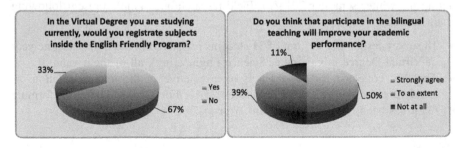

Fig. 1 Interest in subjects within the EFP and their connection with the academic performance

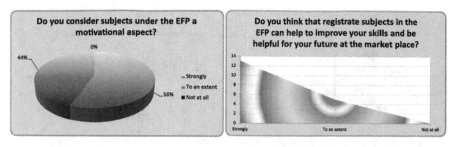

Fig. 2 EFP as motivational aspect for students and benefits for their future at the market place

4 EFP. Motivational Aspect and Innovative Opportunity for the Future

The results depicted in Fig. 2 also suggest a potential link between the EFP and the future benefits. Traditionally Spain has tended to lack skills in languages. Under these circumstances, have these minimized opportunities at the market place? As far as the survey results show, students felt that this opportunity to enhance language skills involved a good performance in their curriculum. In fact, all of them agreed with it. Among them, 72 % strongly agreed EFP helpful for their future work and 28 % to an extent respectively. In any case it has been found a negative answer.

To fully elucidate action measurements to achieve the aim of this research, we should analyze the actual use of the bilingual teaching and potential actions to cover the student's necessities without excluding any opinion. To support this main idea, [5] points out that "the careers of all of the students will be global ones, in which they will need to function effectively in multi-national teams. They will need to understand the cultural differences and historical experiences that divide us, as well as the common values and humanity that unite us". Nowadays there are a wide range of excellent universities and it should make us consider what else we can do to improve ours and classify the EFP as an innovative strategy to achieve our students' goals.

5 EFP. Exploring Benefits on Transversal Competences

It is alleged that transversal competences have been an excellent guide for teachers. [6] assess that "transversal and generic competences, has contributed to multi-disciplinarily and to enriching the qualification and professional perspectives of thousands of graduates" The inclusion of the transversal competences in the university programs is a fundamental pillar for training aspects in a changeable society. The survey conducted explored the impact of EFP in five competences of the Bologna Program: self-learning, capacity for analysis and synthesis, creativity, social skills, initiative and entrepreneur abilities and language skills.

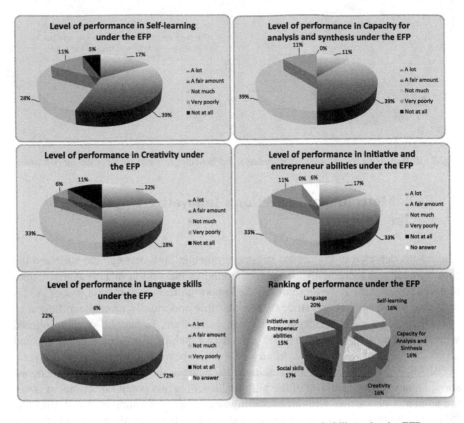

Fig. 3 Representative results about performance on the transversal skills under the EFP

Figure 3 shows some representative results about the performance on the transversal skills under the EFP. More than 50 % of respondents feel that four out of the five transversal competences would improve under the EFP. However, around 40 % remain skeptical about it. In addition, just a few of them hardly admitted any benefits in it (5 % in self-learning and 11 % in creativity). Perhaps the most significant finding is 94 % of the respondents acknowledge the importance of their benefits in language skills. In addition, respondents consider the EFP as a positive academic system to improve not only language skills but also social skills (20 and 17 % respectively). Self-learning, capacity for analysis and synthesis, creativity, and initiative and entrepreneur abilities are underestimated around 16 %.

Results suggest developing a practical and effective framework about bilingual teaching in virtual degrees in spite of the fact that some outcomes are not oriented in this kind of learning. Simultaneously, the framework should contain key distinguishing rules in two optional cognitive foundations of learning, Spanish and English. This concept has been previously defined as cooperative learning when [7] displayed the six main benefits of it: "(…) two linguistic, two curricular and two social: increased frequency and variety of second language practice through different types of interaction, possibility for development or use of the first language in ways that support cognitive development and increased second language skills, opportunities to integrate language with content instruction, inclusion of a greater variety of curricular materials to stimulate language use as well as concept learning, freedom for language teachers to master new professional skills, particularly those emphasizing communication, and opportunities for students to act as resources for each other and, thus, assume a more active role in learning.

6 Conclusions

In September 2014 the University of Burgos launched the Virtual Degree in Computer Science Engineering. In step with the increasing importance of English skills [8], this research has pursued both in parallel. Although it is claimed the difficulties a second language entails [9], the survey outcomes demonstrate that a majority of respondents are strongly motivated by the bilingual learning in virtual degrees. Moreover, they feel there is a correlation between this learning style and their success in future. Among others, the transversal competences enhanced are, language and social skills, self-learning, capacity for analysis and synthesis, creativity and initiative and entrepreneur abilities. Needless to say, this line of research is still very immature and will be required to carry out more detailed analyses of new survey results which will be conducted in the next academic years.

Transnational education is a concept that points out that the academic training has to be alive in a window of knowledge transference and wisdom sharing. Its impact on society, the student and the institutions will be, for sure, highly positive and enriching [10]. All in all, this paper pictures the proposal one day born at the University of Burgos with the aim of going ahead through the transnationalization and the university excellence.

References

1. Cámara J.M., Sánchez, P.L., Represa, C., Zaldívar, C.: Experiencias de Movilidad en la Escuela Politécnica Superior de la Universidad de Burgos. In: Proceedings of the 17th Congreso Universitario de Innovación Educativa en Enseñanzas Técnicas (CUIEET 2009) (2009)
2. (Declaración de Bolonia) The European Higher Education Area. Joint declaration of the European Ministers of Education, Convened in Bologna on the 19th of June 1999 (2000)
3. Cámara, J.M., Lara Palma, A.M., Jimenez, A.: A comprehensive approach to tackle the lack of English language skills of engineering students. In: Proceedings of the International Conference on Engineering Education (ICEE 2011) (2011)
4. Lininger, C., Warwick, D.: La encuesta por muestreo: teoría y práctica. CECSA, México (1985)

5. Goodman, P.: (Institute of International Education): Open Doors 2013. International Students in the United States and Study Abroad by American Students are at All-Time High (2013)
6. Rose, L.F., Haug, G.: Programme Profiles and the Reform of Higher Education in Europe: The Role of Tuning Europe (2013)
7. McGroarty, M.: The benefits of cooperative learning arrangements in second language instruction. NABE J. **13**(2) (2013)
8. Lara, A.M., Stuart, K., Karpe, J., Faeskorn-Woyke, H., Poler, R., Brotóns, R.: Competitive universities need to internationalize learning: perspectives from five European universities. J. Ind. Eng. Manage. 299–318 (2008)
9. Lara Palma, A.M., Brotóns, R., Karpe J., Faeskorn-Woyke, H.: Buenas prácticas en nuestras aulas universitarias. Chapter Book: La internacionalización de las universidades como factor clave para lograr la eficiencia y calidad del aprendizaje. Estudio de un caso. Instituto de Formación e Innovación Educativa. Universidad de Burgos, pp. 195–212 (2010)
10. Giacinto, R., Lara Palma, A.M.: Openplexus: distributed knowledge-sharing in virtual teams. In: Proceedings of the 10th International Conference on Semantic Systems (ICSS 2014) (2014)

A New Way to Improve Subject Selection in Engineering Degree Studies

J.A. López-Vázquez, J.A. Orosa, J.L. Calvo-Rolle,
F.J. de Cos Juez, J.L. Castelerio-Roca and A.M. Costa

Abstract In recent years, different Spanish degree studies are being changed and rearranged in accordance with the European Education System (EES) and the Bologna declaration. At that time there was no tool or methodology to define the manner in which different subjects of a degree had to be studied, for better improvement of knowledge. At present, research studies a new method to define the order in which each subject must be studied in different engineering degrees, with the objective of improving the actual studies to a new type, a type more adjusted to real situations. To do this, different statistical studies on electrical and electronic engineering from the University of A Coruña were evaluated. The main results showed a clear relationship between the marks when accessing the degree and the final average marks after completion of the entire degree. Furthermore, different models were obtained to define the relationship between the marks in the first and second courses as compared with the last one. Finally, it was concluded that more professional subjects depend not only on the previous subjects, but also on the same subjects in the last course, as in the case of electrical machines II. This new procedure had to be proposed in future degree adaptation processes, to improve the quality of university studies.

Keywords Academic performance · Correlation of contents · Historical analysis · Prediction

J.A. López-Vázquez · J.L. Calvo-Rolle (✉) · J.L. Castelerio-Roca
Department of Industrial Engineering, University of A Coruña,
Avda. 19 de Febrero, S/N, 15405 Ferrol, A Coruña, Spain
e-mail: jlcalvo@udc.es

J.A. Orosa · A.M. Costa
Department of Energy and Marine Propulsion, University of A Coruña,
Paseo de Ronda, 51, Ferrol, A Coruña, Spain

F.J. de Cos Juez
Mining Department, University of Oviedo, C/Independencia 13, 33004 Oviedo, Spain

© Springer International Publishing Switzerland 2015
Á. Herrero et al. (eds.), *International Joint Conference*, Advances in Intelligent
Systems and Computing 369, DOI 10.1007/978-3-319-19713-5_47

533

1 Introduction

The Bologna Process started in 1999 [1–3], when ministers from 29 European countries, including Spain, signed the Bologna Declaration, which aimed to establish, by 2010, a European Higher Education Area (EHEA), to achieve convergence and comparability in the European University Systems, facilitating employability, mobility and recognition of university degrees across Europe. Therefore, it is also known as a process of convergence in a European Higher Education Area.

Each university had to design their curricula and set the topics that always conformed to the required minimum quality criteria and that were agreed to by all countries of the EHEA.

All qualifications were grouped into five areas of knowledge: Arts and Humanities, Science, Health Sciences, Social Sciences and Law and Engineering and Architecture.

The curricula are structured according to the European Credit Transfer System (ECTS), in accordance with the Royal Decree 1125/2003 [4], of September 5 (BOE No 224, 18.09.2003), by establishing the European Credit System and Grading System in the official university qualifications, valid throughout the national territory.

According to Spanish law, the guidelines for the design of graduate degrees are defined in the Royal Decree 1393/2007 [5–10] of 29 October (BOE n 260, 30/10/2007), which establishes the official planning of university lessons, and Royal Decree 861/2010, of July 2 (BOE n° 161, 3/7/2010), which partially amends the previous one. In this Royal Decree, fixing the basic structure of undergraduate and graduate degrees as well as the process of verification and accreditation of qualifications must ensure the quality of the degrees.

The curricula leading to the award of the Bachelor's degree must contain a minimum of 60 core credits, of which at least 36 must be linked to some of the materials listed in Annex II of R. D. 1393/2007, and to the branch of knowledge that seeks to ascribe the title. The materials for the branch of knowledge of Engineering and Architecture are: Company, Graphic Expression, Physics, Computing, Mathematics and Chemistry.

In addition, the curriculum must include at least the training modules listed in the Order CIN/351/2009 of 9 February (BOE No 44, 20/2/2009), laying down the requirements for the verification of official university degrees that qualify the exercise of the profession of an Industrial Engineer. These modules are:

1. Sixty credits of the core subjects
2. Sixty ECTS materials of the industrial branch
3. Forty-eight ECTS credits, specific technology (Electricity, Industrial Electronics, Mechanical, Industrial, and Textile Chemistry)
4. Twelve ECTS for the Bachelor Thesis

Given the aforementioned legislation, universities can freely set any evidence, taking into account the definition, training and assessment of the learning outcomes to be acquired from the title, for it to be of "professional" or "social" relevant.

The universities, in exercising their autonomy, design their titles and their own internal systems to ensure their quality. In turn, before the new titles can provide education they have to undergo a verification process by the University Council to check that they have reached the required quality levels. Titles that passed the accredited verification are considered for a period of six years, after which they must again undergo an evaluation process to maintain the accreditation.

In short, the educational structures and content of the curriculum are developed by the universities themselves. In many cases, the design of the curriculum becomes a "negotiation between departments" on what each one wants to impart, without paying too much attention to the skills that the students should acquire or the best location for each subject within the own plan.

In the year 2009, the Bolognia Declaration was introduced in Spanish studies by means of a new distribution of degree studies. In this sense, the old engineering studies were divided into three years for the degree and two more for the master, and the more specific competencies were obtained during the degree and master studies, allowing the engineer to reach a wider number of competencies.

Furthermore, at that point of time, a change in the structure of the academic studies was conducted without any kind of tool to identify a better procedure or to select a more adequate subject for each course. In particular, different European degree studies were taken as references, and agreement between professors was the main tool to develop this task.

In the present research study, the subject of investigation is the detection of the main procedures to do this so that an important and unusual task and a few references are obtained. In consequence, researchers from different institutions of the University of a Coruña have investigated as to how this process must be carried out, based on the statistical analyses of marks obtained in the previous academic studies, and how it can be modelled with the aim of developing a method for optimizing and improving the engineering curricula. In particular, this study is based on the analysis of academic results obtained by the Industrial Engineering students (from 1996/1997 to 2009/2010). In consequence, if a better subjects selection of Electrical Engineering degree studies is done better marks and depth knowledge will be obtained. Furthermore, a time reduction of the number of years per student needed to finish the degree, which is extremely high in Spanish engineering degree studies, can be obtained with this new procedure. Finally, with this methodology common degree studies with other European degrees will help to establish a European Higher Education Area (EHEA), to achieve convergence and comparability in the European University Systems, facilitating employability, mobility and recognition of university degrees across Europe.

Another objective of this study is to design proposals centered on student self-learning, in order to achieve rising graduation rates and lowering dropout rates and securing efficiency in a new Grade curriculum implemented during 2010/2011.

2   Materials and Methods

Different research studies have been carried out on student's academic efficiency [11] based on a new index, defined as a function of the students qualification and the number of calls (1 ordinary and 2 extraordinary) in accordance with Eq. 1.

$$R = \frac{\sum_j 0.8^{c-1} \cdot M_j \cdot C_j \cdot 10}{\sum_j C_j} \tag{1}$$

where;
C is the number of calls (1 ordinary 2 extraordinary)
M represents the marks obtained

With this index, a reduction of 20 % in the students' efficiency is expected, where they pass the subject at a second call.

First of all, the teaching guide was taken as a reference for each subject due to it shows the main contents that must be understood and the new ones. Despite this, these indications were quite ambiguous and variables like time needed to revise the previous knowledge were needed. In this sense, a lot of time was reduced and simultaneously the related contents in the related subjects could be difficult. On the other hand, there were other kinds of contents that could be difficult, and the student would need enough time to be apply it to another course.

As it was explained before, the aim of this study is to determine the main subjects, their relation in accordance to the student's marks and how to improve the organisation of academic studies in the future.

To solve these problems, an initial study of the marks and students' source was made, with 1400 students, to define a representative value of average marks from each different source. Furthermore, a study about the relationship between the input and output marks in the degree would be analysed and then be modelled in accordance with the final marks, with and without corrections, in accordance with the number of calls.

In a similar manner, other studies [12] tried to relate the marks obtained in a subject as a function of the marks obtained in the first course. Despite this, due to the unknown marks of the more important subjects that could be inter-related, a poor curve fitting of the marks in all the subjects in the first course was done, with a posterior analysis of the coefficients. Furthermore, in related studies [13], a relationship was found between the marks obtained when gaining access to an engineering degree and its origins (vocational education and training or secondary school) and more studies showed the influence of previous knowledge on the final marks [14].

A minimum number of 25 students were employed in this study and different parameters were considered, such as, the corrected average in accordance with the number of years needed to pass a subject.

In consequence, in the second stage, a statistical study of the professional subjects in the degree would show the influence of the marks from the previous course subjects. Furthermore, a model could be obtained.

To do this, an initial statistical study was done, based on the Analysis of Variance (ANOVA), to define the relationship between the average marks of each student obtained in the entrance exam of the degree of electrical and electronic engineering of the University of A Coruña and the final average marks of the same students on completion of the entire degree.

Following this, the ANOVA study was done with the aim of defining which subjects showed the same tendency compared to the others, and finally, different regression models were employed to obtain the mathematical relationship between the marks in the subjects in the final courses with respect to the marks obtained in the previous ones by SPSS 22.0.0.0.

In previous studies [15–17] it was only possible to define the design of the curriculum in the master studies by means of questionnaires. Our case study was an example of the degree curriculum design based on the obtained models that could help to develop a self-control system for degree studies [18, 19].

3 Results and Discussion

In this section the main conclusions of this research will be commented on, based on the results obtained.

3.1 Input Groups of Students

As it was explained before, in these two degrees, 1400 students from 1996 to 2011, were selected in accordance with their academic origin. This is an interesting study period due it corresponds with a pre-bologna academic degree plan. In this sense, students from the secondary school (1) and students from the vocational education and training (2) were analysed.

An initial result from the ANOVA study for a significance level of 0.05 showed that both groups of students were clearly different (significance of 0.000). In particular, as we can see from Table 1 and Figs. 1 to 2, the average marks of the students from the secondary school were about 6.0, while for the vocational education and training students this average mark was about 6.5, with a higher standard deviation. Despite this, the maximum and minimum marks were similar.

Table 1 Statistical study of input marks

Group	No.	Average	Standard deviation	Minimum	Maximum
1	1307	5.9925	0.65179	4.95	8.57
2	49	6.5282	1.14547	5.00	8.50

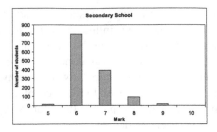

Fig. 1 Histogram of secondary school marks

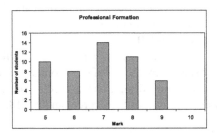

Fig. 2 Histogram of vocational education and training school marks

3.2 Relationship Between the Input and Output Marks

A simple correlation study shows the relationship between the degree input marks and the average degree output marks. At the same time, as there are two different ways to access the degree course (vocational education and training and Secondary Studies) an ANOVA study has been done. Its results show two groups in accordance with their kind of previous studies. At the same time, it is possible with Matlab R2015a to define a model whose access marks are related to the undergraduate final average, as shown in Fig. 3 and Eq. 2. In particular, it is the simple average that shows a better approach. That which does not reduce the average

Fig. 3 Relationship between degree input and output marks

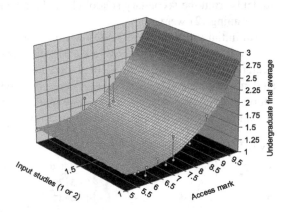

values in accordance with the number of years needed to pass the subject is called simple average. It is common that the final course students need more than one year to pass some of the subjects.

$$z = a + bx^{2.5} + cx^3 \tag{2}$$

$r^2 = 0.8214585098$ | $a = 1.756335034$ | $b = -0.03457759$ | $c = 0.01201803$

At the same time, it is possible to define a model that relates the different degree of access manner to the average marks in the final degree, as we can see in Eq. 3.

$$y = a + bx^{0.5} \ln(x) + c \ln(x)^2 \tag{3}$$

$r^2 = 0.95$ | $a = 4.308497748$ | $b = 9.557320399$ | $c = -7.47343974$

Finally, the One-Way ANOVA results are shown in Table 2 for a significance level of 0.05. In particular, how the manner of access is related to the access mark, but not to the duration of the degree study or the final average mark is shown. This result is of special interest for students that need to choose degree studies.

Table 2 One way ANOVA results with respect to access way (1 or 2)

Variable	Sig.
Access marks	0.046
Years needed to finish studies	0.558
Final average marks (with corrections)	0.096
Final average marks (without corrections)	0.088

3.3 Relationship Between Subjects and Courses

It is well known and usually expected that a good student in vocational education and training or secondary school will get good marks during the degree studies. To analyze and model this effect, more representative subjects from the degree of electrical and electronic engineering were selected.

These two degrees showed common initial subjects during the first and second courses and only few different subjects were needed to obtain the proper contents of each degree study, as we can see in Table 3.

If now we consider the order and year in which each subject is received by students we can conclude from the One-Way ANOVA analysis the more representative subjects of the electrical engineer degree, see Table 3. These subjects are defined by the groups of similar variance which cannot be considered with the same behavior than the other subjects. It is interesting to highlight that the ANOVA test was performed over the final marks of all the students at the same time. For example, it is the case of electrometry. It can be observed that transmission of electricity is close enough to be considered to be with a different evolution.

Table 3 The academic studies of Electrical and Electronic Engineering

University of A Coruña	
Degree in electrical engineering	Degree in electronic engineering
First course	First course
Circuit theory (9c.)	Circuit theory (7.5 c.)
Second course	Second course
Electrometry (3 c.) ⏐ Electrical circuits (6 c.) Automatic regulation (7.5 c.) ⏐ Electrical Machines I (4.5 c.)	Analog electronics (7.5 c.) ⏐ Dig. electronics (7.5 c.) ⏐ Aut. adjustment (10.5 c.) ⏐ Electrical Systems (10 c.)
Third course	Third course
Elec. Machines II (7.5 c.) ⏐ Transmission of elec. (10.5 c.) ⏐ App. of elec. machines (6 c.)	Power electronics (6 c.) Electronic instrumentation (11 c.)

Table 4 Analysis of variance in electrical eng. degree with respect to circuit theory (CT)

Abbreviation	Subjects	Sig.
El	Electrometry	0.036
EC	Electrical circuits	0.216
AR	Automatic regulation	0.953
EM II	Electrical machines II	0.668
TE	Transmission of electricity	0.066
AEM	Application of electrical machines	0.387
EM I	Electrical machines I	0.851

From Table 4 it can be concluded that electrometry is the only one that can be considered different form the others, as its significance level is very low. It is normal, as this subject has a lower number of teaching hours as also content, when compared with the other subjects. As a consequence, this subject can be considered without special revelation for this degree. Despite this, the initial result validates the procedure employed and its understanding. If we now do a correlation study of each subject Table 5 can be obtained. This table shows the correlation between each subject and the final average marks (FAM) and their determination factors.

Interesting conclusions could be obtained from this initial result. For example, it was seen that there was a clear correlation between circuit theory (a subject difficult in the first course) and the final average mark of the entire degree, as we can see in Fig. 4. This result could be plotted and modelled in Fig. 4.

3.4 Modelling the Relationship Between Subjects: Electrical Engineers Analysis

In this section we will try to define the relation between subjects by means of a mathematical model obtained by a polynomial regression of each degree subject. In

Table 5 Determination factor between marks in electrical engineering

	CT	El	EC	AR	MEI	ME2	TE	AEM	FAM
CT	1								
El	0.3306	1							
EC	0.5658	**0.7491**	1						
AR	−0.2424	0.0001	0.0795	1					
EMI	−0.1631	−0.1123	−0.4725	−0.0753	1				
EM2	0.2167	0.2497	0.4498	0.3210	−0.1871	1			
TE	0.5368	0.2421	0.3379	−0.1095	−0.2759	0.3213	1		
AEM	0.3311	0.3607	0.2045	−0.2577	0.1433	0.3171	0.4844	1	
FAM	**0.8035**	0.6722	**0.7290**	0.0197	−0.1598	0.5164	0.6349	0.5313	1

Fig. 4 Final average marks versus circuit theory marks

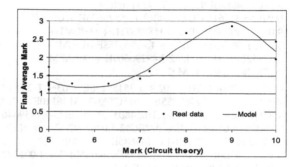

this sense, we needed to employ the marks of the different subjects shown in Tables 4 and 5, as we can see in Eqs. 4 to 7 of Table 6.

As we can see, Eq. 4 shows the relationship between the final average marks and the subjects, with an adequate determination factor. If now we do the same with the subjects of the last courses with the first and second courses (see Table 2) we can define a model for the marks in the subject transmission of electricity with respect to circuit theory, electrometry, automatic regulation, Electrical Machines I and II, and Application of electrical machines, as we can see in Eq. 5. Despite this initial result, it was not possible to define an adequate model between electrical machines II and subjects of the previous courses with an adequate correlation factor, as we can see in Eq. 6.

If now we try to relate the marks of these subjects with the marks of the previous subjects and the marks of the subjects that are taught in the same course, it is possible to define an adequate model, as we can see in Eq. 7.

Table 6 Models in electrical engineer studies

Equation	Determination factor (%)	Eq.
FAM = −1.57934766E + 01-1.11720443E + 00*CT + 7.39511281E-02*CT^2 + 6.50227790E-01*El-4.73196614E-02*El^2-6.54448179E-01*CE + 5.50612694E-02*CE^2 + 2.73053023E + 00*RA-1.93605719E-01*RA^2 + 1.26532882E + 00*MEI-9.44285307E-02*MEI^2 + 1.09123454E-01*ME II-1.29611975E-03*ME II^2 + 9.40681233E-01* TE-5.06288415E-02*TE^2 + 8.25911640E-01* AEM-4.75660529E-02*AEM^2	100.00	4
TE = 5.78260539E + 01 + 4.02696806E + 00*CT-2.13709646E-01*CT^2-7.67936525E + 00*El + 5.83684048E-01*El^2 + 7.24731460E + 00*EC-5.41261582E-01*EC^2-1.25575675E + 01*AR + 9.03922140E-01*AR^2-4.15888371E + 00*EMI + 3.22131906E-01*EMI^2 + 4.78590758E-01*EM2-5.33152136E-02*EM2^2-3.49290669E + 00*AEM + 2.17842874E-01*AEM^2	93.02	5
EMII = 5.23009837E + 02 + 1.75159176E + 03*CT-2.58123916E + 02*CT^2 + 1.21482955E + 01*CT^3-3.83136140E + 02*El + 4.35868924E + 01*El^2-1.33054805E + 00*El^3-7.18596463E + 02*EC + 1.05368444E + 02*EC^2-5.02625926E + 00*EC^3 + 1.21105766E + 02*AR-1.63743685E + 01*AR^2 + 5.94242256E-01*AR^3-7.88691368E + 02*EMI + 1.00139711E + 02*EMI^2-4.12898466E + 00*EMI^3	66.29	6
EMII = 7.53675953E + 01 + 8.96285661E-01*CT-7.12958284E-03*CT^2-1.54907373E + 01*El + 1.03421568E + 00*El^2 + 9.90955776E + 00*EC-6.74333840E-01 *EC^2-1.28005482E + 01*AR + 9.11100157E-01 *AR^2-5.66940514E + 00*EMI + 4.33702894E-01*EMI^2 + 1.18341225E + 01*TE-9.44002307E-01*TE^2-8.32073103E + 00*AEM + 5.88372510E-01*EM^2	95.29	7

3.5 Modelling the Relationship Between Subjects: Electronic Engineer Analysis

Similar to the previous sections, the same procedure was employed to define the relationship between subjects in the electronic engineer degree. In this case, the final average marks could not be modelled as a function of the specific subjects of the degree, as can be seen in Eq. 8. Despite this, its determination factor is quite high reaching values over 77 %.

Table 7 Models in electronic engineer studies

Equation	Determination factor (%)	Eq.
FAM = $-1.37163166E$-$01 + 1.05030968E$-$01*CT$-$1.03257572E$-$02*$ CT^2-$1.77249181E$-$01*AE + 1.47084417E$-$02*AE^2 +$ $4.43763066E$-$01*DE$-$3.43081489E$-$02*DE^2$-$1.51417812E + 00*AR$ $+ 1.16544757E$-$01*AR^2 + 1.53321914E + 00*ES$-$1.31295480E$-$01*ES^2$-$1.87422009E$-$01*PE + 1.73431695E$-$02*PE^2$ $+ 2.99186772E$-$01*EI$-$1.51732813E$-$02*EI^2$	77.26	8
PE = $8.57092524E + 02 + 2.57206450E + 01*CT$-$3.27074623E +$ $00*CT^2 + 1.35598798E$-$01*CT^3$-$1.20091914E + 02*AE +$ $1.90002835E + 01*AE^2$-$9.77986590E$-$01*AE^3$-$1.25013138E +$ $02*DE + 1.77607097E + 01*DE^2$-$8.19520325E$-$01*DE^3$-$1.68837030E + 02*AR + 2.75443137E + 01*AR^2$-$1.47397361E +$ $00*AR^3$-$1.60783482E + 01*ES + 1.71899387E +$ $00*ES^2$-$4.71270282E$-$02*ES^3$	83.61	9
PE = $4.27493050E + 02 + 7.00232869E +$ $00*CT$-$1.78670903E$-$01*CT^2$-$2.92138129E$-$02*CT^3$-$1.36925943E$ $+ 02*AE + 1.99086252E + 01*AE^2$-$9.38978873E$-$01*$ AE^3-$1.56812280E + 02*DE + 2.18380369E +$ $01*DE^2$-$9.86724517E$-$01*DE^3$-$2.08138723E + 02*AR +$ $3.24887550E + 01*AR^2$-$1.66950543E + 00*AR^3 + 2.33267914E +$ $02*ES$-$3.54867071E + 01*ES^2 + 1.77155526E + 00*ES^3 +$ $7.23624452E + 01*EI$-$1.13026200E + 01*EI^2 +$ $5.95124547E$-$01*EI^3$	99.03	10

From this result we can conclude that the final average marks of a student are clearly not influenced by the specific subjects of the degree, as we can see in Table 7.

Once again the relationship between the marks of the more specific subjects in the last course of the degree with respect to the previous ones was analyzed and modelled in accordance with Table 2.

When we try to relate power electronics to the subjects of the previous courses, a low determination factor of 83 % is obtained. Once again this correlation factor can be increased if we include the analysis of the subjects of the last course, in particular the subject of electronic instrumentation, as can be seen in Eq. 9.

Finally, it is interesting to highlight that in the previous studies [15–17] it has only been possible to define the design of the curriculum in the Master's studies by means of questionnaires and our case study is an example of a degree curriculum design, based on the obtained models that can help to develop a self-control system for the degree studies [18, 19].

4   Conclusions

In the present research different but significant conclusions were obtained:

- There is a clear homogeneity in the marks of each of the two students' origins (vocational education and training and secondary school). Despite this, the average marks are different between them.
- In the present research a clear relationship was obtained between the input marks from the previous studies and the final average marks once degree studies were completed.
- It was possible to obtain a model between marks of subjects in the third course with respect to the subject marks from the second and first courses.
- The more difficult subjects showed a better determination factor when their models were improved with subjects of the same course. For instance, it could be related to the fact that some contents of this subject were needed to pass the exams of electrical machines II.
- From these results, future academic studies must be defined considering the need to change this subject in a future course, as it has been done in the new degree studies.
- Another parameter that must be analyzed is the need to consider the years that a student needs to pass each subject. Despite this, it is clear that with this new procedure better marks and depth knowledge will be obtained. In consequence, the number of years per student needed to finish the degree is expected to be reduced.
- Finally, with this methodology common degree studies with other European degrees will help to establish a European Higher Education Area (EHEA), to achieve convergence and comparability in the European University Systems, facilitating employability, mobility and recognition of university degrees across Europe.

References

1. European Commission (Education & training Higher education). http://ec.europa.eu/education/higher-education/bologna_en.htm. Accessed October 2013
2. The Bologna Declaration on the European space for higher education: an explanation. http://ec.europa.eu/education/policies/educ/bologna/bologna.pdf. Accessed Oct 2013
3. University of Navarra. http://www.unavarra.es/conocerlauniversidad/the-university-today/ehea-european-higher-education-area/legislation?submenu=yes&languageId=1
4. Royal Decree 1125/2003, of 5th of Sept., establishing the European credit system and the grading system used in official university degrees valid throughout national territory
5. Royal Decree 1393/20074, of 29th of October, establishing the organization of official university courses
6. Royal Decree 49/2004, of 19th of January, on the harmonization of curricula and official degrees valid throughout national territory

7. Royal Decree 1044/2003, of 1st of August, establishing the procedure by which the European Diploma Supplement is to be issued by universities
8. Royal Decree 55/2005, of 21st of January, establishing the structure of university courses and regulating official university First degree courses
9. Royal Decree 56/2005, of 21st of January, regulating official university Postgraduate degree courses
10. Royal Decree 189/2007, of 9th of February, modifying certain provisions of R.D. 56/2005, of the 21st of January, regulating official university Postgraduate degree courses
11. Alcover, R., Benlloch, J., Blesa, P., Calduch, M.A., Celma, M., Ferri, C., Hernández-Orallo, J., Iniesta, L., Más, J., Vicent, M. J., Zúnica, L. R.: Análisis del rendimiento académico en los estudios de informática de la Universidad politécnica de Valencia aplicando técnicas de minería de datos. XV JENUI. Barcelona, 8-10 de julio de 2009 ISBN: 978-84-692-2758-9
12. Zúnica, L., Alcover, R., Más, J., Valiente, J., Benlloch, J.V., Blesa, P.: Relación entre el rendimiento de dos asignaturas de segundo curso y las asignaturas de primer curso en Ingenierías Técnicas de Informática de la Universidad Politécnica de Valencia. Libro de resúmenes del I Simposio Nacional de Docencia en la Informática (SINDI 2005). Granada, pp. 9–16 (2005)
13. Más, J., Valiente, J.M.: Estudio de la influencia sobre el rendimiento académico de la nota de acceso y procedencia (COU/FP) en la E.U. de Informática. Actas de la VIII Jornadas de Enseñanza Universitaria de Informática JENUI 2002. Cáceres, Spain
14. Esteve Faubel, J.M., Molina Valero, M.A., Espinosa Zaragoza, J.A., Esteve Faubel, R.P.: Autoaprendizaje en el EEES. Una experiencia en Magisterio, especialidad Musical. Revista de Investigación Educativa 27–2, 337–351 (2009)
15. Gharaibeha, K., Harba, B., Salameha, H.B., Zoubia, A., Shamalia, A., Murphyb, N., Brennanb, C.: Review and redesign of the curriculum of a Masters programme in telecommunications engineering – towards an outcome-based approach. Eur. J. Eng. Educ. 38(2), 194–210 (2013)
16. Gavin, K.G.: Design of the curriculum for a second cycle course in civil engineering in the context of the Bologna framework. Eur. J. Eng. Educ. 35(2), 75–185 (2010)
17. Henning, V., Bornefeld, G., Brall, S.: Mechanical engineering at RWTH Aachen University: professional curriculum development and teacher training. Eur. J. Eng. Educ. 32(4), 387–399 (2007)
18. Sáez, Cuesta, de Tejada, J.D., Fuensanta Hernández, P., Luís, Sales, de Fonseca Rosário, P.J.: Impacto de un programa de autorregulación del aprendizaje en estudiantes de Grado. Revista de Educación 353, 571–588 (2010)
19. Dehinga, F., Jochemsb, W., Baartmanc, L.: Development of an engineering identity in the engineering curriculum in Dutch higher education: an exploratory study from the teaching staff perspective. Eur. J. Eng. Educ. 38(1), 1–10 (2013)

ICEUTE 2015: Information Technologies for Transnational Learning

Quality Uncertainty and Market Failure: An Interactive Model to Conduct Classroom Experiments

María Pereda, David Poza, José I. Santos and José M. Galán

Abstract We present an interactive simulation game designed to teach the market effects of quality uncertainty. The instructor can conduct experiments in a virtual classroom, in which students using a computer are embedded in an online market playing the role of buyers. Many industrial engineering programs set aside these market effects because the impact of poor quality in customer behavior is very difficult to evaluate. This work complements traditional classroom approaches to quality improvement and standardization giving engineering students a clear justification for techniques to control and reduce variability in industrial and manufacturing processes. We propose a parameterization for a game and discuss the expected dynamics. Buyers with enough bad experiences form biased quality estimations and stop buying, which can make the market collapse. The game also allows exploring the influence of social networks as mechanism to enhance market performance. The game has been implemented in Netlogo and Hubnet platform.

Keywords Asymmetric information · Quality uncertainty · Classroom experiments · Engineering education · Social networks

M. Pereda (✉) · J.I. Santos · J.M. Galán
INSISOC, Universidad de Burgos, Edif. "La Milanera". C/Villadiego,
S/N, 09001 Burgos, Spain
e-mail: mpereda@ubu.es

J.I. Santos
e-mail: jisantos@ubu.es

J.M. Galán
e-mail: jmgalan@ubu.es

D. Poza
E.T.S. Ingenieros Industriales, INSISOC, Universidad de Valladolid, Pº Del Cauce,
S/N, 47011 Valladolid, Spain
e-mail: poza@insisoc.org

© Springer International Publishing Switzerland 2015
Á. Herrero et al. (eds.), *International Joint Conference*, Advances in Intelligent
Systems and Computing 369, DOI 10.1007/978-3-319-19713-5_48

1   Introduction

Many industrial engineering programs set aside the market effects of poor quality focusing on more tangible quality costs and the corresponding methods and techniques that try to improve quality [1], partly because of the intrinsic difficulty of measuring them. However, in our opinion, their inclusion in quality subjects would help engineering students to get a better vision of quality problems and understand the need of techniques to control quality variability. The simulation game we present pretends to solve this learning deficit. Like other simulation games applied to engineering education [2], our game exploits the characteristics of a simulation and modeling environment.

There is an important tradition, mainly in Economics, in using classroom games for teaching concepts [3, 4]. The strategic nature of many economic phenomena makes their explanation difficult by means of analytics models in the conventional blackboard. In contrast, games can improve the understanding of students when they participate and interact in a less abstract context and much closer to the problem being studied [5, 6]. Simulation games that take advantage of the computer power and the communication possibilities of the Internet do not only motivate students, much more familiar with new information technologies, but also facilitate the labor of the instructor of planning and running game sessions and analyzing and discussing results. Like other simulation games applied to economics and engineering education [7], our game exploits the characteristics of a simulation and modeling environment.

In this work, we have developed an interactive version of the Izquierdo and Izquierdo's model [8], specially designed to teach the effects of quality uncertainty in markets. The instructor can conduct experiments in a virtual classroom, in which students using a computer are embedded in an online market playing the role of buyers. An experiment consists of a series of trade sessions; in each session, students must decide their bidding price for the good without information about its quality distribution. The market is cleared using a fixed customizable supply function, and the students who finally purchase the good find out the actual quality of the unit they have acquired. Moreover, the instructor can explore the effect of social networks to improve market efficiency, letting students share information with their neighbors in the network.

The agent-based model of the quality uncertainty effects on markets by [8] was developed in Netlogo [9]; our version of this model has been implemented in Netlogo too. Netlogo is an application specially designed for developing and simulating agent-based models. The readable but powerful language that allows to express complex agents' behaviors in a few code lines, the interface that facilitates build sophisticated models quickly and the ease of use for researches, instructors and students, have become Netlogo an application widely used within the social simulation community. We have taken advantage of the HubNet technology that is integrated in Netlogo. HubNet lets Netlogo models run in an interactive way with people, who can individually control each of the agents of the model by a networked computer. This is particularly interesting when, as in our case, you want to conduct economic experiments in classroom.

In the next section, we explain briefly the theoretical background of the simulation game. Then in Sect. 3, we explain the different parts of the tool and its purposes. Finally, we briefly discuss the expected dynamics of a trade session conducted by students in a classroom and conclude the paper.

2 Theoretical Background

What is the cost of poor quality perceived by a customer? Quality management subjects address this central issue in any industrial engineering curriculum. A first approach is to quantify the impact of these issues on the costs of goods and services [10–12]. Unlike other quality costs easily measured, such as inspection and testing, scrapping and reworking, or warranty charges and recalls, the impact of poor quality in customer expectation and behavior is very difficult to evaluate. The failure to meet customer expectations resulting in customer defection, i.e. dissatisfied customers who had a bad experience with a purchase will not repurchase, is hidden quality costs which can be very important for firms [13].

Because removing deficiencies and controlling quality variability diminish, among other quality costs, these negative effects of poor quality in customer behavior, organizations implement quality standards and apply statistical techniques, i.e. control charts, six sigma, Taguchi, in order to reduce quality uncertainty in industrial processes and transmit confidence signals to consumers [14]. In some goods and services, this confidence signals can also be accomplished by means of marketing techniques, as buyers claim that high sales are due to superior quality [15].

On the other hand, the problems that quality uncertainty and asymmetric information cause in markets are well known in Economics. Traditionally, quality uncertainty phenomena have been studied assuming that one party, i.e. buyers or sellers, has better knowledge about the value of a good and exploits this information asymmetry to its own advantage. Using the automobiles market, [16] showed that market incentives can trigger adverse selection processes that can ruin the market.

However it is not necessary to assume asymmetric information: quality uncertainty without asymmetric information can lead to market failure, as shown by Izquierdo and Izquierdo [8]. In their model, sellers trade a commodity with a fixed quality distribution, which is a priori unknown to buyers. Buyers estimate the expected quality based on past purchases. Crucially, buyers with sufficiently many bad experiences stop buying the good (and do not update their bad expectations anymore). This leads to biased quality estimations, which can make the market collapse. In connection with this, recent studies on real markets confirm that the fact of quality uncertainty affecting both buyers and sellers results in a loss of efficiency for both parties. Some examples are the case of illicit drug transactions [17] and online markets [18].

3 Description of the Game

The game can be downloaded from https://www.openabm.org/model/4214/.

The application we have developed has two different roles: the instructor that controls the game, and the players.

3.1 The Instructor's Interface

The objective of the instructor's interface (Fig. 1) is twofold. First, it allows defining the parameters of the experiment (e.g. supply function, timing, parameters related to the uncertainty of the quality, etc.). Secondly, it allows the instructor to monitor the evolution of the experiment (number of elapsed rounds, number of buyers, market clearing price in the current round, etc.).

Server buttons (1)
The Server buttons (Sect. (1) in Fig. 1) allow the instructor to start and to finish the experiment. The instructions for connecting the clients (students) to the HubNet server are explained in the manual that is distributed with the application.

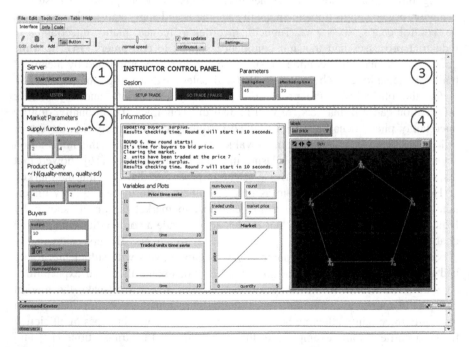

Fig. 1 Instructor's interface. The boxes and numbers inside circumferences are not part of the interface, they have been created to allow for a better explanation

Market parameters (2)

Whereas the demand side of the market depends of the students' decisions (number and value of bids), the supply side of the market is controlled by the instructor.

- The implemented supply function is linear with parameters: intercept 'y0' and slope 'a'.
- The quality of the items sold in the market is uncertain. It is assumed to follow a Gaussian distribution with mean 'quality-mean' and standard deviation 'quality-sd'. Each time a student purchases an item, the quality of the item is sampled from this Gaussian distribution.
- The 'budget' box represents the maximum amount of money that a buyer can spend in the item.

Instructor control panel (3)

Once all the students have connected to the server, the instructor must click on the SETUP TRADE button. This will create a virtual market according to the parameters chosen by the instructor. When the instructor is ready to start the session, she may click on the GO TRADE/PAUSE button. The first round will start.

The experiment consists of several rounds, and each round is divided into two phases (i.e. trading-time and after-trading-time). During the trading-time (configurable), the students have the possibility of bidding (i.e. if a student is interested in an item, she has to post the price she is willing to pay at this time). In the second phase, which duration also configurable, the system calculates the market clearing price and the student's interface shows her benefit in the current round, which depends on the price she proposed during the first phase, the market clearing price and the item quality (see Eqs. 1 and 2).

$$\text{Surplus} = \text{Budget} + \text{Item quality} - \text{Market price (if buyer purchases)} \quad (1)$$

$$\text{Surplus} = \text{Budget (otherwise)} \quad (2)$$

During this phase, the students may assess the results of the current round and maybe reflect on a strategy for the next rounds.

Information (4)

The purpose of this part of the interface is to provide the instructor with useful information during the execution of the experiment. Under the display, there are several information boxes: current round, number of buyers (i.e. the number of students that are currently connected to the server), number of units that were traded during the last round and market clearing price. The two plots on the left show the evolution of the market over the last few rounds: prices and traded units. There is another plot that shows the intersection of the supply function (defined by the instructor) and the demand function (which depends on the number and the value of the students' bids), which yields to the market clearing price. Once the experiment is over, the instructor and the students may discuss the data of these graphs.

The panel on the right displays a window with the buyers (students) that are taking place in the experiment. If the network switch is activated, the instructor will see who each student's neighbours are. If the network switch is not activated, the students will be represented as isolated agents. The color of each agent depends on whether or not the student purchased an item in the previous round (green/red respectively). The label selector on top of this black window allows the instructor to choose what information to display next to each agent (last price, last quality or none).

3.2 The Student's Interface

This interface provides information to help the student to decide whether or not to buy an item and, if so, what price to bid (Fig. 2).

The boxes and numbers inside circumferences are not part of the interface, they have been created to allow for a better explanation.

ID information (1)

This box displays the student's ID, that is, the name introduced by the student when she connects to the server.

Fig. 2 Student's interface. The boxes and numbers inside circumferences are not part of the interface, they have been created to allow for a better explanation

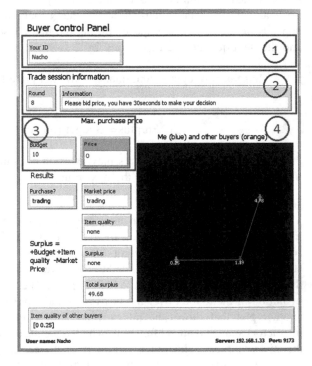

Trade session information (2)
This section shows the current round and reminds the student what she is expected to do at every step along the experiment.

Maximum purchase price (3)
This section consists of two boxes: the one on the left ('Budget') is the maximum amount of money a student can spend in an item. This value is the same for all the students and is defined by the instructor. The one on the right ('Price') allows the student to input a bid for the item she wants to buy. The student must press return after typing any bid.

Results information (4)
The window on the right shows several agents representing the student (in blue) and her neighbors (in orange). Depending on the instructor's decision, an informative number may appear next to each agent (price or quality in the previous round). The information panels on the left indicate whether or not the student has bought an item. If so, the market price, the item quality and the surplus will be displayed. The surplus is calculated as the difference between the student's budget minus the market clearing price plus the resulting quality of the item. The total surplus is the sum of the cumulative surpluses along the experiment. The box entitled 'Item quality of other buyers' will display a vector with the quality of the items acquired by the student's neighbors in her social network (if a network has been created).

4 Expected Dynamics

We propose a parameterization for a simulation game with 20 students in the Table 1. Although it is not possible to assure the final result of the experiment, because it depends on the sequence of decisions taken by students and the inherent stochasticity of the game, we can discuss briefly the expected dynamics. A complete theoretical study of the simulation model can be found in Izquierdo and Izquierdo [8].

Table 1 Parameterization for a simulation game with twenty students

Parameter	Value	Parameter	Value
y0	2	quality-mean	6*
a	4	quality-sd	2
budget	10	num-neighbors	2

* The instructor should inform students of this value

When students play the game on their own, they only update their expected quality when purchasing a good. In this situation, if there were no quality variability, the market would achieve an equilibrium. On the contrary, the dynamics that are expected to happen as a result of buyers trading in a market with quality uncertainty are the ones known as market failure, that is, the good is no longer bought after a number of periods in which the buyers (the students) have accumulated a set of bad purchase experiences causing the underestimation of the good's price.

The inclusion of the network structure serves as a mechanism of countervail the quality uncertainty in the market, and so preventing the market to fail. The students now can count with no only their own purchase experience, but with the quality information of goods bought by the members of their network. This mechanism does not secure the market from collapsing, but can be an approach of preventing it. As a result, different game results can take place, but it is expected a lower damage due to quality variability.

5   Concluding Remarks

Interactive exercises used to illustrate key economic and management issues can facilitate their understanding. We have presents in this work a tool to develop an online classroom experiment that helps to understand relevant aspects about quality management -the market effect of quality uncertainty and standardization (or the lack of it)-. Besides, the application can also be used to explore the effect of social networks as information diffusion mechanisms.

We think that the approach is useful to understand relevant concepts in quality management, to induce interest and motivation in the students by means of a participatory process and to help instructors to conduct online classroom experiments not only in simulation rooms but also in e-learning and transactional educational contexts.

Although the current development is designed to be used by means of computers, future research can be focused on migrating the application to smartphones as part of the movement "meet students where they are".

Acknowledgments The authors would like to thank Dr. Luis R. Izquierdo for his advice and comments on this paper. The authors acknowledge support from the Spanish MICINN Project CSD2010-00034 (SimulPast CONSOLIDER-INGENIO 2010).

References

1. Pyzdek, T.: The Handbook for Quality Management: A Complete Guide to Operational Excellence, 2nd edn McGraw-Hill (2012)
2. Deshpande, A.A., Huang, S.H.: Simulation games in engineering education: a state-of-the-art review. Comput. Appl. Eng. Educ. **19**(3), 399–410 (2011)
3. Brauer, J., Delemeester, G.: Games economists play: a survey of non-computerized classroom-games for college economics. J. Econ. Surv. **15**(2), 221–236 (2001). doi:10.1111/1467-6419.00137
4. Hazlett, D.: Using classroom experiments to teach economics. Teaching Economics: More Alternatives to Chalk and Talk

5. Frank, B.: The impact of classroom experiments on the learning of economics: an empirical investigation. Econ. Inq. (1997)
6. Durham, Y., McKinnon, T., Schulman, C.: Classroom experiments: not just fun and games. Econ. Inq. **45**, 162–178 (2007). doi:10.1111/j.1465-7295.2006.00003.x
7. Deshpande, A.A., Huang, S.H.: Simulation games in engineering education: A state-of-the-art review. Comput. Appl. Eng. Educ. **19**(3), 399–410 (2011)
8. Izquierdo, S.S., Izquierdo, L.R.: The impact of quality uncertainty without asymmetric information on market efficiency. J. Bus. Res. **60**(8), 858–867 (2007)
9. Wilensky, U.: "NetLogo (and NetLogo User Manual)" Center for Connected Learning and Computer-Based Modeling, Northwestern University. http://ccl.northwestern.edu/netlogo/ (1999)
10. Campanella, J., Principles of quality costs: principles, implementation, and use, Milwaukee, Wis. Am. Soc. Qual. **3** Sub edit (1999)
11. Ali, Hassan: Classical model based analysis of cost of poor quality in a manufacturing organization. Afr. J. Bus. Manage. **6**(2), 670–680 (2012)
12. Schiffauerova, A., Thomson, V.: A review of research on cost of quality models and best practices. Int. J. Qual. Reliab. Manage. **23**(6), 647–669 (2006)
13. Juran, J.M., Godfrey, A.B.: Juran's Quality Handbook, vol. 1, issue 2, p. 1872. McGraw-Hill, New York (1999)
14. Jones, P., Hudson, J.: Standardization and the costs of assessing quality. Eur. J. Polit. Econ. **12**(2), 355–361 (1996)
15. Miklós-Thal, J., Zhang, J.: (De)marketing to manage consumer quality inferences. J. Mark. Res. **50**(1), 55–69 (2013)
16. Akerlof, G.A.: The market for 'lemons': quality uncertainty and the market mechanism. Q. J. Econ. **84**(3), 488–500 (1970)
17. Ben Lakhdar, C., Leleu, H., Vaillant, N.G., Wolff, F.C.: Efficiency of purchasing and selling agents in markets with quality uncertainty: the case of illicit drug transactions. Eur. J. Oper. Res. **226**(3), 646–657 (2013)
18. Dimoka, A., Hong, Y., Pavlou, P.A.: On product uncertainty in online markets: theory and evidence. MIS Q. **36**(X), 395–A15 (2012)

Public Access Architecture Design of an FPGA-Hardware-Based Nervous System Emulation Remote Lab

Gorka Epelde, Andoni Mujika, Peter Leškovský
and Alessandro De Mauro

Abstract This paper presents a public access architecture design of a remote lab developed for remote experimentation with an emulation of the nervous system of C. elegans nematode. The objective of this system is to emulate the neural system and the virtual embodiment of a C. elegans worm and to let the biologists' and neuroscientists' community remotely evaluate different neuronal models, and observe and analyse the behaviour of the virtual worm corresponding to the neuronal model being evaluated. This paper first analyses related remote lab technologies which then is followed by the description of the adopted architecture design, selected cloud deployment option and remote user interface approach.

Keywords Virtual laboratories · Remote laboratories · Nervous system emulation · Neuro-computational response models on Field-Programmable gate arrays (FPGAs)

1 Introduction

The *Si elegans* project aims at providing a comprehensive artificial Caenorhabditis elegans (C. elegans) emulation system from which the principles of neural information processing, related to the behaviour of the C. elegans worm, can be derived.

G. Epelde (✉) · A. Mujika · P. Leškovský · A. De Mauro
eHealth and Biomedical Applications, Vicomtech-IK4, Donostia - San Sebastian, Spain
e-mail: gepelde@vicomtech.org

A. Mujika
e-mail: amujika@vicomtech.org

P. Leškovský
e-mail: pleskovsky@vicomtech.org

A. De Mauro
e-mail: ademauro@vicomtech.org

© Springer International Publishing Switzerland 2015
Á. Herrero et al. (eds.), *International Joint Conference*, Advances in Intelligent
Systems and Computing 369, DOI 10.1007/978-3-319-19713-5_49

The C. elegans hermaphrodite, a soil-dwelling worm, is well known worm regarding its cell composition. The morphology, arrangement and connectivity of each cell, including neurons, has been completely described [1]. Despite the extensive information on the worm's composition and behaviour [2], there is a lack of knowledge regarding the neural information processes that lead to the identified behaviours.

The *Si elegans* project is replicating the nervous system of the C. elegans on a highly parallel, modular, user-programmable, reconfigurable and scalable FPGA hardware architecture. It embodies the worm in a virtual environment for behavioural studies, taking the sensory-motor loop and realistic body physics into account. The resulting computational platform will be provided through an open-access web portal to the scientific community to test their neuronal models and different hypothesis.

This paper focuses on the architecture design defined to provide the scientific community with remote access to the neuro-computational research platform being developed.

2   Related Work Remote Labs Technology

A remote laboratory is commonly defined as an experiment which is conducted and controlled remotely through internet [3]. The main objective of *Si elegans* is to offer the scientists the possibility to remotely conduct behavioural experiments on an FPGA network-based hardware emulation system of the C. elegans worm. As such, *Si elegans* presents a remote laboratory designed for neuro-biological studies.

Given the benefit in terms of availability, administration and cost, remote labs have had a large proliferation during the last two decades. Two reference papers analyse and report on the state of art of remote laboratories [3, 4]. The research work in [5] takes these two reference papers, and provides an analysis from a design and development point of view for industrial electronics applications. The *Si elegans* project belongs to the industrial electronics remote laboratory category, considering that the core of the experiment is emulated in an FPGA network.

2.1   Existing Remote Lab Review

The analysis done in [5] concludes that most industrial electronics remote laboratories share a common architecture design. The main components of the conceptual architecture shared among many remote laboratories are: User interface, Web-server, Lab server, Instruments, Controller and Controlled Objects.

Our analysis of the remote labs has been mainly based on [3–5]. From the analysis we distinguished two different approximations: (i) toolkit initiatives for remote laboratories, (ii) FPGA specific remote laboratory implementations, and

(iii) **biological system simulation platforms**. The identified remote labs have been examined by their software distribution approach (open source vs. commercial), last developer activity (for the open source projects) and basic functionalities, such as login and experiment booking.

Toolkits Initiatives for Remote Laboratories

iLab is a remote laboratory solution developed by the Massachusetts Institute of Technology [6]. It provides solutions for lab servers, clients, and a service broker (identified as web server in the generic architecture introduced above) to share different lab servers among clients in remote locations. Other toolkit initiatives provide gateways to integrate their solutions under the iLab service brokering. The solution has to be deployed using Microsoft based technologies (Windows Server, Structured Query Language (SQL) Server, NET Framework). An Application Programming Interface (API) is provided in order to extend it freely.

iLab's software distribution approach is based on an open source license. The code is being distributed via subversion (SVN) code repository, manifesting frequent and recent updates.

WebLab-Deusto is a remote laboratory solution developed by the University of Deusto [7]. It offers solutions similar to iLab in regard to the lab servers, the clients, and the service broker. One of the benefits of WebLab-Deusto is the amount of extensions developed to make it interoperable with other remote laboratories (e.g. iLab) and to integrate it with learning management and authentication systems (i.e. LDAP, Oauth 2.0, and OpenID). Moreover, it offers seamless cross platform deployment (GNU/Linux, Mac OS X and Microsoft Windows systems).

For the client side, it allows the use of the Google Web Toolkit (GWT), JavaScript, Java, and Flash, although recommends the use of GWT to achieve maximised cross-browser interoperability. The lab server software is written in Python programming language. It differentiates between core and laboratory/experiment machine, allowing for a flexible distribution model (all in one machine, each of them in one machine) simply by changing related configuration files.

WebLab-Deusto's software distribution approach is based on an open source license. The code is being distributed via GitHub, also manifesting frequent updates.

Sahara Labs is a software suite developed by UTS Remote Labs that helps to enable remote access to computer controlled laboratories. It is designed to be a scalable and stable platform that enables the use and sharing of a variety of types of remote laboratories and maximises remote lab usage by implementing queuing and booking (or reservations) for users over a group of identical laboratories. The main language used for Sahara Labs development purposes is Java. The project is hosted at Sourceforge, and is being developed at GitHub. It has frequent updates.

Open Collaborative Environment for the Leverage of Online insTrumentation (**OCELOT**) is a complete solution framework for projects of remote collaborative interactions [8]. OCELOT software aims to provide a framework to quickly bring remote instrumentation solutions to the user. Its user interface is based on mixed reality and interactive multimedia.

OCELOT is open-source based. The implementation is based on the open-source implementation of OSGi enterprise server. The core components of a generic remote laboratory (User Interface, Web server, Lab Server and Controller as identified in the introduction) are launched in a centralised way on top of the OCELOT middleware.

The distribution is Git based, with logs showing recent activity only in internationalisation. The core development activity seems to have been abandoned. The last activity was registered on February 2012.

Laboratory Virtual Instrument Engineering Workbench (**LabVIEW**) is a system-design platform and development environment for a visual programming language by National Instruments (NI) [9]. LabVIEW allows for automation of processing and measuring equipment. LabVIEW provides integration of NI reconfigurable I/O (RIO) hardware targets through its FPGA module. In the introduced conceptual architecture the LabVIEW software provides user interface, web-server and lab-server component functionalities.

With regard to remote users, LabVIEW includes the support for remote panels, through which the developer can publish his new applications easily on a web-server. However, remote panels require a plug-into run which is not available for mobile devices. Moreover, not all browsers support LabVIEW remote plugin, and the plugin approach breaks with the ongoing web development trends which use HTML5.

FPGA Specific Remote Laboratory Implementations

For the study of specific remote laboratory implementations we have focused on those tackling FPGA hardware configuration, learning and research. Following, two representative FPGA specific remote laboratory implementations are introduced.

Remote FPGA Lab is a web-based remote FPGA lab, developed by the National University of Ireland in Galway [10]. The main contribution of this development is the offer of novel web-based FPGA hardware design capability, and real-time control and visualization interface to a remote, always-on FPGA hardware implementation.

For the webserver, it uses the Django web-framework and Python programming languages. For the webpages, HTML and JavaScript is used. Webpage JavaScript supports the addition, placement and update of items on a predefined system block diagram, providing this way the definition of the console interface.

The Remote FPGA Lab is a research project having the design source not open, though licensable. The use of the application is open.

Distance Internet-Based Embedded System Experimental Laboratory (**DIESEL**) was developed by the University of Ulster [11]. The generic architecture consists of a gateway server (laboratory administrator) connected to the Internet and a number of experimental workstations connected to the server via a network-hub. This architecture is functionally composed of three interacting components: (i) a server based booking system, accessible through the web, which allows students to reserve a time slot on any available experimental workstation, (ii) a client application, which the end users install on their PCs to facilitate access to remote experimentation resources, and (iii) a server application, which runs on each remote workstation to facilitate the remote access process.

The DIESEL client–server approach uses a distributed software architecture developed using Microsoft. NET technology. The development is not active anymore.

Biological System Simulation Platforms

Traditionally most simulation platform models such as GENESIS [12], or NEURON [13] have been restricted to neurophysiological aspects. Recently the closed-loop interaction between neural circuits and biomechanics has been addressed. Animatlab software [14] allows to define a neural circuits and biomechanics interaction loop for any type of animal, while the OpenWorm project [15] targets specifically the C.elegans nematode. Most of the aforementioned biological simulation platforms are standalone applications that do not allow for remote multi-user experimentation. The only remarkable solution regarding the remote public access is the Geppetto simulation engine [16] being developed by the OpenWorm project, which offers a modular and web-based multi-scale simulation solution build over Java and OSGi technologies. The drawback of this solution regarding the *Si elegans* platform requirements is that it is targeted towards software-based simulation, and their interfaces and modular approach has not been designed for an implementation of a real time closed-loop between a hardware based neural emulation and a physically-based virtual environment calculations.

Beside the biological simulation platform discussed, the analysed generic remote lab toolkits are focussed on e-learning, whereas the *Si elegans* is focused on a specific research task. The Implementation of the *Si elegans* related research flows (e.g. definition of C.elegans behavioural experiments and the neuronal model and networks, and their conversion for the FPGA based emulation) do not follow a regular remote lab experimentation. In consequence, the actual development is not grounded on any of the analysed remote lab technology as such, but will individually reuse some of the open-source toolkit modules available (e.g. booking system).

3   Architectural Design of the *Si elegans* Platform

Based on the requirement to provide public access to the scientific community, a cloud-based architecture has been defined for the *Si elegans* platform. The aim of defining a cloud-based architecture is to allow for parallel activity of scientists in:

(i) defining new neuronal models and networks, (ii) defining new behavioural experiments, (iii) booking for emulation of any previously defined behavioural experiment run on a defined neuronal model and network, (iv) analysing existing results from previous emulations, and (v) collaborating at distance and contributing to research by sharing and discussing new models and results.

3.1 Si elegans *Platform*

The final design of the *Si elegans* architecture divides the solution into two main blocks: (i) public access services (provided through web technologies relying on cloud services), and (ii) hardware based C. elegans worm emulation with exclusive access. The second part, takes into consideration that the FPGA hardware implementing the nervous system of the worm, although being versatile and reconfigurable, is dedicated only to a single worm at a time. Functionalities that do not require a direct access to the emulation system are provided through the public access part (e.g. definition of new experiments and models, revision of results and the collaboration with other researchers). This way, in case an emulation is already running or a downtime on the hardware based emulation system occurs, the scientists can continue using the public access functionalities. Thus they can continue with their research. Figure 1 depicts the separation of the public access and the hardware access logic layers.

The blocks defined for the *Si elegans* framework have been selected based on the main components (User interface, Web-server, Lab server, Instruments, Controller and Controlled Objects) identified by Tawfik et al. at [5] as part of the conceptual architecture shared among industrial applications remote laboratories. In our design, the Web-server is referred to as Cloud-based server, in order to emphasize its cloud deployment nature. The Interface Manager block corresponds to the Controller

Fig. 1 Public access logic identification over the *Si elegans* top level diagram

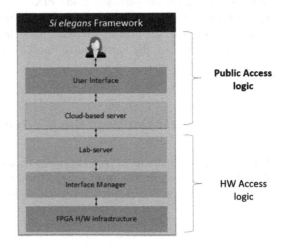

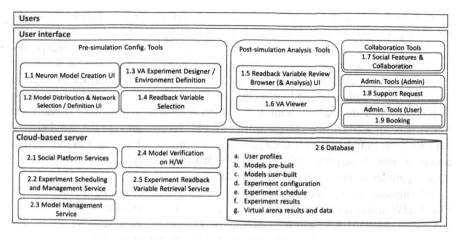

Fig. 2 System level components identified for the public access logic

component in Tawfik et al. paper, being responsible for controlling the FPGA-based neuron network, which in this case would correspond to the Controlled object in Tawfik et al. paper.

Figure 2 represents the system level components identified for the development of the public access logic.

User interface components represent the visual web components that will be shown to the user, while the cloud-based server components pertain to the services that support those UI functionalities and launch HW Access logic actions. Virtual arena (VA) refers to the visualisation of the virtual 3D worm and the environment. Readback relates to monitoring of the neuron model parameters for debugging purposes. For the development of the Cloud-based server logic and to implement the required research flows we are using the Django web-framework and Python programming languages. In addition, we are considering both, the WebLab-Deusto and the Remote FPGA lab solutions for the reuse of specific blocks (e.g. booking and scheduling module).

Figure 3 represents the system level components identified for the development of the hardware public access logic.

Lab-server

| 3.1 Simulation Controller | 3.2 Physics Engine | 3.3 Result Upload Service |

Interface Manager

| 4.1 Interface Manager | 4.2 Result Buffer | 4.3 Experimental Result Dispatcher |

FPGA/connectome H/W infrastructure

| 5.1 HW FPGAs for Neuron Network | 5.2 HW FPGAs for Muscle Network | 5.3 Synaptic Communication | 5.4 Gap Junction Communication |

Fig. 3 System level components identified for the hardware access logic

The HW access logic block is composed of the lab server, interface manager and the FPGA HW infrastructure. The lab server is the responsible for: (i) controlling at a high level the emulation (start, stop, and restore the system from a breakpoint), (ii) hosting the physics engine (for the physics calculation of the virtual worm and its environment), and (iii) uploading results to the cloud. The interface manager is responsible for initialising and managing the FPGA HW infrastructure (in addition to buffering and dispatching neuron variables readback information). Finally the FPGA HW infrastructure is composed of 302 FPGAs dedicated to the simulation of neurons (one FPGA per each C.elegans worm neuron), additional FPGAs used to accelerate muscle response calculations, and hardware with logic controller for the optical parallel communication between neurons.

Information on the implementation of the platform can be found for the FPGA HW infrastructure at [17], for the physics-based simulation of the virtual environment at [18], for neuron modelling to FPGA conversion process at [19], and for the optical interconnection technology being developed at [20].

3.2 Remote User Interaction with the Virtual Environment

An additional challenge of the public access design is the availability of remote interaction with the virtual 3D worm and its environment (whether in real-time or predefined). User interfaces tackling other *Si elegans* platform functionalities (e.g. defining new neuronal models and networks, collaboration with other users) are not part of this analysis since they are foreseen to be provided through regular web development technologies.

Decision must be taken on where and how the experiments are defined and the results of the physical emulation (the worm's behaviour) are represented and accessed.

Tawfik et al. [5] provide a good insight into the user interface technologies used to provide interaction with industrial electronic applications related to remote labs. They have identified the use of the following technologies: Desktop Sharing, Common Gateway Interface (CGI), ActiveX and Java Applets, Rich Internet Applications (RIAs), and Asynchronous JavaScript and XML (AJAX).

For the remote access to the virtual 3D worm of the *Si elegans* project, two possible scenarios have been considered: synchronous access (at the time of emulation), and asynchronous (no direct interaction while emulation is running) experiment definition and results analysis.

We have considered to offer the synchronous interaction through end-user tools like Virtual Networking Computing (VNC) and mixed web approaches [21]. The asynchronous interaction was considered to be provided through web 3D technologies, i.e. the experiment will be predefined, then carried out in the FPGA network and then visualized in a 3D environment with no real-time interaction.

We finally promote the asynchronous approach due to the following issues:

- The necessity of an exclusive access to the FPGA hardware in the *Si elegans* project limits the availability of the emulation system for multiple researchers at a time.
- Large communication latencies of the web limit the synchronous interaction with the 3D Virtual Arena. Basic advice suggesting acceptable response times of interactive applications [22, 23] of 100 ms latency (considered to be the limit for users to experience **direct object manipulation** in an UI), together with the necessity of an immediate response in sensory/neural experiments make the synchronous access in overseas internet communication prohibitive.

We are developing the experiment configurator as a web 3D application, providing a 3D GUI, and the simulation results renderer as a distinct web 3D application, accessible once the hardware emulation engine and the physics engine finish computing the experiment results. The planned remote UI technologies are based on HTML5, AJAX and WebGL technologies.

3.3 Cloud Deployment

In the *Si elegans* project a cloud based approach has been planned, given: (i) most existing close-loop (between neural circuitry and biomechanical simulation) biological simulation platforms are standalone and to allow remote users to emulate their experiments and collaborate with other scientists, and (ii) ensure availability of as many platform functionalities as possible and to reduce maintenance tasks as possible.

The aim of this subsection is to introduce the different cloud technology service levels, and to decide the best entry point (selected service level) for the *Si elegans* public access platform development.

Cloud services are designed to provide services at different architecture layers [24], dividing it to hardware, infrastructure, platform and application services. Starting at the infrastructure level, each layer can be implemented as a service for the next, higher level layer. Gray [25] considers the following three service based business models:

- Infrastructure as a Service (IaaS): IaaS provides hardware and software infrastructure resources over the Internet. These resources are offered as on-demand services and are usually based on virtual machines assignment.
- Platform as a Service (PaaS): PaaS provides platform level resources on-demand. Making use of services at this level, the application developer does not need to interact at the virtual machine level but rather interacts with the higher level APIs provided by the cloud service.
- Software as a Service (SaaS): SaaS offers end-users applications as on-demand services through the Internet.

In our cloud-based approach, the PaaS technologies have been selected as an entry point. Nevertheless, within *Si elegans* we will develop a SaaS level service to provide public access to the hardware emulation system. The selection of PaaS entry level instead of the IaaS level has been motivated by the effortless system maintenance, maximising the system's uptime, the moderate maintenance costs and the facilitated scalability of the final platform. Event though, the project might consider moving a VHDL compiler solution (software to compile FPGA design file to FPGA programming binaries), or moving the physics engine (visual 3D worms physics simulation) to the cloud to take advantage of the reduced maintenance tasks. To move these modules to the cloud IaaS access would be required. Therefore, PaaS service provider selection focusses on solutions that leave the door open to future access at a lower level (IaaS) in order to integrate more specific services (e.g. Amazon Beanstalk, Google Managed VMs).

4 Conclusions

In this paper we present the analysis done and the decisions taken with regard to the public access architecture design of an FPGA-hardware-based nervous system emulation remote lab.

First, the main toolkits used for remote labs and existing FPGA based remote lab development experiences have been analysed. Furthermore, following the main components identified by Tawfik et al. [5] for remote laboratories for industrial electronics applications, we have defined our high level platform architecture. Based on the high level platform architecture, the components required for the public access logic have been identified.

Following, the approach for the remote access to the experiment configuration and results visualisation user interface have been analysed. The asynchronous experiment definition and results revision approach have been selected. The planned remote UI technologies will be based in HTML5, AJAX, and WebGL technologies.

For the deployment of our solution on the cloud, a PaaS approach has been selected, but with a possibility for IaaS level access to allow integrating services (e.g. VHDL compiler) that require a direct deployment on the virtual machine.

Actual work is focussed on the development of the system level components for the public access logic and the definition of the communication protocol between the public access logic and the hardware access logic.

Acknowledgments This work was partially funded by the EU 7th Framework Programme under grant FP7-601215 (*Si elegans*).

References

1. White, J.G., Southgate, E., Thomson, J.N., Brenner, S.: The structure of the nervous system of the nematode caenorhabditis elegans. Philos. Trans. R. Soc. Lond. B Biol. Sci. **314**, 1–340 (1986)
2. Hart, A.: Behavior. The C. elegans Research Community, WormBook (2006)
3. Gomes, L., Bogosyan, S.: Current trends in remote laboratories. Ind. Electron. IEEE Trans. On. **56**, 4744–4756 (2009)
4. Gravier, C., Fayolle, J., Bayard, B., Ates, M., Lardon, J.: State of the art about remote laboratories paradigms-foundations of ongoing mutations. Int. J. Online Eng. **4**, 1–9 (2008)
5. Tawfik, M., Sancristobal, E., Martin, S., Diaz, G., Castro, M.: State-of-the-art remote laboratories for industrial electronics applications. In: Technologies Applied to Electronics Teaching 2012. Pp. 359–364. IEEE Press, New York (2012)
6. Massachusetts institute of technology: iLab Project, http://ilab.mit.edu/wiki
7. University of Deusto: WebLab-Deusto, http://www.weblab.deusto.es/website/
8. Ocelot : Opensource software, http://ocelot.telecom-st-etienne.fr/
9. National Instruments: LabVIEW, http://www.ni.com/labview/
10. Morgan, F., Cawley, S., Newell, D.: Remote FPGA lab for enhancing learning of digital systems. ACM Trans. Reconfigurable Technol. Syst. **5**, 1–13 (2012)
11. Callaghan, M.J., Harkin, J., McColgan, E., McGinnity, T.M., Maguire, L.P.: Client–server architecture for collaborative remote experimentation. J. Netw. Comput. Appl. **30**, 1295–1308 (2007)
12. Genesis: The genesis simulator, http://www.genesis-sim.org/GENESIS/
13. Neuron: Neuron simulator, http://www.neuron.yale.edu/neuron/
14. Cofer, D., Cymbalyuk, G., Reid, J., Zhu, Y., Heitler, W.J., Edwards, D.H.: AnimatLab: a 3D graphics environment for neuromechanical simulations. J. Neurosci. Methods **187**, 280–288 (2010)
15. OpenWorm: The OpenWorm project, http://www.openworm.org/
16. OpenWorm: Geppetto simulation engine, http://www.geppetto.org/
17. Machado, P., Wade, J., McGinnity, T.M.: Si elegans: FPGA hardware emulation of C. elegans nematode nervous system. In: 6th World Congress on Nature and Biologically Inspired Computing. pp. 65–71. IEEE Press, New York (2014)
18. Mujika, A., de Mauro, A., Robin, G., Epelde, G., Oyarzun, D.: A physically-based simulation of a caenorhabditis elegans. In: 22nd International Conference in Central Europe on Computer Graphics, Visualization and Computer Vision—WSCG 2014, Vaclav Skala—Union Agency, Plzen (2014)
19. Krewer, F., Coffey, A., Callaly, F., Morgan, F.: Neuron models in FPGA hardware a route from high level descriptions to hardware implementations. In: 2nd International Congress on Neurotechnology, Electronics and Informatics. pp. 177–183, Scitepress, Lisbon (2014)
20. Petrushin, A., Ferrara, L., Liberale, C., Blau, A.: Towards an electro-optical emulation of the c. elegans connectome. In: 2nd International Congress on Neurotechnology, Electronics and Informatics. pp. 184–188, Scitepress, Lisbon (2014)
21. webtoolkit: Wt, C ++ Web Toolkit, http://www.webtoolkit.eu/wt
22. Bochicchio, M.A., Longo, A.: Hands-on remote labs: collaborative web laboratories as a case study for IT engineering classes. IEEE Trans. Learn. Technol. **2**, 320–330 (2009)
23. Card, S.K., Robertson, G.G., Mackinlay, J.D.: The information visualizer, an information workspace. In: Proceedings of the SIGCHI Conference on Human Factors in Computing Systems. pp. 181–186. ACM, New York (1991)
24. Zhang, Q., Cheng, L., Boutaba, R.: Cloud computing: state-of-the-art and research challenges. J. Internet Serv. Appl. **1**, 7–18 (2010)
25. Gray, M.: Cloud computing: demystifying IaaS, PaaS and SaaS, http://www.zdnet.com/news/cloud-computing-demystifying-iaas-paas-and-saas/

Education Apps. One Step Beyond: It's Time for Something More in the Education Apps World

Francisco J. Cardenal and Vivian F. López

Abstract In this publication the key parts of the process undertaken to develop an educational fun math app-centric world are collected. The key points in the design, idea, analysis, development and testing of an App that combines learning with fun.

Keywords App · Application · Mobile · Apps · Education app · Blended education · Math

1 Introduction

Education Applications are one of the categories that more increase in the Apple App Store or App Store. It's a growing market that allows us to gain knowledge of many subjects, including mathematics. To prove it, you just have to find the Education category in the App Store. The market of education Apps is a continuously growing industry and becomes one of the few sectors whose numbers improved every year. According to Forrest Research forecasts [1], the mobile device market will triple in 2017. Focusing on the education section, there are several schools that use mobile devices to base their learning. The first time we heard of it was a Scottish school, but now only the public schools of San Diego uses 20,000 mobile devices. According to Apple itself, in late 2013 the company of the bitten apple sold a whopping 8 million mobile devices exclusively for educational institutions [2]. The latest data indicate that there are more than 80,000 education apps on the market. The education Apps accounted for more than 9 % of the total [3]. Over 80 % of the applications in the Education category of the iTunes Store were aimed at children. In 2009, 47 % of the

F.J. Cardenal (✉) · V.F. López
Departamento Informática Y Automática, University of Salamanca,
Plaza de La Merced S/N, 37008 Salamanca, Spain
e-mail: pacocardenal@pacocardenal.com

V.F. López
e-mail: vivian@usal.es

© Springer International Publishing Switzerland 2015
Á. Herrero et al. (eds.), *International Joint Conference*, Advances in Intelligent
Systems and Computing 369, DOI 10.1007/978-3-319-19713-5_50

best selling Apps were intended for children. Now, that percentage is 72 % [4]. The percentage of Apps for children continues to increase, while the adult market declines. The Apps for Preschoolers occupy 58 % of all categories and have increased by 23 % [5].

The main objective of this work is the development of a specific application that covers a significant part in mathematical terms of the current formal education: working with decimals. The whole process to reach that goal will be discussed in this article in the following sections.

In Sect. 2 we present educations apps, Sect. 3 provides a short overview of the actual market of education apps. In Sect. 4 we present our education app. Section 5 explains the conclusions of the paper.

2 Education Apps. The Study

The target of the App is a child of about 9 or 10 years, beginning with the use of fractions and decimals. Choosing the iPhone as target platform has been made by the boom that mobile devices are having on our society, using them, among other fields, for education. This is also the cause that sparked our interest in the software development for these devices.

After deciding the development of this application, the first step was to communicate with people who could guide us both in education (and more specifically children education), and developing education Apps. Specifically, we communicated with the creators of education applications Teachley [9] (Apple Design Award winners at WWDC 2014), who gave us very useful notions and concepts to develop our App. They have confirmed what we have indicated earlier: learning is more effective if the student, in this case the child, is involved. This is what stimulates their attention and the reason that teaching a particular topic or subject becomes part of their leisure and approaches it more like a fun as an obligation. After several meetings with them, the following conclusions were drawn:

- The application should be entertaining. This is the main point of the whole project and we wanted to develop an application that does not reach the level of game, but it will be very close to it.
- The user interface should be simple. The third pillar on which the application would be based was the simplicity and clarity of the user interface. The display should show only the essentials and make users agree to it clearly and quickly as shown in Fig. 1.

On the development of education applications, the following lines were drawn to follow in the application:

- Simplify. The less the better.
- Increasing the degree of difficulty of the App from a below average user level to a higher level.

Fig. 1 The game inside the App

- Perform any readily observable modification when a change in a difficulty level is performed.

The first tip led us to simplify the interface as simple as possible, the second to manage the complexity of the questions asked in the application as we will discuss below, and the third to create a separate file for modifications of each difficulty level present in the project. Thus, children are offered a new way of teaching and a new way of learning. The concept of learning through play forms the basis of the entire project and how to improve education, children education in this case.

3 The Actual Market of Education Apps

All parents are concerned about the poor quality or lack of interest in their children's early school stages. Ludic education applications can be a way of improving education today. To achieve this, these applications must be a real mix of play and learning. Many of the applications currently on the market are little more than colored numbers that help the education of the children of the house calling his attention.

Most educational Apps that can be found on websites like Eduapps [6] or eduTecher [7] are very simple Apps without a defined goal that can be applied to formal education. Likewise occurs with the iCuardenos App [8] or the awarded Teachley Apps [9]. Neither offers to cover several courses comparable to current child formal education levels in Spain.

With all the above facts, it is not unreasonable to say that the market of educational fun apps for kids is an emerging market, offering a good opportunity for the future and values of stability and more than enough benefits. The education market needs a change. Parents need a change. Children need a change. And education Apps could be this expected change in educational quality and quantity.

4 Our Education App

The App developed is an educational fun application designed for the little ones who
start to work with decimals within the mathematical universe. The recommended age
for use is between about 9 or 10 years. Our leading software requirements was:

- Conduct educational games applications
- Adapt the difficulty level of the application
- Generate an addictive environment for the child to learn while playing
- Joining the characteristics of video games and board games
- Generate different questions randomly
- Adapt the application to mobile devices
- Maximize the interaction between the user and the application
- Include music and sounds
- Manage movements and animations
- Develop the game in 2D top-down perspective
- Discard the focus on the lighter side or in the educational side of the game. Both
  of them should be mixed.

The App has been developed for iOS operating system, but, with a little effort,
we can port it to Android to reach a bigger market.

Game goal
The goal is to reach the goal with your racing car before the opponent. To do this,
you must respond as quickly as possible, stating whether the statement is displayed
is True or False. If the answer is right, your car advance an amount of meters
relating to the response speed. The sooner the better.

If you are wrong, your car will not move, and one of the three attempts that you
have to overcome all levels will be removed. If the enemy reaches the goal before
you, you must start over at one of the unlocked levels as shown in Fig. 2.

Fig. 2 Difficulty levels

4.1 Application Features

Unified Process

The Unified Process is a generic framework for software development. Its main advantages are: standardization, flexibility and adaptation. As fundamental concepts emphasize discipline, which is used to organize key management and development activities, and artifacts from the results. The various stages of that process are listed below: initiation, development, construction and transition. In the requirements phase different use case diagrams have been created to reflect the system's functionality. Figure 3 shows the use case diagram *Home Menu Selection*.

Also, these phases are combined to form an engineering stage and another stage of production. Returning to the term of discipline, used to organize the management and development activities, the various disciplines are: requirements, analysis, design, implementation and testing.

Sprite Kit

Sprite Kit is a framework [10] that offers both graphical capabilities like animation and sound to deal with all matters relating to the graphic elements of a game developed for mobile devices that have the Apple iOS 7 operating system or higher. This way we can create a fairly complex game or application in a simple, rapid and standardized way.

Sprite Kit is much easier and faster to use than engines like Cocos2D and Unity, and becomes the perfect place to develop applications that require high interaction from the user, games especially, by new developers and programmers. Sprite Kit is available for both iOS 7 and OS X 10.9 and above and supports different types of items.

Game design

In terms of design, the first step was to create the class diagram. Most classes descend from one of the most widely used classes in Sprite Kit: the Scene class [11], which is the basis of many of the screens appearing in the App. The class that focuses so much time spent with the application as the time spent on its

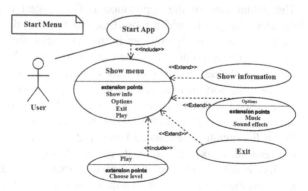

Fig. 3 Use case diagram "Home Menu Selection"

development is the Game class, which will be carrying most of workload game included in the app. To help complete the design we also developed the activity diagram where the activities performed by the application are observed.

The design has been one of the project star phases, with the longest run. The reason has been the quest for the best graphical interface and the most suitable items and pictures optimized for the device screen. The design goal was to make an application with a simple look but both attractive and intuitive for use again. In terms of design, we have two distinct parts: interface design and game design.

Interface application design

The application screens should show all-important information without data that overload the interface. Due to the size of the device, everything shown on the screen should be large enough to display it. We wanted to design an interface that was attractive but not too overdone because, although there is a game running underneath, the area where the user should focus is the panel where the questions appear. An important aspect of the interface was interactivity. Although the user should not press more than true or false button, he simply must feel that his choice was used to control his character, in this case his vehicle. It is that sense of connection between a typical touch application, where only a series of buttons are pressed, and a game controlled by movements of the device, which has satisfied us with the result achieved.

As for the different parts of the interface, we have two very distinct parts: the game running underneath, and the application of true false that sits atop. To leave the maximum free space available on the screen to ask questions and the characters roam the greatest possible route, we have placed all the controls in the lower parts (buttons True-False) and higher (indicating question number, time and remaining lives). Furthermore, to give the game more movement, we decided, in addition to vehicles, encourage the bottom of each level to give a greater sense of mobility, so that the bottom of each level occupies 50 % more height screen device to which it is intended. Through a photographic image processing used with small changes and tweaks, an acceptable enough interface to enjoy the lighter side of the application was achieved.

Interface game design

The lighter side of the application is the biggest part of the design stage and therefore the part in which more time has been used. At each stage of this design have been tested using children in that age range for errors, problems or potential upgrades. Other point within the project has been the digital image processing, a key part in the development of games and it has led us to acquire new knowledge in this field. Report all images used in the application by modifying original images were created with free license, but not all of them allow distribution of the application for benefits. To get the feel of movement, we need a sequence of still images that will generate an animated image [12]. In turn was due to take extra care with background, colors and contrasts as seen in Fig. 4, due to the small screen of the device, which could lead to a bad user experience. Indicate that, in addition to graphics, the game includes both music and sound effects that acclimate the activity and provide feedback to the user.

Fig. 4 Color and contrast on screen

Architecture

The application was developed using the Model View Controller pattern MVC, which isolates presentation and data. With Sprite Kit and iOS 7 is very easy to follow this pattern. The MVC [13] consists, as its name indicates, of three pillars:

- Model: responsible for the business logic, the part of algorithms and data. The model can be a single object or be a multiple structure. Represents the information or knowledge model.
- View: responsible for presenting the view or interface. The view is the visual representation of the model. The view is related to the model through the controller, and it is thanks to this through which they can communicate.
- Controller: driver intermediary between Model and View and responsible for all notifications. It is the real link between the user and the system, and is responsible for managing the inputs and outputs of the application.

Implementation

Due to the novelty of the Sprite Kit framework, the available documentation is hard to find and the proposed examples are not as frequent as with other advanced graphics engines. For this reason, we have relied on examples found in the bibliography appended to implement a project consisting of automatic generation questions for each difficult level, changing parameters of each level, sprite animation, position on screen and tap detection. Furthermore, it has different levels adapted to play a major difficulty. To provide the application addictiveness feature, it was tried to adjust the difficulty as possible in relation to the speed of the opponent vehicle and the category of mathematical operations. This adjustment was made based on the evidence we commented afterwards.

Evolution

The application has been developed following an iterative model, so that changes and increases in the application have been many and varied. Project started focusing on the lighter side of the application. After get it, we started to develop a graphic interface that allows its proper use. After that, the different application screens were incorporated, linked and provided with new graphics. Finally both logic and presentation were refined and the most interesting questions were selected and adapted for each difficulty level.

We want to reiterate that the application was initially designed to develop using Cocos2D [14], but the App evolved to Sprite Kit, for being, in our opinion, a more suitable environment, and for being Apple's own environment. Once it was decided to use this tool, the first thing to be achieved was to unite a specific learning and a fun experience. To achieve it, we searched both educational and recreational applications and different material of studies on this subject, which led us to the foundation of the application: to face an opponent and reach the finish line before him using our knowledge.

After considering several options such as multiple choice answers, direct confrontations with other types of games or eliminating wrong answers, we chose to make an application based on true-false, but with wide penalty of random correction, thus avoiding fast and ignorant response. This true-false application was very simple in the beginning, showing a series of questions consecutively, and default iOS buttons. After this first model, we proceeded to customize buttons, expanding the number of questions and generate them in random order. Finally, we implemented the load of questions from an external file.

The next phase would determine the game associated with this other game of true false. We had to develop a layer that run below the principal and maintained addiction and learning. After evaluating different options as horse racing games or board games, we chose a drag car race, but customizing them with formula 1 cars, for being a known sport, adapting their colors to red and blue representing the Ferrari teams (with Fernando Alonso as a representative of our country) and several of its rivals teams. In this game our character was only showed at first. A small portion of the screen represented a road on which our car was circulating to reach goal. The implementation of the victory was not done by position, but right questions because they always represented the same distance that the car run on the screen.

Then we proceeded to parameterize the distance traveled depending on how quickly the user selected the correct answer. It thus had a variation in the distance that moved the character and didn't allowed us to determine their position by the number of correct answers, so this algorithm was modified to depend on the position on the screen. To complete this application layer, adding an opponent character to play against increased the thrill of the game. The first time it was added, the opponent is moved when the user answer a question, moving the two vehicles at a time. However, it was understood that it would be a more challenging situation to see how the enemy was moving across the screen while the user was trying to solve the proposed question. This was the technique chosen for the final version of the car race layer.

Upon completion of both layers, we had to relate them. To do this, we linked the appropriate action to each of the buttons with the movement and distance traveled by the vehicles due to information of response time remaining and the speed of the opponent that resides in the LevelData.plist file. Furthermore both this distance as that traveled by our vehicle was optimized so that the maximum number of possible questions were displayed on the screen.

4.2 Experiments and Results

Tests

The tests were conducted at an early stage in the simulator itself provided by the Apple programming environment, and then we installed the application on a real device (iPhone 4) to be tested by different people. Before going into the test section, we would like to stress that in order to test the application on a physical device and not just into a simulator, you need to have an Apple's iOS developer license, which was obtained by integrating us into the iOS development program of the University of Salamanca and the friendliness and helpfulness of José Rafael García-Bermejo Giner (Coti).

As we said before, we had a number of children to test the application and the support of their parents to learn from a more adult perspective. In these tests several bugs and small errors in the game were discovered. In tests we had specifically 4 children with the next ages: 8, 9, 10 and 12 years old. Recall that the target user of the application is a child between 9 and 10 years old. The choice of these ages between 8 and 12 years was conducted to have a clear idea of whether both interface as difficulty was adjusting correctly.

Difficulty tests

In the first tests only a mathematical question appeared on screen and you should answer true or false. They performed a total of ten questions and we determined both the number of correct answers and the time taken to resolve them. The questions were chosen at random from a set of 30 different levels of low, medium difficulty, where addition and subtraction of whole numbers and decimal numbers was included. These tests were basic to adapt the difficulty of the different levels in the application.

Interface tests

The second objective of this test group was helping us to made a sufficiently attractive and intuitive App. Everybody should access to any of the screens and options naturally and play the game without explanation. During these tests there was no help screen, so the children had no way of knowing how it worked except for practice.

The first product delivered to them, had only the game screen with a similar interface to the final one, and where they should always answer the same 10 questions that appeared in random order. The result was more than satisfactory because, thanks to addiction caused by the game, they wanted to continue using the application until it was not necessary, and they evolved largely in their response times.

As to what concerns us, the interface, initially it had a true or false buttons similar to accept or cancel options present in the iOS 7 mobile operating system.

However, these buttons were not sufficiently intuitive and we had to explain several times what that meant and how they should respond in each case. In addition, the game screen in which our vehicle was moving (then we had not an opponent vehicle), was between True-False buttons and the text of the question, which appeared at the top of the screen. Being such a short distance, the look of the game was not as great as in the final version.

The third set of tests was performed with the modified interface very close to the final version. It was found that the user response was good in both management and addictiveness, so it was decided to keep this interface to the final product.

Results

This results was taken from a group of ten children between 10 and 12 years old who answered 30 of 100 random math adding and subtracting operations. In first place in a notebook and, in second place, within the App.

The children answered correctly to 65 % of the questions in the traditional way (only with the notebook) and 83 % with the App. Then, we used the App first, and the results were 90 % of correct answers using the App and 76 % with the hand-writing way.

We repeated this tests ten times, and the results with the App were an average of 14 % better than the same kind of operations answered on a notebook, without any game pushing the children to improve their results. Also, the response time was 30 % better with the use of the App.

5   Conclusions

In this paper, we have presented an App that combines learning with fun. The children's education Apps that are now on the market are all aimed at improving small educational or psychological aspects of the children to whom they are addressed.

We have described the current state of the work but there are many open lines of work. We have finished the touch and feel of the App, defined the structure of the cases and the game models, and we have been working on alternatives of storage and variations of the main racing game.

Current lines of work are focused on new cognitive science research. We want to capture meaningful data for teachers about what to teach next. Our intention is to help students in their learning process and to help teachers in their teaching process as well.

We have seen improvement in children's mental calculus and we think that this kind of Apps will be an important part of formal education in a future. We want it and we will work hard to achieve it.

Acknowledgments This work has been supported by the Spanish Government through the project iHAS (grant TIN2012-36586-C01/C02/C03) and FEDER funds.

This App was awarded with the TCUE-5 prize [15] at University of Salamanca.

References

1. Gownder, J.P., et al.: Global business and consumer tablet forecast update, 2013 to 2017. Forrest Research forecast. http://bit.ly/1gujliI. Accessed 16 Mar 2015
2. Apple Inc. Apple education web page. https://www.apple.com/es/education/. Accessed 16 Mar 2015
3. Universia España. Ya existen más de 80.000 aplicaciones educativas. http://bit.ly/1L5ES2T. Accessed 16 Mar 2015
4. Universidad de La Rioja. Aplicaciones educativas para dispositivos móviles. http://bit.ly/1GMfDCY. Accessed 16 Mar 2015
5. Common Sense Media Inc. Best Preschool Apps. https://www.commonsensemedia.org/lists/best-preschool-apps. Accessed 16 Mar 2015
6. Raúl Santiago y Alicia Díez. Eduapps, aplicaciones educativas. http://www.eduapps.es. Accessed 16 Mar 2015
7. Bellow, A.: Educational technology web tools. http://www.edutecher.net/. Accessed 16 Mar 2015
8. Rubio. iCuadernos by Rubio. http://www.icuadernos.com/. Accessed 16 Mar 2015
9. Teachley, LLC. Research, learning, fun. http://www.teachley.com. Accessed 16 Mar 2015
10. iUridium. Sprite Kit tutorials. http://www.sprite-kit.com/. Accessed 16 Mar 2015
11. Volevodz D.: iOS 7 Game Development, ISBN:978-1-78355-157-6 (2014)
12. Penn, J., et al.: Build iOS Games with Sprite Kit, ISBN: 978-1-94122-210-2 (2014)
13. Conway, J., et al.: iPhone Programming, ISBN: 978-0-32194-205-0 (2014)
14. Cocos2d-x.org. Open Source Game Development. http://www.cocos2d-x.org/. Accessed 16 Mar 2015
15. Fundación Universidades y Enseñanzas Superiores CYL. Transferencia de conocimiento Universidad empresa. http://www.redtcue.es/. Accessed 16 Mar 2015

Self-Assessment Web Tool for Java Programming

Bruno Baruque and Álvaro Herrero

Abstract Self-assessment capabilities, that enables the student to have an insight on his own learning process, are a very desirable skill on any higher education student. It is even more important for transnational students as it allows them to successfully adapt to their new international learning environment. Present work proposes an online tool to help computer science students to develop their self-assessment skills while learning Java programming. It consists on a plug-in for the widely popular Moodle learning management system to work with source code for the Java programming language. The developed plug-in lets students to upload source files, analyzes the code and presents a report to the student using industry standard tools. The report includes both errors, points of improvement and a general comparison with the rest of his classmates, from the software quality standpoint. This way, the student is provided with a framework against to which compare the correctness of the solution he/she has programmed before it is delivered to the teacher for evaluation. The teacher is able to access all this information too, facilitating an insight of how the class, as well as individual students, is progressing. It is expected that the shortening on the classical loop of the student problem solution and teacher feedback will enhance the student self-awareness and will improve his overall performance in programming courses.

1 Introduction

Formative assessment can be defined as employing appropriate activities to provide feedback to enhance student motivation and achievement during instruction—as students learn. Providing helpful information while learning occurs contrasts with providing feedback solely after instruction (examination).

B. Baruque (✉) · Á. Herrero
Civil Engineering Department, University of Burgos, Burgos, Spain
e-mail: bbaruque@ubu.es

Á. Herrero
e-mail: ahcosio@ubu.es

© Springer International Publishing Switzerland 2015
Á. Herrero et al. (eds.), *International Joint Conference*, Advances in Intelligent Systems and Computing 369, DOI 10.1007/978-3-319-19713-5_51

583

According to [6]

"Self-assessment is more accurately defined as a process by which students (1) monitor and evaluate the quality of their thinking and behavior when learning and (2) identify strategies that improve their understanding and skills. That is, self-assessment occurs when students judge their own work to improve performance as they identify discrepancies between current and desired performance."

This ability of being capable to understand how learning is taking place and what can oneself do to progress in this task is very valuable for everyone, but specially for people on their formative years. This can be seen as a crucial ability for students of higher education and more precisely, those pursuing STEM-related (science, technology, engineering and mathematics) degrees [3]. This kind of studies are characterized for being rich in complex problem-solving tasks involving also abstract concepts and multiple ways of tackling the same problem. This has been recognized by the European Higher Education Area (EHEA), which stresses the need for students to develop key cross-curricular skills such as "problem solving", "information management" or "critical thinking" [4].

This is especially true in computer science studies, where students are taught how to build automated solutions to real-life problems or to manage already existing complex systems to perform these tasks on a daily basis. This area of knowledge is particularly prone to this kind of self-assessment activities, since usually students can test whether the solution they have designed works or not by running a test on a regular computer system. Present work focuses on computer programming courses, but the underlying principles could be extended to other IT disciplines.

Usually, programming practical lessons follow a similar scheme: the teacher proposes a problem which will be solved by writing code in a certain programming language. A very well defined set of requisites is delivered to students for their solutions to comply with. Students then design an algorithm needed for the solution and implement it. After that, the teacher collects all the proposed solutions and checks them (according to the requisites). Finally, feedback is given to the students (probably along with a numerical mark) by the teacher.

Since students in this area—and increasingly in many others nowadays—are used to have access to interactive ways of learning, it seems quite natural to have a system to guide them when completing the proposed tasks, rather than waiting for the teacher to either give direct feedback during class sessions or ultimately mark their exercises to assess the quality of their work.

To overcome this limitation, the aim of the proposed system is to automate the feedback loop for the students and therefore reducing the time frame when they receive assessment on the quality of their work. This would make it more flexible, as students do not need to be in the same classroom as the teacher. It will also add a way to compare their work with that of their classmates in order to provide a context to reinforce the self-awareness of the students [8, 10]. The main characteristics of the proposed web tool are described in the following sections.

2 Related Work

There are many other efforts being developed and tested in order to help university students master complex concepts through the help of technology [10]. This trend has been even emphasized in later years, with the wake of MOOCs (Massive Open Online Courses) systems, that provide access to courses of different disciplines to hundreds or even thousands of students at the same time.

Some of those systems are used for disciplines such as Physics [11], Chemistry [2] or Mathematics [3], among others. There are other ones that are aimed at lowering the input barrier for beginner students to programming courses [12]. Some efforts even go one step further, creating adaptive paths of exercises for students according to their progress on previous tasks in the platform [5].

In present case, the proposed tool has some specific characteristics that make it different from the others. First of all, the tool is aimed at students who have mid-level expertise in programing. That means that, probably, they have already mastered the basics of the topic and are learning more advanced concepts. Therefore, the output of the tool is based on industrial standards, rather than simplifying the information for everyone to understand. That way, students start to relate with the outputs offered from professional tools, while they are shielded to a certain degree from all its complexity. Additionally, it is specifically programmed to enable the tasks to be completed by a group of students, automatically sharing the results of a test with all the group members. This is meant to encourage collaborative work among students. Another advantage of the proposed tool is that it enables students to compare their performance with the rest of the course (in an aggregated anonymized way). Again, this feature presumes some maturity from students in order to interpret the obtained results. This intends to reinforce the aspect of self-assessment, indicating if the student should make an effort to reach the same progress state where the rest of students are or not.

3 Proposed Automated Self-Assessment Web Plug-In

3.1 Objectives

The main objectives of the tool designed for each student to perform self-assessment are:

From the teacher point of view: The tool must be able to detect and highlight errors or malfunctions in the code written by the student. It must be flexible enough to let the teacher decide which are the errors that he/she wants to highlight for the students, including the feedback given to the students. It would be also interesting to spot other errors in an automated way, to help the teacher with the task.

The tool should be able to present the teacher different levels of information detail about the assignments. The tool should present the teacher with an overview of the classroom progress per assignments. It should be able to objectively measure this according to standard measures of software quality. It should also be able to disaggregate this information for different assignments and individual students (or group of students).

From the student point of view: The information presented to the student should be clear and informative. The process of self-assessment should be as straightforward as possible for the student. The student should see the application as a helpful tool for his learning and not the other way round.

The students should see the tool as a help and guide for his own self-assessment, but let him complete the progress by himself. It is important for the student to internalize the process of reflecting about his work, finding errors or weak spots in his reasoning... as a summary, to let him engage in the problem-solving dynamic. The tool should not appear to him as a trial-and-error way of completing the assignments.

The tool will also let the students (or group of students) to compare his progress to the rest of his classmates.

3.2 Tool Description

The tool is implemented as an extension of the well-known Moodle Learning Management System [7]. This has mainly two advantages. Firstly, as it is one of the most widely spread LMS, both teacher and students are probably familiar with the basic workflow of the tool. Second, the user authentication system, assignment setting and many other aspects of LMS are already implemented, letting us focus on the specific task of programming self-assessment.

The software is then programmed using standard languages and tools for web programming (PHP and MySQL mainly). It is divided into several modules, which relate to different aspects of the teaching-learning management. In present work, authors have extended the module of "Tasks" (specifically the "assignment-upload"), which is the part that implements the workflow for delivering assignments to be graded by the teacher.

What the extension does is altering this basic structure by performing some automated actions when the student uploads the source code for the assignment, providing both the student and the teacher with complete information about the quality of the proposed solutions. Up to now, the extension is devised for the assessment of programming assignments implemented in the Java programming language. In order to do so, it relies on the results obtained from two of the most used libraries for testing the correctness and quality of source code: jUnit [13] and PMD [9]. The first is used to check the functionality of the program proposed by the student (unitary tests), while the second one is used to analyze problems or suggest best practices on the codification of the program (static code tests). The use of JUnit implies that it is the teacher the one responsible for defining the exact behavior that is required on the students program and the feedback that will be given to them when the behavior presented does not match the requested specifications. Students will see the task as any other upload assignment task in the course. The difference is that, as soon as they upload their answer (implemented as Java source code); they will be provided with an exhaustive feedback automatically obtained from the teacher's jUnit tests and PMD analysis results. This process is limited to a certain number of tries (set by the teacher), in order to prevent students to engage in a trial-and-error process; rather than a reasoned solution to the weakest aspects of the assignment.

As depicted in Fig. 1, the typical workflow for using the proposed tool would be the following:

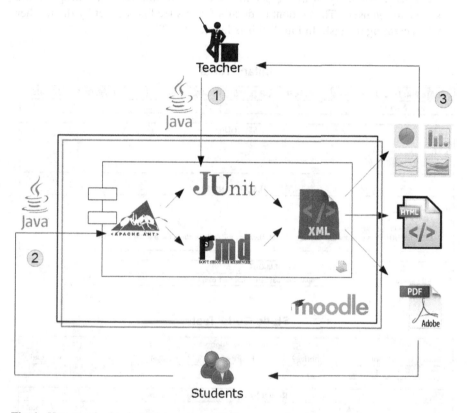

Fig. 1 User (teacher/students) interaction with the proposed system

1. The teacher designs a new programming assignment and prepares the jUnit tests that will be used to check its correctness along with its description; in the way Test Driven Development (TDD) [1] is performed in the industry.
2. The teacher creates a new programming assignment task in the Moodle system, and sets the task as usually. For this task, a suite of jUnit tests is also required. Additionally, the maximum number of tries for each student to use the auto-correct tool can be set. In Fig. 1, this is labeled as "1".

3. The students works on the assignment and finally get the requested program. They are advised to do their own tests while working on the assignment. When they think the program is complete enough an want to check it, they can upload a copy into the correspondent task space (as a zip file). In Fig. 1, this is labeled as "2".
4. The upload is analyzed in the server by using the ANT [14] library to automatically perform the tasks of compiling, running tests, saving the results, calculating statistics and formatting all the information to be presented to the user (teacher and student). If the uploaded files complies with some rules and can be compiled correctly, a summary of the errors and/or advise to improve their program are shown to the users. The amount of details for this feedback is set by the teacher when creating the task. In Fig. 1, this is labeled as "3".

Unitary Tests

DATE AND TIME OF TESTS	TOTAL NUM. OF TESTS	NUM. OF FAILURES	NUM. OF EXCEPTIONS
2015-03-2613:55:01	6	1	1

ASSERTIONS
TEST: testMeEngañaste   DESCRIPTION:*Third Test. The value returned does not match with the one expected.*

EXCEPTIONS

TEST: testExplota
    ERROR:
    TYPE:java.lang.Exception
    DESCRIPTION:This exception is thrown as a test. Here you will find the stack trace of the exception.

SYSTEM-OUT	SYSTEM-ERR
First test passed. Second test passed. Fourth test passed. Fiftht test passed.	

Static Code Tests

Results by Priority						
Files	Priority 1	Priority 2	Priority 3	Priority 4	Priority 5	Total
1	1	0	3	0	0	4

Results by Type of Problem				
Possible Bugs	Suboptimal Code	Overcomplicated Expressions	Design	Total
4	0	0	0	4

Duplicated Code

Duplicated				

Fig. 2 Overview of the results obtained from the unitary and static code tests that are displayed to students. The user can select any group of errors for further inspection. Automatic feedback for each individual error, included by the teacher, is also shown (on demand) to the student

5. Students can visualize both an overview of results or each one of them individually, for further details. A high-level comparison with the rest of the classmates is also offered to students, comparing the errors of their assignment with an average of the class. They can choose to read this on the Moodle web application or download a report in PDF format for later analysis (Figs. 2 and 3).

While students are uploading their answers, the teacher can access at any time to a summary or snapshot of how the whole class is working on the assignment, presented as an aggregation of the individual errors from each one of the students. For a deeper knowledge of students performance, the teacher is able to select a given student and analyze in detail the source code uploaded and the exact errors produced. To do this, access to both the student uploaded file (with the source code) and the same report the student has received as feedback (on web and PDF form) is given

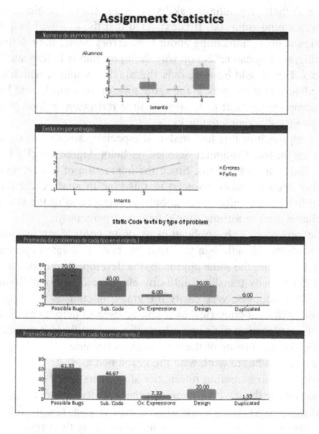

Fig. 3 Overview of the summary of errors of the whole class for a given task, when accessed by the teacher. Details about how many errors each group of the class has is also included in the data shown. The teacher can also obtain details about each individual error in the delivery of each student/group

to the teacher. This is a great insight into the progress of the class and can be used to prepare the next class sessions to better help their students by focusing on those issues where students are having problems.

4 Planned Experiences and Expected Results

The presented software is a completely functional tool, that has passed all software testing needed in this type of software. For the first field testing to be performed, the tool will be used to assist students with the assignments of programming classes on a second-year subject from the bachelor studies in the Computer Science Degree at the University of Burgos. This subject is one of the most difficult ones for students, since it requires them to master complex concepts and students are still not proficient enough with their programming skills (although they have already completed two basic programming subjects). It is expected that the tool will help them complete their assignments by informing about their errors before they deliver the final code for marking, so they can reflect on why their program is failing and how to correct these errors. This should improve over the usual dynamic in which the students receive the feedback on their work at the same time of their marks and have already moved on to complete the next task, so very little reflection is done over the flaws and errors of previously completed tasks.

The on-line approach of this tool makes it especially suitable to be used by the students from the on-line Computer Science bachelor degree already being implemented at the University of Burgos. Since the interaction of the group of students with the teacher becomes more reduced in this type of studies, the advantage of having a mean to automatically assess about the correctness of the tasks in a more flexible way than e-mail or forums should be quite appreciated.

Finally, the proposed web application is quite convenient for international/ transnational students. Moodle is a tool that has been translated into many different languages. Following the same approach, the developed auto-assessment tool is also prepared to be easily translated into any language (already including English and Spanish). This would be very easily integrated in courses that require student to move between different countries, as it can be used to help them follow courses taught in a different location with less effort. If a student finds it difficult to follow the teacher of understand some of the explanations because of the language, it will be much easier for him/her to work with the developed tool.

This can help also to incoming international students. For future incoming students is easy to provide them with a sample of the work local students are required to complete at a certain moment (with the exact same exercises and feedback). For current students, having the choice to receive feedback using their own language with very low effort (or practice a different one) could also be considered an interesting advantage.

5 Future Work

The system is regarded as a very interesting tool for supporting any programming assignment to university students. This means that it may not be the best way of introducing programming to inexperienced students, since it assumes students are familiar with reading requisites and complying with strict programming rules; but it could fit very well with undergraduate students from their second course on to the final courses.

As this was the initial setup for the software, it shapes a bit the main improvements and future work to be done in the system. Up to this moment, the system is not prepared to cope with problems mainly raised with students in their very first programming tasks: infinite loops, compiling problems, very sub-efficient solutions (time and resource wise); that could affect the performance of the system and even blocking access to other users.

Also, as versions of the Moodle system and the other tools (jUnit, PMD, ANT, etc.) are released regularly, the software should be updated, taking into account the requisites that newer versions of Moodle imposes to its modules. At the same time, the latest Moodle versions give the solution for this same problem: Moodle currently complies with the Learning Tools Interoperability (LTI) v.1.1 standard. This means that it includes several modules that enable Moodle to communicate with any other web tool that implements this specification. This way, both pieces of software can evolve at their own pace, maintaining their functionality intact. One of the clearest ways of improving the proposed auto-assessment system would be to implement the system in its own, dedicated server.

Finally, the ultimate goal of the system is not only to help students with the assessment of their work in a rather static way, but also to analyze their answers to different proposed task and guide them into new problems that could help them master concepts that are still missing from their learning, creating a personalized pathway through their programming courses, maximizing their engagement and learning outcomes.

References

1. Beck, K.: Test driven development: by example. Addison-Wesley Longman, London (2002)
2. Dalgarno, B., Bishop, A.G., Adlong Jr, W., Bedgood, D.R.: Effectiveness of a virtual laboratory as a preparatory resource for distance education chemistry students. Comput. Educ. **53**, 853–865 (2009)
3. Hotard, D.J.: The effects of self-assessment on student learning of mathematics. Ph.D. thesis, Louisiana State University (2010)
4. Keraudren, P.: New skills and jobs in europe: pathways towards full employment. Technical report, European Comission—Directorate-General for Research and Innovation (2012)
5. Marcus, N., Ben-Naim, D., Bain, M.: Instructional support for teachers and guided feedback for students in an adaptive elearning environment. In: Eighth International Conference on Information Technology: New Generations (ITNG), pp. 626–631 (2011)

6. McMillan, J.H., Hearn, J.: Student self-assessment: the key to stronger student motivation and higher achievement. Educ. Horiz. **87**(1), 40–49 (2008)

7. Moodle Pty Ltd: Moodle web site. https://moodle.org/. Accessed 15 Jan 2015

8. Panader, E., Alonso-Tapia, J.: Self-assessment: theoretical and practical connotations. when it happens, how is it acquired and what to do to develop it in our students. Electr. J. Res. Educ. Psychol. **11**(2)(30), 551–576 (2013), http://dx.doi.org/10.14204/ejrep.30.12200

9. Pelisse, R., Dangel, A.: PMD web site, http://pmd.sourceforge.net/. Accessed 15 Jan 2015

10. Pirker, J., Riffnaller-Schiefer, M., Gütl, C.: Motivational active learning: Engaging university students in computer science education. In: ITiCSE '14, Proceedings of the 2014 Conference on Innovation and Technology in Computer Science Education. pp. 297–302. ACM, New York (2014), http://doi.acm.org/10.1145/2591708.2591750

11. Prusty, G.B., Russell, C., Ford, R., Ben-Naim, D., Ho, S., Vrcelj, Z., Marcus, N., McCarthy, T., Goldfinch, T., Ojeda, R., Gardner, A., Molyneaux, T., Hadgraft, R.: Using intelligent tutoring systems in mechanics courses, a community of practice approach. In: The 22nd Annual Australasian Association of Engineering Education (2011)

12. Ramos, J., Trenas, M.A., Gutiérrez, E., Romero, S.: E-assessment of matlab assignments in moodle: application to an introductory programming course for engineers. Comput. Appl. Eng. Educ. **21**(4), 728–736 (2013), http://dx.doi.org/10.1002/cae.20520

13. Saff, D., Cooney, K., Birkner, S., Philipp, M.: jUnit web site. http://junit.org/. Accessed 15 Jan 2015

14. The apache software foundation: the apache ANT project. http://ant.apache.org/. Accessed 15 Jan 2015

ICEUTE 2015: Teaching and Evaluation Methodologies

Building Augmented Reality Presentations with Web 2.0 Tools

Krzysztof Walczak, Wojciech Wiza, Rafał Wojciechowski,
Adam Wójtowicz, Dariusz Rumiński and Wojciech Cellary

Abstract Augmented reality enables superimposing various kinds of computer generated content, such as interactive 2D and 3D multimedia objects, in real time, directly on a view of real objects. One of the application domains, which can particularly benefit from the use of AR techniques, is education. However, to enable widespread use of AR in education, simple and intuitive methods of preparation of learning content are required. In this paper, we propose an easy-to-use method of augmenting printed books with arbitrary multimedia content. A popular Moodle-based educational platform has been extended with a plugin enabling creation of AR presentations. In addition, an associated mobile application is provided, which helps teachers assign the presentations to book pages and enables students to access the content.

Keywords Augmented reality · AR learning · E-Learning · Web 2.0 · Moodle

K. Walczak (✉) · W. Wiza · R. Wojciechowski · A. Wójtowicz ·
D. Rumiński · W. Cellary
Department of Information Technology, Poznań University of Economics,
Poznań, Poland
e-mail: walczak@kti.ue.poznan.pl

W. Wiza
e-mail: wiza@kti.ue.poznan.pl

R. Wojciechowski
e-mail: rawojc@kti.ue.poznan.pl

A. Wójtowicz
e-mail: awojtow@kti.ue.poznan.pl

D. Rumiński
e-mail: ruminski@kti.ue.poznan.pl

W. Cellary
e-mail: cellary@kti.ue.poznan.pl

© Springer International Publishing Switzerland 2015
Á. Herrero et al. (eds.), *International Joint Conference*, Advances in Intelligent
Systems and Computing 369, DOI 10.1007/978-3-319-19713-5_52

1 Introduction

Augmented reality (AR) technology enables superimposing various kinds of computer generated content, such as interactive 2D and 3D multimedia objects, in real time, directly on a view of real objects. Widespread use of the AR technology has been enabled in the recent years by the remarkable progress in consumer-level hardware performance, in particular, in the computational and graphical performance of mobile devices and quickly growing bandwidth of mobile networks. Entertainment, e-commerce and tourism are examples of application domains in which AR-based systems become increasingly used.

One of the application domains, which can particularly benefit from the use of AR techniques, is education. AR can provide entirely new, intuitive methods of presentation of learning content. Learning content in the form of interactive, animated, three-dimensional virtual scenes, which – in addition – can be presented directly in the context of a real environment, can significantly improve students' comprehension and involvement [1]. The main problem, however, is that AR presentations are relatively difficult to create in comparison to typical learning content widely used in classrooms. As a consequence, pre-designed, ready-to-use AR presentations dominate in this domain. This significantly limits their utility in the learning process.

In this paper, we propose a solution to this problem by providing an easy-to-use method of augmenting printed books with arbitrary multimedia content. A popular Moodle-based learning platform has been extended with a plugin enabling creation of AR presentations. Also, a custom mobile application is provided, which helps teachers to assign the presentations to book pages and enables students to access them.

The remainder of this paper is organized as follows. Section 2 provides an overview of the state of the art in the domain of AR-augmented books. In Sect. 3, requirements for effective use of AR techniques in the learning process are discussed and the AR Web 2.0 learning system is introduced. In Sects. 4 and 5, the two main components of the system are presented: the AR Web 2.0 Platform and the Mobile AR Learning Application. Section 6 concludes the paper and indicates directions of future works.

2 State of the Art

The augmented reality technology enables presentation of learning content in the form of computer-generated virtual objects appearing directly in the context of real objects [2]. Recent research indicates that AR may significantly increase students' motivation for learning [1, 3, 4]. Also, a few studies reported on a positive influence of the use of AR techniques on the cognitive attainment of learners in different domains [5–9].

One of the possible applications of AR in education is the creation of augmented reality books – AR books – which are paper books augmented with computer-generated content. AR books provide physical means of operation, which enables users to leverage their prior experience from the real world for natural and intuitive interaction with the electronic content [10].

Until recently, AR books relied on fiducial markers printed on their pages to control the position of virtual objects. Typically, such markers have the form of a black square frame surrounding a special pattern enabling its identification. An AR book is captured with a camera and the captured video is augmented with overlaid virtual objects aligned with the markers. The most recognized software library for tracking fiducial markers is ARToolKit [11].

The approach to book augmentation based on pre-printed fiducial markers causes a number of disadvantages:

- markers must be placed on pages in advance – during the publishing process, which restricts the set of books that can be augmented only to new titles;
- the set of pages with markers cannot be modified, which excludes from augmentation the pages which have not been intended for augmentation during the publishing process;
- markers themselves do not provide significant learning value, so they take up valuable space on book pages;
- artificial markers in a book can be unacceptable for aesthetic reasons.

Nowadays, the problems with explicit markers have been mitigated with the advent of software libraries for tracking real objects based on their natural features. The most common software packages which enable tracking 2D images based on their inherent characteristics are: Vuforia [12], Layar [13], Metaio [14], and D'Fusion [15]. These solutions can be applied to tracking 2D images presented on any plane in a real environment; in particular they can be used for tracking book pages (or fragments of book pages). These software libraries open up new opportunities for the book augmentation. It is no longer necessary to include additional artificial markers in books, since the book pages may act as markers themselves. In this way, it becomes possible to enrich both new and already published books with additional multimedia content.

However, not every page is suitable for tracking to the same extent. For example, it is difficult to recognize and track pages that contain mostly text written in small print. Due to limitations of contemporary cameras, it is difficult to detect differences and distinguish between such pages. On the contrary, richly illustrated pages are more suited for tracking, particularly, if they contain pictures with high-contrast geometric areas. Nowadays, most of the textbooks for primary and secondary schools contain many illustrations, so they are well suited to be enriched with additional content.

Another advantage of solutions based on the natural feature-based tracking is that the pictures on book pages may be semantically related to the overlaid synthetic content. For example, a picture showing a chemical molecule can be used as a marker to display 3D model of the molecule. In this way, learners may receive consistent learning material, which comprises the content of traditional textbooks and related multimedia content. Such content may provide valuable add-on material – 3D model of a molecule is much easier to understand than a drawing, movie may better represent dynamic objects than an image, sound accompanying a book may significantly enrich learning experiences (esp. in the case of learning foreign languages), etc.

A number of studies on AR books have been undertaken in the past. The forerunner of AR books based on fiducial markers was MagicBook [16]. Books designed using the MagicBook concept were normal books with text and pictures on each page. However, some pictures surrounded with thick, black borders were used as fiducial markers for positioning add-on content attached to the book pages. An extended version of the MagicBook metaphor for learning cultural heritage was proposed in [17]. In this solution, users could not only browse virtual objects representing cultural artefacts by turning pages of the book, but could also take actions on the content using dedicated interaction markers attached to the side of the book. The MagicBook concept has been applied in the educational context in the AR-Jam project [10] for early literacy education and in GEN-1 book for learning construction engineering [18].

A novel type of AR books, called mixed reality books, has been introduced in [19]. The motivation of this study was the observation that considering an AR book as a physical container for markers demonstrated lower learning effectiveness of the book. Therefore, the main intention of the mixed reality book was to provide users with a less disruptive reading experience with a book without fiducial markers. The authors presented an AR version of an existing book entitled The House That Jack Built.

Another example of an AR book that does not contain fiducial markers is The Thirsty Crow. This is an interactive AR book for preschool children aimed at learning numbers using old folklore literature [20]. Users can view virtual objects on a mobile device using a mobile application. The same approach was applied by Carlton Books Ltd. which published a series of books entitled An Augmented Reality Book [21].

The main limitation of the above solutions is that the additional multimedia material accompanying a book is limited to the content designed by the creators of a given book or a specific AR application. Teachers are not able to modify the content based on the learning context, for example, in order to adapt the content to diverse cognitive abilities of different student groups.

3   Web 2.0 AR Learning System

Efficient use of AR techniques to support learning process requires meeting the following requirements:

- It should be possible to augment arbitrary visual markers, e.g., book pages or selected images or drawings, also in books already published;
- Creation of AR presentations should be as easy as possible, so that teachers without advanced technical skills could perform this task. Ideally, creation of AR presentations should be performed in the same way as creation of other types of educational content;
- It should be possible to use different kinds of content as the additional learning material – including 3D models, animations, images, sounds and video sequences;
- The AR content should be easily accessible, in various contexts, without the need of specific hardware;
- The solution should not impose additional costs – neither on the schools, nor on the teachers, nor on the students.

To meet the above-listed requirements, the AR Web 2.0 learning system has been designed. The system consist of two components: AR Web 2.0 Platform, which is a learning content management system with the support for AR content and Mobile AR Learning Application, which enables easy access to the AR content on mobile devices. The two components are described in the following sections.

4 The AR Web 2.0 Platform

In [22], a Moodle-based [23] educational platform has been proposed that provides an easy-to-use and secure environment designed to facilitate employment of various Web 2.0 tools, such as blogs, social bookmarking, e-portfolios, wikis, friend IMs and video embedding, by educationalists. The platform provides distributed communities of teachers with a secure system designed for decentralized creation and sharing of learning content.

To enable efficient use of AR content in the learning process, the abovementioned platform has been extended with an additional, custom-designed plugin *AR Marker&Content*, presented in Figs. 1 and 2. The plugin permits adding, removing and modifying AR markers and multimedia content, which should constitute an AR scene. The AR markers are freely selected 2D images, which can represent parts of real book pages or covers – there is no need for specific markers, such as QR-codes. The Vuforia third-party software library is used to recognize and track the markers within the scene [12]. In turn, multimedia content that can appear within the AR scene can be 2D images, 3D images, video sequences, sound sequences, text or embedded web pages. To enable free composition of AR scenes from arbitrary multimedia content, various scene templates encoded in the CARL (Contextual Augmented Reality Language) language are used [24, 25].

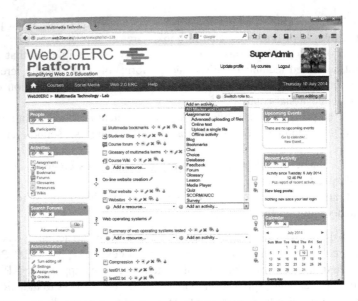

Fig. 1 Adding a new AR Marker&Content plugin instance in the activity mode

Fig. 2 AR Marker&Content
plugin in the block mode

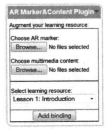

The plugin can be used in two modes: the activity mode and the block mode. The activity mode allows *AR marker–AR content* pairs to be created and placed in the context of a given course section. In Fig. 1, it is illustrated, how the AR Marker&Content plugin can be used to add an AR activity within a course available on the AR Web 2.0 Platform. The process of creating an *AR marker–AR content* pair requires uploading an image file representing the AR marker, i.e., a digital image of the paper book page or cover, and some multimedia content to be presented on the marker, e.g., a 3D model representing a chemical molecule to be shown on the book page.

In the block mode presented in Fig. 2, the proposed plugin allows *AR marker–AR content* pairs to be bound with the Web 2.0 learning resources already created within the course. For instance, an AR marker can be associated with video content in the context of a wiki-related augmented learning scene, while – at the same time – the marker can be associated with sound sequences in the context of a blog-related augmented learning scene. Also, with this approach, large multimedia objects can be uploaded once and effectively reused with different markers or different Web 2.0 resources.

The initialization of an augmented learning scene on a mobile device can be performed in two ways: either by browsing the learning content and clicking on the link representing an AR resource or by selecting an AR resource from the list of available AR resources provided by the server through Web Services. Based on the available set of AR scene templates designed for presentation of different types of multimedia content, the plugin creates XML-based specification of a particular AR scene in the CARL language. The CARL specification of an AR scene contains descriptions of the markers to be tracked, the content to be shown, and the interface used for manipulation and control of the AR scene.

The CARL specification is then used by the Mobile AR Learning Application as the input. An example of a trackable marker specification in CARL is presented in

```
<Trackables uri="http://platform.web20erc.eu/t/covChem.dat">
  <Trackable id="chemistry">
    <Begin>
      <ObjectBegin id="moleculeID"/>
    </Begin>
  </Trackable>…
```

List. 1 CARL – specification of a trackable marker and an associated content object

List. 1. The *Trackables* object is used to specify the binary representations of AR markers that can be detected and tracked within the 3D space by the mobile application. Moreover, it indicates which object will be used as augmentation content (initialized in the Begin section).

In turn, List. 2 shows an example of multimedia content description generated by the plugin. It provides a location of the multimedia content (in this case a 3D model representing a chemical molecule) indicated by a URI address and actions that can be executed on this object by the mobile application. There is only one default action declared, named *init*, which is responsible for setting the position, scale and orientation of the object in a 3D space. The action also activates the multimedia content and sets its state.

```
<ContentObjects>
  <ContentObject id="moleculeID" initState="hidden">
    <Resources>
      <Component id="molecule">
        <Location detail="med" uri="http://web20erc.eu/mol"/>
      </Component>
    </Resources>
    <Actions>
      <Action name="init" state="hidden">
        <SetPosition comp="molecule" x="0" y="0" z="0"/>
        <SetScale comp="molecule" scale="5"/>
        <SetOrientation comp="molecule" axis="x" angle="0"/>
        <Activate comp="molecule" active="true"/>
        <ObjectState value="shown"/>
      </Action>
    </Actions>
  <ContentObject>
<ContentObjects>
```

List. 2 CARL specification of a content object to be presented on a marker

5 Mobile AR Learning Application

The Mobile AR Learning Application has been designed to enable convenient access of students to the AR learning material and to support teachers in the process of creating AR resources. To run the application, a mobile device equipped with a camera and the Android operating system is required. The application can be configured to work with a particular AR Web 2.0 Platform server – eliminating the need of manual insertion of connection details.

Students may access the AR content within the Mobile AR Learning Application in two ways:

- by retrieving all the AR content for a course, or
- by accessing a single AR activity during a course.

In the first case, when a student logs-into the platform, he/she selects a desired course from the list of available courses. The application then retrieves CARL specifications of all the available AR activities. The CARL specifications contain both the descriptions of visual markers to track and the AR content to present on the markers. When a student points his/her mobile device towards a concrete book page or illustration, the AR activity assigned to the currently visible visual pattern appears on the screen of the mobile device.

In the latter case, while following a particular topic and browsing the related course on-line with all its resources and activities, a student can select the *AR Marker&Content* activity on a particular topic and then can point the camera of a

Fig. 3 Augmenting a chemistry book with a 3D model (left) and a video (right)

mobile device towards a designated page in the book. The Mobile AR Learning Application retrieves CARL specification of the AR activity to be used by the student and starts presentation of the AR content. This approach may be connected with game-like scenarios, where a student has to find appropriate content in a book, e.g., performing a kind of a quiz.

After activation of an AR activity, a student can interact with the presented multimedia content using gestures such as pinch or pan to scale or rotate a 3D object or tap to start playing a video sequence. Figure 3 (left) shows augmentation of a chemistry book with an interactive 3D model of a molecule. Another example is shown in Fig. 3 (right), where instructional video is presented on the book cover.

AR presentations may be configured directly within the Mobile AR Learning Application, in the teacher mode. A teacher makes a photo that will be used as an AR marker – an image of the book cover or a book page, part of the page, etc. Then, using the application, the teacher connects to the AR Web 2.0 Platform, selects a course and a topic and creates a new *AR Content&Marker* resource. The platform uses an AR marker service (e.g., Vuforia) to convert the image into an AR marker binary file, which can be later directly used during the tracking process. Then, the teacher may attach the AR content to the marker. The content may be selected from the list of multimedia files available on the mobile device used by the teacher. If the teacher does not have access to the content (e.g., the content is stored on a computer), the application attaches the original image that may be later replaced using the edit mode on the AR Web 2.0 Platform.

6 Conclusions and Future Works

Nowadays, mobile devices, such as smartphones and tablets, are powerful and widely available to both adults and children. It is important to exploit the potential of these devices beyond typical web browsing, social networking and gaming – which constitute vast majority of their usage. AR learning applications provide a good opportunity to benefit from the advanced technical capabilities offered by available mobile devices, to make the process of learning more effective and appealing.

The system presented in this paper, which extends the popular learning platform Moodle and couples it with a mobile AR application, makes the creation of AR content almost as simple as the creation of other types of learning content. Therefore, teachers without advanced technical skills are able to perform this task, so their students may benefit from the AR technology. Using the AR Web 2.0 learning system, teachers sharing a common course on a server may jointly prepare AR content for their students. Moreover, as the plugin follows security policy of the platform, a teacher may precisely specify which users can access the AR plugin functionality in which mode (e.g., editing or non-editing students or teachers).

Students can access the AR activities using available mobile devices – either in the classroom or at home. AR provides an intuitive and convenient interface to access various types of interactive multimedia content. Moreover, since the content is associated with printed books that the students are familiar with, the content is well organized and easy to access – being always at hand.

For the proof-of-concept implementation, the Android OS has been selected, as it is currently the most popular mobile operating system. However, other mobile platforms, such as iOS and Windows Mobile, could also be supported.

There are several directions for future works. First, support for other types of mobile devices should be provided to enable the use of the system by a vast majority of teachers and students. Second, a classroom evaluation of the system should be performed to verify students' and teachers' acceptance of the system and educational value of this kind of additional learning content. Third, possibilities of gamification of the AR activities should be explored to further increase students' motivation – through the use of engaging scenarios – performed by students both separately and in groups.

Acknowledgments This research work has been supported by the Polish National Science Centre Grant No. DEC-2012/07/B/ST6/01523 and by the European Commission within the project "Ed2.0Work – European network for the integration of Web2.0 in education and work".

References

1. Wojciechowski, R., Cellary, W.: Evaluation of learners' attitude toward learning in aries augmented reality environments. Comput. Educ. **68**, 570–585 (2013)
2. Azuma, R.: A survey of augmented reality. Presence: Teleoperators Virtual Environ. **6**(4), 355–388 (1997)

3. Di Serio, A., Ibáñez, M.B., Kloos, C.D.: Impact of an augmented reality system on students' motivation for a visual art course. Comput. Educ. **68**, 586–596 (2013)
4. Martín-Gutiérrez, J., Contero, M.: Improving academic performance and motivation in engineering education with augmented reality. Commun. Comput. Inf. Sci. **174**(2), 509–513 (2011)
5. Cheng, K.H., Tsai, C.C.: Children and parents' reading of an augmented reality picture book: analyses of behavioral patterns and cognitive attainment. Comput. Educ. **72**, 302–312 (2014)
6. Enyedy, N., Danish, J.A., Delacruz, G., Kumar, M.: Learning physics through play in an augmented reality environment. Int. J. Comput. Support. Collab. Learn. **7**(3), 347–378 (2012)
7. Ibáñez, M.B., Di Serio, A., Villarána, D., Kloos, C.D.: Experimenting with electromagnetism using augmented reality: impact on flow student experience and educational effectiveness. Comput. Educ. **71**, 1–13 (2014)
8. Lin, T.J., Duh, H.B.L., Li, N., Wang, H.Y., Tsai, C.C.: An investigation of learners' collaborative knowledge construction performances and behavioral patterns in an augmented reality simulation system. Comput. Educ. **68**, 314–321 (2013)
9. Westerfield, G., Mitrovic, A., Billinghurst, M.: Intelligent augmented reality training for motherboard assembly. Int. J. Artif. Intell Educ. **25**(1), 157–172 (2015)
10. Dünser, A., Hornecker, E.: Lessons from an AR book study. In: Proceedings of the First International Conference on Tangible and Embedded Interaction (TEI'07), pp. 179–182. ACM, New York (2007)
11. ARToolKit, http://artoolkit.sourceforge.net/ Accessed 12 Mar 2015
12. Qualcomm Vuforia, https://www.qualcomm.com/products/vuforia Accessed 12 Mar 2015
13. Layar, https://www.layar.com/ Accessed 12 Mar 2015
14. Metaio, http://www.metaio.com/ Accessed 12 Mar 2015
15. D'Fusion Studio Suite, http://www.t-immersion.com/products/dfusion-suite Accessed 12 Mar 2015
16. Billinghurst, M., Kato, H., Poupyrev, I.: The magic book: a transitional AR interface. Comput. Graph. **25**, 745–753 (2001)
17. Walczak, K., Wojciechowski, R., Cellary, W.: Dynamic interactive VR network services for education. In: Proceedings of the ACM Symposium on Virtual Reality Software and Technology (VRST 2006), pp. 277–286. ACM, New York (2006)
18. Behzadan, A.H., Kamat, V.R.: Enabling discovery-based learning in construction using telepresent augmented reality. Autom. Constr. **33**, 3–10 (2013)
19. Grasset, R., Dünser, A., Billinghurst, M.: Edutainment with a mixed reality book: a visual augmented illustrative children's book. In: Proceedings of the 2008 International Conference on Advances in Computer Entertainment Technology, pp. 292–295. ACM, Yokohama (2008)
20. Tomi, A.B., Rambli, D.R.A.: An interactive mobile augmented reality magical playbook: learning number with the thirsty crow. In: Proceedings of International Conference on Virtual and Augmented Reality in Education 2013. Procedia Computer Science, vol. 25, pp. 123–130. Elsevier (2013)
21. Carlton Publishing Group, http://www.carltonbooks.co.uk/ Accessed 12 Mar 2015
22. Wójtowicz, A., Walczak, K., Wiza, W., Rumiński, D.: Web Platform with Role-based Security for Decentralized Creation of Web 2.0 Learning Content. In: Proceedings of the 7th International Conference on Next Generation Web Services Practices (NWeSP), pp. 523–529. IEEE (2011)
23. Moodle, https://moodle.org/ Accessed 12 Mar 2015
24. Rumiński, D., Walczak, K.: CARL: a language for modelling contextual augmented reality environments. In: Camarinha-Matos, L.M., Barrento, N.S., Mendonça, R. (eds.) Technological Innovation for Collective Awareness Systems. IFIP Advances in Information and Communication Technology, vol. 432, pp. 183–190. Springer, Berlin (2014)
25. Rumiński, D., Walczak, K.: Dynamic composition of interactive AR scenes with the CARL language. In: Proceedings of the 5th International Conference on Information, Intelligence, Systems and Applications (IISA 2014), pp. 329–334. IEEE (2014)

Illustrating Abstract Concepts by Means of Virtual Learning Artifacts

David Griol and José Manuel Molina

Abstract Education is one of the most interesting applications of virtual worlds, as their flexibility can be exploited in order to create heterogeneous groups from all over the world who can collaborate synchronously in different virtual spaces. In this paper, we describe the potential of virtual worlds as an educative tool to teach and learn abstract concepts by means of programmable 3D objects. We describe the main experiences carried out recently in the application of these technologies in transnational educational activities that combine the Moodle learning resources and programmable 3D objects in the Second Life virtual world.

Keywords E-Learning · Intelligent environments · Immersive virtual worlds · Manageable 3D objects · Second Life · OpenSimulator · Sloodle

1 Introduction

The development of so-called Web 2.0 has also made possible the introduction of a number of Internet applications into many users' lives, which are profoundly changing the roots of society by creating new ways of communication and cooperation. The popularity of these technologies and applications has produced a considerable progress over the last decade in the development of social networks increasingly complex.

Among them, we highlight virtual social worlds, which are simulated graphic environments in which humans, through their avatars, "cohabit" with other users [1, 2]. Thanks to the social potential of virtual worlds, they are becoming a useful tool in the teaching-learning process [3, 4]. This way, virtual environments currently enable the creation of learning activities that provide an interactivity degree that is

D. Griol (✉) · J. Manuel Molina
Computer Science Department, Carlos III University of Madrid,
Avda. de la Universidad, 30, 28911 Leganés, Spain
e-mail: david.griol@uc3m.es

J. Manuel Molina
e-mail: josemanuel.molina@uc3m.es

© Springer International Publishing Switzerland 2015
Á. Herrero et al. (eds.), *International Joint Conference*, Advances in Intelligent
Systems and Computing 369, DOI 10.1007/978-3-319-19713-5_53

607

often difficult to achieve in a traditional classroom, encouraging students to become protagonists of the learning process and also enjoy while they are learning.

However, most of the virtual campus and educational applications in these immersive environments have only been created to replicate real world places without providing benefits from, for instance, accessing these applications in a classical webpage [5]. To address this problem, several initiatives and re-search projects currently focus on the integration of virtual worlds and virtual learning environments.

One of the most important initiatives is Sloodle (Simulation Linked Object Oriented Dynamic Learning Environment),[1] a free and open source project which integrates the multi-user virtual environments of Second Life[2] with the Moodle learning-management system.[3] This way, Sloodle provides a range of tools for supporting learning and teaching to the immersive virtual world which are fully integrated with the Moodle web-based learning management system, currently used and tested by hundreds of thousands of educators and students worldwide.

There are also different tools and programming languages, like the Linden Scripting Language (LSL) [6] or the Scratch tool[4] that make possible creating manageable 3D representations of abstract entities, which are usually very difficult to learn. The objects creates can also react to the user inputs and modify their main properties.

Our paper focuses on three key points for the creation of enhanced learning activities using immersive virtual worlds. Firstly, we promote the use of open source applications and tools for the creation of educative environments in virtual worlds, such as the tools and applications provided by means of the combination of the Open-Simulator virtual worlds[5] and the Moodle learning management system. Secondly, we emphasize the benefits of working in immersive environments to create visual objects that can clarify concepts that are difficult to understand due to their abstraction level. Thirdly, we show that it is possible to use these technologies for pedagogic purposes in transnational education and show a practical application of the integration and evaluation of these functionalities to carry out educative activities in the Second Life virtual world.

2 State of the Art

The benefits of virtual worlds for teaching and learning have fostered different research projects which aim is to help to use virtual environments in education. For example, the AVATAR Project (Added Value of teAching in a virtuAl woRld) [7] improves the quality of teaching and education in secondary schools through an innovative learning environment using a virtual world. The NIFLAR Project (Networked Interaction in Foreign Language Acquisition and Research) [8] is aimed at

---

[1]http://www.sloodle.org.

[2]http://secondlife.com/.

[3]http://moodle.org/.

[4]http://scratch.mit.edu/.

[5]http://opensimulator.org

enriching, innovate and improve the learning process by using video conferencing and virtual worlds for interaction among students from Spain and Holland. V-LeaF (Virtual LEArning platForm) [9] is an educative platform developed by Universidad Aut?noma de Madrid, which promoted cooperative and collaborative learning versus traditional learning.

Another project, 3D Learning Experiences Services[6] included several educative projects for language learning developed with OpenSimulator for secondary education in the Netherlands. Similarly, the AVALON Project (Access to Virtual and Action Learning live Online) was aimed to language teachers and learners http://avalonlearning.eu/. The project sought to develop best practices in teaching and learning of languages in multi-user environments (MUVEs) like Second Life or OpenSimulator.

The River City Project [10] simulated a city besieged by health problems. Students were organized in small research groups trying to find why residents were getting sick, using technology to track clues and figure out the causes of the disease, developing and testing hypotheses through experiments and extracting conclusions from the collected data.

At elementary level, the Vertex Project [11] involved students from 9 to 11 years old, who worked in groups to combine traditional activities such as writing a story, draw a scene or make a collage with the plan and design of their virtual worlds, with the educative goal of developing their creativity and imagination. They observed that within a multi-user world, students improve their communication skills, learn to cooperate and improve their self-esteem and confidence.

Additionally, WiloStar3D[7] is a virtual school that offers distance education using virtual worlds to improve reading comprehension, problem-solving ability, and creativity. Students learn to experience and interact in the virtual world, and take different roles, according to the activity, participating in media projects with other students.

Also some projects have worked on creating virtual communities to share knowledge. For example, the Impending Gale was a project developed by Game Environment Applying Real Skills (GEARS) [12]. They created an online community called the Social and Educational Virtual World (VSEW) that contains avatars, chat, Voice Over Internet Protocol (VoIP) communication, areas for tutoring with teachers and various social objects that can interact with avatars. In a similar way, Euroland [13] was a project based on collaborative activities carried out among several classrooms in the Netherlands and Italy, which implied the creation of a 3D virtual world that was initially empty and was progressively populated with culture houses.

However, although the pedagogical background has been carefully developed for these projects, the lack of interaction modalities in these environments may have a negative impact on the students' learning outcomes. Recently, Mikropoulos and Natsis presented a ten-year review on the educative applications of virtual reality

---

[6]http://www.3dles.com/en/.
[7]http://www.wilostar3d.com.

covering more than 50 research studies, and have pointed out that, although virtual worlds support multisensory interaction channels, visual representations predominate [3].

3 Our Proposal to Illustrate Abstract Concepts by Means of Educative Virtual Environments

In order to exploit Second Life and OpenSimulator technologies to develop a fully equipped virtual activities for collaborative learning, we propose the use of Sloodle activities and manageable learning 3D objects.

Sloodle activities make it possible to create and control the collaborative sessions [14]. The main objective is to take advantage of the Moodle learning management system for the creation of virtual courses and the possibility of using these functionalities through the 3D tools described in the previous section. In addition, the different activities requiring the Sloodle tools have been designed to make the participants be able to collaborate and express their ideas and make decisions, with the help of the material provided, such as documents, meeting minutes, presentations, and so on. This can be achieved thanks to the possibilities for user control and multimedia resources location in Sloodle.

Sloodle provides a range of tools for supporting learning and teaching to the immersive virtual world. Firstly, it allows controlling the user registration and participation in a course thanks to an *Access Checker*. Also new users can register in a course using the so-called *EnrolnBooth*. Secondly, there are several tools to create surveys in Sloodle, such as *Choice Horizontal, Quiz Chair* or *Quiz Pile on. Choice horizontal* allows instructors to create and show surveys in the virtual world, compile the information and show the results in a course. With *Quiz Chair* an avatar can answer questionnaires of a course in SL, while with *Quiz Pile On* provides a similar functionality with a more amusing format. This way, questions are presented in the form a text that floats over a pile, the students seat over the correct answer and if the answer is wrong, he falls over.

Thirdly, the *Sloodle Presenter* tool allows creating presentations in Second Life, which can combine images, web pages and videos and may be configured so that any avatar or only the owner of the corresponding sim controls the display of the presentation. Finally, there are other interesting tools for object sharing such as the *PrimDrop*, which allows students to deliver their works by sending objects in Second Life, or the *Vending Machine*, which can be used to deliver object to the students. Figure 1 shows different images corresponding to the described Sloodle tools.

Our proposal to create manageable learning 3D artifacts is based on the Linden Scripting Language (LSL) to create objects that react to the user inputs by changing their properties (position, volume, color, name, description, owner, etc.) or specific behavior. This language allows creating objects that react to the user inputs by changing their properties (position, volume, color, name, description, owner, etc.) or specific behavior.

Fig. 1 Sloodle tools. Top *Access Checker* activated by entering a login zone (left), *Access Checker* with a door (middle) and *Enrol_Booth* (right). *Middle* Choice *Horizontal*, Quiz Chair and Quiz Pile On. Bottom PrimDrop (left) and Vending Machine (right)

There are different tools that facilitate the use of this language. For instance, Scratch is a graphical programming language based on constructing programs by snapping together graphical blocks, which was developed by the Lifelong Kindergarten group at the MIT Media Lab. Scratch can be installed and freely redistributed and its source code is available under a license that allows modifications for non-commercial uses. This way, it is possible to create custom objects and applications.

LSL also allows controlling primitives, objects, and avatars by both physical events (touch, collide, listen) and logical events (temporizer, sensors). LSL includes a set of built-in functions, events and states definition and handling, different types and operators, and structures for flow control. Before their usage, the 3D objects are designed, developed and stored in the metabot's inventory. The *llRezAtRoot* function allows extracting the object and places it in SL. Then, the objects wait for commands that activate them and modify their behavior.

Figure 2 shows different images of the activities developed by the student using the LSL programming language. The images in the figure show important aspects of our proposal. The first image (right up corner) shows one of the panel of the scratch tool, which is employed to assign functionalities to the objects placed in the virtual world. Thanks to this tool, the students can access to the actions list and develop scripts with the help of an assistant. The second image (left up corner) shows several students paying attention to the instructions of the work they must develop and accessing the resources of the subjects using the *Sloodle Presenter* utility while they can also interact with the automatic bot, which is the avatar dressed in black in the image. In the bottom left image there is a student moving around in the virtual world to enter a work space of a subject, while the bottom center image shows the initial materials that the students had to carry out one of the proposed activities. Finally, the bottom right image shows several students interacting with a set of objects described previously in Sect. 3, with the objective to develop an essay about one of the units of the subject. Active behaviors of objects are developed with Scratch.

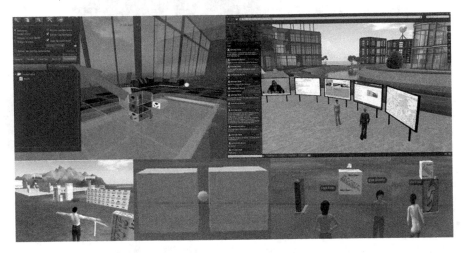

Fig. 2 Activities designed with the LSL programming language

4 Practical Application of the Proposal

We have used our proposal for creating a virtual learning environment supporting synchronous and collaborative learning at the Computer Science Department of the Carlos III University of Madrid (Spain) and the Institute of Information Technology at the University of Ulm (Germany). In particular, we present the different activities we have designed and how we have extended SL through the design of ad-hoc objects to enhance collaborative learning in small groups and integrate SL with Sloodle for managing collaborative sessions and multimedia contents.

The participants in the experiment were 70 students attending Language Processors and Formal Languages and Automata Theory courses at the described universities. These subjects cover theoretical and practical contents about Finite Automata, Push-Down Automata, Turing Machines, and the design and analysis of programming languages.

Based in other works such as [15], we developed a collection of practical activities for the development and programming of abstract entities (Finite Automata,

Push-down Automata, Turing Machines) and algorithms (bottom-up syntax analysis, grammar definition and usage). The main objective was to take advantage of the graphical possibilities of the virtual worlds to clarify complex concepts that students usually find difficult to understand due to their high level of abstraction. This way, graphical simulation can be useful to illustrate their behavior and facilitate the student learning fostering group work to collaboratively build knowledge in order to reach common predefined objectives.

One of the practices consisted in the implementation of a virtual simulator of a simplified Turing Machine. The system simulates a finite state machine with a single reading header and a single tape from which bits are read sequentially one after the other. The input to the system is contained in a text file, which also contains the machine transition function. The developed Turing Machine shows step by step the intermediate values until the analysis is completed. Internally, the simulator is also a finite state machine that includes four states and the events raise the transition from one to the other. At the beginning of the simulation, the machine is in a predefined state. If the avatar offers the notecard to the machine, it accesses an initial state in which it starts analyzing the inputs. At this point, the machine is listening in the control channel for a signal from one of the buttons: run or debug. When a signal arrives from any of these objects, the simulator starts working.

Each student had at his/her disposal a computer and was invited to interact with the other students and with the teacher through the virtual world. Before the lectures started, the tutors guided the students through the virtual campus and provided them with a little training session on the available communication features and avatar basic movements. The students had a basic knowledge of 2D computer graphics, and image and video editing.

The teacher gave the initial lectures in a virtual classroom of the virtual campus. To perform collaborative sessions the students were aggregated in small groups of 2 or 3 members, each of which had at its disposal a separate environment where it was possible to discuss, visualize and record information. The conversations and decisions were automatically saved for later references in a database using the Moodle plug-in.

We have carried out a subjective evaluation through an opinion survey based on the evaluation methodologies for educative virtual environments proposed in [2], which were adapted from [16, 17]. Three main aspects have been evaluated: perception of presence, communication and perceived sociability, and specific evaluation of the virtual environment. The responses to the questionnaire were measured on a five-point Likert scale ranging from 1 (for nothing/strongly disagree) to 5 (very much/strongly agree).

The questionnaire results corresponding to the perception of presence are summarized in Table 1. As can be observed, the presence perception of the students was very high. The control factors (CF) values reveal that the user perceived a good sense of control on the environment and proposed activities, as the median of all questionnaire items is 3.7. Similar results have been obtained for the sensory, distraction and realism factors (SF, DF, and RF), even if with the dispersion was higher.

Table 1 Results related to the perception of presence

	Total	Scaled	CF	SF	DF	RF
Mean score	117.2	3.7	3.9	3.7	3.9	3.4
Std. Dev.	9.8		0.5	1.1	0.8	0.6
Max. value	147.0	4.6	5	5	5	5
Min. value	89	2.8	4	3	3	2

Table 2 shows the results of the communication (COM), awareness (AW), and perceived sociability (PS) questions. Also in this case, the median is high, but higher data variability denotes a heterogeneous perception of the communication features. A deeper analysis confirmed that the lower average score was obtained by the last question, which is the question concerning the avatar gestures. This indicates the students had some difficulties in communicating by using the avatar gestures. On the whole, students felt the offered communication appropriate to favor discussion with others and speech communication improved the interaction with other students. Let us note that, even if the median is 3.9, a very good result, there were some negative judgments by users who found it more difficult to figure out what was happening at some moments.

Table 2 Results related to communication, awareness and perceived sociability

	COM	AW	PS
Mean score	4.1	3.9	3.8
Std. Dev.	1.3	1.1	1.0
Max. value	5	5	4
Min. value	2	3	3

In Table 3 we summarize the results corresponding to the evaluation of the virtual environment and the results concerning the additional questions referring to the perception of learning and global satisfaction. As can be observed, the mean of all the items of this category is high (4.5) and the low standard deviation indicates that most subjects agreed in assigning a very positive score to the activities in the virtual environment and were satisfied with the experience.

5 Conclusions

The development of social networks and virtual worlds brings a wide set of opportunities that can be exploited for education purposes, as they provide an enormous range of possibilities for evaluating new learning scenarios in which the students can explore, meet other residents, socialize, participate in individual and group activities, and create and trade virtual objects and services with one another.

In this paper, we propose the use of the Moodle Learning Management System and the development of programmable objects to create enhanced learning activi-

Table 3 Results related to the virtual environment

	Virtual environment	Learning	Satisfaction
Mean score	4.5	4.4	4.7
Std. Dev.	0.4	0.9	0.5
Max. value	3	2	3
Min. value	5	5	5

ties in this kind of immersive virtual environments, thus complementing traditional educative methodologies for explaining abstract concepts that are usually very difficult to teach and learn. A practical application of our proposal in several Computer Science subjects of two European universities has been presented and evaluated, obtaining very satisfactory results in terms of presence, commitment, performance and satisfaction.

As a future work we want to extend the experience during this academic year, including in our study several functionalities to adapt and extend the proposed activities taking into account student's specific needs, considering their evolution during the course as one of the main aspects to perform this adaptation. We are also evaluating the benefits of using the described technologies, tools and programming languages in combination with open-source metaverses like the ones that can be created using OpenSimulator and the OSgrid Project.

Acknowledgments This work was supported in part by Projects MINECO TEC2012-37832-C02-01, CICYT TEC2011-28626-C02-02, CAM CONTEXTS (S2009/TIC-1485). We want also to thank the help of the ERASMUS-PDI program at the UC3M to make possible the collaboration between the Carlos III University of Madrid and the University of Ulm.

References

1. Arroyo, A., Serradilla, F., Calvo, O.: Multimodal agents in Second Life and the new agents of virtual 3D environments. In: Proceedings of the IWINAC'09, pp. 506–516 (2009)
2. Lucia, A., Francese, R., Passero, I., Tortora, G.: Development and evaluation of a virtual campus on Second Life: the case of SecondDMI. Comput. Educ. **52**(1), 220–233 (2009)
3. Mikropoulos, T., Natsis, A.: Educational virtual environments: a ten-year review of empirical research (1999–2009). Comput. Educ. **56**(3), 769–780 (2011)
4. Andrade, A., Bagri, A., Zaw, K., Roos, B., Ruiz, J.: Avatar-mediated training in the delivery of bad news in a virtual world. J. Palliat. Med. **13**(12), 1–14 (2010)
5. Girvan, C., Savage, T.: Identifying an appropriate pedagogy for virtual worlds: a communal constructivism case study. Comput. Educ. **55**(1), 342–349 (2010)
6. Rymaszewski, M., Au, W., Ondrejka, C., Platel, R., Gorden, S.V., Cézanne, J., Cézanne, P., Batstone-Cunningham, B., Krotoski, A., Trollop, C., Rossignol, J.: Second Life: The Official Guide. Sybex (2008)
7. Santoveña, S., Feliz, T.: El Proyecto added value of teaching in a Virtual World (AVATAR). Didáctica, innovacián y multimedia **18**, 1–9 (2010)

8. Jauregi, K., Canto, S., Graaff, R., Koenraad, T., Moonen, M.: Verbal interaction in Second Life: towards a pedagogic framework for task design. Comput. Assist. Lang. Learn. J. **24**(1), 77–101 (2011)
9. Rico, M., Camacho, D., Alamán, X., Pulido, E.: A high school educational platform based on Virtual Worlds. In: Proceedings of the 2nd Workshop on Methods and Cases in Computing Education (2009)
10. Ketelhut, D., Dede, C., Clarke, J., Nelson, B., Bowman, C.: Assessment of problem solving using simulations. In: Baker, E., Dickieson, J., Wulfeck, W., O'Neil, H. (eds.) Studying Situated Learning in a Multi-user Virtual Environment. Lawrence Erlbaum Associates (2007)
11. Bailey, F., Moar, M.: The vertex project: Children creating and populating 3D virtual worlds. J. Art Des. Educ. **20**(1), 19–30 (2001)
12. Barkand, J., Kush, J.: GEARS a 3D virtual learning environment and virtual social and educational world used in online secondary schools. Electron. J. e-learning **7**(3), 215–224 (2009)
13. Ligorio, M., Talamo, A., Simons, R.: Euroland: a virtual world fostering collaborative learning at a distance. In: Proceedings of the First Research Workshop of EDEN Research and Innovation in Open and Distance Learning (2000)
14. Griol, D., Molina, J.: Using virtual worlds and Sloodle to develop educative applications. International workshop on evidence-based technology enhanced learning. Adv. Intell. Soft Comput. **152**, 99–106
15. Gasperis, G.D., Maio, L.D., Mascio, T.D., Niva, F.: Il Metaverso open source: Strumento Didattico per Facoltá Umanistiche. In: Proceedings of the Didamatica. pp. 1–10 (2011)
16. Witmer, B., Singer, M.: Measuring presence in virtual environments: a presence questionnaire. Presence: Teleoperators Virtual Environ. **7**(3), 225–240 (1998)
17. Kreijns, K., Kirschner, P., Jochems, W., van Buuren, H.: Measuring perceived sociability of computer-supported collaborative learning environments. Comput. Educ. **49**, 176–192 (2007)

Mining Skills from Educational Data for Project Recommendations

Pavel Kordík and Stanislav Kuznetsov

Abstract We are focusing on an issue regarding how to actually recognize the skills of students based on educational results. Existing approaches do not offer suitable solutions. This paper will introduce algorithms making possible to aggregate educational results using ontology. We map the aggregated results, using various methods, as skills that are understandable for external partners and usable to recommend students for projects and projects for students. We compare the results of individual algorithms with subjective assessments of students, and we apply a recommendation algorithm that closely models these skills.

Keywords Educational data mining · Data transformation · Data warehouse · Student skills · Ontology

1 Introduction

The cooperation of educational facilities with industries has always been a very important field. Companies have always been significantly interested in research that is offered by academic institutions. Being it contractual research or mutual projects. This cooperation enables the university to acquire extra funds, new technologies and increase its publication activities.

The requirements of industrial companies have changed in the past few years. Companies have begun focusing on extensive cooperation with university students. They are looking for very talented individuals as well as "reliable and tried" students who can offer employment during or even after their studies. Even though, this is not a mainstream custom, it encourages people at universities to begin with education data mining and to discover methods to meet this demand [1]. Another reason to

P. Kordík (✉) · S. Kuznetsov
Faculty of Information Technology, Czech Technical University in Prague,
Czech Republic, Europe
e-mail: pavel.kordik@fit.cvut.cz

S. Kuznetsov
e-mail: kuznesta@fit.cvut.cz

© Springer International Publishing Switzerland 2015
Á. Herrero et al. (eds.), *International Joint Conference*, Advances in Intelligent
Systems and Computing 369, DOI 10.1007/978-3-319-19713-5_54

begin mining student data on universities is to increase the quality of the education process and assessment reliability thanks to acquired knowledge [2].

As described here [3, 4], education data mining is experiencing a boom. This means, the field is branching out into many directions. However, some of these directions stay unchanged within some communities [5], e.g. EDM[1] and LAK.[2] One of the important directions is the inventing of new methods, processes, and algorithms that can better assess and classify the experiences of students. Many works trying to estimate the level of student?s knowledge (skills) by processing of original educational data without further transformation. For example, Brijesh Kumar Baradwaj [6] applied the decision tree as a classification method for evaluating students performance using information like attendance, class tests, seminar and assignment marks [7]. Used thousands of actions in Cognitive Tutor curriculum, covering a diverse selection of material from the middle school mathematics curriculum to show method that is effective for distinguishing between the relative impact of state and trait explanations for student behavior. At [8] using fine-grained skill models, a skill model is a matrix that relates questions from the online tutoring system to the skills needed to solve the problem, to fit student performance with Bayesian Networks [9]. Compare the performance of the three estimates of student skill knowledge using simulated data from the DINA[3] model using common clustering methods: hierarchical agglomerative clustering, K-means, and model-based clustering.

Our approach is different. We try to estimate the skills of students based on evaluation from university courses. We first use a data mining process to create a skill tree, and then we use algorithms to transform these data into final standardized skills. These skills represent various levels of abstraction and have been simplified to meet the requirements of our industrial partners. This process has made it possible to automatically generate student profiles that we may compare with each other. Then, we recommend the most suitable students for specific projects.

In this paper, we present our ontology (skill tree) and describe mapping of student subjects. This will be followed by a student skill generating process where we present 4 algorithms and compare these algorithms with a student reference (i.e. subjective profile). Finally we present the direction of our future work.

2 Data and a Data Warehouse

We created a data warehouse to consolidate data from our university systems. The system is designed according to the Inmona [10] architecture and is tailored to our needs. The data warehouse makes it possible to automatically collect data, consolidate and visualise it in the form of dashboards and ad-hoc reports. Further on, the data warehouse enables us to store historical, data and so we are able to undertake analyses containing extensive periods of time.

---

[1] Educational Data Mining.

[2] Learning Analytics and Knowledge.

[3] The deterministic-input noisy-AND model.

This paper focuses on data from the study results of students. These are the data containing the final results of the students. Our faculty uses the ECTS grading scale. We work with 6 grades from A to F. In order to achieve an E grade, students have to show at least a 50 % knowledge of the subject matter, otherwise they will receive an F.

3 Ontology

Our ontology represents a simplified "skill tree" inspired by the ACM tree.[4] We simplified the tree by removing skills that are not taught on the faculty. We did not want the tree to be excessively deep(a maximum of 2 levels), so that clients (people without experience in the education system) can easily understand it. The higher levels of the tree contain skills that represent abstract levels of knowledge (centriol) and which are made up of more than one specific skill. Around each of these skills, we have a cluster of specific skills. The representation of skill tree is shown in Fig. 1.

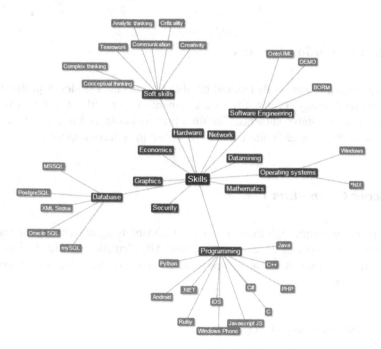

Fig. 1 Skills are represented by tree so higher level nodes generalise skill of leaf nodes

---

[4]http://www.acm.org/about/class/class/2012.

3.1 Subject Mapping

We have to map every subject taught on our faculty to the skill tree. Because we are a technical faculty, most of the subjects are divided according to subjects of study, these are in compliance with ACM ontology. We began with automatic mapping based on the description of the subject in our system, and then we consulted this with the lecturer in order to increase the accuracy of the mapping. Besides a list of skills that cover a given subject, we assigned a value to each skill (from 0 to 1), this is the level that the given subject covers a specific skill. This resulted in a matrix that represents the core of the entire system.

Example 1 shows mapping of subjects with code MI-PDD[5] to skills.

Example 1 MI-PDD:

- Datamining weight $= 1$
- Mathematics weight $= 0.2$
- Database weight $= 0.4$
- ...

4 Skill Calculation Process

First, we created a process that would be able to generate a student's profile based on the student's study results. Then we compare these profiles against each other and recommend suitable students. The final process contains 3 phases. We created several algorithms for each phase that have shown to be most applicable.

4.1 Score Calculation Phase

In this phase, we work with the grades of the student using algorithms, we transfer them into absolute values which we call scores. All calculations made are done individually for each student, i.e. no comparison is made. We use 2 different algorithms for this calculation.

4.1.1 The SimpleCompute Algorithm

This was the first algorithm to be implemented thanks to its intuitive nature. We use it to calculate the basic score for each individual skill of a student.

---

[5]Data Preprocessing.

Algorithm steps:

- We acquire the grades of each student for all subjects and transform it into a value (coefficient) according to this rules - A is 1 point, B is 0.8 point, C is 0.6 point, D is 0.4 point and E 0.2 point.
- To calculate the induvidial skills of each student we use the Formula 1.

$$\text{Skill} = \sum (\text{subject weight} * \text{grade coefficient}) \tag{1}$$

4.1.2 CumeDistCompute Algorithm

It often happens on academic grounds that subjects with a number of identical credits are not identically challenging. Sometimes the same subjects may be slightly different. The main idea of the algorithm is the assumption that if a subject is easy, it will mean a better total assessment of the students and vice versa. We used the CUME_ DIST() function to be able to react to this occurrence and to calculate scores. Algorithm steps:

- We acquire a list of all students and their subjects. For every subject, we select a subset of students that have finished the subject with a grade better than F
- We calculate an intermediate value for each student using the CUME_ DIST() function
- We multiply the result with the value of the skill for the given subject
- The algorithm runs 2 times in order for results of a tree leaf to get into the parent node

4.2 Comparison and Normalization Phase

In this phase, we compare the students and then calculate the number of stars using a normalization function. The SSP portal has two main groups of students, bachelor's degree students and master's degree students, we call it programs.The algorithms compare two students that are in the same programs of study and in the same year.

4.2.1 The CompareMinMax Algorithm

The main feature of this program is the use of Min Max normalization. Algorithm steps:

- The entire set of students is taken for every program of study (BI and MI)
- A minimum and maximum skill value is calculated for each set
- Using MinMax we calculate a value from 0 to 1 for each student and each of his/her skills and transfer this value into stars according to the following Formula 2

$$\text{Number of stars} = \frac{\text{Student score} - \text{Min score}}{\text{Max score} - \text{Min score}} * 5 \qquad (2)$$

4.2.2 CompareSurvivalFunc Algorithm

The algorithm to calculate individual "stars" uses a so called survival function. This function tries to eliminate the dependency of a set of students on one specific individual and tries to reflect the entire spread of student skills.
Algorithm steps:

- Identically with the previous algorithm, students are divided into sets according to their program of study (BI-MI) and year
- Each set is arranged in descending order and we calculate the number of stars for each skill according to the following Formula 3

$$\text{Number of stars} = \text{Round up}(\frac{\sum_{i=1}^{n}(\text{Value i to value n})}{\sum(\text{All values})} * 5) \qquad (3)$$

5 Practical Apllications

In the last phase of our process, we assign students to an assignment. Any assignment may contain a random number of positions, e.g. team researcher, database analyst, C++ programmer, etc. The client may select the necessary amount and level of skills of every positions. We are then able to calculate the most suitable students based on these values.

We select a vector of required skills for every position of the assignment. For each student, we calculate a similarity value in regards to the required vector of skills using Euclidean distance, the folowing Formula 4. Finally, we arrange the values. If the assignment does not state otherwise, we send an e-mail to the first 50 students, informing them of their match

$$\text{Similarity value} = \sqrt{\sum(\text{StudentValue[n]} - \text{SkillValue[n]})^2} \qquad (4)$$

5.1 SSP Portal

The Student Cooperation Portal (SSP) is an information system that has been designed and developed by the Faculty of Information Technology, CTU, Prague. This portal provides a platform for industrial partners (referred to as "clients") who input their projects into the system and for students who look through these projects and apply if interested.

Because a large number of students is involved in the portal, it is very important to correctly recommend the "correct" students for positions in the project. This applies vice versa as well, it is necessary to offer the projects to the correct students. Following these assumptions ensures a high success rate for these projects, as it is important that only students who are capable of solving the assignment apply for the assignment.

The calculated skills also enable us to automatically generate students CV and their profile specialization. Each students then has the opportunity to anonymously compare with other students in the portal.

6 Evaluation

We implemented two groups of skills that every student may have. The first group consists of so called "Objective skills". These are skills that have been calculated from study data using our own algorithms. The second group consists of so called "Subjective skills". These are skills that every student enters into his/her profile by himself/herself.

When comparing students, we worked on the assumption that students that have entered their subjective skills into their profile know more about their level of skills then our algorithms, and so we may use them as reference. So, in order to compare our algorithms, we have chosen a method where we compare how closely our algorithms are able to copy the subjective skills of the students.

Because the portal has several types and versions of algorithms, we present the results of all the combinations.

6.1 Initial Settings and Results

We selected 409 active students for analysing. These students were registered at the portal for at least one semester and show signs of medium activity. Because we selected a method of comparing our algorithm with a subjective number of students, we may only select those students who filled out both of their profiles. Table 1 shows all algorithm combinations.We estimated student profiles for each combination and then compared them with their subjective profiles. The calculations were performed for all students and all skills. The result of comparison is on Fig. 2.

The chart shows the result of how different algorithms hit subjective values. The size of the bubbles depends on the number of hits. The best result was achieved by algorithms that well estimate the subjective value and have the biggest bubble on the main diagonal. The best result was achieved by a combination of CumeDistCompute & CompareMinMax algorithms because they have the largerst bubbles ont the main diagonal and their neighborhoods.The second best result was achieved by a combination of CumeDistCompute & CompareSurvivalFunc. From the results, we can conclude that is better to use CumeDistCompute algorithm in the first phase of skill calculation process. Since the algorithms using SimpleCompute for the most part deviated from the diagonal.

Table 1 Proposed algorithms for skill estimation can be combined from components proposed above

Score calculation phase	Comparison and normalization phase
SimpleCompute	CompareMinMax
SimpleCompute	CompareSurvivalFunc
CumeDistCompute	CompareMinMax
CumeDistCompute	CompareSurvivalFunc

This result may be affected by the fact that the current combination of algorithms we use in portal is just a winning combination. So, the students may vote their skills depending on the generating dates. Long-term analysis in the future can varify this assumption.

The result implies that automatic algorithms can estimate the student skills with sufficient accuracy to provide comparison between students. The biggest advantage of the entire solution is the fact that the profiles are generated automatically after the first time student login into system, and therefore students need no fill/update.

7 Discussion

This paper presents the method we use to map individual subjects to our created ontology. We also presented a tree of skills. There are many more methods of how to describe the relation between subjects and skills. We have developed our method based on cooperation with lecturers of our university and industrial partners in order to present an optimal compromise between comprehensiveness and simple use. The tree itself will be continuously researched and developed. The university, similar to a living organism, is constantly developing, and so we have to review our solutions and offer new solutions in order to better describe the current focus of the university.

This paper also presents processes that transfer the final grades of the students to scores and then the transformation of these scores into levels of skills (stars). The entire process is divided into 3 separate phases. This kind of separation is beneficial as it enables us to develop each phase separately without influencing the behaviour of the entire process.

We presented two algorithms in phase one. The SimpleCompute algorithm represents a very intuitive and direct approach. Its disadvantage is the fact that every

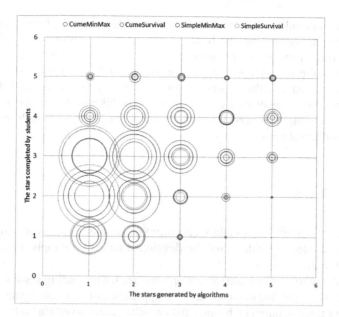

Fig. 2 Evaluation of proposed algorithms with regard to subjective evaluations. Chart shows skills of 184 students that both entered subjective value and the value provided by an algorithm was positive. Error in the graph is caused by the absence of data first-year students

subject has a different level of difficulty and so it is necessary to take into account the spread of grades within the subject. The second algorithm is CumeDistCompute and reacts better to the spread of grades and has shown to be the better choice of the two.

In the second, phase we presented two algorithms that compare the score of individual students and transform it into the final level of skill. The CompareMinMax algorithm is a simple algorithm that compares the score of each student with the largest score in the group. The algorithm does not react to the total spread of students and if there happens to be a brilliant individual, the skill value of the set will not decrease. Another algorithm is the CompareSurvivalFunc which uses for transformations the so called Survival function. The algorithm reacts to the total spread of students in a set and partially eliminates the effect that a brilliant individual would have on the set. The differences in the algorithms are not big, but the CompareSurvivalFunc is better.

On the results at Fig. 2 is interesting that most of the students with lower score (1, 2, 3 stars) tend to enter higher value of skills—some of them overestimate their skills and some have not finished enough courses to qualify for more stars. The third phase uses an algorithm based on "Euclidean distance". This algorithm is very simple, but it is not able to precisely define the absolute position of a student. We try to solve these and other issues using information technologies as well as with various approaches to data.

Comparison results imply the direction, we should take in the future. This means to find algorithms that would be better in evaluating the complexness of subjects. This will however mean to begin using more data, e.g. to monitor who is assessing the student, if the assessment of a given lecturer correlates with the assessment of other lecturers, and how this is different from the mean, and to use this information to adjust assessment evaluations, etc. Another possibility to begin using data from individual end-of-term papers, individual tests or skill tests for which the students could get additional scores.

8 Future Work

Our future work shall focus mainly on improving results of automatic algorithms. More sophisticated algorithms will be developed which will apply advanced data mining methods.

We continue to work on alternative approaches for presenting results. An interesting fact, we would like to mention, is the approach based on using game elements where every student, based on his/her study results, receives certain "achievements", takes part in "challenges" and thanks to these new elements can improve his/her profile.

Currently, we develop a tool for the portal to undertake various analyses, e.g. user behaviour analysis, etc., including visualisation and reporting of current and historical data. We do this in order to better understand the needs of our users (students and clients) and to offer them better services for mutual cooperation.

9 Contributions and Conclusion

The main contribution of this paper is that it shows new approaches to the calculation and presentation of student skills. This approach is based on the mapping of achieved student results and on a standardized tree of skills. Thanks to this, student profiles are created that can be presented to industrial partners (clients) and are well applicable when recommending students for a specific project. This process is now automated and using relatively simple algorithms we achieve good recommendation results when compared to reference student skills.

The advantage of our approach is that computed skills are updated automatically when new grades are available and students do not need to adjust their skills manually. We believe that computed skills are also more objective than skills entered by student—those may be strongly biased.

References

1. Edmondson, G.: Making industry-university partnerships work, lessons from successful collaborations. 2012 Sci. Bus. Innov. Board AISBL **1**, 1–50 (2012)
2. Chalaris, M., Gritzalis, S., Maragoudakis, M., Sgouropoulou, C., Tsolakidis, A.: Improving quality of educational processes providing new knowledge using data mining techniques. In: Procedia—Social and Behavioral Sciences (0) (2014), pp. 390–397. 3rd International Conference on Integrated Information (IC-ININFO)
3. Romero, C., Ventura, S.: Educational data mining: a survey from 1995 to 2005. Expert Syst. Appl. **1**, 135–146 (2007)
4. Romero, C.: Handbook of Educational Data Mining. CRC Press, Boca Raton (2011)
5. Siemens, G., Baker, R.S.J.d.: Learning analytics and educational data mining: towards communication and collaboration. In: Proceedings of the 2Nd International Conference on Learning Analytics and Knowledge. LAK '12, pp. 252–254. ACM, New York (2012)
6. Brijesh Kumar Baradwaj, S.P.: Mining educational data to analyze students' performance. CoRR (2012)
7. Baker, R.S.J.D.: Is gaming the system state-or-trait? Educational data mining through the multi-contextual application of a validated behavioral model. In: vol. 3 (2007)
8. Pardos, Z.A., Heffernan, N.T., Anderson, B., Heffernan, C.L., Schools, W.P.: Using fine-grained skill models to fit student performance with bayesian networks. Handbook of Educational Data Mining, pp. 417–426 (2010)
9. Ayers, E., Nugent, R., Dean, N.: A comparison of student skill knowledge estimates. In: Educational Data Mining—EDM 2009, Cordoba, Spain, 1–3 July 2009. Proceedings of the 2nd International Conference on Educational Data Mining, pp. 1–10 (2009)
10. Inmon, W.H., Strauss, D., Neushloss, G.: DW 2.0: The Architecture for the Next Generation of Data Warehousing. Morgan Kaufmann Publishers Inc., San Francisco (2008)

Virtual Teams in Higher Education: A Review of Factors Affecting Creative Performance

Teresa Torres Coronas, Mario Arias Oliva, Juan Carlos Yáñez Luna
and Ana María Lara Palma

Abstract Many studies have shown how teams improve their effectiveness and efficiency when a composition and skills balance is properly managed. Creativity has been an ignored variable, especially in virtual contexts. Contextual factors provoke significant differences between traditional team work and the virtual team work. Some authors suggest that virtual environment can achieve higher levels of creativity due to greater openness, flexibility, diversity and access to information compared to traditional conditions. However, building trust and team cohesiveness in virtual structures could be more difficult and creative performance can be affected, decreasing innovative solutions. We focus our study in critical aspects of creative performance that should be taking into consideration in emerging eLearning collaborative processes in higher education.

Keywords Virtual teams · Creativity · Team working · Higher education

1 Introduction

Facing technological revolution is a challenge for most higher education institutions. Internet and expressions like ".com" are phenomena that have transformed the old workplaces in modern "e-workplaces" (Walker [52]). The pressure of the

T. Torres Coronas (✉) · M.A. Oliva
Universitat Rovira i Virgili, Tarragona, Spain
e-mail: teresa.torres@urv.net

M.A. Oliva
e-mail: mario.arias@urv.net

J.C.Y. Luna
Universidad Autónoma de San Luis Potosí, San Luis Potosí, Mexico
e-mail: jcyl@uaslp.mx

A.M.L. Palma
Universidad de Burgos, Burgos, Spain
e-mail: amlara@ubu.es

© Springer International Publishing Switzerland 2015
Á. Herrero et al. (eds.), *International Joint Conference*, Advances in Intelligent
Systems and Computing 369, DOI 10.1007/978-3-319-19713-5_55

new economy is forcing to most of the organizations to be more dynamic and to have innovative approaches to improve competitiveness and survive. Flexible organizational structures are a keystone to success. The new virtual structures break traditional physical ways of organizing work, being necessary to develop a borderless network organization [12, 19]. In these virtual teams, instead of traditional face-to-face and same time and space coordination methods, we can create virtual teams. In such teams, people geographically dispersed overcome time and space barriers using new technologies [22, 28, 29]. Technology integration is creating a hybrid model where distinction between virtual communication tools and face-to-face communication is less meaningful as time goes on [8].

In this new context, many authors have examined the emergence of virtual teams, analyzing their characteristics and factors affecting their performance compared with traditional teams [38, 40]. But there are still few studies focusing on creativity as a critical factor.

This article defines an integrative framework of creativity and teams, studying key variables that affect the creative performance in this new way of organizing work in learning environments. Traditional research covers performance variables such as communication, clarity of roles, diversity and trust in virtual environments; but creative performance is not included.

Our study begins with creative fundamentals in order to understand its importance for organizations and their teams. We continue introducing classical variables that affect team's performance and the ecological model of effectiveness, integrating creativity in the analyzed model. Finally we focus on creativity development environments, adapting traditional studies conducted in face-to-face environments.

2 Nature an Importance of Creativity

2.1 An Approach to Creativity

Smolensky and Kleiner [47] define creativity as the process that requires the integration of the left brain (logic) and the right (imagination) thinking to solve problems or produce something new. Other definitions of creativity do not emphasize the process, but the result (Prince [41]).

Most of the accepted definitions of creativity points out the developing new ideas and products. Ideas should be innovative and useful, should be appropriate; and should influence the way we do business [3, 4]. It is difficult to explain why groups are formed, but there are two reasons widely accepted in the literature [18]: (i) a functional reason who considers that people come together to perform a task, and (ii) a psychological reason that is related to the perception of a shared identity and a need for affiliation or recognition.

A successful working team exists when teamwork improves performance. A well-designed team is capable to obtain synergies that exceed the capabilities of each individual team member [13, 33]. The participation of students in working teams has a positive impact on learning and performance. For instance, development of decision making skills, increased confidence, or improved implementation of technology, improved quality of learning, lower rates of absenteeism and desertion among others advantages.

2.2 Determinants of Performance Work Teams (Face-to-Face)

There are different models to study teams work performance, as the proposed by Gladstein [20] cited in Jex and Britt [21]. The group effectiveness model is an accepted and holistic framework to study teamwork. According to Gladstein the effectiveness of the efforts of a team on a given task depends on inputs (group level and organizational level), process and output as we can see in Fig. 1:

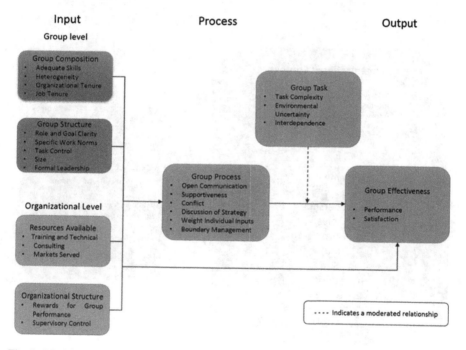

Fig. 1 Model of group effectiveness [20]

The effectiveness of work teams is influenced by external and internal factors. The basis of the ecological model of the behavior places teamwork within the organizational context, as we can see in Fig. 2.

According to the ecological model, the behavior of a working team is a subsystem within a larger system, the organization [49] also cited in [50]. The team performance will depend on organizational strategy, authority structures, formal regulations, organizational resources, human resources selection, performance evaluation and reward systems, organizational culture and atmosphere of physical work [44]. Proper management of these processes will be the key to encouraging performance.

Moreover, for a better productivity of team working, it is necessary to ensure the existence of a balance of skills, abilities and personalities among its members (Belbin [9]). These factors are those that have received more attention as determinants of the performance of a team because the evidence indicates that individuals with job skills and intellectuals are involved and have better contribution, helping to improve team performance (Mathieu et al. [30]).

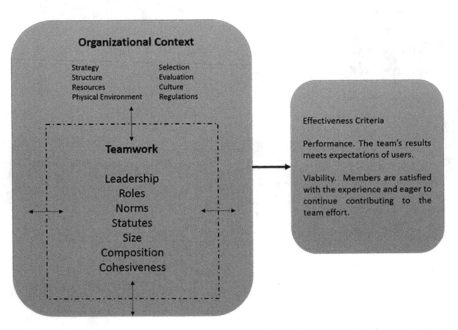

Fig. 2 Ecological model of the effectiveness of teamwork (adapted from Sundstrom et al. [49])

2.3 Determinants of the Performance of Virtual Teams

In virtual teams it have been identified different aspects that determine performance, such as the method and type of communication [40], building a culture of collaboration and trust [42], technical project management [23], a correct objectives setting [40], leadership style and roles of its members [43, 48] or transformational leadership [11] among others.

Factors related to technology have been studied as well, as the selection of appropriate technology [37] or the role of communication in trust [28, 29, 46]. Managing virtual teams has important challenges such as trust, cohesion, and team identity and eradicates the isolation of its members [26]. Maznevski and Chudoba [31] work analyses the key factors for the effectiveness in Virtual Teams. These authors look at how structural features such as task, culture and group characteristics interact with technology and decision outcomes (quality, commitment and team cohesion). Building commitment requires a high level of involvement. Because of that complexity, decision making meetings face to face can be more effective than a simple e-mail. The creation of trust is as well enhanced in face-to-face environments.

Any team should follow a structured process of making decisions that encourage critical thinking [16]: (1) identify and explore the problem, (2) generate possible solutions [14]; (3) refine and critique possible solutions and (4) implement the solution. During the virtual team decision making process, [16] points out that creativity must be a constant. Individuals have to try to see the problem differently, to restructure their thinking; to generate as many ideas as possible before any assessment is done.

3 Creative Team Development

Structures centered in teams have improved organizational performance, but its future growth and development depends on creativity [17, 34]. Nemiro [36] examined the importance of the social environment on creativity, identifying six factors to increase groups creativity (Table 1):

Table 1 Necessary factors environment to encourage creativity of teams (adapted from Nemiro [35])

Factor	Description
Autonomy and freedom	To give responsibility to individuals to launch new ideas, make decisions, and feel in control of their own work [51]
Challenge	A challenging and meaningful work, the sense of working on important and challenging tasks [5]
Clearly direction	The objectives that facilitate creativity must be clear, negotiated, and realistic, shared and valued [2, 55]
Diversity, flexibility and tension	Diversity in terms of assigning work to people with whom we interact and tolerance for differences [51]. To be flexible with differences is required [1]. Diversity and flexibility provoke creative tension (Ekvall et al. [15])
Supporting creativity	Create an environment to support creativity [5, 10]
Trust and safe participation	Emphasize the promotion of participation in a non-threatening environment or evaluative [55]

All this research has shown that from the point of view of the task, clear direction and commitment are very important; and from the point of view of interpersonal relations, trust and participation are the most important factors to increase creativity. Management based on creativity must lead all identified environmental factors. In that way, it will be possible to enhance employees' internal drive to perceive every project as a new creative challenge [7].

Organizations cannot leave the creativity flow as a spontaneous process. Organizations must work rigorously to manage the world of ideas.

A literature review can see that the most important factors in the field of Virtual Teams are the problems of nonverbal communication and social context that ICT originates (Table 2).

Table 2 Factors to encourage creativity in virtual environments

Factor	Description
Communication	Loss of communication in a face to face environment, lower quality interaction [6] and sense of social anonymity [25]
Clarity of objectives and roles	Having a purpose and clearly goals [28, 29, 39], and well-defined roles [28, 29]
Trust	Creation of trust environment [28, 29] and Social relationship processes [27]

The theory of McGrath [32] suggests that a team with no history faces a complex problem and high uncertainty in the environment (such as Virtual Teams). Short et al. [45] also cited in [24] and his theory of social presence, that also consider the development of trust in teams, a phenomenon that particularly affects Virtual Teams. These authors suggest based communication technologies harm a good development of interpersonal relationships. Other authors as Walther [53, 54] and his theory of social information processing, consider that electronic communication does not differ from face to face in terms of the ability to exchange information. Walther [54] and Young and Nabuco de Abreu [56] found that social discussion based computer can reach higher quality in communication, even more than traditional communication channels. The implication to be drawn from all these studies is that some teams will struggle to develop an acceptable level of confidence, while for other teams geographic dispersed teams with intensive and appropriate use of information technologies will overcome any problem.

4 Conclusion

The dynamics of virtual teams and improving their creative performance reveals major unresolved issues. In this paper we have tried to address these issues, integrating the work of related fields—creative performance and virtual-teams, seeking theoretical evidence to support the need for further study of these two variables. The analyzed demonstrate, first, that new virtual work environments are still being explored and analyzed to find effective and efficient designs. This statement is evident in the framework of the implementation of creative management of virtual work environments. Although a marked and growing concern about applying creative management to virtual teamwork, the number of studies in this field is not yet manifested significant. Future research should take further explore the variables that affect the creation of a creative environment for virtual teams. Secondly, it shows the clear relationship between the variables that have traditionally been studied as determinants of the effectiveness of team work and conditioning the creative performance in both traditional and virtual environments. Thirdly, it is important to emphasize the correct design of a social system that encourages creative team performance, to clarify the conflicting results reported in many studies, such as those listed on the confidence level. Developing appropriate for quantitative analysis measures are also needed to strengthen the qualitative observations that reflect the relationships among the characteristics of virtual environments and creative performance of virtual machines.

Finally, note that creativity is not only a result of applying techniques or to follow a model of problem solving, creativity is mostly the result of a complex system in which many variables interact: people, processes and context. Virtual environments are revolutionizing these variables, research needs to move forward to address this revolution. This could help to show that we may not be dealing with two very different ways of organizing work.

References

1. Abbey, A., Dickson, J.W.: R&D work climate and innovation in semiconductors. Acad. Manag. J. **26**, 362–368 (1983)
2. Amabile, T.M.: A model of creativity and innovation in organizations. Res. Organ. Behav. **10**, 123–167 (1988)
3. Amabile, T.M.: Creativity in context: update to "the social psychology of creativity", xviii edn. Westview Press, Boulder (1996)
4. Amabilie, T.M.: Componential theory of creativity. Harvard Business School, Boston (2012)
5. Amabile, T.M., Gryskiewicz, S.S.: Creativity in the R&D laboratory. Greensboro, NC (1987)
6. Andres, H.P.: A comparison of face-to-face and virtual software development teams. Team Perform. Manag. **8**, 39–48 (2002)
7. Andriopoulos, C., Lowe, A.: Enhancing organisational creativity: the process of perpetual challenging. Manag. Decis. **38**, 734–742 (2000)
8. Arnison, L., Miller, P.: Virtual teams: a virtue for the conventional team. J. Work. Learn. **14**, 166–173 (2002)
9. Belbin, R.M.: Team roles at work, 2nd ed. Butterworth-Heinemann, Oxford (2010)
10. Brand, A.: Knowledge management and innovation at 3M. J. Knowl. Manag. **2**, 17–22 (1998)
11. Bryman, A.: Leadership and organizations. Routledge Library Editions, London (2013)
12. Davidow, W.H., Malone, M.S.: The virtual corporation. Harper Collins, New York (1992)
13. De Pillis, E., Parsons, B.: Implementing self-directed work teams at a college newspaper. Coll. Student J. Staff. **47**, 53–63 (2013)
14. Edmondson, A.C., Bohmer, R.M., Pisano, G.: Speeding up team learning. Harv. Bus. Rev. **79**, 125–132 (2001)
15. Ekvall, G., Arvonen, J., Waldenstrom-Lindblad, I.: Creative organizational climate: construction and validation of a measuring instrument (1983)
16. Farkas, M.: A note on team process. Harvard Business School, Boston (2001)
17. Feurer, R., Chaharbaghi, K., Wargin, J.: Developing creative teams for operational excellence. Int. J. Oper. Prod. Manag. **16**, 5–18 (1996)
18. Fincham, R., Rhodes, P.: Principles of organizational behavior, 4th edn. Londres, Oxford (2005)
19. Galbraith, J.R.: Designing organizations: strategy, structure, and process at the business unit and enterprise levels, 3rd ed. Wiley, New York (2014)
20. Gladstein, D.L.: Groups in context: a model of task group effectiveness. Adm. Sci. Q. **29**, 499–517 (1984)
21. Jex, S.M., Britt, T.W.: Organizational psychology: a scientist-practitioner approach, 3rd ed. Wiley, New York (2014)
22. Johnson, P., Heimann, V., O'Neill, K.: The "Wonderland" of virtual teams. J. Work. Learn. **13**, 24–29 (2001)
23. Kayworth, T., Leidner, D.: Leadership effectiveness in global virtual teams. J. Manag. Inf. **18**, 7–40 (2002)
24. Kear, K., Chetwynd, F., Jefferis, H.: Social presence in online learning communities: the role of personal profiles. Res. Learn. Technol. **22**, 1–15 (2014)
25. Kiesler, S., Siegel, J., McGuire, T.W.: Social aspects of computer-mediated communication. In: Dunlop, C., Kling, R. (eds.) Computerization and controversy: value conflicts and social choices, pp. 330–349. Harcourt Brace, Boston (1991)
26. Kirkman, B.L., Rosen, B., Gibson, C.B., Tesluk, P.E., McPherson, S.O.: Five challenges to virtual team success.pdf. Acad. Manag. Exec. **16**, 67–79 (2002)
27. Lewicki, R.J., Bunker, B.B.: Developing and maintaining trust in work relationships. In: Kramer, R.M., Tyler, T.R. (eds.) Trust in organizations: frontiers of theory and research, pp. 114–139. SAGE, Thousand Oaks (1996)
28. Lipnack, J., Stamps, J.: Virtual teams: reaching across space, time and organizations with technology. Wiley, New York (1997)

29. Lipnack, J., Stamps, J.: Virtual teams: the new way to work. Virtual teams new W. to Work. Strateg. Leadersh. Vol. 27, pp. 14–19 (1997b)
30. Mathieu, J.E., Tannenbaum, S.I., Donsbach, J.S., Alliger, G.M.: A review and integration of team composition models: moving toward a dynamic and temporal framework. J. Manag. (2014)
31. Maznevski, M.L., Chudoba, K.M.: Bridging space over time: global virtual-team dynamics and effectiveness. Organ. Sci. **11**, 473–492 (2000)
32. McGrath, J.E.: Time, interaction, and performance (TIP): a theory of groups. Small Gr. Res. **22**, 147–174 (1991)
33. Miner, F.: Group vs. individual decision making: an investigation of performance measures, decision strategies, and process loss/gains. Organ. Behav. Hum. Perform. **33**, 112–124 (1984)
34. Naghavi, M.A.S., Nekoo, A.H., Molladavoodi, A.: Methodology for software development as organizational creativity factor. African J. Bus. Manag. **6**, 8050–8054 (2012)
35. Nemiro, J.E.: Connection in creative virtual teams. J. Behav. Appl. Manag. **2**, 92–95 (2001)
36. Nemiro, J.E.: The creative process of virtual teams. Creat. Res. J. **14**, 69–84 (2002)
37. Nemiro, J.E.: Creativity in virtual teams: key components for success. Jossey-Bass/Pfeiffer, San Fran-cisco (2004)
38. Nunes, S.T., Consultant, S., Osho, G.S., Nealy, C.: The impact of human interaction on the development of virtual teams. J. Bus. Econ. Res. **2**, 95–100 (2004)
39. O'Hara-Devereaux, M., Johansen, R.: Global work: bridging distance, culture, and time. Jossey-Bass, San Francisco (1994)
40. Potter, R.E., Cooke, R.A., Balthazard, P.A.: Virtual team interaction: assessment, consequences, and management. Team Perform. Manag. **6**, 131–137 (2000)
41. Prince, G.M.: The practices of creativity. In: Thorne, P. (Ed.) Organizing genius. Blackwell Business, Oxford (1992)
42. Raghuram, S., Garud, R., Wiesenfeld, B., Gupta, V.: Factors contributing to virtual work adjustment. J. Manag. **27**, 383–405 (2001)
43. Rickards, T., Chen, M.-H., Moger, S.: Development of a self-report instrument for exploring team factor, leadership and performance relationships. Br. J. Manag. **12**, 243–250 (2001)
44. Robbins, S.P.: Comportamiento organizacional, 10th edn. Pearson Educación, México (2004)
45. Short, J., Williams, E., Christie, B.: The social psychology of telecommunications. Wiley, New York (1976)
46. Smagt, T. van der: Enhancing virtual teams: social relations v. communication technology. Ind. Manag. Data Syst. **100**, 148–156 (2000)
47. Smolensky, E.D., Kleiner, B.H.: How to train people to think more creatively. Manag. Dev. Rev. **8**, 28–33 (1995)
48. Solomon, C.M.: Managing virtual teams. Workforce **80**, 60–64 (2001)
49. Sundstrom, E., De Mouse, K.P., Futrell, D.: Work teams. Am. Psychol. **45**, 120–133 (1990)
50. Tohidi, H.: Teamwork productivity and effectiveness in an organization base on rewards, leadership, training, goals, wage, size, motivation, measurement and information technology. Procedia Comput. Sci. **3**, 1137–1146 (2011)
51. VanGundy, A.B.: Idea power. AMACOM, New York (1992)
52. Walker, J.W.: E-leadership? Hum. Resour. Plan. **23**, 5–6 (2000)
53. Walther, J.B.: Computer-mediated communication: Impersonal, interpersonal, and hyperpersonal interaction. Hum. Commun. Res. **23**, 1–43 (1996)
54. Walther, J.B.: Group and interpersonal effects in international computer-mediated collaboration. Hum. Commun. Res. **23**, 342–369 (1997)
55. West, M.A.: The social psychology of innovation in groups. In: West, M.A., Farr, J.L. (eds.) Innovation and creativity at work, pp. 309–322. Wiley, New York (1990)
56. Young, K.S., Nabuco de Abreu, C.: Internet addiction: a handbook and guide to evaluation and treatment (2011)

Author Index

© Springer International Publishing Switzerland 2015
Á. Herrero et al. (eds.), *International Joint Conference*, Advances in Intelligent
Systems and Computing 369, DOI 10.1007/978-3-319-19713-5

Printed in the United States
By Bookmasters